Rattan Lal
Carlos C. Cerri
Martial Bernoux
Jorge Etchevers
Eduardo Cerri
Editors

Carbon Sequestration in Soils of Latin America

Pre-publication
REVIEWS,
COMMENTARIES,
EVALUATIONS . . .

"This book is a timely and important addition to the global discussions on soil carbon sequestration. Although focusing on soils of Latin America, the book brings forth considerable valuable information on the science of soil carbon sequestration and procedures to assess soil carbon in tropical, temperate, and desert regions. It provides solid, scientific evidence on the potentials for soil carbon sequestration in different biomes in Latin America, as well as impacts of different land management systems. Although soil carbon is not yet recognized for carbon credits under the current Kyoto agreement, the book will contribute significantly to discussions to ensure improvement of Kyoto for the next reporting period.

In addition to assembling all evidence on soil carbon sequestration in Latin America, the added value of the book is the additional picture of land degradation in the region. This ensemble of evidence contributes considerably to understanding the opportunities and challenges of soil carbon sequestration and environmental management in the Latin America region."

Julian Dumanski, PhD
Retired, Senior Scientist,
Government of Canada;
Consultant, World Bank

More pre-publication
REVIEWS, COMMENTARIES, EVALUATIONS . . .

"Latin America makes up about 15 percent of the world's land area. The authors have looked at this region in great detail and have presented a wealth of data and other information about the carbon status of soils in the region. More important, they have looked at different biomes in the region and various management systems to determine when they are serving as a source or sink of carbon in the atmosphere. The book should be a valuable resource to researchers and policy makers concerned with carbon sequestration. Although the book focuses on Latin America and only detailed data from this region are presented, many of the authors have presented methodologies and discussions that apply to soils anywhere in the world. Therefore, even though the book will have its greatest value to workers in Latin America, it will be of interest and help to anyone concerned with carbon sequestration."

B. A. Stewart, PhD
Distinguished Professor of Soil Science,
West Texas A&M University,
Canyon

"Carbon sequestration in soils through judicious land use and management is an important strategy to offset emissions and mitigate the climate change both regionally and globally. Latin America—with its large landmass and widely differing soils, biomes, and ecoregions—provides a large potential to sequester soil carbon. This book comprised of twenty-four chapters contributed by internationally recognized experts provides a state-of-the-art discussion on different aspects of soil carbon sequestration in Latin America. Not only the soil carbon sequestration potential of different biomes of the region is extensively covered but also some of the advanced methods of assessment of soil carbon are well described.

Methods suggested for viable land-use change to improve carbon sequestration in soils of different biomes and ecoregions in the second section of the book could be useful tools for making a carbon accounting system under the Kyoto agreement. Furthermore, the conversion of degraded lands in the continent to various land-use systems, especially to productive new grass-legume pastures and agroforestry of silvopastoral systems and adoption of direct seeding mulch based cropping systems (DMC), provide a large potential of carbon sequestration.

The book also emphasizes that despite many uncertainties and limitations in soil carbon assessment, this region could be a good candidate for carbon trading under the CDM through suitable market structure. Although the book specifically reports carbon sequestration potential in Latin American soils, methodological and management approaches presented could be equally applicable to other regions and hence the book could be of wider benefit."

Bal Ram Singh, PhD
Professor,
Department of Plant
and Environmental Sciences,
Norwegian University
of Life Sciences (UMB)

NOTES FOR PROFESSIONAL LIBRARIANS AND LIBRARY USERS

This is an original book title published by Food Products Press®, an imprit of The Haworth Press, Inc. Unless otherwise noted in specific chapters with attribution, materials in this book have not been previously published elsewhere in any format or language.

CONSERVATION AND PRESERVATION NOTES

All books published by The Haworth Press, Inc., and its imprints are printed on certified pH neutral, acid-free book grade paper. This paper meets the minimum requirements of American National Standard for Information Sciences-Permanence of Paper for Printed Material, ANSI Z39.48-1984.

DIGITAL OBJECT IDENTIFIER (DOI) LINKING

The Haworth Press is participating in reference linking for elements of our original books. (For more information on reference linking initiatives, please consult the CrossRef Web site at www.crossref.org.) When citing an element of this book such as a chapter, include the element's Digital Object Identifier (DOI) as the last item of the reference. A Digital Object Identifier is a persistent, authoritative, and unique identifier that a publisher assigns to each element of a book. Because of its persistence, DOIs will enable The Haworth Press and other publishers to link to the element referenced, and the link will not break over time. This will be a great resource in scholarly research.

Carbon Sequestration in Soils of Latin America

FOOD PRODUCTS PRESS®

Crop Science

Amarjit S. Basra, PhD
Editor in Chief

The Lowland Maya Area: Three Millennia at the Human-Wildland Interface edited by A. Gómez-Pompa, M. F. Allen, S. Fedick, and J. J. Jiménez-Osornio

Biodiversity and Pest Management in Agroecosystems, Second Edition by Miguel A. Altieri and Clara I. Nicholls

Plant-Derived Antimycotics: Current Trends and Future Prospects edited by Mahendra Rai and Donatella Mares

Concise Encyclopedia of Temperate Tree Fruit edited by Tara Auxt Baugher and Suman Singha

Landscape Agroecology by Paul A. Wojtkowski

Concise Encyclopedia of Plant Pathology by P. Vidhyasekaran

Molecular Genetics and Breeding of Forest Trees edited by Sandeep Kumar and Matthias Fladung

Testing of Genetically Modified Organisms in Foods edited by Farid E. Ahmed

Fungal Disease Resistance in Plants: Biochemistry, Molecular Biology, and Genetic Engineering edited by Zamir K. Punja

Plant Functional Genomics edited by Dario Leister

Immunology in Plant Health and Its Impact on Food Safety by P. Narayanasamy

Abiotic Stresses: Plant Resistance Through Breeding and Molecular Approaches edited by M. Ashraf and P. J. C. Harris

Teaching in the Sciences: Learner-Centered Approaches edited by Catherine McLoughlin and Acram Taji

Handbook of Industrial Crops edited by V. L. Chopra and K. V. Peter

Durum Wheat Breeding: Current Approaches and Future Strategies edited by Conxita Royo, Miloudi M. Nachit, Natale Di Fonzo, José Luis Araus, Wolfgang H. Pfeiffer, and Gustavo A. Slafer

Handbook of Statistics for Teaching and Research in Plant and Crop Science by Usha Rani Palaniswamy and Kodiveri Muniyappa Palaniswamy

Handbook of Microbial Fertilizers edited by M. K. Rai

Eating and Healing: Traditional Food As Medicine edited by Andrea Pieroni and Lisa Leimar Price

Physiology of Crop Production by N. K. Fageria, V. C. Baligar, and R. B. Clark

Plant Conservation Genetics edited by Robert J. Henry

Introduction to Fruit Crops by Mark Rieger

Generations Gardening Together: Sourcebook for Intergenerational Therapeutic Horticulture by Jean M. Larson and Mary Hockenberry Meyer

Agriculture Sustainability: Principles, Processes, and Prospects by Saroja Raman

Introduction to Agroecology: Principles and Practice by Paul A. Wojtkowski

Handbook of Molecular Technologies in Crop Disease Management by P. Vidhyasekaran

Handbook of Precision Agriculture: Principles and Applications edited by Ancha Srinivasan

Dictionary of Plant Tissue Culture by Alan C. Cassells and Peter B. Gahan

Handbook of Potato Production, Improvement, and Postharvest Management edited by Jai Gopal and S. M. Paul Khurana

Carbon Sequestration in Soils of Latin America edited by Rattan Lal, Carlos C. Cerri, Martial Bernoux, Jorge Etchevers, and Eduardo Cerri

Carbon Sequestration in Soils of Latin America

Rattan Lal
Carlos C. Cerri
Martial Bernoux
Jorge Etchevers
Eduardo Cerri
Editors

Food Products Press®
An Imprint of The Haworth Press, Inc.
New York • London • Oxford

For more information on this book or to order, visit
http://www.haworthpress.com/store/product.asp?sku=5755

or call 1-800-HAWORTH (800-429-6784) in the United States and Canada
or (607) 722-5857 outside the United States and Canada

or contact orders@HaworthPress.com

Published by

Food Products Press®, an imprint of The Haworth Press, Inc., 10 Alice Street, Binghamton, NY 13904-1580.

PUBLISHER'S NOTE
The development, preparation, and publication of this work has been undertaken with great care. However, the Publisher, employees, editors, and agents of The Haworth Press are not responsible for any errors contained herein or for consequences that may ensue from use of materials or information contained in this work. The Haworth Press is committed to the dissemination of ideas and information according to the highest standards of intellectual freedom and the free exchange of ideas. Statements made and opinions expressed in this publication do not necessarily reflect the views of the Publisher, Directors, management, or staff of The Haworth Press, Inc., or an endorsement by them.

Cover design by Marylouise E. Doyle.

Library of Congress Cataloging-in-Publication Data

Carbon sequestration in soils of Latin America / Rattan Lal . . . [et al.], editors.
p. cm.
Includes bibliographical references and index.
ISBN-13: 978-1-56022-136-4 (case 13 : alk. paper)
ISBN-10: 1-56022-136-4 (case 10 : alk. paper)
ISBN-13: 978-1-56022-137-1 (soft 13 : alk. paper)
ISBN-10: 1-56022-137-2 (soft 10 : alk. paper)
1. Soils—Carbon content—Latin America. 2. Carbon sequestration—Latin America. I. Lal, R.
S592.6.C35C38 2006
631.4'22—dc22

2006015831

CONTENTS

PART IV: CONCLUSION

ABOUT THE EDITORS

Rattan Lal, PhD, is a professor of soil physics in the School of Natural Resources and director of the Carbon Management and Sequestration Center, FAES/OARDC at Ohio State University. He was a soil physicist for 18 years at the International Institute of Tropical Agriculture in Ibadan, Nigeria. In Africa, he conducted long-term experiments on land use, watershed management, methods of deforestation, and agroforestry. He is a fellow of the Soil Science Society of America, American Society of Agronomy, Third World Academy of Sciences, American Association for the Advancement of Sciences, Soil and Water Conservation Society, and Indian Academy of Agricultural Sciences. Dr. Lal is the recipient of the International Soil Science Award, the Soil Science Applied Research Award, and several others. He has authored and co-authored nearly 1,100 research publications and nine books and has edited or co-edited 43 books.

Carlos C. Cerri, PhD, is a professor at the Centro de Energia Nuclear na Agricultura of the Universidade de São Paulo, Brazil, where he teaches graduate students and performs research on soil carbon sequestration and trace gas mitigation on tropical conditions. He was director of CENA between 1991-1997 and coordinator of 51 national and international scientific projects on environmental science. His present functions are as Brazilian coordinator of GEF Project (GEFSOC—2740-02-4381) "Assessment of soil organic carbon stocks and change at national scale," coordinator of NASA Project "Biogeochemical consequences of agricultural intensification in the Amazon basin," representative scientific member of IPCC since 1995 until now, member of French Academy of Agriculture, "Chevalier dans l'Ordre des Palmes Académiques" of French and Commend of the Ordem Nacional do Mérito Científico—Brazilian President Decree.

Martial Bernoux, PhD, is research soil scientist at the French Institut de Recherche pour le Développement (IRD) since 1999, now member of research unit "Soil carbon sequestration and soil bio-functioning." In 1992 he received an agronomist engineer degree and a master degree of agronomic sciences, and in 1998 PhD in soil science from the University of Orléans (France) and the University of Sao Paulo (Brazil). He has lived and worked in France and mainly in Brazil, and also developed research activities in Martinique—Lesser Antilles. Since 1999, he has been involved in the eval-

Carbon Sequestration in Soils of Latin America

doi:10.1300/5755_a

uation of the role of soils and agricultural systems in carbon sequestration in tropical conditions focusing onto soil carbon temporal and spatial dynamics and the integration of spatial and temporal dimensions. He participated in the elaboration of the first Brazilian National Communication submitted to the United Nation Framework Convention on Climate Change. He also served as contributing author of the Intergovernmental Panel on Climate Change Good Practice Guidance for Land Use, Land-Use Change and Forestry. He participated to the 2006 edition of the *Global Environment Outlook Year Book* produced by the United Nations Environment Programme. He has authored and co-authored 110 research publications, including 30 papers indexed at the ISI Web of Science.

Jorge Etchevers, PhD, is a professor of soil science at the Colegio de Postgraduados, Mexico, and a permanent visiting professor of the Universidad of Concepción, Chile. In recent years, his main field of research has been soil carbon, carbon sequestration, and recuperation of degraded soils. In addition, he has continued working on several aspects of the nutrients in the soils and plant systems. He is a member of the Mexican Academy of Science and the National Academy of Agricultural Science of Mexico. He has been distinguished with several awards and recognitions in Mexico and Latin America. He is a past President of the Chilean Society of Soil Science, and directed for nearly 10 years, *Terra* (today *Terra Latinoamericana*) which is the journal of the Mexican Soil Science Society. He is a member of the Editorial Committee of five scientific journals of the region, and has published nearly 150 refereed papers, authored or edited several book and book chapters, contributed with more than 400 presentations in congresses, scientific and technical meetings, and graduated nearly 150 students in the last 40 years. He is a member of the Soil Science Society of America, the American Society of Agronomy, the International Union of Soil Science Societies, the Mexican Society of Soil Science, and the Latin American Society of Soil Science.

Eduardo Cerri, PhD, is a post-doc at the Centro de Energia Nuclear na Agricultura (CENA-USP), Brazil. In 2003 he received his PhD in soil science from the University of Sao Paulo. He worked for the GEFSOC project financed by the Global Environment Facility for three years. His work focuses on spatial variability of soil properties and modeling organic matter dynamics in tropical soils. He received awards from University of Sao Paulo, Soil Science Society of America and START as young scientist.

CONTRIBUTORS

Dimas Acevedo, Universidad de Los Andes, Instituto de Ciencias Ambientales y Ecológicas, Merida, Venezuela.

M. Acosta, Instituto Nacional de Investigaciones Forestales Agrícolas y Pecuarias, Mexico.

A. Albrecht, Institut de Recherche pour le Développement (IRD), Laboratoire Matière Organique des Sols Tropicaux (MOST), France.

Alfredo Alvarado, Department of Soil Science, University of Costa Rica, San Jose, Costa Rica.

Bruno J. R. Alves, Embrapa-Agrobiologia, Antiga Rodovia Rio– São Paulo, Rio de Janeiro, Brazil.

Edgar Amézquita, Tropical Soil Biology and Fertility (TSBF) Institute, Centro Internacional de Agricultura Tropical (CIAT), Cali, Colombia.

M. C. Amézquita, Centro Internacional de Agricultura Tropical (CIAT), Cali, Colombia.

D. Arrouays, Institut National de la Recherche Agronomique (INRA), Unité Infosol, France.

Miguel Ayarza, Tropical Soil Biology and Fertility (TSBF) Institute, Centro Internacional de Agricultura Tropical (CIAT), Cali, Colombia.

C. Balbontín, Instituto de Recursos Naturales, Colegio de Postgraduos, Montecillo, Mexico.

Edmundo Barrios, Tropical Soil Biology and Fertility (TSBF) Institute, Centro Internacional de Agricultura Tropical (CIAT), Cali, Colombia.

E. Blanchart, Institut de Recherche pour le Développement (IRD), Laboratoire Matière Organique des Sols Tropicaux (MOST), France.

Robert M. Boddey, Embrapa-Agrobiologia, Antiga Rodovia Rio– São Paulo, Rio de Janeiro, Brazil.

Daniel E. Buschiazzo, Instituto Nacional de Tecnología Agropecuaria (INTA), Consejo Nacional de Investigaciones Cientificas y Técnicas

doi:10.1300/5755_b

(CONICET), and Dept. Agronomía, Universidad Nacional de La Pampa, La Pamapa, Argentina.

Mercedes M. C. Bustamante, Departamento de Ecologia, Universidade de Brasília, Brasília, Brazil.

P. Buurman, Soil Science and Geology, Department of Environmental Sciences, Wageningen University, the Netherlands.

Y. M. Cabidoche, Institut National de la Recherche Agronomique (INRA), Laboratoire Agronomie-Pédologie-Climatologie (APC), Guadeloupe, France.

T. Chevallier, Institut de Recherche pour le Développement (IRD), Laboratoire Matière Organique des Sols Tropicaux (MOST), France.

C. Clermont-Dauphin, Institut National de la Recherche Agronomique (INRA), Laboratoire Agronomie-Pédologie-Climatologie (APC), Guadeloupe, France.

Marc Corbeels, Département des Cultures Annuelles, Département des Cultures Annuelles, Centre de Coopération Internationale en Recherche Agronomique pour le Développement (CIRAD), France, and Centro da Empresa Brasileira de Pesquisa Agropecuária (EMBRAPA)–Cerrados, Brasília, Brazil.

David A. Cremers, Chemistry Division, Los Alamos National Laboratory, Los Alamos, New Mexico.

Octávio C. de Oliveira, Instituto Brasileiro de Geografia e Estatística (IBGE), Rio de Janeiro, Brazil.

Martín Díaz-Zorita, Díaz-Zorita, Duarte & Asoc., Nitragin Argentina, Consejo Nacional de Investigaciones Cientificas y Técnicas (CONICET), Dept. Agronomía, and Universidad de Buenos Aires, Buenos Aires, Argentina.

Michael H. Ebinger, Earth and Environmental Sciences Division, Los Alamos National Laboratory, Los Alamos, New Mexico.

C. Feller, Institut de Recherche pour le Développement (IRD), Laboratoire Matière Organique des Sols Tropicaux (MOST), France.

R. F. Follett, Soil Plant Nutrient Research, U.S. Department of Agriculture, Fort Collins, Colorado.

Juan A. Galantini, Comision de Investgaciones Científicas-CERZOS, Dept. Agronomía, Universidad Nacional del Sur, Bahía Blanca, Argentina.

D. Gómez, Departamento de Suelos, Universidad Autónoma Chapingo, Chapingo, Mexico.

Ronny D. Harris, Material Science Division, Los Alamos National Laboratory, Los Alamos, New Mexico.

Rosa M. Hernandez, Universidad Simón Rodríguez, Centro de Agroecología Tropical, Instituto de Estudios Científicos y Tecnológicos (IDECYT), Miranda, Venezuela.

R. C. Izaurralde, Joint Global Change Research Institute, College Park, Maryland.

Claudia P. Jantalia, Departmento. de Fitotecnia, Universidade Federal Rural do Rio de Janeiro, Rio de Janeiro, Brazil.

J. J. Jiménez, Carbon Management and Sequestration Center, The Ohio State University, Columbus, Ohio.

John Kimble, U.S. Department of Agriculture, National Resources Conservation Service, National Soil Survey Center, Lincoln, Nebraska.

M. C. Larré-Larrouy, Institut de Recherche pour le Développement (IRD), Laboratoire Matière Organique des Sols Tropicaux (MOST), France.

Michele O. Macedo, Departmento. de Solos, Universidade Federal Rural do Rio de Janeiro, Rio de Janeiro, Brazil.

M. Martínez, Centro de Investigación en Ecosistemas, Universidad Nacional Autónoma de Mexico, Morelia Michoacán.

R. Martínez, Centro de Investigación en Ecosistemas, Universidad Nacional Autónoma de Mexico, Morelia Michoacán.

O. Masera, Centro de Investigación en Ecosistemas, Universidad Nacional Autónoma de Mexico, Morelia Michoacán.

G. W. McCarty, Environmental Quality Lab, U.S. Department of Agriculture, Beltsville Agricultural Research Center East, Beltsville, Maryland.

F. F. C. Mello, Centro de Energia Nuclear na Agricultura (CENA), Universidade de São Paulo, Piracicaba, Brazil.

Clifton M. Meyer, Earth and Environmental Sciences Division, Los Alamos National Laboratory, Los Alamos, New Mexico.

A. Monterroso, Departamento de Suelos, Universidad Autónoma Chapingo, Chapingo, Mexico.

E. Murgueitio, Centro para la Investigación en Sistemas Sostenibles de Producción Agropecuaria (CIPAV), Cali, Colombia.

C. Ortiz, Instituto de Recursos Naturales, Colegio de Postgraduos, Montecillo, Mexico.

Idupulapati Rao, Tropical Soil Biology and Fertility (TSBF) Institute, Centro Internacional de Agricultura Tropical (CIAT), Cali, Colombia.

J. B. Reeves III, Animal Manure and By-Products Laboratory, U.S. Department of Agriculture, Beltsville Agricultural Research Center East, Beltsville, Maryland.

Alexander S. Resende, Centro da Empresa Brasileira de Pesquisa Agropecuária (EMBRAPA)–Agrobiologia, Antiga Rodovia Rio–São Paulo, Rio de Janeiro, Brazil.

C. W. Rice, Department of Agronomy, Kansas State University, Manhattan, Kansas.

Mariela Rivera, Tropical Soil Biology and Fertility (TSBF) Institute, Centro Internacional de Agricultura Tropical (CIAT), Cali, Colombia.

Mandius Romero, Corporación Colombiana para la Investigación en Agricultura (CORPOICA), Departamento de Agroecología, Tibaitata, Colombia.

Marco A. Rondón, Tropical Soil Biology and Fertility (TSBF) Institute, Centro Internacional de Agricultura Tropical (CIAT), Cali, Colombia.

Renato Roscoe, Centro da Empresa Brasileira de Pesquisa Agropecuária (EMBRAPA)-Agropecuária Oeste, Dourados, Mato Grosso, Brazil.

Ramón A. Rosell, Consejo Nacional de Investigaciones Cientificas y Técnicas (CONICET), Departmento Agronomía, Universidad Nacional del Sur, Bahía Blanca, Argentina.

Yolanda Rubiano, Rural Innovation Institute, Centro Internacional de Agricultura Tropical (CIAT), Cali, Colombia.

Lina Sarmiento, Universidad de Los Andes, Instituto de Ciencias Ambientales y Ecológicas, Merida, Venezuela.

Eric Scopel, Département des Cultures Annuelles, Centre de Coopération Internationale en Recherche Agronomique pour le Développement (CIRAD), France, and Centro da Empresa Brasileira de Pesquisa Agropecuária (EMBRAPA)-Cerrados, Brasília, Brazil.

Segundo Urquiaga, Centro da Empresa Brasileira de Pesquisa Agropecuária (EMBRAPA)–Agrobiologia, Antiga Rodovia Rio–São Paulo, Rio de Janeiro, Brazil.

C. Venkatapen, Institut de Recherche pour le Développement (IRD), Laboratoire Matière Organique des Sols Tropicaux (MOST), France.

Boris Volkoff, Institut de Recherche pour le Développement (IRD), France.

Lucian Wielopolski, Brookhaven National Laboratory, Environmental Sciences Department, Upton, New York.

Preface

Latin America (LA) comprises a large land mass of 50 political entities and is home to 535 million inhabitants, or 7 percent of the 2004 world population. The region has vast natural resources comprising 1,945 Mha of total area, and 1,909 Mha of land area. Predominant land uses in LA comprise 753 Mha of agricultural land, 140 Mha of cropland, 17 Mha of permanent crops, 595 Mha of permanent pastures, and 1,752 Mha of forest, savanna, and other land uses. The total area under tropical rainforest in LA is about 900 Mha, and is being deforested at the rate of 2 to 3 percent per year. Tropical deforestation and soil cultivation contribute 1.7 to 2.0 Gt C emission annually on a global basis.

The LA region also comprises rapidly developing economies (e.g., Brazil, Mexico). Net electricity consumption in LA is projected to grow by 3.2 percent per year, from 668 billion kwh in 2001 to 1,425 billion kwh in 2025. Hydropower and other renewable resources account for nearly three-fourths of the total energy consumed in the region. However, their share is projected to fall to 57 percent in 2025 and is being replaced by natural gas. The natural gas markets in the region constituted 3.9 percent of the world's consumption in 2001. At the beginning of 2004, LA held 4.1 percent of the world's natural gas reserves, or about 250 trillion cubic feet. Natural gas consumption in the region increased from 2.0 trillion cubic feet in 1990 to 3.5 trillion cubic feet in 2001, and is projected to grow at an average rate of 3.8 percent per year to 8.5 trillion cubic feet in 2025.

The CO_2 emissions from fossil fuel consumption are increasing for the region. Total emissions were 1.1 Tg C per year in 1900, 45.0 Tg C per year in 1950, 181.1 Tg C per year in 1975, and more than 400 Tg C per year in 2000. Fossil fuel emissions have grown more than ninefold since 1950 at the rate of 4.3 percent per year from 1997 onward. Two countries (Brazil and Mexico) are among the 20 highest fossil fuel–CO_2 emitting countries of the world.

Agriculture is an important industry in the region. It can be a source or sink for atmospheric CO_2 depending on the land use, management, and the

Carbon Sequestration in Soils of Latin America

doi:10.1300/5755_c

extent of adoption of recommended management practices. Therefore, a workshop was organized to discuss the impact of land use change and soil/crop/pasture management on soil C stock and fluxes. The workshop was held from June 1-5, 2004, in Piracicaba, Brazil. It was jointly organized by The Ohio State University, University of São Paulo, and Institute de Recherche pour le Dévelopment (IRD) of France. The workshop was financed by the U.S. Department of State, the Inter-American Institute (IAS), USAID, U.S. Department of Energy, U.S. Department of Agriculture (USDA), and the Nature Conservancy.

The objectives of the workshop were to (1) collate and synthesize available information on soil carbon dynamics under natural and managed ecosystems, (2) estimate the potential of soil C sequestration in different ecoregions, (3) address land use and management options to realize the potential, (4) identify land use and management options that will render agricultural ecosystems as sink for atmospheric CO_2, (5) identify knowledge gaps and prioritize researchable issues, (6) develop a scientific network to implement a research program to validate rates of soil C sequestration for diverse land uses and management systems, and (7) establish mechanisms of scientific exchange and create training opportunities.

This book is organized into four sections. Part I is titled "Physiography and Background," and contains four chapters. One chapter is devoted to principal biomes or ecoregions of Latin America. The remaining 3 chapters address issues pertaining to soil C sequestration, opportunities, and challenges. Part II is titled "Soil Carbon Sequestration in Different Biomes of Latin America." This section has 13 chapters dealing with 7 principal biomes: (1) western mountain regions and deserts of Mexico and Central America, (2) western mountain regions and deserts of South America, (3) the Caribbean and Savanna regions including Llanos, (4) Amazon tropical wet forest, (5) South American savanna or Cerrados, (6) the Atlantic Forest region, and (7) the Pampas or Meadows. These 13 chapters provide a state-of-the-knowledge synthesis of soil C sequestration for different biomes of Latin America. Part III is titled "Soil Carbon Assessment Methods," and comprises 6 chapters describing recent advances in methods of measurement and prediction of soil C pool. The last section has one chapter and is a summary of recommendations compiled by the work groups and discussion panels. This 24-chapter book is a state-of-the-knowledge compendium on soil C sequestration in LA.

The organization of the symposium and publication of the volume were made possible by the cooperation and support of sponsoring and funding organizations listed previously. The editors thank the authors for their outstanding efforts to document and present their information on the current understanding of soil C sequestration in LA. Their efforts have contributed to enhancing the overall understanding of opportunities and challenges of soil C sequestration in LA, and how to better use soil resources of this vast region and make it a sink for atmospheric CO_2. These efforts have advanced the frontiers of soils science with regard to soil C sequestration, while improving agronomic productivity and mitigating climate change.

Special thanks must be given to Mr. Andrew Dowdy of the U.S. State Department for his help, support, and initiative in organizing the workshop, providing funding support and liaison with administration on the one hand and the scientific community in LA on the other. This workshop was made possible by his initiative, vigor, and dedication. Thanks are also due to Mr. Gustavo Necco, of the IAI, for his help and support. In addition, valuable contributions were made by numerous faculty, staff, and students of OSU and USP. We thank Ms. Brenda Swank for her help in preparing the manuscripts. Thanks are also due to the staff of The Haworth Press, Inc., for their timely efforts in publishing this information and making it available to the scientific community, policymakers, and land managers. The efforts of many others were also very important in publishing this important scientific information in a timely manner.

PART I:
PHYSIOGRAPHY AND BACKGROUND

Chapter 1

Soil Ecoregions in Latin America

Boris Volkoff
Martial Bernoux

INTRODUCTION

Large soil units generally reflect bioclimatic environments (the concept of zonal soils). Soil maps thus represent summary documents that integrate all environmental factors involved. The characteristics of soils represent the environmental factors that control the dynamics of soil organic matter (SOM) and determine both their accumulation and degradation. Soil maps thus represent a basis for quantitative studies on the accumulation processes of soil organic carbon (SOC) in soils in different spatial scales. From this point of view, however, and in particular if one is interested in general scales (large semicontinental regions), soil maps have several disadvantages.

First, most soil maps take into account the intrinsic factors of the soils, thus the end results of the formation processes, rather than the processes themselves. These processes are the factors that are directly related to environmental conditions, whereas the characteristics of the soils can be inherited (paleosols and paleoalterations) and might no longer be in equilibrium with the present environment.

Second, soils seldom are homogenous spatial entities. The soil cover is in reality a juxtaposition of several distinct soils that might differ to various degrees (from similar to highly contrasted), and might be either genetically linked or entirely disconnected. This spatial heterogeneity reflects the conditions in which the soils were formed and is expressed differently according to the substrates and the topography. The heterogeneity also depends on the duration of evolution of soils and the geomorphologic history, either regional or local, as well as on climatic gradients, which are particularly obvious in mountain areas. The heterogeneity is visually expressed in detailed maps and is implicit in medium-scale maps, as it is often taken into account in the definition of cartographic units. However, it is commonly masked in

Carbon Sequestration in Soils of Latin America

doi:10.1300/5755_01

general maps, which might group territories with dissimilar characteristics within the same clusters. Thus, parameters that are linked to the environment and that could affect the dynamic of the SOM might not be apparent in soil maps.

It is thus necessary to consider the soils and the biophysical environment simultaneously and to supplement the data given by soil maps with information on the biophysical environment (climate, vegetation, landscape).

For this purpose it seems convenient to work with soil synthetic geographical entities that must be defined as a given combination of soil, climate, and vegetation within a specified physiographic (geologic, topographic, and geomorphologic) context.

An elementary landscape with its specific soils may repeat itself regularly, forming a large regular unit, as occurs in some wet tropical plains. But normally it changes because of the variation in the biophysical environment. The changes generate a definite spatial frame of the soils (Fridland, 1974). At a regional scale this frame is described in terms of soil macrostructure. Zonal macrostructures, horizontal in plains or vertical in mountains, result from gradual changes in the bioclimatic (climate and vegetation) environment. More complex spatial macrostructures are found when different types of landscapes are spatially ordered, as in hydrographic basins, or appear spatially disordered when they are strongly controlled by the lithology, for example.

On the basis of the study of soil associations and the soil spatial macroorganizations, using available soil maps and assessing the general biophysical characteristics of Latin America, there are a limited number of distinct soil regions, or "soil ecoregions."

This regionalization can facilitate the delineation of the soil C storage and help in the selection of representative sites for the study of the mechanisms of this storage.

CONTINENTAL PARAMETERS

Latin America represents more than one-eighth of the earth's land surface. It encompasses a great latitudinal span, with its broadest expanse within the tropical zone. It sprawls across 83 degrees of longitude, from 35°W at the northeastern coast of Brazil to 118°W at the California-Mexico border (Figure 1.1). Its latitudinal extent is 90 degrees, from 34°N at the California-Mexico border to 56°S at Cape Horn into the sub-Antarctic. The land area is about 20,000,000 km^2, under the jurisdiction of more than 45 countries (Figure 1.2). The region's 2000 population was estimated at 400 million people.

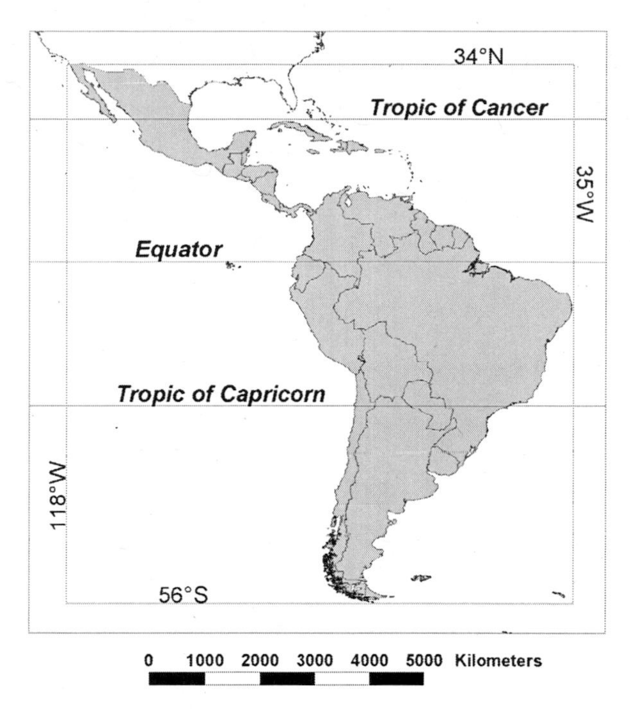

FIGURE 1.1. Geographic location of Latin America.

THE BIOPHYSICAL OR NATURAL ENVIRONMENT

Pattern of Landforms

A comprehensive description of the natural environment, especially the landforms and the geological history, together with an extended bibliography, can be found in SI-WWF-IUCN (2001). Only the most relevant issues will be presented here.

Latin America includes the South American continental plate, the southern portion of the North American plate, and the independent Caribbean plate. South America and North America were widely separated through most of their geological histories, and they became connected via the Isthmus of Panama during the Pleistocene period. The Antillean plate with its collection of islands formed only during the Cenozoic period.

The outstanding geological feature of South America is the Andes, which extends in a nearly straight line over 7,000 km from the north to the southern tip of the continent. The southern Andes are the oldest, with signif-

FIGURE 1.2. Countries of Latin America.

icant uplift already present in early Cenozoic times, prior to the Oligocene period. Most of the uplift of the central Andes was in the Miocene period or later, whereas most of the uplift of the northern portion of the cordillera occurred in the Plio-Pleistocene period. To the north the Andes become more complex, breaking into three separate cordilleras on the Ecuador-Colombia border.

Much of the rest of the South American continent consists of two great crystalline shields. The northeastern portion of the continent constitutes the Guayana shield, whereas much of Brazil south of Amazonia is underlain by the Brazilian shield. They consist of Precambrian igneous basement rocks overlain by ancient and much-eroded Precambrian sediments (Figure 1.3).

The Guayana region has been the most heavily eroded, with basement elevations mostly below 500 m interrupted by massive table mountains, rising to 2,000 or 2,500 m. The peak of the highest of these, Pico da Neblina on the Venezuela-Brazil border, reaches an altitude of 3,015 m and is the high-

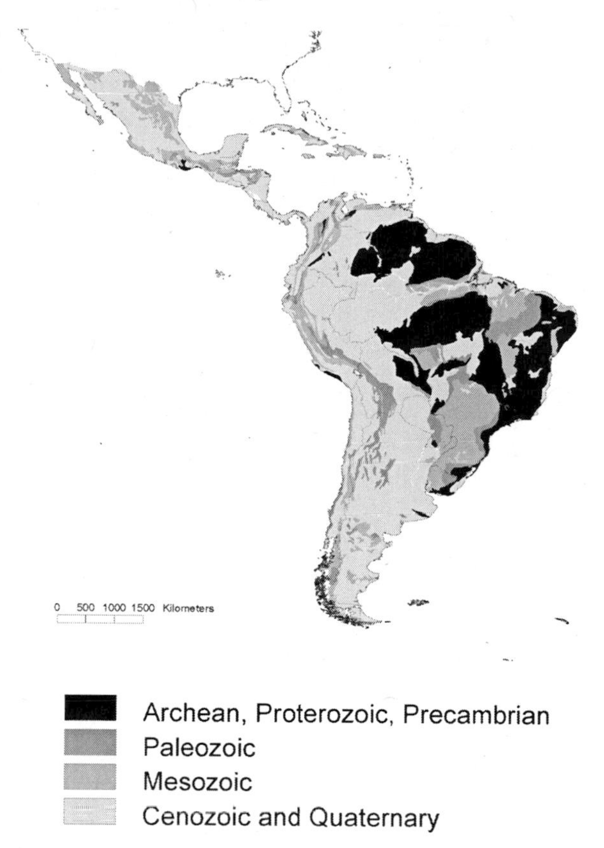

FIGURE 1.3. Geology of Latin America: General stratigraphy. (*Source:* Created from data compiled from CGMW-UNESCO, 2000.)

est point in South America outside the Andes. The Brazilian shield is generally higher and less dissected, with much of central Brazil having an elevation of 800 to 1,000 m.

In contrast to these ancient shields, the central Amazonian basin is low and geologically young. Prior to the Miocene period, most of Amazonia constituted a large inland sea opening to the Pacific. With uplift of the central Andes, this sea became a giant lake that gradually filled with Andean sediments. The region remains low and flat, such that Iquitos, Peru, is only 110 m above sea level and most of Amazonian Ecuador, Peru, and Bolivia are below 200 m in elevation.

Like Amazonia, some other distinctive geological features of the South American continent are relatively low, flat, and geologically young, such as the Chaco/Pantanal/Pampa region to the south, the Venezuelan-Colombian llanos to the north, and the trans-Andean Chocó region of Colombia and Ecuador to the west. Large portions of all of these regions are seasonally inundated.

Middle America is more complex geologically than South America. Toward south to central Nicaragua, Central America is an integral part of the North American continent. The region from southern Nicaragua to the Isthmus of Darién in Panama is geologically young and presents recent volcanism, uplift, and associated sedimentation.

Like South America, Middle America has a mountainous spine that breaks into separate cordilleras in the north. In general, the Middle American cordilleras are highest to the north in Mexico, and lowest in Panama to the southeast. In Mexico, the geology is complicated by a band of volcanoes that bisects the continent from east to west at the latitude of Mexico City.

In southern Central America, volcanism has been most intensive mainly in central Costa Rica and western Panama. The Yucatán Peninsula area of Mexico and Guatemala is a flat limestone formation more like the Greater Antilles than the mountainous terrain and volcanic soil of most of Middle America.

The Antillean islands constitute the third geologic unit of Latin America. The Antilles have a complex geological history. Some parts of the Greater Antilles island (Jamaica) are connected to Central America. Most of the Greater Antilles (Cuba, Hispaniola, Puerto Rico) has significant areas of serpentine and other ultra basic rocks. The Antilles have extensive areas of limestone. The Lesser Antilles are actively volcanic.

Major Physiographic Regions

Several major physiographic regions can be defined on the basis of terrain features, geological structure, hypsography, and watershed organization.

One broad region is composed of the Western steep-sided mountain ranges. It comprises Middle America, Central America, and the Andean cordilleras and includes the Middle America and Andean intermountain high plateaus (Figure 1.4).

Just to the east of the mountain ridge, much of the South American continent is a vast area of gentle relief. It can be divided into Guyana, central and eastern Brazilian and Patagonian uplands, and Orenoque, Amazon, and Paraná-Uruguay (or Plata) basin plains (Figure 1.5). Eastward, Central America and the Caribbean and Antillean islands constitute the last region.

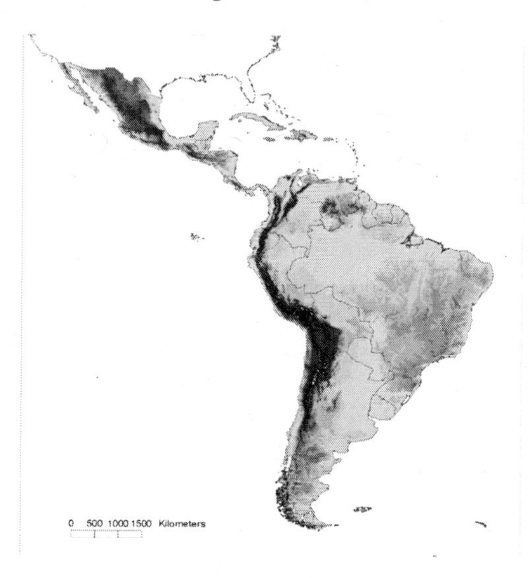

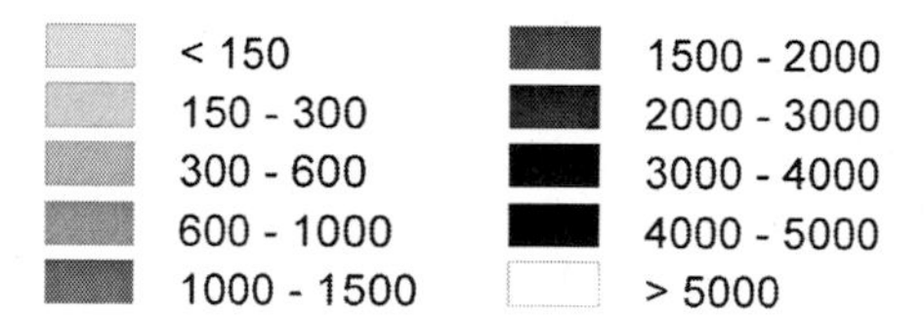

FIGURE 1.4. Hypsometry of Latin America. (*Source:* Created from data extracted from USGS, 1996.)

Glaciation

Pleistocene-Quaternary glaciations have affected only the southernmost part of the continent (Clapperton, 1993). Glacial, fluvio-glacial, and loess deposits, remains of the last glacial period, can be found on the foot slopes of the hills, the plateaus, and the low plains of the eastern side of the southern Andes (Figure 1.6).

Paleosoils

The South American inner lowlands and highlands, which have not undergone any quaternary glacial processes, exhibit a thick mantle of strongly weathered materials elaborated during the Cenozoic and Early Quaternary wet tropical climate periods (Tardy, 1997).

FIGURE 1.5. Hydrographic features of Latin America. (*Source:* Created from data extracted from ESRI, 1996, ArcAtlas database.)

The successive global long warm periods of the Cretaceous period and the warm and wet conditions of the Cenozoic period were periods of intense weathering. They produced thick saprolites and simultaneously strongly leveled the topography. Furthermore, periods of arid climate alternated with periods of wet climate. Consequently, episodes of erosion, transport, and deposition of the weathered materials followed periods of intense weathering and deepening of the profiles. The landscape resulting from such cyclic geomorphic evolution is typically a stepped landscape composed of ordered sequences of leveled land surfaces developed as well on residual saprolites as on transported continental sediments (Thomas, 1994).

FIGURE 1.6. Quaternary glacial and recent volcanic deposits. (*Source:* Created from data compiled from ESRI, 1996, ArcAtlas database.)

Two main features characterize the central and eastern parts of the South American continent: (1) remnants of the mid-Cenozoic summit surface, the South American surface (King, 1962), scattered throughout the whole zone and forming the highest plateaus, and (2) the "Barreiras formation" the Late Cenozoic-Pleistocene continental deposit that is the foundation of many coastal and inner low plateaus.

All the plateaus are uniformly covered by typical Ferralsols (or Latosol according Brazilian soil classification). The map of Ferralsols in Latin America (Figure 1.7) gives a good indication of the extent of the South American ancient landscape.

Climate

The climate of Latin America is controlled by two main factors: the latitude and the topographic patterns. North-south trends in climatic zones

Ferralsols

FIGURE 1.7. Ferralsols in Latin America. (*Source:* Created from data compiled from FAO, 2001, SOTERLAC database.)

reflect the fundamental impact of the latitude. The Tropic of Cancer in Mexico and the Tropic of Capricorn define major thermal demarcations. Tropical climatic regimes span nearly the entirety of Latin America. Only in the southern part of South America is the climate cold. Temperatures decrease with increasing elevation, such that the tropical climate of the lowlands and lower slopes changes to subtropical and temperate climates at intermediate elevations of the Andes and the interior uplands, and finally to cold climate at the top of the mountain ridges. The barrier effect of the western cordilleras oriented N-S and ocean currents are other climatic controls with great regional importance, mainly on the annual rainfall amount and rainfall seasonal distribution (Figure 1.8).

The Köppen Climate Classification System is the most widely used system for classifying the world's climates. Its categories are based on the an-

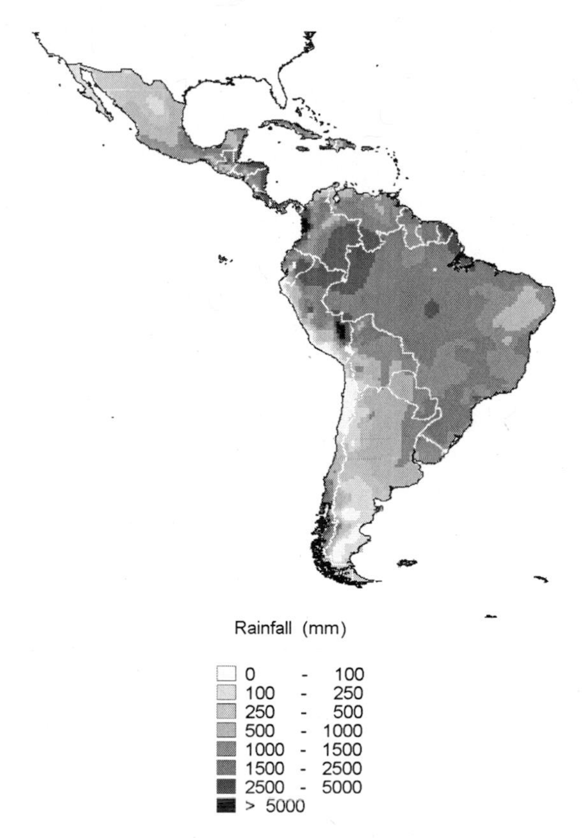

FIGURE 1.8. Annual rainfall in Latin America. (*Source:* Created from data extracted from FAO-SDRN, 1997.)

nual and monthly averages of temperature and precipitation. Each type is designated by a capital letter. Four major climatic types of the Köppen system are recognized in Latin America: (1) tropical moist climates (A) where all months have average temperatures above 18°C, (2) dry climates (B) with deficient precipitation during most of the year, (3) temperate (midlatitude) climates (C), and (4) polar climates (E) (Figure 1.9). A fifth category, the cold climate (D), has no significant area (FAO-SDRN, 1997).

Tropical moist climates (A) are controlled by equatorial and tropical air masses. They extend northward and southward from the equator to about 15 to 25° of latitude. All months have average temperatures greater than 18°C. Annual precipitation is greater than 1,500 mm. Three minor Köppen climate types exist in the A group, and their designation is based on seasonal distribution of rainfall.

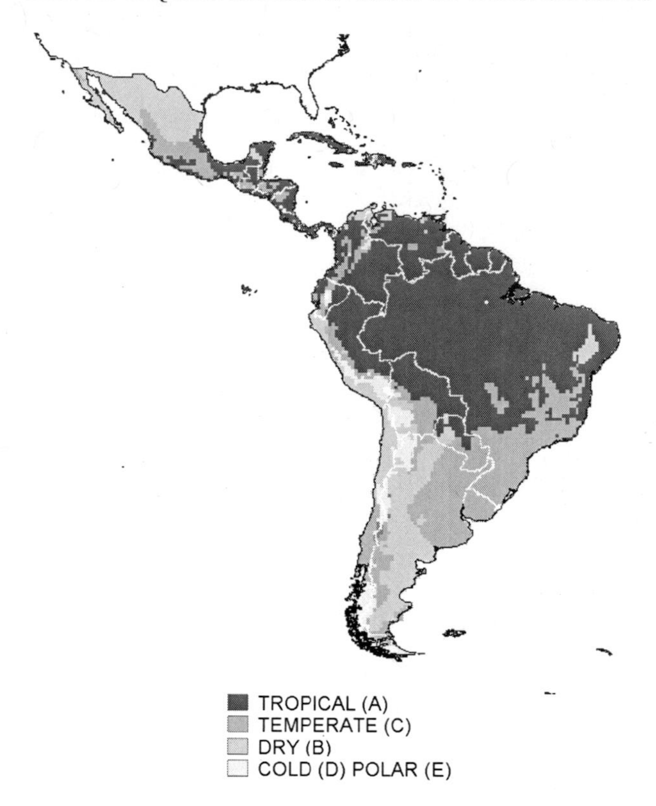

FIGURE 1.9. Climates of Latin America (Koeppen types). (*Source:* Created from data extracted from FAO-SDRN, 1997.)

The tropical wet (Af) is a tropical climate where precipitation occurs all year round and total rainfall is 2,000 mm or greater. The monthly temperature averages vary from 24 to 30°C. It is the typical Amazonian climate (Figure 1.10a).

The tropical wet and dry climate (Aw) has a distinct dry season, with at least one month with precipitation <60 mm. The total rainfall is normally lower than in the Af type. It is the climate of the southwestern side of the Amazonian basin.

The tropical monsoon climate (Am) has an annual rainfall lower, equal to, or greater than Af. Most of the precipitation falls over seven to nine months. The seasonal pattern of moisture is due to the migration of the intertropical convergence zone. The wet season is synchronous with the high sun and the presence of the convergence zone. During the rainy season, the climate is warm and humid, similar to the tropical wet climate. During

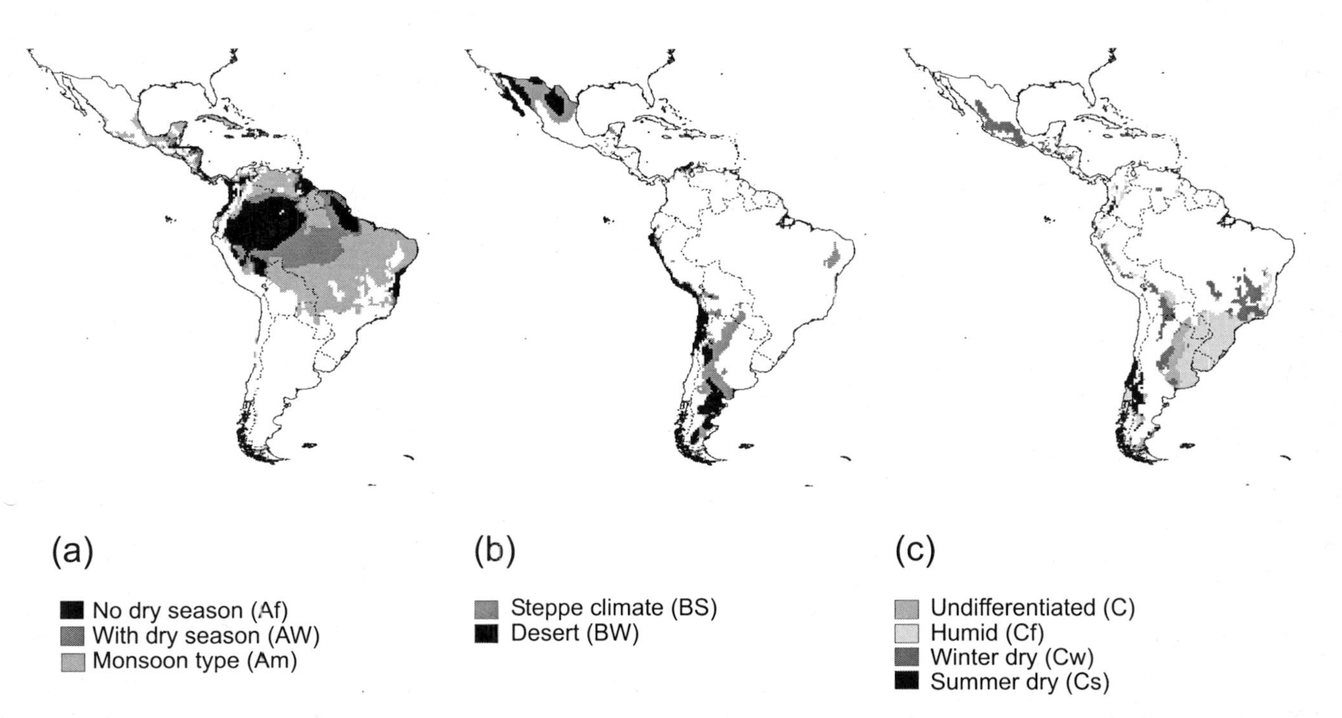

FIGURE 1.10. Climate subtypes: (a) tropical climate subtypes, (b) dry climate subtypes, (c) temperate climate subtypes. (*Source:* Created from data extracted from FAO-SDRN, 1997.)

the dry season very little rainfall occurs. It is the climate of central Brazil, northeastern Brazil, the coast of Venezuela, and central Mexico.

The dry climates (B) extend north and south of the equator and in large continental regions of the subtropics and midlatitudes. The most obvious feature of these climates is that annual evaporation exceeds annual precipitation. The two main subtypes are the dry semiarid climates (BS) and the dry arid climates (BW).

The dry semiarid climate (BS) is characterized by steppe vegetation. It receives more precipitation than the BW. With less rain, the steppe would be classified as an arid desert. With more rain, it would be classified as a tall-grass prairie. The semiarid steppe climates cover considerable parts of northern and western Mexico and western Argentina (Figure 1.10b).

The dry arid (BW) is a true desert climate. It is dominated by xerophytic vegetation. The dry tropical desert climate predominates in low-latitude deserts approximately between 18 to 28° in both hemispheres. It has major expanses in the east of the Andes and in narrow regions in southern South America (between 20°S and 30°S from the coastal desert of Atacama to the Argentine Chaco and Patagonia) and in northern Mexico where aridity is universal except at higher elevations.

The temperate climates (C) are mainly found between 30 to 50° latitude. They have a seasonal regime characterized by a cold or mild winter and a warm summer. The average temperature of the coldest month is <18°C and that of the warmest month is >10°C. They are generally moist climates with mean annual rainfall ranging from 500 to 5,000 mm, and have warm and humid summers with mild winters.

In the Cs type (Mediterranean climate) the summer is a dry season. Aridity may extend for up to five months. It rains primarily during the winter season. Locations in Latin America are California, central Chile, and central-western Argentina (Figure 1.10c).

In the Cw type winter is a dry season with at least ten times less precipitation in the driest month of winter than in the wettest month of summer. This climate type is centered on the tropics of Cancer and Capricorn. It coincides with the intermediate elevations of the highlands of central Mexico and central-eastern Brazil. It also occurs in central and northeastern Argentina.

The humid Cf type (humid subtropical climate) is characterized by at least 30 mm precipitation in the driest month, and the difference between the wettest and driest months is less than between Cw and Cs. It occurs mainly on the eastern border of the continent. The humid subtropical climate has hot humid summers and mild winters. It is located in southeastern Brazil-Uruguay-northeastern Argentina with annual precipitation ranging from 2,000 mm in Brazil to less than 1,000 mm in northern Argentina. It occasionally occurs in Central Mexico.

The polar climate (E), or subboreal, is the climate of the southeastern tip of the continent. The average temperature of the warmest month is greater than 10°C, while the coldest month is less than –30°C. It extends northward from southern Chile through most of the Andean highlands of Chile, Peru, and Colombia. On the southern part of South America it is wet, temperate-cold, and very cold at high elevations. Northward in mountains and Andean highlands, the general mean annual temperatures are between 3°C in the south and 6°C in the north, and mean annual precipitation varies, decreasing from west to east.

The Pattern of Natural Vegetation

Broad vegetation classes according Fedorova et al. (1993) and JRC (2000) were extracted from ArcAtlas data (ESRI, 1996). At the very broadest level the lowland vegetation types of South America may be summarized as follows:

I. Forest vegetation (Figure 1.11)
  A. *Evergreen forest* (or evergreen rain forest) in Amazonia, the coastal region of Brazil (from Bahia to Serra do Mar), the Choco and the lower Magdalena Valley, and along the Atlantic coast of Central America to Mexico;
  B. *Semideciduous forest* (or semievergreen rain forest) close to the same regions, mainly the Guayana shield in the Amazon region, and also along the Pacific side of Mexico and Central America. The semideciduous forest also comprises the Brazilian Atlantic forest ("Mata Atlantica") and the central Brazilian Forest ("Cerradão").
  C. *Deciduous forest* (or dry forest) extends in a discontinuous strip from northwestern Argentina to northeastern Brazil, encompassing Chaco, eastern Bolivia and Caatinga, northern Colombia and Venezuela, coastal Ecuador and adjacent Peru, central and western Mexico, with scattered smaller patches elsewhere (also the Mediterranean climate region of central Chile);
II. Nonforest vegetation (Figure 1.12)
  D. *Wet savanna* of the cerrado in central Brazil, the Llanos de Mojos and adjacent Pantanal of Bolivia and Brazil, the llanos of Colombia and Venezuela, and the Grand Sabana and Sipaliowini savanna in the Guyana region;

E. *Grassland* in the Pampas region of northeastern Argentina and adjacent Uruguay and southernmost Brazil;
F. *Desert and xeric shrubland* on northern Mexico, the dry Sechura and Atacama regions along the western coast of South America, and the Monte and Patagonian steppes of southeastern South America.

Montane formations occur along the Mexican-Central American cordillera system with subtropical coniferous and mixed forests, tropical and equatorial montane forests, and the Andean Cordillera with tropical and equatorial montane forests, subboreal forest in southern Chile, montane

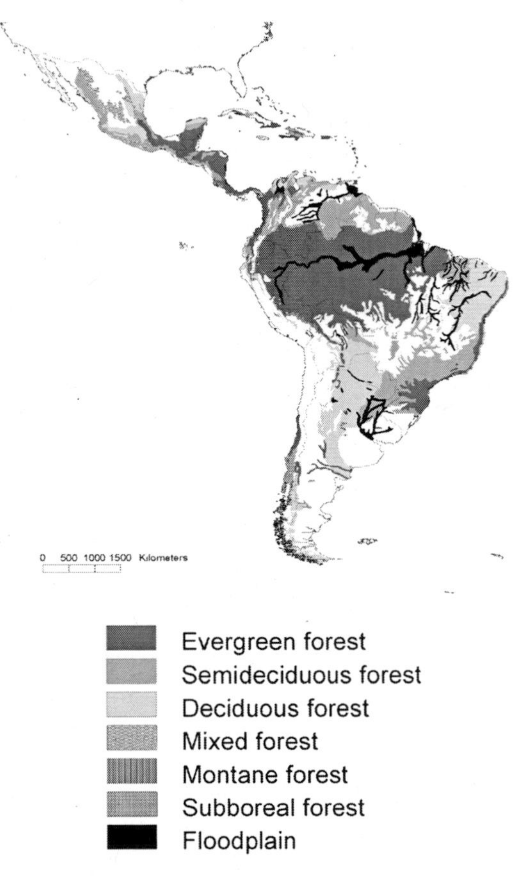

FIGURE 1.11. Forest vegetation in Latin America. (*Source:* Created from data compiled from ESRI, 1996, ArcAtlas database.)

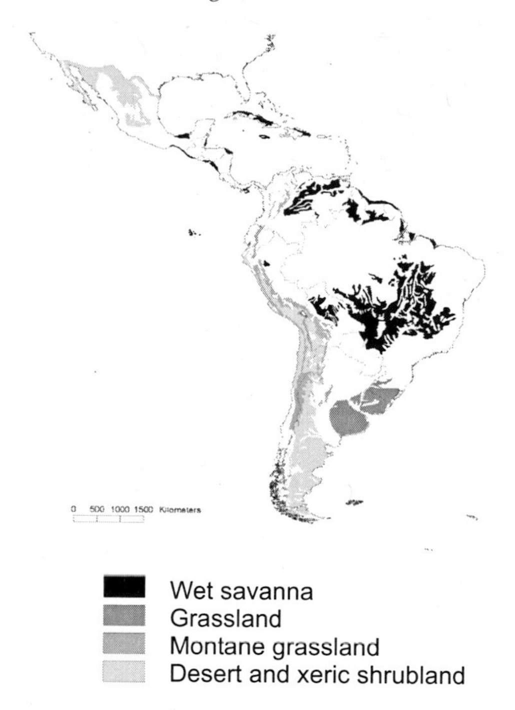

FIGURE 1.12. Nonforest vegetation in Latin America. (*Source:* Created from data compiled from ESRI, 1996, ArcAtlas database.)

grasslands (Puna, Paramos), and in the tepuis of the Guayana highland region.

Detailed descriptions of these vegetal formations can be found in NSG-WWF-ESRI (2001).

Pattern of Soil

The sources of information are the published large-scale soil maps (1:1,000,000 to 1:5,000,000) synthesized on the SOTERLAC map (FAO, 2001). This map applies the Revised Legend FAO (FAO, 1990) founded on observable soil characteristics (diagnostic soil horizons). For the purpose of a broad generalization on the scale of the whole of Latin America, we have integrated the original soil units in larger units using genetic criteria (soil-forming conditions and processes) instead of diagnostic horizons as in the former soil classifications (Marbut, 1928; Baldwin et al., 1938; Aubert and

Duchaufour, 1956). The resulting distribution pattern is easier to understand and reflects better environmental relationships.

Classically, soils can be grouped into two broad categories: zonal soils directly related to the regional bioclimatic environment and azonal soils determined by local factors such as topography or the parent material.

Zonal soils are directly related to the broad bioclimatic zones. Five main zonal types are found in Latin America.

Ferrallitic Soils (Kaolinic Soils)

Ferralltic soils (Robinson, 1949) group. These are the Ferralsols (or Latosols), Acrisols, Lixisols, Plinthosols, Nitosols, and part of the Cambisols. Ferrallitic soils are formed under wet tropical conditions. The wet tropical climate promotes strong chemical weathering of rocks. Feldspars and ferromagnesian minerals are substituted by clay minerals, mainly kaolinite and sesquioxides, the sand content reflecting the content of coarse quartz in the original parent rock. After long periods of chemical weathering the "saprolite" (weathered rock) may extend down to a depth of more than ten meters. Ferrallitic soils are thoroughly weathered and extensively leached soils. They are red or yellow in color, deep, finely textured, contain no more than traces of weatherable minerals, strongly leached, and have low-activity clays. They are prevalent in large parts of the eastern South American Precambrian shields, the sedimentary plains comprising the Amazon basin, and Guyana's coast in regions with wet tropical climate. But they are also largely represented in tropical regions with a pronounced dry season, as well as in regions with subtropical warm climate or with semiarid climate (Figure 1.13). Many of these last soils were most likely formed under a previously more humid climate. Ferrallitic soils are also sparsely found in the mountain folded belts of Central America and the equatorial Andes and Central America,

Ferralsols occur on old, stable geomorphic surfaces, typically in level to undulating terrain. They are particularly well represented on old erosional or depositional surfaces in the Amazon Basin and central and eastern South America (Figure 1.7). They are found in the semiarid regions in northeastern Brazil, where they must be considered as fossil soils (Volkoff, 1985).

Acrisols and Lixisols are ferrallitic soils in which clay has washed out of an eluvial horizon down to an argic subsurface horizon that has low-activity clays. Acrisols have a low base saturation level. Lixisols have a moderate to high base saturation level. Acrisols and Lixisols are found together with Ferralsols, generally with Acrisols or Lixisols on slopes. In South America Acrisols are prevalent in the humid Amazon basin and southeastern coastal

FIGURE 1.13. Ferrallitic, fersiallitic, and siallitic soils in Latin America. (*Source:* Created from data compiled from FAO, 2001, SOTERLAC database.)

regions. Lixisols occur in their subhumid periphery, extending mainly into northeastern Brazil. On a continental scale there is no apparent zonality for Ferralsols, but there is a clear climatic zonality of Acrisols and Lixisols.

Elsewhere in regions with wet tropical climates Nitosols may occur on basic rocks, and Plinthosols are frequently related to depressional areas. In these regions the typical setting of ferrallitic Cambisols is on the eroding slopes of low hills and uplands and on the steeper slopes of mountain areas.

Fersiallitic Soils

These soils are considered to be zonal soils in subtropical climates (Botelho da Costa, 1959; Volkoff, 1998). Some fersiallitic soils are neutral or eutrophic soils, for example, ferric Luvisols in northern Brazil, while others are acidic, distrophic soils. Alisols represent the fersiallitic acid soils.

Neutral fersiallitic soils (ferric Luvisol) are common in warm subtropical regions with distinct dry and wet seasons (e.g., Mediterranean climate). In Latin America they occur on gently sloping young land surfaces. They are moderately weathered soils and have high-activity clay mineral (hydromica, smectite). They typically have a brown to dark brown surface horizon over brown to strongly brown or red subsurface horizon. Soft powdery lime may occur in the subsoil horizon in the drier climatic zones. They are usually associated with Cambisols. Neutral fersiallitic soils are found in central coastal Chile and northeastern Brazil.

Acid fersiallitic soils (Alisols) are formed on strongly but incompletely weathered materials, normally under moist subtropical climates. Secondary high-activity clays (vermiculite and smectite) dominate the clay complex. They are strongly acidic, red or brown-yellow in color, and they are most common in old land surfaces with a hilly or undulating topography in southeastern Brazil and on eroding steep slopes in the foothills of the Andes (Paraguay and western Amazon basin).

Siallitic Soils or Haplic Luvisols (Brown Soils)

They are soils of wet temperate climates and have a small extent. They occur in temperate central Argentina. Vertisols, commonly associated with siallitic soils, have a greater extension. They are found in Uruguay, southern Argentina, and northeastern Mexico (Figure 1.13).

Steppic Soils

Steppic soils are Phaeozem, Kastanozem, and Chernozem. They are associated with a semiarid climate and steppe vegetation, and are characterized by a thick, dark surface layer that is rich in organic matter and bases. Their agricultural potential is generally high. Phaeozems are soils of prairie regions and occur under subhumid conditions. They are dusky red soils with high base saturation. Kastanozems have a brownish surface layer and carbonate and/or gypsum accumulation at some depth. They occur in the driest parts of the steppe zone. Chernozems have a deep, very dark surface layer and carbonate enrichment in the subsoil. Steppic soils are the principal soils of Uruguay and northeastern Argentina.

Xeric Soils or Aridisols (Calcisols and Gypsisol)

These are mineral soils with low organic matter content. Redistribution of calcium carbonate and gypsum is an important mechanism of horizon differentiation in soils. They are associated primarily with arid climates. Calcisols are soils with secondary carbonate enrichment. Gypsisols are soils with a horizon of secondary gypsum enrichment. Xeric soils are extensively found in northern Mexico, central Argentina, and central Andean regions (northwestern Argentina and central Chile) (Figure 1.14).

The zonation of fersiallitic, steppic, and xeric soils is commonly north-south. It is independent of the main morphostructural units, which on a continental scale indicate the role of the climate and the vegetation in soil formation.

Azonal soils are divided in two groups: soils conditioned by the topography and those conditioned by the parent material.

Soils Conditioned by Topography

The soils conditioned by topography include (1) incompletely developed and eroded soils (Cambisols and Regosols), and (2) hydromorphic wetland

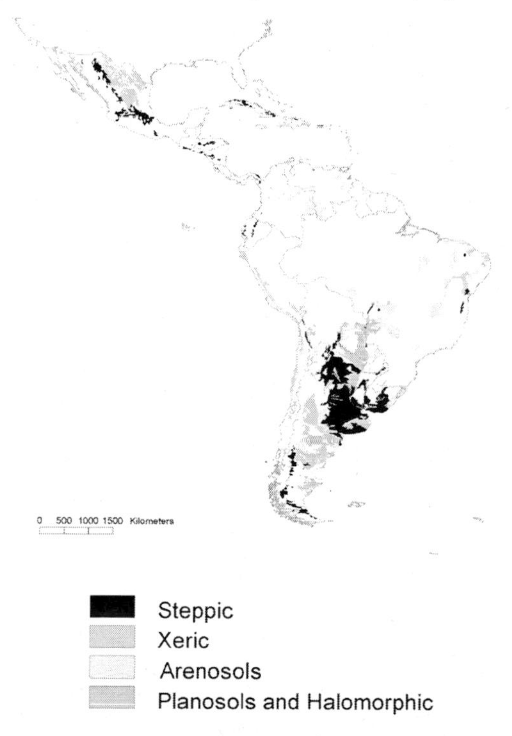

FIGURE 1.14. Arid soils in Latin America. (*Source:* Created from data compiled from FAO, 2001, SOTERLAC database.)

soils (hydromorphic soils, Solonetz, Planosols) and alluvial soils (Fluvisols). Cambisols occur predominantly at medium altitudes in hilly and mountainous regions under relatively moist climates. They are especially present in the wet tropical zone (Figure 1.15). Regosols occur in widely differing environments. They are very shallow soils over hard rock (Leptosols or Lithosols) or also deeper soils that are extremely gravelly and/or stony in unconsolidated materials and which have only surficial profile development (Regosols). They are particularly common in mountain regions and in arid regions (Figure 1.15). Fluvisols are soils developed in alluvial deposits along rivers and lakes, in deltaic areas. Hydromorphic soils (Gleysols) are soils of the wetland, of depression areas and low-landscape positions with shallow groundwater. Hydromorphic soils are associated with halomorphic soils with a high content of sodium and/or magnesium ions (Solonetz in semiarid, temperate, and subtropical regions; Solonchaks in arid and semiarid regions), and Planosols, soils with a degraded eluvial surface horizon lying abruptly over dense subsoil, mainly in subtropical and temperate

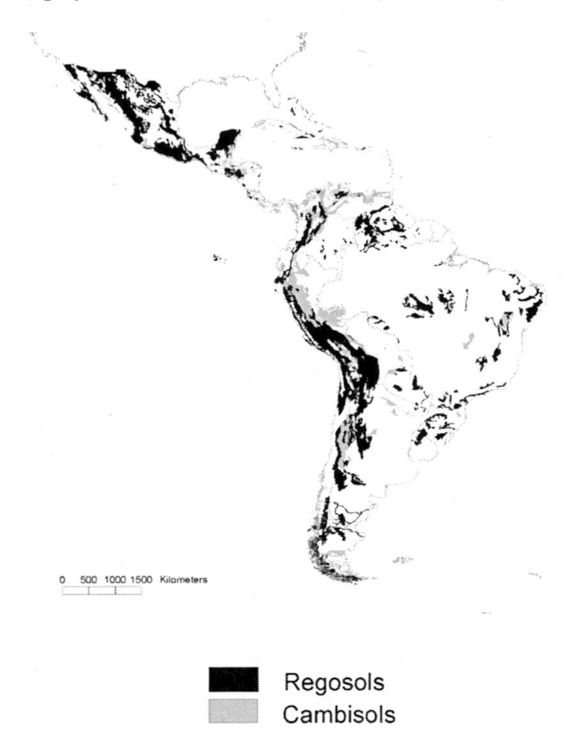

FIGURE 1.15. Cambisols and Regosols in Latin America. (*Source:* Created from data compiled from FAO, 2001, SOTERLAC database.)

and subhumid regions of southern Brazil, Paraguay, and Argentina (Figure 1.14).

Soils Conditioned by Parent Material

Arenosols are developed on sands (Figure 1.14), Andosols on weathered volcanic ash (also landscapes with fall of ash) (Figure 1.7), and Rendzina on soft limestone. Tropical Podzols, where leaching is the predominant soil-forming process, can be linked to the category of soils conditioned by sandy parent material. Some Vertisols are also directly related to sedimentary clay deposits.

SOIL REGIONS IN LATIN AMERICA

The broad geographic belts determined by climate and vegetation with specific types of genetic zonal soil are subdivided into soil regions. A soil region is defined by one or several soil associations and a distinctive spatial distribution of these soils. Soil associations and the spatial structures (macrostructures of the soil cover) are determined by the regional characteristics of the environment, including landscape, drainage pattern, and soil parent materials (bedrock, superficial sedimentary cover, present formed or inherited saprolites), and local variations of the vegetal cover.

Following Glazovskaya (1984), Latin America was divided into 13 regions (Figure 1.16).

Amazonian Region

This region constitutes the core of the South American continent. It lies on both sides of the equator, and on crystalline basement and sedimentary rocks. It matches approximately with the Amazon basin and a large part of the Orenoque basin. It is mainly a low plain. The central lowlands area is a broad sedimentary plain, composed of Cenozoic deposits, with altitudes below 200 m. The river valleys isolate a set of well-drained low plateaus. The bordering Guyana and Brazilian shields have a smooth, hilly relief. The altitude increases gradually toward the periphery of the region where, particularly in the Brazilian shield, the sloping hilly relief alternates with vestiges of perfectly leveled, old geomorphic surfaces. The only noticeable elevations are the Guyana highlands.

Within this extended area the climate ranges from wet tropical without a dry season to tropical monsoon. The evergreen forest, the Amazonian "rain forest," covers most of the region. Wet savanna occurs in both the north and

FIGURE 1.16. Soil regions in Latin America. (*Source:* Adapted from Glazovskaya, 1984.)

the south. Colombian llanos were provisionally included in the region because of their climatic environment being similar to the Amazonian climate.

Ferrallitic soils are almost exclusive to the Amazonian region. Ferralsols are closely associated with Acrisols and predominate on the well-drained plateaus of the low plain and on the higher geomorphic surfaces. Plinthosols occur in the lower, poorly drained areas. Acrisols-Cambisols associations characterize the hilly areas.

At the scale of the total Amazonian region the distribution of the soil associations is determined by bioclimatic factors that infer a large zonal soil macrostructure. At a lower scale watershed-organized macrostructures prevail, as can be observed in the main secondary basins, for example in the Rio Negro, Tapajós, or Purús basins.

Andean Equatorial Region

It is predominantly a mountainous region following north from the equator to Colombia. The mountains rise above 5,000 m, the highest peaks being volcanoes. There are also leveled surfaces lying at 2,500 to 3,000 m. The climate is mainly wet tropical with very high annual precipitation in the northern Pacific coastal parts. The vegetation is the montane rain forest on the eastern Pacific and Caribbean-Colombian coasts, mountain semideciduous forests in the midaltitude zones, and montane grasslands ("paramo") on the high-mountain zones.

The region is mainly characterized by zonal mountain bioclimatic soil macrostructures. Specific macrostructures related to the volcanoes and to their present and past activity bring an additional soil pattern. It can be assimilated to a "volcanic macrostructure." Acrisols and ferrallic Cambisols are formed below 2,000 m, Cambisols on midaltitude zones, and Regosols and Leptosols on paramo above 3,000 to 4,000 m. Andosols are formed on the volcanic ashes that cover the steep slopes of volcanoes and some levelled high surfaces.

Central American Region

It comprises parts of Mexico, Guatemala, El Salvador, Honduras, Nicaragua, Costa Rica, and Panama. This region has a complex topography. In the north the transvolcanic belt is a large mountain range running from west to east in the central portion of Mexico. The Sierra Madre del Sur runs parallel to the Pacific coast in a northwestern direction. The Sierra Madre de Chiapas runs parallel to the Pacific coast. In southern Panama the elevation is less than 500 m. The Yucatan Peninsula is a relatively lowland region.

The climate is mainly wet tropical. In the south, rainfall ranges from about 2,500 mm per year in central Panama to over 5,000 mm per year in southern Nicaragua. In the north of the region the climate is wet temperate (subtropical). The maximum rainfall is in summer; the winter is dry. Annual precipitation is between 1,200 to 2,000 mm on the mountain slopes and 600 to 800 mm in the interior parts.

In the south and on the eastern mountains and plains the vegetation is an evergreen tropical forest. The ferrallitic soils predominate and include Acrisols and Nitosols. Regosols-Leptosols (Rendzina) are found on Yucatan limestone.

In the central, western, and northern parts, soils and vegetation vary widely both on the mountain slopes, according to their orientation, and on the intermontane plateaus and depressions. Vegetation ranges from montane evergreen forest to montane deciduous and mixed forests (pine-oak forests). Cambisols are found on the steeper slopes. Fersiallitic soils, which can be either neutral or acidic, are formed in the interior regions. Andosols are present on volcanic ashes, and basaltic lava is present on volcano slopes as well as on uplands and intermontane depressions.

The main soil spatial distribution pattern is zonal mountain bioclimatic with occurrence of volcanic macrostructures.

Brazilian Atlantic Region

This region is located between 7°S and 27°S. It covers the easternmost elevated (800 to 2,000 m) part of the Brazilian highland and its eastern slopes facing the Atlantic Ocean. The eastern slopes receive at least 1,500 mm per year of rainfall, with either a uniform precipitation distribution throughout the year or the occurrence of a short dry season.

It is divided into two subregions: the northeastern Brazilian soil subregion and the southeastern Parana soil subregion.

The northeastern Brazilian soil subregion extends from the Serra do Mar in Rio de Janeiro and Sao Paulo states to Bahia state along the Atlantic coast. The climate is typically wet tropical. The coastal forests are evergreen tropical forest. The evergreen forests are bordered westward through Minas Gerais and Bahia states, by semideciduous and deciduous forests (Bahia interior forest). The soil cover is mainly composed of Ferralsols and Acrisols.

The southeastern Parana soil subregion is located on a basalt plateau. The subregion has a humid temperate (subtropical) climate with no dry season. Plateaus are occupied by coniferous *Araucaria* forests with sparse areas of tallgrass prairie. Tropical evergreen forests are found along the river valleys. The *Araucarias* forest is bordered on the west by a semideciduous forest (Parana/Parnaiba semideciduous forest). The soil cover is composed of Alisols in association with Ferralsols on plateaus and Acrisols on the slopes.

The latitudinal zonality attributable to the climatic change from the north, tropical, to the south, and subtropical is partly masked by paleoclimatic influences. Most of the Ferralsols covering the inner and border plateaus are remnant of an ancient landscape that is probably not in equilibrium with the present bioclimatic environment. Geomorphology and lithology are the factors controlling the soil spatial patterns. Paleoclimatic and disordered lithologic soil macrostructures characterize this soil region.

Central Brazilian Region

The region represents the area of the Brazilian central plateau. The climate is typically tropical wet and dry. The dry period, which occurs from May through September or October, coincides with the coldest months of the year. The average annual rainfall varies between 1,250 and 2,000 mm, and the average annual temperature ranges between 20 and 26°C.

The Central Brazilian soil region is characterized by wet savanna vegetation (Campo cerrado) that is a woodland savanna. Areas of grassland (Campo limpo) and semideciduous forests occur within the cerrado.

Soils of this region are essentially ferrallitic soils. The main soil type is a Ferralsol that is very similar to the Ferralsol of the Amazonian region. They are strongly leached and have a stable microaggregate structure. Many soils have high contents of iron oxides, which accounts for the very low cation exchange capacity. Ferralsols are associated with Acrisols (Podzolico Vermelho Amarelo), and those of the Central Brazilian soil region have developed on old, highly weathered materials that are the result of a long weathering process probably initiated in the beginning of Cenozoic era. Ferralsols are found on extended residual plateaus, and Acrisols occupy the dissected lower areas between them. Soils associated with products of the weathering of basalts, basic rocks (dolerites), and sandstones are widespread. Dark red Ferralsols and Nitosols are found on basalts, mainly in the southern parts. Arenosols are distributed throughout the area. The soil spatial organization is controlled by geomorphology and lithology. Paleoclimatic and disordered lithologic macrostructures are characteristics of this soil region.

East Brazilian Region

This region occupies northeastern Brazil (Piauí, Ceará, Rio Grande do Norte, Paraíba, Pernambuco, Sergipe, Alagoas, Bahia, and northern Minas Gerais) on crystalline basement or on sedimentary deposits. It is a gently undulating plain whose continuity is broken by isolated plateaus, remnants of several geomorphic cyclic old surfaces.

Annual rainfall ranges from 250 to 1,000 mm, and the average annual temperature is between 24 and 26°C. The climate is tropical with five to six dry months. It is a dry climate (subarid dry climate) with 6 to 11 dry months in the interior regions.

The natural vegetation is a deciduous forest that ranges from dry tropical forest to "caatinga," a shrubby sclerophyllous vegetation. A cerrado-like vegetation covers the residual plateaus of the region. The "agreste" is a deciduous forest along the coastal wetter areas. The soil cover is represented by several associations: fersiallitic soils, Regosols, Planosols on crystalline lowlands, ferrallitic soils (Lixisols) and Arenosols on sedimentary lowlands, and ferrallitic soils (Ferralsols and Lixisols) on residual plateaus. Soil cover pattern is characterize by paleoclimatic (Ferralsols covering the residual plateaus must be considered as Paleosols) and disordered lithologic macrostructures.

Paraguay-Preandean Region

This region extends meridionally between 15°S and 40°S and occupies mostly flat interior plains and foothills of the eastern slopes of the Andes. It encompasses the eastern lowlands of Bolivia situated at the southern limit of Amazonian forests (Chiquitano), the Pantanal located near the borders of Brazil, Bolivia, and Paraguay, the Chaco, east of the Andes in southeastern Bolivia, in the western and center of Paraguay, and northwestern and northeastern Argentina, the Córdoba Montane and "Pampa occidental," transitional vegetation to Monte in central and southern Argentina.

The annual precipitation in this region decreases from east to west and from north to south. The Chaco has a mean annual rainfall of 950 mm and is characterized by a strong dry season during the winter. The north and the east are characterized by a mean annual rainfall of 1,000 to 1,400 mm and by a strong dry season during the winter. At the southern and western boundaries the climate is dry (dry semiarid), with an annual rainfall of 350 to 650 mm, and an average temperature of 12 to 28°C.

At the limit of Amazonian forests, deciduous forest and wet savanna mark the transition to drier thorny scrub forests that extend farther south in the Chaco. In the east the vegetation is characteristic of deciduous xerophytic forests, palm groves, and grassy savannahs. In lower areas that are easily subject to flooding, there are grasslands and bogs. The Chaco vegetation consists of xerophile forests mixed with palm savannas. The forests are composed of quebracho colorado *(Schinopsis balansae)* and quebracho blanco *(Apidosperma quebracho-blanco).*

Three soil subregions can be distinguished: a northern gently undulating plain with ferrallitic soils; a central, leveled plain characterized by recent intracontinental sediments with fersiallitic and halomorphic soils; and a southern area frequently covered by loess with steppic soils.

The ferrallitic soils of the northern subregion are Ferralsols and Acrisols, most of which are Paleosols. The main soils of the central subregion are chromic Cambisols and chromic Luvisols grading to Phaeozems. They are associated with Solonetz and Gleysols. Soils of the southern subregion are predominantly typical Phaeozems. Regional soil cover is characterized by zonal, plain, bioclimatic macrostructure and watershed-organized macrostructure.

Caribbean Region

The region comprises a continental part, the northern plains of Venezuela, the northeastern Andes, and the islands of the Greater and Lesser Antilles.

The general climate is tropical wet and dry. The dry season, three to five months long, occurs between December and April. Total annual rainfall ranges from 1,000 to 1,600 mm. It is lower (300 to 1,000 mm per year) in the northeast Venezuelan coast, and higher on the southwestern slopes of the northeastern Andes (2,500 mm per year). The mean annual temperature is 27°C.

The natural vegetation is a complex mosaic of evergreen tropical forest, deciduous forest, and wet savanna. Xeric shrublands are found on the coastal cordillera in the northern part of Venezuela. A uniform savanna (llanos) is located along the Orinoco River. The mosaic of vegetation is even more complex in the islands. However, the evergreen tropical forest is the most significant Antillean vegetation.

The most representative soils of the continental parts are the ferrallitic soils (Acrisols) in llanos and fersiallitic soils (Luvisols and Alisols) associated with Cambisols in the drier parts of the Venezuelan coast. In the Caribbean islands (Cuba, Haiti-St. Domingue) very diversified soil mosaics occur, with the frequent occurrence of Acrisols. It has a disordered lithologic soil macrostructure

South American Meadow Region

The region extends north and south from the estuary of the La Plata River. It is relatively flat, ranging from sea level to elevations of about 500 m in some areas. The complex geology includes Precambrian, Cretaceous, and Jurassic rocks, as well as more recent sedimentary rocks (loess, sandstone, limestone). Many freshwater and saltwater lagoons are present.

The climate is humid temperate (subtropical). The annual average rainfall ranges from 1,000 to 1,600 mm. The rainfall is distributed uniformly throughout the year; summers are hot (24 to 27°C) and winters are mild (10 to 16°C).

The vegetation is comprised of a tallgrass meadow ("Pampas"). Xeric vegetation occurs in the southern part.

Soils of the Pampa plain are dominantly Phaeozems. Phaeozems are associated with fersiallitic soils (ferric Luvisols and Alisols) and Vertisols in the northern parts of the region (Uruguay and southern Brazil). A plain bioclimatic soil zonality is observed.

Mexican Region

The Western Sierra Madre separates the two principal dry zones of Mexico: the Chihuahuan Desert in the east and the Sonoran Desert in the west. The Chihuahuan Desert corresponds to the major portion of the central plateau of Mexico. Eastward, the central plateau is bounded by the Eastern Sierra Madre. The coastal gulf plain begins in the east at the base of the Eastern Sierra Madre.

In the Chihuahuan Desert and the Sonora Desert, which comprises the Sonora coastal plain and the major part of Baja California, annual precipitation averages less than 200 mm, and less than 50 mm in some areas.

In the Western Sierra Madre temperatures and rainfall fluctuate widely due to great variations in elevation. Mean annual rainfall is around 500 mm, with the western sides generally receiving more rainfall than other regions. The temperature varies between extremes of –3°C and 28°C. Summers are wet and winters are mild. Pine-oak forests grow on elevations of 1,500 to 3,500 m. The southwestern slopes on the Western Sierra Madre are covered by semideciduous tropical forests. In the Eastern Sierra Madre, the average annual rainfall ranges from 250 to 300 mm in the north, and 900 to 1,500 mm in the southern parts. The climate is temperate humid on the northeastern slope, and temperate subhumid on the western slope and highest portions. Mixed pine-oak forests cover most of the mountains.

In the eastern coastal plain, the climate is dry and hot, with precipitation levels below 500 mm per year. Precipitation levels increase gradually toward the south. The native vegetation type covering much of northeastern Mexico and parts of southern Texas is mesquite-grassland. The vegetation then grades to a deciduous forest to the south

Regosols, Leptosols, and xeric soils (Calcisols) are the dominant soils of the deserts. Vertisols occur along the coastal gulf plain, and steppic soils (Phaeozems) occur along the eastern base of the Western Sierra Madre on the Mexican central plateau. There is predominantly a mountain bioclimatic soil zonality.

Central Andes Region

This meridionally extended region includes the Pacific coast, with latitudes ranging from the equator to 38°S, and the central Andes from Colombia, Peru, and Bolivia to northern Argentina. The coastal area is an almost uninterrupted desert (the Atacama Desert and the Sechura Desert) that occupies an extended continuous strip. The average width of the coastal area is less than 100 km. Close to the border of Peru and Bolivia, the Andean mountain range is divided into two mountain systems, the Cordillera Occidental and the Cordillera Oriental, with a large plateau in between called the Altiplano.

This coastal desert is virtually rainless. The xeric conditions extend up to 1,500 m on the western drier slopes. In Andean Peru and eastern Bolivia the average precipitation in the Altiplano ranges from 500 to 700 mm per year. The average annual temperature is low, ranging from 5 to 7°C. In southwestern Peru, precipitation is lower and varies between 250 and 500 mm per year. Eastern slopes of the Andes have a wet and humid climate, and rainfall typically exceeds 2,500 mm per year.

The Pacific coast is a desert. The vegetation of the high plateaus is the "Puna." It is a xeric shrubland that grades into montane grassland in humid areas. The eastern Andean slopes are covered with evergreen or deciduous forests (southern Andean and Peruvian yungas and Bolivian montane dry forests).

Regosols (Leptosol and Lithosols) are the dominant soils of the region, and are associated with Cambisols in areas less prone to erosion. Fersiallitic soils occur in association with Cambisols. They are neutral (chromic Luvisols) under deciduous forests and acidic (Alisols) under evergreen forests. The mountain bioclimatic soil zonality is the prevailing pattern. A latitude bioclimatic zonality occurs in the coastal area.

South Argentine Region

This soil region is connected with the Central Andes region and is restricted to the pre-Andean zone of western Argentina ranging from 30° S to the Tierra del Fuego in Chile and Argentina. The region covers the piedmont plains and intermontane depressions of the first range of the Andes. The northern and central parts correspond to the "monte desert." It is a plain that lies at an altitude of 1,000 to 1,500 m, the altitude decreasing toward the east. A number of closed internal drainage basins are present. In the south, Patagonia is a stepped plateau sloping from the foothills of the Andes to the east. The maximum altitude is 2,000 m.

The climate of the monte part is dry, semiarid, and arid; rainfall is 200 to 300 mm. The climate of Patagonia is very dry and cold; the average annual precipitation normally does not exceed 200 mm. The climate of southern Patagonia is cold and humid, with 200 to 300 mm of rainfall per year and an average temperature below 8°C.

Vegetation of the region consists of three main formations: the scrub monte desert, the Patagonian steppe, and the Patagonian Tierra del Fuego grassland. In the monte desert the dominant vegetative formation is xeric scrubland with evergreen bushes and cactus scrub. The Patagonian steppe is also xerophytic. It consists of resinous evergreen bushes and herbaceous species. In the Patagonian grasslands the dominant vegetation is a grass-steppe mixed with shrubs.

Most soils of the monte desert are Calcisols. Solonetz occur in relief depressions. In the northern and central Patagonia, soils are xeric Calcisols associated with halomorphic soils (Solonetz and Solontchak) in river valleys. Soils with steppic characteristics occur in southern Patagonia. In the Patagonian steppe zone, Kastanozems are associated with calcic and haplic Luvisols, and in the Patagonian Tierra del Fuego grassland zone, Phaeozems are associated with Cambisols. The general spatial soil pattern is zonal, bioclimatic plain-zonal.

South Chilean Region

The region stretches from 35°S to 56°S, i.e., to the southern tip of Tierra del Fuego. It is a mountainous region. In its northern part, the Andean cordillera rises to 4,000 m or more, whereas in the south, altitudes do not exceed 2,000 m. In the south, glaciers fill the valleys, whereas the northern part displays several active volcanoes. Intermediate depressions and central valleys are covered by volcanic ash and glacial moraines fields.

Most of the region receives 2,000 to 5,000 mm rainfall per year. On the eastern slopes facing Patagonia, the rainfall decreases to 1,000 mm per year or less. Rainfall is distributed regularly around the year. The climate is "moist mid-latitude type with cold winters." It is a temperate, or subboreal, climate, temperate-cold in the south. The vegetation is a temperate (or subboreal) broadleaf and mixed forest: the Valdivian temperate rain forest in the north, the Magellanic subpolar forest in the south.

The mountain bioclimatic zonality of the soils is not clearly apparent because Regosols (Leptosols or Lithosols) are the predominant soil types of the region. Only slightly developed soils (Cambisols) are found on the less eroded parts of the topography. Andosols are largely represented in the north.

CONCLUSION

Three groups of soil ecoregions are found in Latin America.

1. Arid soil ecoregions where the predominant soils are undeveloped soils and slightly developed soils directly correlated to the present arid climate. They consist of the Mexican, the Central Andes, and the South Argentina regions. Within each ecoregion, variations in the soil characteristics are associated with climatic changes. A range of climatic subzones is determined by either geographic or orographic factors.
2. Mountain soil ecoregions where limited soil development is associated with the steep topography. Ecoregions in this group are differentiated according to their position in latitude, which determines the bioclimatic environment as well as the main characteristics of their soils and of their soil associations. These regions are the Central American, the Andean Equatorial, and the South Chilean. Within each ecoregion, soil characteristics vary according to the altitude (vertical bioclimatic zonation).
3. Other soil ecoregions, where wetter climates coupled with a fairly leveled topography allow soil development and conservation of an ancient soil cover. Although this is not a general rule, the boundaries of these regions can occasionally match the limits of natural physiographic units, as is the case in the Amazon basin. In general, the boundaries are gradual, thus relatively imprecise, the definition of the soil ecoregions being based on criteria that do not account for clear geographical limits, such as climate-vegetation combinations, soil associations, and soil cover macrostructures. Soil distribution rules are specific to each region. Spatial patterns are determined by geomorphologic and lithologic factors as well as by the climate-vegetation combinations. Therefore, the Amazonian region is a wet tropical forest region, with a relatively homogeneous soil cover dominantly composed of Ferralsol; the Paraguay pre-Andean region has a tropical wet and dry climate that grades to dry climate toward the south, and contains a parallel soil zonation that ranges from Acrisol to Luvisol and Phaeozem; the Eastern Pampas region has a subtropical wet (or temperate) climate and a relatively homogeneous soil cover composed of Phaeozem; the Caribbean region has tropical wet and dry climate and a complex soil cover mainly due to a heterogeneous lithology; the Central Brazilian region has a tropical wet and dry climate and a relatively homogeneous soil cover mainly composed of inherited Ferral-

sol; the East Brazilian region has a tropical wet and dry climate and a contrasted soil cover composed of Luvisol and inherited Ferralsol; the Brazilian Atlantic region has a subtropical wet (or temperate) climate with contrasted soils consisting of Alisol and Ferralsol. A number of regions, including the Paraguay pre-Andean and the Brazilian Atlantic regions, consist of two or more overlapping climatic zones and need to be further subdivided.

The major characteristics of the soil ecoregions are summarized in Tables 1.1 and 1.2.

In conclusion, organizing soil and environmental data in soil ecoregions significantly reduces the spatial variability. By working at this scale, it will be possible to assess the potential of soil C sequestration on the basis of the dominant soil types, and then to build spatial models that could be used to establish more accurate evaluations.

REFERENCES

Aubert, G. and P. Duchaufour. 1956. Projet de classification des sols. In C.R. 6th Cong. Inter. Sci. Sol, Paris, E, pp. 597-604.

Baldwin, M., C.E. Kellog, and J. Thorp. 1938. Soil classification. In *Soil and men: Yearbook* (pp. 979-1001). USDA, Washington, DC.

Botelho da Costa, J.V. 1959. Ferrallitic, tropical fersiallitic and tropical semi-arid soils. Definition adopted in the classification of soils of Angola. In C.R. 3rd Congr. Interafr. Sols, pp. 317-319.

CGMW-UNESCO. 2000. Geological map of the world at 1:25000000. CD-ROM. UNESCO, Paris.

Clapperton, C. 1993. *Quaternary geology and geomorphology of South America.* Elsevier, Amsterdam.

Deckers, J., F. Nachtergaele, and O. Spaargaren. 2003. Tropical soils in the classification systems of USDA, FAO and WRB. http://www.fao.org/ag/agl/agll/wrb/doc/KAOWDeckerscorr280203.doc.

ESRI. 1996. *ArcAtlas: Our earth.* CD-ROM. ESRI, Redlands, California.

FAO. 1990. *Soil map of the world, revised legend.* World Soil Resources Report 60. FAO, Rome.

FAO. 2001. A soil and terrain digital database (SOTER) for Latin America and the Caribbean (SOTERLAC) at 1: 5000000 scale. Land and Water Digital Media Series 5. CD-ROM. FAO, Rome.

FAO-SDRN. 1997. Global climate maps. http://www.fao.org/waicent/faoinfo/sustdev/EIdirect/climate/EIsp0002.htm. FAO, Rome.

Fedorova, I.T., Y.A. Volkova, and D.L. Varlyguin. 1993. Legend to the world vegetation cover map. http://www.ngdc.noaa.gov/seg/cdroms/ged_iib/datasets/b01/fvv.htm.

TABLE 1.1. Characteristics of the soil regions: Environment.

Region name	Topography	Climate type	Climate subtype	Vegetation
Amazonian	Plain	Tropical	Af, Aw, Am	Evergreen forest and wet savanna
Andean Equatorial	Mountain	Tropical	Af, Aw, Am	Montane forest and montane grassland
Central American	Mountain	Tropical	Am, Aw	Montane and mixed forest
Brazilian Atlantic N	Plateau	Tropical	Af	Evergreen and semideciduous forest
Brazilian Atlantic S	Plateau	Temperate	Cf, Cw	Evergreen and semideciduous forest
Central Brazilian	Plateau	Tropical, temperate	Am, Cw, Cf	Wet savanna and semideciduous forest
East Brazilian	Plain and plateau	Tropical, dry	Am, BS	Deciduous forest
Paraguay Preandean N	Plain	Tropical	Am	Deciduous forest and wet savanna
Paraguay Preandean C	Plain	Dry, temperate	BS, Cw, Cf	Deciduous forest
Paraguay Preandean S	Plain	Dry	BS	Deciduous forest
Caribbean	Low mountain	Tropical	Af, Aw, Am	Wet savanna, evergreen forest, decidous forest
Eastern Pampas	Plain	Temperate	Cf, C	Grassland
Mexican	Mountain	Dry	BS, BW	Desert and xeric shrubland
Central Andes	Mountain	Dry, polar	BW, E	Desert, xeric shrubland, and montane grassland
South Argentina	Plain and plateau	Dry	BS, BW	Desert and xeric shrubland
South Chilean	Mountain	Polar	E	Subboreal forest

TABLE 1.2. Characteristics of the soil regions: Soils.

Region name	Main soils[a]	Other soils[a]	Soil macrostructure
Amazonian	Ferralsol, Acrisol	Cambisol, Plinthosol	Zonal plain bioclimatic and watershed organized
Andean Equatorial	Cambisol, Acrisol	Andosol	Zonal mountain bioclimatic
Central American	Cambisol, Luvisol	Andosol	Zonal mountain bioclimatic
Brazilian Atlantic N	Ferralsol	Acrisol	Paleoclimatic and disordered lithologic
Brazilian Atlantic S	Alisol	Ferralsol, Acrisol, Cambisol	Paleoclimatic and disordered lithologic
Central Brazilian	Ferralsol	Acrisol	Paleoclimatic and disordered lithologic
East Brazilian	Ferralsol, Regosol, Luvisol	Lixisol, Planosol	Paleoclimatic and disordered lithologic
Paraguay Preandean N	Ferralsol, Acrisol	Plinthosol, Solonetz	Zonal plain bioclimatic and watershed organized
Paraguay Preandean C	Phaezem, Solonetz	Luvisol, Gleysol	Zonal plain bioclimatic and watershed organized
Paraguay Preandean S	Phaeozem	Solonetz	Zonal plain bioclimatic and watershed organized
Caribbean	Acrisol	Luvisol, Alisol, Cambisol	Disordered lithologic
Eastern Pampas	Phaeozem	Alisol, Vertisol, Solonetz	Zonal plain bioclimatic
Mexican	Regosol, Calcisol	Leptosol, Phaeozem, Vertisol	Zonal mountain bioclimatic
Central Andes	Regosol	Cambisol, Chromic Luvisol	Zonal mountain bioclimatic
South Argentina	Calcisol, Luvisol	Solonetz, Kastanozem	Zonal plain bioclimatic
South Chilean	Regosol	Cambisol, Andosol	Zonal mountain bioclimatic

[a]Soils are named according to the Soil Map of the World legend (FAO, 1990); correlations with USDA Soil Taxonomy can be found in Deckers et al. (2003).

Fridland, V.M. 1974. Structure of the soil mantle. *Geoderma,* 12(1-2):35-41.
Glazovskaya, M.A. 1984. *Soils of the world: Soil geography.* Moscow University Publishing House, Moscow. English translation, Oxonian Press, New Delhi.
JRC. 2000. Global 2000 land cover (GLC). http://www.gvm.jrc.it/glc2000/default GLC2000.htm.
King, L.C. 1962. *The morphology of the earth.* Oliver and Boyd, London.
Marbut, C.F. 1928. A scheme for soil classification. Proc. 1st Int. Congr. Soil. Sci. 1, 1927, Washington, 4, pp. 1-31.
NSG-WWF-ESRI. 2001. Global 2000 ecoregions. http://www.nationalgeographic.com/wildworld/terrestrial.htm.
Robinson, G.W. 1949. *Soils, their origin, constitution and classification,* 3rd ed. Murby & Co., London.
SI-WWF-IUCN. 2001. CPD: The Americas. http://www.nmnh.si.edu/botany/projects/cpd/index.htm.
Tardy, Y. 1997. *Petrology of laterites and tropical soils.* Balkema, Rotterdam.
Thomas, M.F. 1994. *Geomorphology in the tropics.* John Wiley & Sons, Chichester, UK.
USGS. 1996. GTOPO30: Global 30 arc second elevation data. CD-ROM. EROS Data Center, Sioux Falls, South Dakota.
Volkoff, B. 1985. Organisation régionale de la couverture pédologique du Brésil: Chronologie des différenciations. *Cahiers ORSTOM* (Série Pédologie), 21(4): 225-236.
Volkoff, B. 1998. Red and lateritic soils: World scenario. In *Red and lateritic soils,* Vol. 1, J. Seghal, W.E. Blum, and K.S. Gajbhiye (eds.) (pp. 57-75), Oxford & IBH Publishing Co., New Delhi.

Chapter 2

Challenges and Opportunities of Soil Carbon Sequestration in Latin America

C. C. Cerri
M. Bernoux
C. E. P. Cerri
R. Lal

INTRODUCTION

Latin America (LA) covers an area of about 20.5 million km^2 containing 1 country in North America (Mexico), 7 countries in Central America, 20 countries in the Caribbean, and the 13 countries that comprise South America. The population of LA was estimated to be 535 million in 2002. Anthropogenic disturbances in LA started in the 1500s, mainly by Spanish and Portuguese colonists. As a result of colonization, a wide range of crops were adopted in different regions of LA to adapt to specific soil and climatic conditions. The best agricultural systems (including crops, varieties, fertilizer type and amount, control of diseases, etc.) to obtain high yields with better economic return, mainly for export, are being adopted in several regions of LA, yet large areas remain under native vegetation, of which some have been cleared for agricultural use and others are being managed under shifting cultivation. For these newly cleared areas, few, if any, research data indicate whether the methods used for conversion to agricultural ecosystems are appropriate, and whether indeed this is a suitable region for the specific crops or pasture being established. Yet answers to these questions are extremely important because of the drastic impacts of land-use change and cropping/farming systems on the environment in general and climate change in particular.

Agricultural land use in 2004 comprised about 38 percent of total area of LA, including 30 percent of pastures and 8 percent of arable land and permanent crops (FAO, 2004). In some regions, natural ecosystems are rapidly

Carbon Sequestration in Soils of Latin America

doi:10.1300/5755_02

being changed to pasture, cropland, and plantations with large emissions of CO_2, CH_4, and N_2O into the atmosphere. Therefore, the overall goals of this chapter are to (1) assess soil carbon (C) stocks under native vegetation without human intervention for the entire LA region, (2) estimate the original soil C stocks in soils presently under agricultural land use, and (3) discuss the challenge and opportunities for sequestering C in soils.

SOIL CARBON STOCK ESTIMATION UNDER NATIVE VEGETATION

The terrestrial biosphere contains about 1,500 to 1,600 Pg of C in the top 1 m of soil (Eswaran et al., 1995; Batjes and Sombroek, 1997; Lal, 2002) and an additional 470 to 655 Pg of C in the vegetation (Schimel, 1995). Together, C stocks in the terrestrial ecosystem are three times the amount of C in the atmosphere (Figure 2.1). The soil C stock in the top 30 cm is about 800 Pg C, which is almost the same amount as that stored in the atmosphere. It is important to realize that soils are fragile ecosystems. With improper land use, soil organic carbon (SOC) can be mineralized and transferred to the atmosphere as CO_2. With judicious management, however, the system can sequester C from the atmosphere into the soil, through plant litter decomposition and humification. The process of soil C sequestration, through

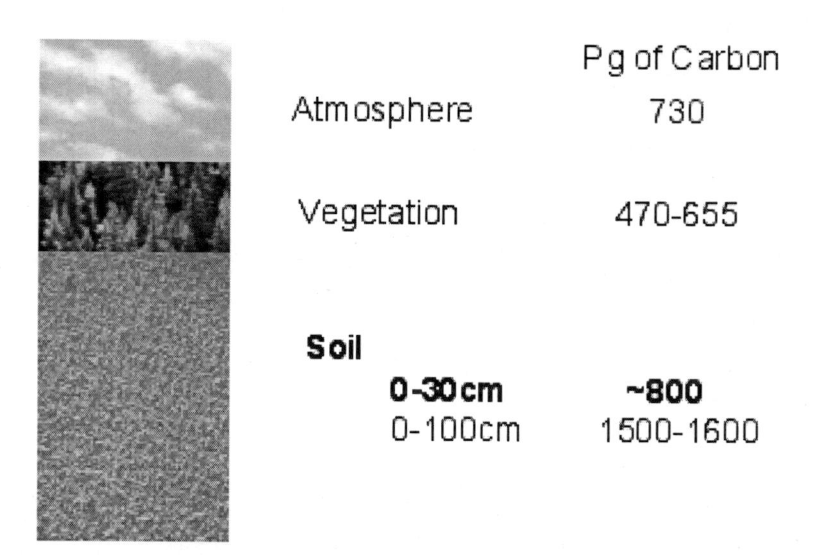

FIGURE 2.1. Global carbon stocks in the atmosphere and terrestrial ecosystems. Values are in Gt of C, where 1 Gt = 10^9 t = 1 Pg.

adoption of judicious land use and recommended management practices, is an important regional and global strategy to offset fossil fuel emissions and mitigate climate change.

Soil C stocks under undisturbed conditions in Latin America (Figure 2.2) were estimated by using natural vegetation maps in association with their soil C data available in SOTERLAC. As a first approximation, about 100 Pg of C was stored in the top 30 cm depth before human intervention in Latin America This stock corresponds to one-eighth of the global soil C stocks (Figure 2.2). This ratio of soil C stock of LA:world is similar to that between LA area and the world land area (1:7).

The LA region is favorably endowed with high soil C content and its geographical distribution. It contains average values of soil C stocks; in other words, there are no exceptionally high C concentrations as observed in the temperate zones, and there are no extremely low concentrations as observed in desert areas of the world. Under both conditions, edaphoclimatic aspects are limiting to agricultural land use. Therefore, LA is a promising region for expansion of agriculture land use since the desert area is very small and a temperate climate is not pronounced in the region. Despite this, the region has a medium value of soil C stock at the global scale. The

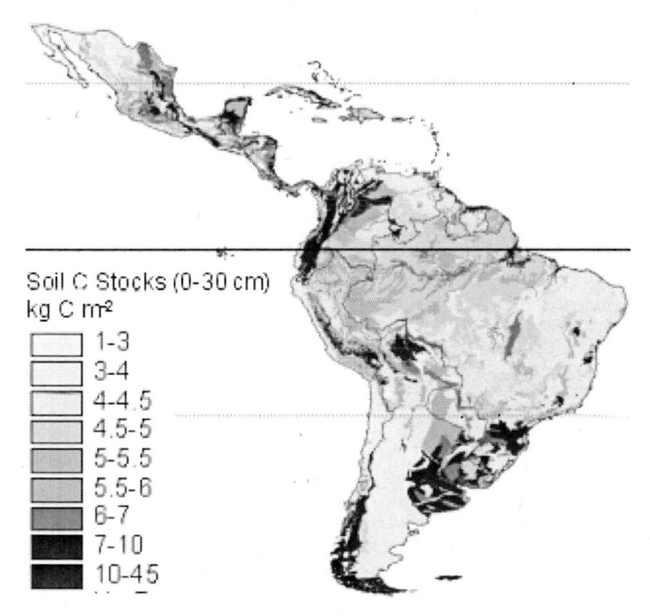

FIGURE 2.2. Distribution of soil C stocks (0 to 30 cm) under native vegetation in Latin America.

SOC stock is distributed in different ecoregions within LA. Those different ecoregions are described in Chapters 1 and 4 of this volume.

CHANGES OF SOIL CARBON STOCKS UPON LAND-USE CHANGE

Anthropogenic impacts on natural ecosystems of LA are relatively recent compared to other world ecoregions. The first colonization occurred about 500 years ago, but more intensive human intervention for agricultural use started only during the eighteenth century. Despite this more intense land-use change, only 25 percent of the total area has been used for cultivation. As an example, almost one-half of the world's undisturbed tropical evergreen forest is located in LA. Of this large forest area, only 12 percent had been deforested by 2000.

Figure 2.3 shows the areas impacted by human intervention, mainly for agricultural land use. The agricultural areas and its distribution in Latin America were obtained from the "Global Land Cover Map for the Year 2000" (European Commission Joint Research Centre, 2002).

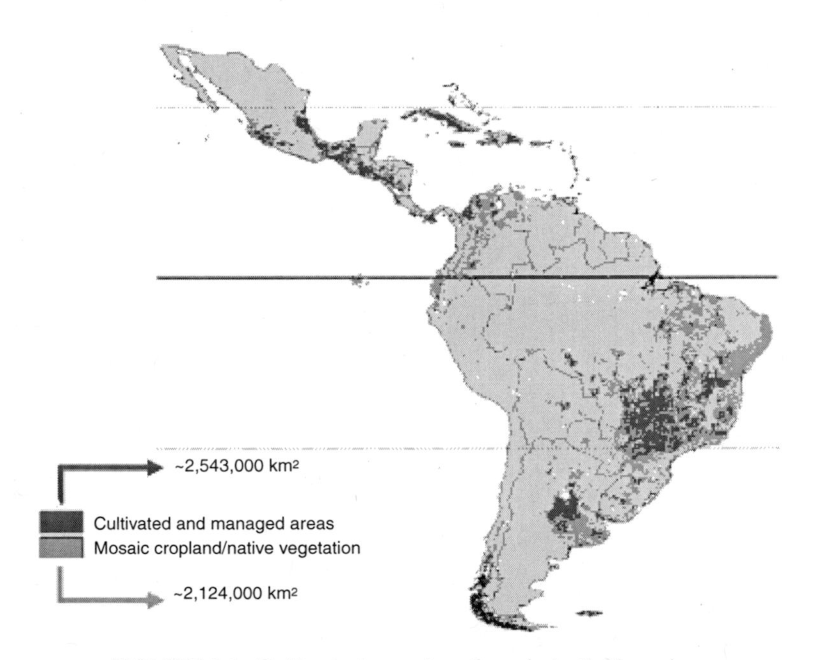

FIGURE 2.3. Cultivated area locations in Latin America.

For the purpose of the present chapter, the map shown in Figure 2.3 was subgrouped into three main land-use categories: (1) cultivated and managed areas, (2) mosaic croplands, and (3) undisturbed and other land uses. Of the 25 percent of total agricultural land use, 2,543,000 km^2 is used for cultivated and managed areas, and 2,124,000 km^2 for mosaic cropland. These land uses are spread over the LA continent. The undisturbed area is concentrated in three main ecoregions: the deserts of Mexico, the Amazonian tropical rain forest, and the temperate zone in the south.

The dispersed cultivated areas are located under different soil ecoregions classified by Bernoux et al. (see Chapter 4 in this volume). Moreover, present cultivated areas are located in soils with different carbon stocks. For instance, using a geographic information system (GIS), when a land-use map was overlaid on the soil total C, it showed that a total area of 4,667,000 km^2 of agricultural land contains 23.7 Pg C in the top 30 cm depth. Using the two main land-use categories, it is estimated that 13.0 and 10.7 Pg C are stocked in the top 0 to 30 cm depth of present cultivated and managed areas and mosaic cropland, respectively. With the assumption that the mean soil C density for the entire LA region is 4.88 kg $C \cdot m^{-2}$, the areas under agricultural use are slightly higher in SOC content (i.e., 5.04 kg $C \cdot m^{-2}$). This is expected because land-use conversion to agriculture is preferentially done on fertile soils.

In general, conversion of natural to agricultural ecosystems leads to depletion of the SOC stock. The depletion is caused by the use of inappropriate agricultural management practices, such as disking, plowing, minimum residue return, accelerated soil erosion, and so forth. In addition, the prevailing climatic conditions of the tropics accentuate the process of mineralization rather than humification. Such a trend can be reversed by adoption of new management practices, such as minimum and no-till systems of seedbed preparation. These management practices, besides maintaining or even increasing agronomic productivity, are economically and environmentally compatible, because they have the potential to sequester C from the atmosphere into the soil. These are the reasons for an exponential increase in area under those improved management practices in LA observed between 1990 and 2004. It is also widely reported that well-managed pastures at low animal stocking rates allow pasture regeneration and increases soil C stocks.

FINAL CONSIDERATIONS AND CONCLUSIONS

Estimating the potential of soil C sequestration in a vast region with diverse soils, climates, land uses, and ethnic and economic groups is a major challenge. The region is located between latitudes 30°N and 50°S, and comprises diverse ecoregions. Landowners adopt a wide range of site-specific appropriate management practices. Thus, there exists a wide range of mineralization and humification rates and management practices that are leading to a change in the ratio of mineralization/humification. Since the 1980s, farmers have adopted new management practices that have changed the mineralization:humification ratio.

A strategy to better estimate the potential of soil C sequestration for LA is to stratify the region. One criteria for stratification involves a matrix with data on soils, climates, vegetations, and geology to establish soil ecoregions. The soil ecoregions have been adopted as a basis for estimating the potential of soil carbon sequestration in Latin America reported in this volume.

The LA region has a vast potential of soil C sequestration because of its large land area and favorable climate. In addition to these natural conditions, national and international institutes have been developing and transferring specific agricultural technologies that are being continuously adopted by landowners. Despite the scientific advances, numerous landowners still practice subsistence agriculture. Such lands have a large potential of soil C sequestration because these soils are severely depleted of their SOC stocks. Adoption of improved agricultural technologies on these lands has a large potential to enhance productivity, improve soil quality, and mitigate the greenhouse effect.

In this regard, two main concepts must be emphasized, independent of the ecoregion. The first concept is related to the fact that the potential of soil C sequestration is finite. Once the SOC stock attains a threshold level, gains and losses of C in a soil reach a new steady state and there is no more potential of additional C sequestration. The time to reach this new equilibrium depends on soil, climate, land use, and management practices. Besides mitigating the greenhouse effect, increases in SOC stock also enhance agronomic productivity. The second concept is related to the emissions of non-CO_2 gases (e.g., CH_4 and N_2O) from the soil to the atmosphere because several agricultural practices influence the flux of these gases with a high global warming potential. The choice of appropriate land-use and soil management practices can reduce the emission of these gases expressed in C equivalent (CE) to a level lower than the CE sequestered in the soil.

REFERENCES

Batjes NH and Sombroek WG. 1997. Possibilities for carbon sequestration in tropical and subtropical soils. *Global Change Biology,* 3, 161-173.

Eswaran H, Van den Berg E, Reich P, and Kimble J. 1995. Global soil carbon resources. In Lal R, Kimble J, Levine E, and Stewart BA (eds.), *Soil and global change,* pp. 27-43. CRC Lewis Publishers, Boca Raton, Florida.

European Commission Joint Research Centre. 2002. A vegetation map of South America. EUR20159.

FAO. 2004. *Production yearbook*, FAO, Rome, Italy.

Lal R. 2002. Soil carbon dynamic in cropland and rangeland. *Environmental Pollution,* 116, 353-362.

Schimel DS. 1995. Terrestrial ecosystem and the carbon cycle. *Global Change Biology,* 1, 77-91.

Chapter 3

Soil Carbon Sequestration in Latin America

R. Lal

INTRODUCTION

It is widely recognized that soil carbon (C) sequestration is a win-win strategy. This is especially true in developing countries and emerging economies where soil resources are often degraded, land-use change and agricultural intensification are imminent, and the need to sequester carbon in soil is more important than ever before because of the necessity to restore degraded soils and ecosystems, improve water quality, enhance biodiversity, and increase agronomic productivity to achieve food security. Yet reliable information on the attainable and potential rate of soil carbon sequestration in relation to appropriate land use and recommended management practices (RMPs) is not known. The energy demand is also rapidly changing in many regions of Latin America (USDOE, 2004), leading to more total and per capita emissions in the future. Therefore, the objective of this chapter is to discuss the potential of terrestrial carbon sequestration in general and soil C sequestration in Latin America in particular. The chapter also outlines some economic, policy, and human dimension issues of soil C sequestration.

TERRESTRIAL CARBON POOL

There are five large global C pools: (1) an oceanic pool estimated at 38,000 Pg (1 Pg = petagram = 10^{15} g = 1 billion metric tonnes or 1 billion Mg), (2) a geologic pool estimated at about 5,000 Pg comprising 4,000 Pg of coal and 500 Pg of each of oil and gas, (3) a pedologic or soil C pool comprising 1,550 Pg of soil organic C (SOC) and 950 Pg of soil inorganic C (SIC), (4) an atmospheric pool estimated at 750 Pg and increasing annually at the rate of 3.3 Pg C per year, and (5) a biotic pool of 620 Pg including

Carbon Sequestration in Soils of Latin America

doi:10.1300/5755_03

60 Pg of detritus material (Lal, 2004). These five pools are interconnected. The atmospheric pool is increasing due to three anthropogenic activities: fossil fuel combustion, deforestation and soil cultivation, and cement production.

The terrestrial C pool has two principal components: the soil C pool and the biotic C pool (Figure 3.1). Both of these components are interconnected, depend on each other, and are influenced by atmospheric conditions. Together, the terrestrial C pool of 3,120 Pg is 4.1 times the atmospheric pool. The soil C pool by itself (2,500 Pg) is 3.3 times the atmospheric pool. The residence time of C in the terrestrial pool is longer than that in the atmospheric pool. Therefore, transfer of a fraction of C from atmospheric to terrestrial C pools is a prudent strategy of minimizing the CO_2-related effect on climate change.

SOIL CARBON SEQUESTRATION

In the context of the twenty-first century, soil has numerous functions (Exhibit 3.1). The process of transferring atmospheric C into a pedologic/soil C pool is called soil C sequestration. Soil C sequestration is a natural process and involves transfer of atmospheric CO_2 into biomass C via photosynthesis. Transfer of biomass C into SOC occurs through humification of the biomass returned to the soil. The humification efficiency of biomass C is 5 to 15 percent and depends on the quality of the biomass (C:N ratio, lignin and suberin contents), soil properties (amount and mineralogical composition of clay, water retention, and movement), and climate (mean annual temperature, amount and distribution of precipitation). The humification efficiency is high for biomass containing a high concentration of suberin and a low C:N ratio, in heavy-textured soils containing high-activity clays, and in cool and humid climates.

There are two principal components of soil C sequestration: SOC and SIC (Figure 3.2). The sequestration of SOC occurs through formation of stable aggregates, especially microaggregates (Bronick and Lal, 2005; Blanco-Canqui and Lal, 2004).

$$\text{Microaggregates} = [(\text{Cl-P-OM})_x]_y \quad (3.1)$$

This model was proposed by Edwards and Bremner (1967). In this model (Equation 3.1), Cl is clay particle, P is polyvalent cation, OM is long-chain organic molecule, and x and y refer to the number of units joined together. This process of formation of an organo-mineral complex is an important natural reaction which supports life on the earth. The SOC encapsulated

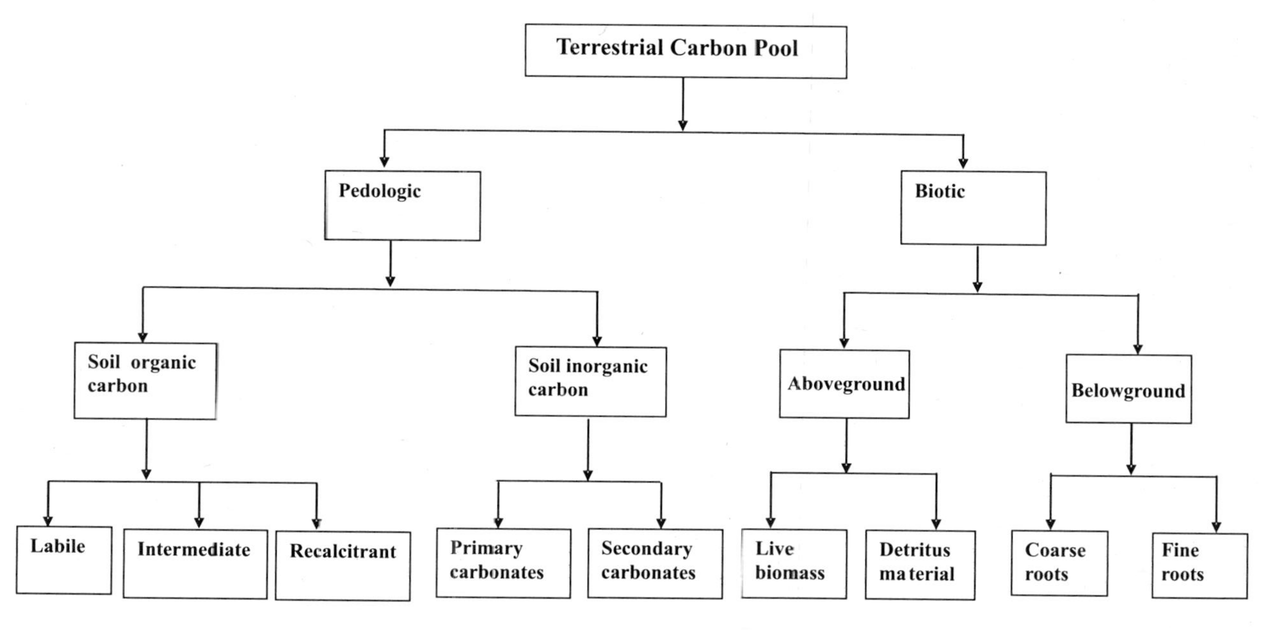

FIGURE 3.1. Components of the terrestrial carbon pool.

EXHIBIT 3.1.
Multiple Functions of Soil for the Twenty-First Century

1. Nutritive value for agriculture
2. Supportive strength for construction
3. Filtering ability for purifying water
4. Sink and buffering capacity for pollutants
5. Beholder and habitat of life (1,000 or more different spp. per g of soil)
6. Repository of germplasm
7. Archive of planetary and human history
8. Moderator of the climate

within stable microaggregates is physically protected against microbial processes.

Other mechanisms of protection involve formation of chemically and biologically recalcitrant fractions. Transfer of SOC from the surface into the sub-soil (below 30 cm depth) is another safeguard mechanism against mineralization. There are two processes of SIC sequestration. Formation of secondary carbonates is an important process in arid and semiarid regions. Carbonates are formed through dissolution of CO_2 in soil into carbonic acid and its reaction with Ca^{+2} or Mg^{+2} to form carbonates (Lal and Kimble, 2000) (Equations 3.2-3.4).

$$CO_2 + H_2OCO_3 \leftrightarrows H_2CO_3 \tag{3.2}$$

$$H_2CO_3 \leftrightarrows H^+ + H\,CO^-{}_3 \tag{3.3}$$

$$2\,H\,CO^-{}_3 + Ca^{+2} \leftrightarrows CaCO_3 + H_2O + CO_2 \tag{3.4}$$

The source of CO_2 in soil air is through root respiration or microbial decomposition of soil organic matter. Biogenic processes are important to the formation of secondary carbonates (Monger, 2002). The rate of SIC sequestration through formation of secondary carbonates is low and ranges from 5 to 15 Kg $C{\cdot}ha^{-1}$ per year.

Leaching of bicarbonates from the surface layer into the groundwater is another mechanism of transfer of SIC from the surface into the subsoil. Leaching of bicarbonates can occur in subhumid and semiarid climates as dissolved load in water percolating through the soil. In irrigated soils, leaching can occur with percolation of irrigation water, especially if it is not

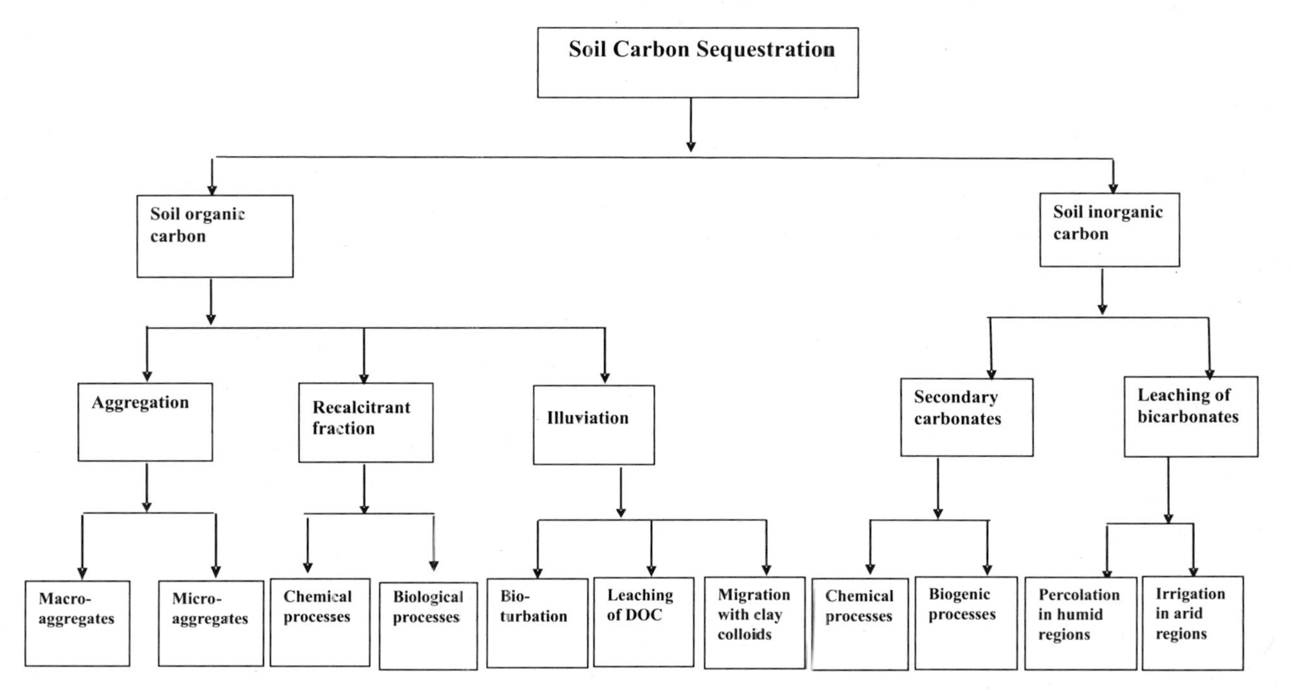

FIGURE 3.2. Components and processes of soil carbon sequestration.

saturated with bicarbonates and is good-quality water. Nordt et al. (2000) outlined a reaction involved in leaching of bicarbonates (Figure 3.3).

The rate of leaching of bicarbonates ranges from 0.25 to 1.0 Mg C·ha^{-1} per year (Wilding, 1999). These rates depend on soil chemical properties, hydraulic conductivity and quality of percolating or irrigation water, and the climatic parameters.

SOIL CARBON SINK CAPACITY

The soil C sink capacity depends on the difference between the present SOC pool under agricultural land use compared to the SOC pool under an undisturbed or a natural ecosystem. Most soils lose the SOC pool upon conversion from natural to managed ecosystems. The magnitude of the loss may be 25 to 75 percent depending on the antecedent pool, land use, management, and climate. The loss of the SOC pool occurs due to a decrease in the amount of biomass returned to the soil, an increase in the rate of mineralization because of change in soil moisture and temperature regimes, and a decrease in the amount of root or belowground biomass returned to the soil. In addition, there is also more loss of the SOC pool due to erosion and leaching in agricultural compared to natural ecosystems. Consequently, most agricultural soils lose 30 to 40 Mg C·ha^{-1} by the time the SOC pool has reached the new equilibrium level. It is this loss or deficit in the SOC pool that creates the SOC sink capacity. Conversion to a restorative land use and adoption of RMPs can resequester 60 to 70 percent of the historic C lost.

The soil C sink capacity thus created depends on a range of interacting factors (Figure 3.4). Important among the soil factors are solum depth, clay

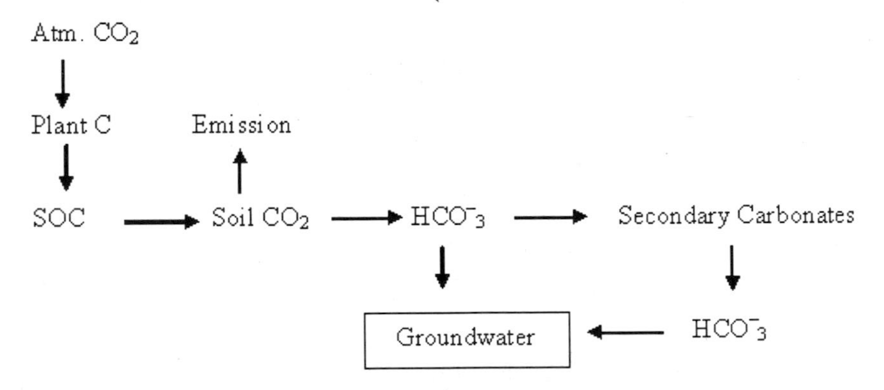

FIGURE 3.3. A reaction involved in leaching of bicarbonates.

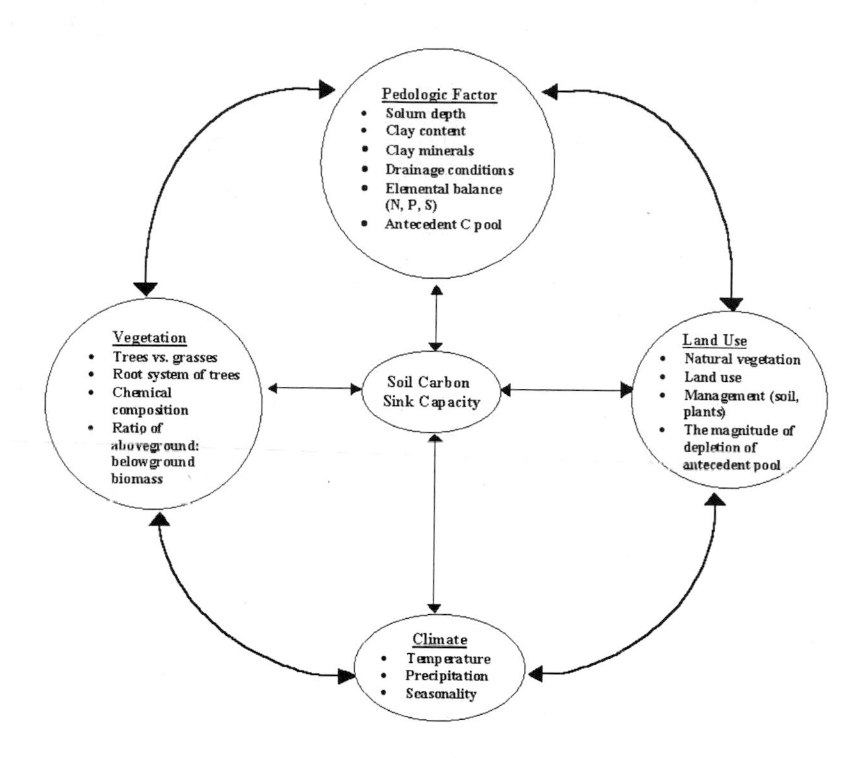

FIGURE 3.4. Factors affecting soil carbon sink capacity.

content and minerology, structural attributes and the ability to form aggregates, soil moisture retention, charge density and surface area, and the antecedent pool. The deficit or loss of the SOC pool is greater in coarse-textured than fine-textured soils, greater in excessively drained than poorly drained soils, and greater in soils with a high rather than low antecedent SOC pool. The loss of the SOC pool is also greater and more rapid in tropical than temperate climates (Lal, 2004). Land use and vegetation are also important factors. The magnitude of SOC loss is greater in those land-use and management systems that involve frequent and intense soil disturbance (e.g., cultivation of seasonal crops with plow tillage) and extractive or fertility mining practices (e.g., subsistence farming with low external input). Thus, the SOC sink capacity is greater in croplands than pasture or forest lands, and greater in soils prone to erosion and other degradative processes as compared to prime lands.

RATE OF SOIL ORGANIC CARBON SEQUESTRATION

Rate of SOC sequestration through conversion to a restorative land use and adoption of RMPs is generally lower than the rate of depletion of the SOC pool upon conversion from natural to agricultural ecosystems (Figure 3.5). The sequestration follows a sigmoid curve with a low rate in the initial phases, attaining the maximum between 10 to 30 years after adoption of RMPs or restorative land use. Although the loss of an SOC pool is rapid and drastic, the gain of an SOC pool is slow and variable. Indeed, the rate of SOC sequestration depends on a wide range of factors (Figure 3.6). Most important among these are the soil type and climate. Predominant soil factors affecting the rate of SOC sequestration are texture, clay minerals, water retention capacity, and aggregation. The soil's ability to form aggregates is also an important factor. Important among the climatic factors are temperature and precipitation. Rates of SOC sequestration are generally higher in cool and humid than in warm and arid climates. Therefore, agricultural soils of loam to clayey textures in temperate climates are likely to have higher rates of SOC sequestration than coarse-textured soils in the tropics and subtropics. The rate of SOC sequestration may be higher in irrigated soils, especially if the water-use efficiency is high and irrigation increases the net primary productivity (NPP) with a large increase in belowground biomass returned to the soil.

The rate of SOC sequestration in humid temperate soils is 300 to 1,000 Kg $C{\cdot}ha^{-1}$ per year. In comparison, the rate of SOC sequestration in arid/semiarid tropical climates may be 0 to 250 Kg $C{\cdot}ha^{-1}$ per year. In some exceptional cases, a high rate of >1.0 Mg $C{\cdot}ha^{-1}$ per year has been reported. Such high rates are often associated with adoption of RMPs which produce a large quantity of biomass, especially the biomass associated with deep root systems (Fisher et al., 1994) and with supplemental irrigation.

The rate of SIC sequestration is generally low, and ranges from 5 to 15 Kg $C{\cdot}ha^{-1}$ per year. In contrast, the leaching of bicarbonates may occur at rates ranging from 0.25 to 1.0 Mg $C{\cdot}ha^{-1}$ per year. Although the rate of formation of secondary carbonates is low, carbonates are more stable and have longer residence time in soil than the SOC pool.

OPPORTUNITIES OF SOIL CARBON SEQUESTRATION IN LATIN AMERICA

Only a fraction of the maximum potential of soil C sequestration is the attainable potential. The attainable potential depends on the extent of adoption of RMPs. Commodification of soil C may be an important factor deter-

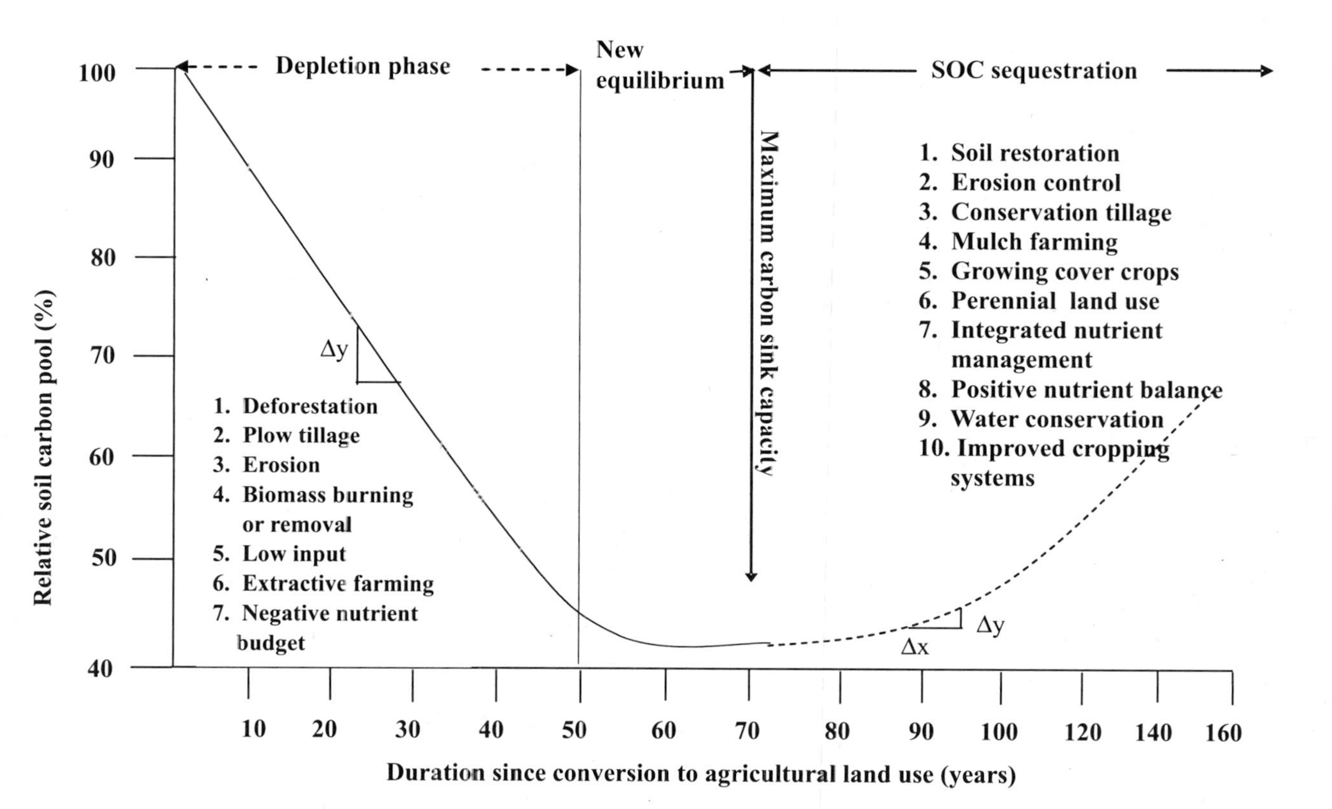

FIGURE 3.5. Rate of depletion and sequestration of the SOC pool in agricultural soils.

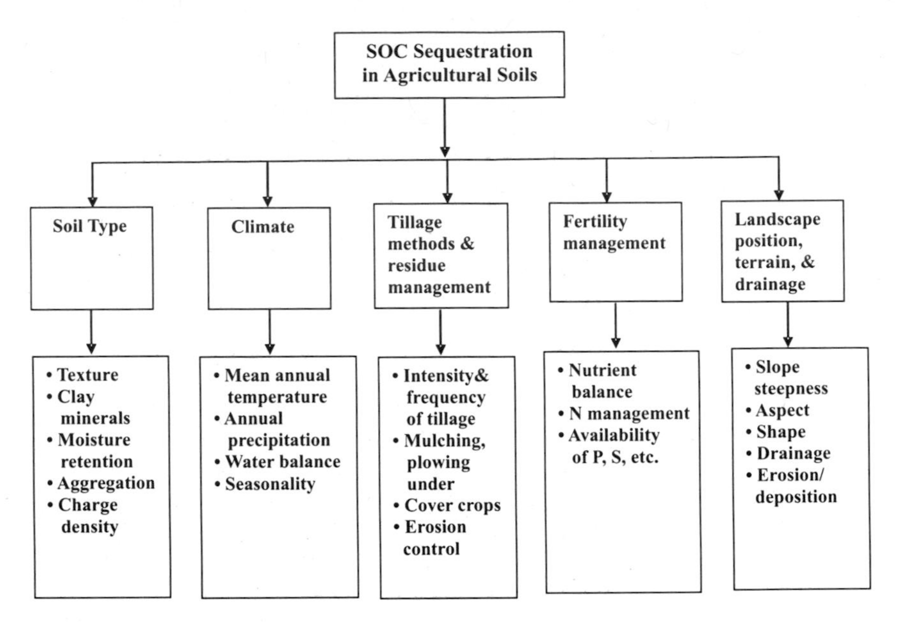

FIGURE 3.6. Factors affecting rates of sequestration of soil organic carbon in agricultural soils.

mining the adoption rate of RMPs. The term *commodification* or *commoditization* implies making soil C a commodity which can be bought or sold (traded) in the domestic or international market similarly to any other farm produce (e.g., soybeans, corn, milk, meat). Soil C is a tradeable commodity that has neither transport costs nor any other expenses in its storage and handling. There are also no quality differences, because soil C is the same whether in temperate or tropical regions, in commercial farms or shifting cultivations, and in cropland or plantations. Trading of soil C credits can occur under the Clean Development Mechanism (CDM) of the Kyoto Protocol, the BioCarbon Fund of the World Bank, or any domestic opportunities similar to the Chicago Climate Exchange in the United States (World Bank, 2003; UNEP, 2002; Johnson and Heinen, 2004; Kiss et al., 2002). Trading C credits provides economic incentives to farmers for restoring degraded soils and improving the environment. This is especially important to small landholders of the tropics and subtropics in remote areas (e.g., the Andean region, the Caribbeans, and Central America). With the current pricing structure and low demand, the economic prospects are rather meager (Butt and McCarl, 2005). It is important, therefore, that a just value of soil C is objectively assessed.

Numerous factors determine the value of soil carbon (Figure 3.7). The value of soil C depends on the benefits to the farmer and to society. Benefits to the farmer are due to improvement in soil quality through adoption of RMPs. These practices increase agronomic yield, enhance water and nutrient use efficiencies, reduce losses and decrease input, and improve farm income and profits. Increase in productivity is also associated with increase in the property value.

Although benefits to the farmer through improvement in soil quality are easy to quantify, assessment of the societal benefits of SOC sequestration is rather complex and confounded by numerous factors. Societal benefits are due to increases in ecosystem services such as a decrease in erosion and sedimentation, improvement in water quality due to reduction in nonpoint source pollution, increase in biodiversity associated with saving of land for nature conservancy, and mitigation of climate change because of reduction in net emission of CO_2 (Figure 3.7).

Regardless of the tradeable value of SOC, the inherent value of soil humus is relatively easy to conceptualize. Principal constituents of soil humus of relevance to agronomic productivity are nutrients (e.g., N, P, K, S, Zn, Cu, Mo) and plant available water. It is generally argued that 1 Mg of C in humus contains nutrients and water worth US$200 at the current price of fertilizers and lifting of irrigation water. In contrast, the price of C is $3.7 per Mg in the Chicago Climate Exchange and about $60 per Mg in the European market. The high price in the European market is due to adoption of the Kyoto Protocol and mandatory reduction of emissions by the industry. The voluntary reduction of emissions by the United States has so far not been as effective in increasing the tradeable value of soil C as it has in the European countries.

Technological adoption also depends on several other factors (Figure 3.8). Important among these are (1) institutions to provide support through practical/relevant research, extension services, and outreach activities, (2) infrastructure with regard to access to markets, availability of essential input needed for adoption of RMPs, and channels of communication with researchers and extension agents, and (3) knowledge about the relevant farming systems in terms of their effectiveness for restoring the SOC pool and soil quality, social acceptability, and adaptability in an ecological niche, and the extension approach with regard to being bottom-up rather than top-down, and a demand-driven and participatory approach. The most important criteria of adoption of improved technology is its impact on enhancing farm income, creating off-farm jobs, and improving the environment by saving land for nature conservancy and mitigating climate change (Figure 3.8). Above all, the program must be community based and self-sustaining.

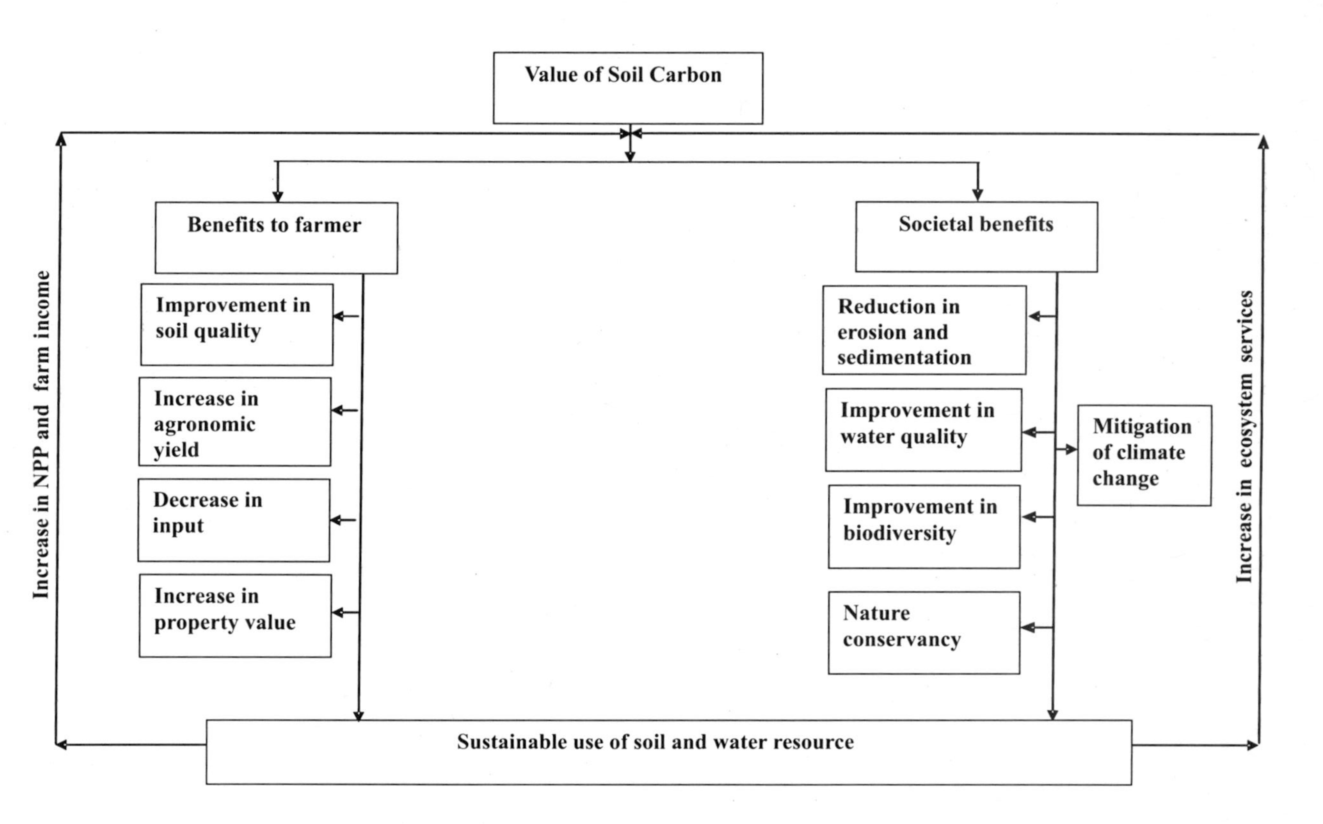

FIGURE 3.7. Factors affecting commodification of soil carbon.

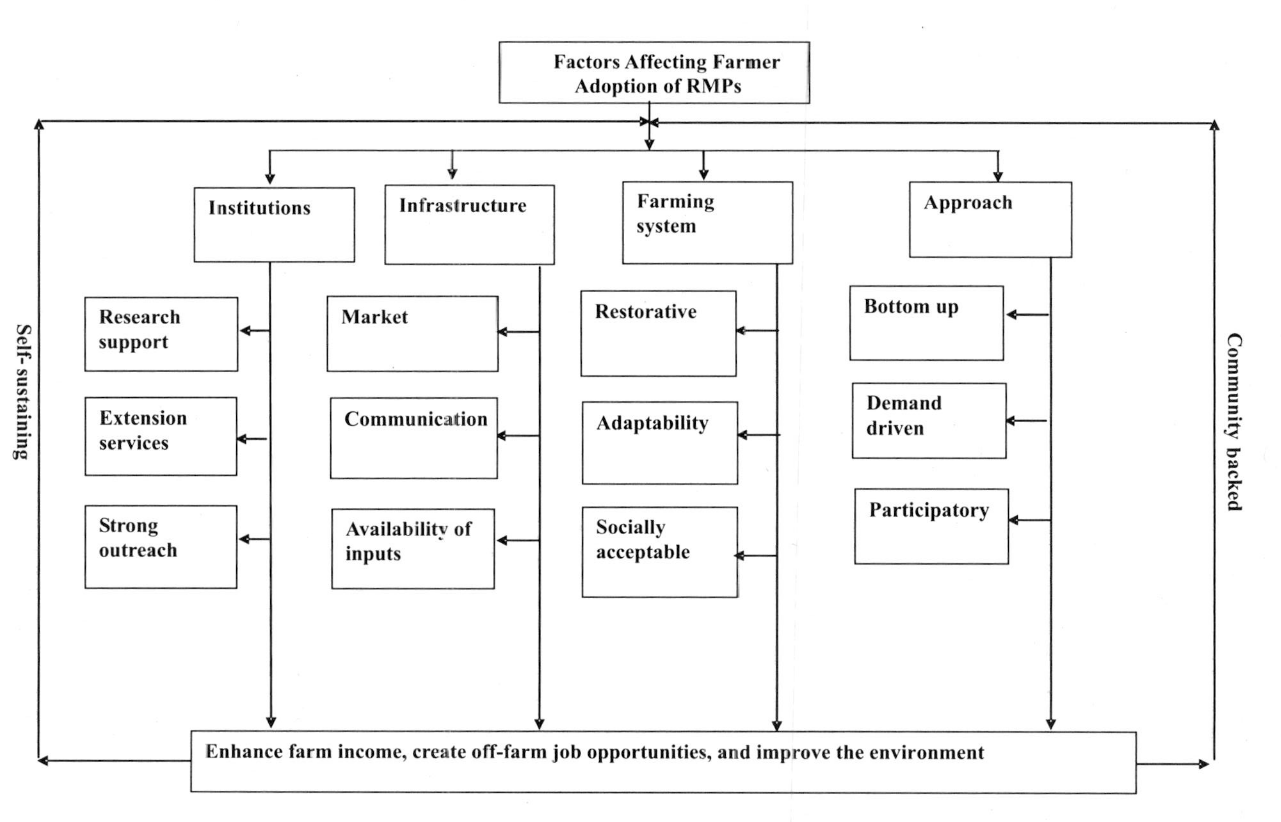

FIGURE 3.8. Adoption of recommended practices by farmers depends on numerous factors.

Although the just reward can be useful, providing subsidies can be counterproductive.

Latin America has a large landmass, diverse climates, and a wide range of soils. Diverse biophysical, socioeconomic, and cultural environments provide an ample opportunity to convert degraded soils to a restorative/perennial land use and adopt RMPs on croplands and pastureland. Important RMPs for croplands include adoption of a no-till system of seedbed preparation, frequent incorporation of a cover crop in the rotation cycle, and maintenance of soil fertility through the strategy of integrated nutrient management (INM). Management of pastures through establishment of RMPs and of tree plantations through stand management and fertility (N, P) enhancement are important strategies of SOC sequestration.

There are large arid and semiarid regions in Latin America, especially along the western coastal regions. These regions have a potential of SIC sequestration through formation of secondary carbonates, and also leaching of bicarbonates in subhumid climates and irrigated agriculture. Credible data on rate of SOC sequestration are available, but the information on SIC sequestration is scanty.

Trading C credits, not yet begun in Latin America, offers tremendous opportunities of commodification of soil C as farm produce. Such opportunities exist in both domestic and international markets and need to be explored and strengthened. In this regard, the importance of measuring and monitoring and verification cannot be overemphasized (see Part III in this book). The potential of C sequestration for three strategies is shown in Table 3.1. For the decade from 2003 to 2012, the total potential of terrestrial C sequestration is 1.37 Pg C. Of this, the potential of soil C sequestration is 93 Tg C, or 9.3 Tg C per year. If the C can be traded, the region will have a net income from farming C at the rate of $1 billion per year (Table 3.1). Realization of this income has numerous implications.

TABLE 3.1. Terrestrial carbon sequestration potential in Latin America from 2003 to 2012.

Strategy	Area (Mha per year)	Carbon sequestration (10^{12}g C)	Net present value (US$$10^6$)
Reforestation rate	1.71	177.9	1,231.4
Sustainable agriculture	11.0	98.1	644.1
Deforestation halted[a]	0.87	1,097.3	8,362.3
Total		1,368.3	10,237.8

Source: Adapted from Niles et al., 2002.
[a]Annual deforestation rate = 5.59 Mha per year.

CONCLUSIONS

Latin America is rapidly developing, with increasing demands for energy and conversion of natural to managed ecosystems. Therefore, carbon sequestration in terrestrial ecosystems offers a unique niche to mitigate climate change by reducing net emissions of CO_2 while restoring degraded soils and improving the environment. The soil carbon pool is a reactive pool and an important component of the global C cycle. Of the two components of soil carbon pool (i.e., soil organic and inorganic components), the organic component is more dynamic. The rate of sequestration of soil organic carbon is more than that of inorganic carbon, although the residence time of inorganic carbon is longer than that of the organic component. Important management practices that enhance sequestration of organic carbon in soil include afforestation, conversion of an agricultural to a perennial (e.g., plantations) land use, adoption of conservation tillage with residue mulching and cover crops, and use of integrated nutrient management strategies based on a judicious combination of inorganic fertilizers and organic manures to achieve a positive nutrient balance.

Soil C is a tradeable commodity and can be traded in domestic and/or international markets. Unlike other farm commodities (e.g., corn, soybeans, milk, meat), however, soil C has no transport costs and is of a uniform quality regardless of farm size, soil type, or climate. Trading C credits is an option to enhance income of small landholders and resource-poor farmers of Latin America. However, the current market price is low and mechanisms are not yet in place to facilitate the trade. Regardless, soil C sequestration is essential to improving the soil quality in order to increase agronomic productivity and achieve global food security. It is in this regard that soil C sequestration is a truly a win-win situation.

REFERENCES

Blanco-Canqui, H. and R. Lal. 2004. Mechanisms of carbon sequestration in soil aggregates. *Crit. Rev. Plant Sci.* 23: 481-504.

Bronick, C. and R. Lal. 2005. Soil structure and management: A review. *Geoderma* 124: 3-22.

Butt, T. A. and B. A. McCarl. 2005. Implications of carbon sequestration for landholders. *Journal of the SFMRA* 68: 116-122.

Edwards, A. P. and J. M Bremner. 1967. Micro-aggregates in soils. *J. Soil Sci.* 18: 64 -73.

Fisher, M. J., I. M. Rao, M. A. Ayarza, C. E. Lascano, J. I. Sanz, R. J. Thomas, and R. R. Vera. 1994. Carbons storage by introduced deep-rooted grasses in the South American savanna. *Nature* 266: 236-238.

Johnson, E. and R. Heinen. 2004. Carbon trading: Time for industry involvement. *Env. Intl.* 30: 279-288.

Kiss, A., G. Castro, and K. Newcombe. 2002. The role of multi-lateral institutions. In I. Swingfield (ed.), *Capturing carbon and conserving biodiversity: The market approach.* Earthscan Publications, London, pp. 90-101.

Lal, R. 2004. Soil carbon sequestration impacts on global climate change and food security. *Science* 304: 1623-1627.

Lal, R. and J. M. Kimble. 2000. Pedogenic carbonate and the global carbon cycle. In R. Lal, J. M. Kimble, and D. A. Stewart (eds.), *Global climate change and pedogenic darbonates.* CRC/Lewis Publishers, Boca Raton, Florida, pp. 1-14.

Monger, H. C. 2002. Pedogenic carbonates: Links between biotic and abiotic $CaCO_3$. 17th World Cong. Soil Sci., August 14-21, 2002, Thailand, Symp. #20, Oral Presentation #891.

Niles, J. O., S. Brown , J. Pretty, A.S. Ball, and J. Fay. 2002. Potential carbon mitigation and income in developing countries from changes in use and management of agricultural and forest lands. *Philos. Transact. (Royal Society)* 360(1797): 1621-1639.

Nordt, L. C., L. P. Wilding, and L. R. Drees. 2000. Pedogenic carbonate transformations in leaching soil systems. In R. Lal, J. J. Kimble, and B. A. Stewart (eds.), *Global climate change and pedogenic carbonates.* CRC Publishers, Boca Raton, Florida, pp. 43-64.

UNEP. 2002. *A guide to emission trading: An emerging market for the environment.* UNEP, Nairobi, Kenya.

USDOE. 2004. *International energy outlook.* U.S. Department of Energy, Washington, DC.

Wilding, L. P. 1999. Comments on paper by R. Lal. In N. Rosenberg, R. C. Izaurraldo, and E. Malone (eds.), *Carbon sequestration in soils: Science, monitoring and beyond.* Battelle Press, Columbus, Ohio, pp. 146-169.

World Bank. 2003. *BioCarbon Fund: Harnessing the carbon market to enhance ecosystems and reduce poverty.* World Bank, Washington, DC.

Chapter 4

Soil Carbon Stocks in Soil Ecoregions of Latin America

Martial Bernoux
Boris Volkoff

INTRODUCTION

Global warming incites the global community and especially scientists to deepen their studies about the global carbon cycle. Rising levels of atmospheric CO_2 have led to attention of C stocks (CS) in main terrestrial compartments, mainly soils and phytomass. The world's mineral soils represent a large reservoir of C, with estimations ranging from 1,115 to 2,200 Pg C (1 Pg = 1,015 g, or 1 Pg = 1 billion tons) in the first meter (Post et al., 1982; Batjes, 1996). Plant biomass is estimated to range between 560 and 835 Pg C (Whittaker and Likens, 1975; Bouwman, 1990). Tropical forest ecosystems would account for 20 to 25 percent of the carbon contained in soil and vegetation (Brown and Lugo, 1982; Dixon et al., 1994). Volkoff and Bernoux (Chapter 1) presented a regionalization of Latin America based on soil and environmental data: the complex landscape of Latin America is divided into 13 soil ecoregions in order to significantly reduce the spatial variability. The objectives of this chapter are (1) to assess the original soil C stocks of Latin America for the different countries and soil ecoregions, and (2) to estimate the part of this total that could have been impacted by past land-use activities.

Map of Soil C Stocks of Latin America Under Initial Native Vegetation

The map of soil C stocks was obtained using the vectorized format of the 1:5,000,000 Digital Soil Map of the World (DSMW, version 3.5, FAO, 1995) as spatial component, and the carbon content database elaborated

Carbon Sequestration in Soils of Latin America

doi:10.1300/5755_04

from the results published in Batjes (1996). Batjes (1996) reported the median soil C content for the 0 to 30 cm, 0 to 50 cm, 0 to 100 cm, and 0 to 200 cm layers. The estimates correspond to the OC present in the layer after correction using the volume percentage of fragments coarser than 2 mm. The estimates are derived from information of 4,353 individual soil profiles contained in the World Inventory of Soil Emission Potentials (WISE) database.

The C stock (in k·gm^{-2}) representative for each soil mapping unit (of the DSMW) was calculated. This estimate was calculated according to the relative proportion of all soil units within each individual soil mapping unit. Then the obtained matrix was combined with the spatial component to derive the maps of the C stored in different soil depths.

In total, the upper 100 cm of the soil of Latin America contains about 185 Pg C (Figure 4.1). Figure 4.1 shows the unequal distribution (in kg C·m^{-2}) of this pool in Latin America. The stocks for the layers 0 to 30 cm, 0 to 50 cm, and 0 to 200 cm are, respectively, 100 Pg, 132 Pg, and 277 Pg of C for Latin America. We also calculated the global soil carbon stocks. These

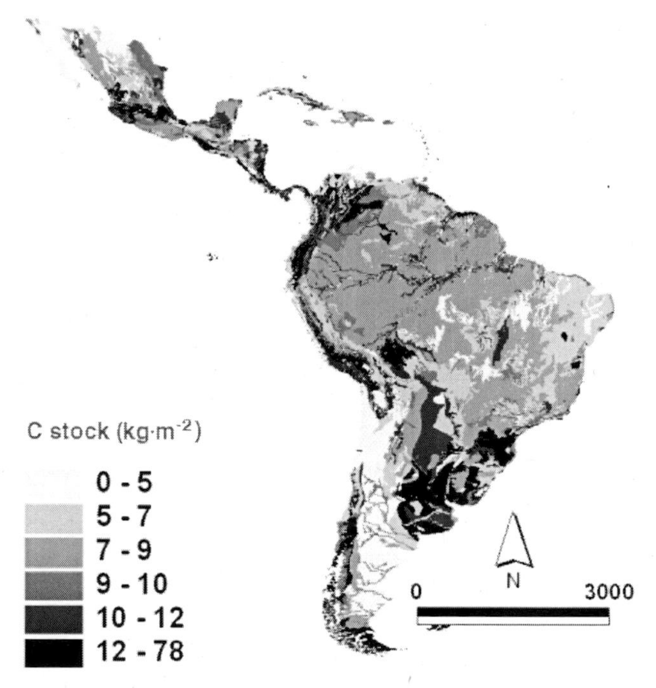

FIGURE 4.1. Soil carbon stock distribution in Latin America for the 0 to 100 cm soil layer under native conditions.

stocks are, respectively, 803 Pg, 1,101 Pg, 1,589 Pg, and 2,521 Pg of C for the layers 0 to 30 cm, 0 to 50 cm, 0 to 100 cm, and 0 to 200 cm.

After digitally intersecting the digital soil map of the world with country boundaries, it is possible to derive for each country the C stock for the different soil depths. Results for each country are reported in Table 4.1.

These C stocks are theoretically stocks under undisturbed conditions because they derive from soil profiles under native vegetation. It is important to highlight that these estimates can be very imprecise, particularly for the small islands of the Caribbean. For instance, Venkatapen et al. (2004) calculated that about 11 Tg of C are stored in the soils of Martinique (French West Indies). The estimate obtained from the DSMW overestimated this value by 37 percent. This is due to imprecision in the boundaries of this territory that led to an overestimation of the area by about 9 percent. Also, at that scale the high spatial variability of soils is reduced to only eight soil mapping units. On the other hand, for a large country such as Brazil, the estimates are more robust. For instance, Bernoux et al. (2001, 2002) calculated that about 36.4 ± 3.4 Pg C was stored in the top 30 cm of the 8.3 million km^2 occupied by soils in Brazil. The value reported for Brazil in Table 4.1 was overestimated by only 7 percent. Schroeder and Winjum (1995) calculated 72 Pg of stored C in the top meter of Brazil, very close to the present estimates reported in Table 4.1.

The soil C map was also overlaid with the boundaries of the different soil ecoregions presented by Volkoff and Bernoux (Chapter 1, Figure 1.16). The corresponding stocks and area for each soil ecoregion are reported in Table 4.2.

Most of the soil ecoregion has a mean C content ranging from 4 to 7 kg $C{\cdot}m^{-2}$. Only the East Brazilian and South Argentina regions presented values below this range, mainly due to semiarid and arid climatic conditions. On the other hand, the South Chilean region shows the highest mean C stock because of the moist midlatitude climate with cold winters.

PRESENT LAND USE AND ESTIMATES OF SOIL CARBON STOCKS IMPACTED BY AGRICULTURAL ACTIVITIES

To estimate the C pool that is potentially affected by the agricultural activities, a digital map of the present land use of the region was overlaid with the C map. The land-use component was derived from the Global Land Cover (GLC) map for the year 2000 (European Commission Joint Research Centre, 2003) The GLC 2000 dataset is a main input dataset to define the boundaries of the different ecosystems, such as forest, grassland, and culti-

TABLE 4.1. Inventory of soil carbon pools for the different countries of Latin America.

Regions and countries or territories	Number of map unit		Area		Soil carbon stock for different layers (Tg of C)			
	Total	With soil	All map unit km²	% with soil	0-30 cm	0-50 cm	0-100 cm	0-200 cm
Caribbean								
Antigua and Barbuda	5	5	438	100.0	3	3	5	7
Bahamas	29	27	12,984	97.3	34	42	66	106
Barbados	3	3	454	100.0	3	4	5	6
British Virgin Islands	5	5	315	100.0	2	2	3	4
Cayman Islands	2	2	89	100.0			1	1
Cuba	98	96	111,349	99.9	710	934	1,316	2,145
Dominica	5	5	786	100.0	5	7	10	13
Dominican Republic	40	37	49,864	99.1	263	330	450	699
Grenada	4	4	328	100.0	2	2	3	5
Guadeloupe (France)	8	8	1,797	100.0	11	13	18	27
Haiti	28	27	27,220	99.8	147	181	246	381
Jamaica	17	17	11,553	100.0	60	78	114	200
Martinique (France)	8	8	1,185	100.0	7	10	15	20
Montserrat	1	1	109	100.0	1	1	1	1
Puerto Rico	17	17	9,141	100.0	52	66	93	150
St Kitts and Nevis	5	5	437	100.0	3	3	4	5
St Lucia	3	3	628	100.0	3	4	6	8
St Vincent and the Grenadines	5	5	449	100.0	2	3	4	6
Trinidad and Tobago	6	6	5,276	100.0	21	29	42	54
Turks and Caicos	7	7	783	100.0	2	2	3	5
US Virgin Islands	1	1	238	100.0	1	1	2	3
Subtotal	297	289	235,424	99.6	1,331	1,717	2,406	3,845

Central America								
Belize	37	33	22,053	99.3	144	182	242	358
Costa Rica	59	59	51,984	100.0	329	451	653	941
El Salvador	24	22	21,047	99.6	100	137	199	281
Guatemala	78	75	109,478	99.8	636	819	1,141	1,677
Honduras	71	69	114,079	98.4	708	928	1,298	1,862
Mexico	463	408	1,959,288	99.5	9,436	12,054	16,479	22,467
Netherlands Antilles	8	8	915	100.0	3	4	5	6
Nicaragua	70	65	130,214	92.8	827	1,117	1,608	2,406
Panama	70	69	75,425	99.3	537	732	1,073	1,686
Subtotal	880	808	2,484,483	99.1	12,719	16 422	22,698	31,683
South America								
Argentina	422	292	2,779,966	98.2	11,998	15,877	22,850	33,561
Bolivia	175	130	1,090,350	97.3	5,555	7,428	10,505	16,156
Brazil	1,070	927	8,547,596	98.8	39,033	51,705	71,531	105,086
Chile	550	478	760,780	94.3	4,805	6,185	8,678	14,914
Colombia	236	215	1,137,588	99.8	6,572	8,659	11,956	19,053
Ecuador	105	105	254,376	100.0	1,692	2,292	3,200	4,604
Falkland Islands (Malvinas)	17	17	11,964	100.0	250	393	646	1,786
French Guiana	79	76	81,454	99.8	457	614	885	1,612
Guyana	166	160	209,729	99.8	1,086	1,405	1,961	3,359
Paraguay	73	59	400,126	99.6	1,816	2,473	3,528	4,708
Peru	406	334	1,292,626	97.7	6,625	8,771	12,031	16,610
Suriname	108	103	145,109	97.8	808	1,122	1,672	3,174
Uruguay	56	45	178,512	98.1	1,075	1,458	2,091	3,486
Venezuela	153	137	927,495	98.0	4,495	5,958	8,340	13,364
Subtotal	3,616	3,078	17,817,670	98.4	86,264	114,340	159,875	241,473
Latin America Total	4,793	4,175	20,537,577	98.5	100,314	132,479	184,979	277,001

TABLE 4.2. Soil C stocks and mean C content for the soil ecoregions defined by Volkoff and Bernoux.

Soil ecoregion	Area (1,000 km^2)	Carbon stock, 0-30 cm (Tg)	Mean C content, 0-30 cm ($kg \cdot m^{-2}$)
Amazonian	6,734.4	33,711	5.0
Andean Equatorial	630.8	4,136	6.6
Brazilian Atlantic	1,133.0	6,267	5.5
Caribbean	655.6	3,370	5.1
Central American	1,137.0	7,140	6.3
Central Andes	1,806.5	7,822	4.3
Central Brazilian	1,882.1	7,986	4.2
East Brazilian	1,012.3	3,806	3.8
Eastern Pampas	825.8	5,419	6.6
Mexican	1,367.6	5,793	4.2
Preandean	1,763.2	8,125	4.6
South Argentina	961.5	3,185	3.3
South Chilean	346.3	3,743	10.8
Total	20,256.1	100,503	5.0

vated systems. The Latin America part of the GLC 2000 was extracted before further analysis and was projected to the Lambert Equal-Area Azimuthal projection that respects areas. The projection was realized in ArcView 3.2a with the extension grid and theme projector (Jenness, 2004) with a pixel size of 0.5 km.

Table 4.3 resumes the extension of the different land-use classes in the entirety of Latin America.

For better readability, the number of original categories was reduced to ten principal categories by merging original categories. Table 4.3 reports the land surface corresponding to the original categories.

The most probable localization where soil C stocks may have been disturbed is under “Cultivated and managed areas,” and the least likely were the two classes corresponding to a mosaic of cropland and other natural vegetation, aggregated under the category “Mosaic cropland/native vegetation” (Figure 4.2). The locations of these areas are detailed in Figure 4.3.

These two categories cover nearly one-fourth of Latin America and certainly have influenced present soil C stocks. The area that in the year 2000 was classified as cultivated and managed areas corresponded, crossing the soil C map (Figure 4.1) and the land-use map (Figure 4.2), to original total C stocks in the 0 to 30 cm layer of about 12,960 Tg C before disturbance. This stock amounted about 10,730 Tg C for areas characterized with a mosaic of croplands and natural vegetation. That means that about 23.7 Pg C

TABLE 4.3. Surface of the land-use categories of Latin America from the Global Land Cover map.

Land-use category	Surface (km^2)
Tree cover, broadleaved, evergreen	7,157,857
Tree cover, broadleaved, deciduous, closed	1,067,491
Tree cover, broadleaved, deciduous, open	313,440
Tree cover, needle-leaved, evergreen	460,324
Tree cover, mixed leaf type	101,374
Tree cover, regularly flooded, fresh water	300,780
Tree cover, regularly flooded, saline water	19,475
Mosaic: tree cover/other natural vegetation	745
Shrub cover, closed-open, evergreen	91,886
Shrub cover, closed-open, deciduous	1,409,855
Herbaceous cover, closed-open	2,018,088
Sparse herbaceous or sparse shrub cover	1,661,480
Regularly flooded shrub and/or herbaceous cover	393,576
Cultivated and managed areas	2,542,855
Mosaic: cropland/tree cover/other natural vegetation	1,178,687
Mosaic: cropland/shrub and/or grass cover	945,525
Bare areas	554,572
Snow and ice	23,845
Artificial surfaces and associated areas	14,216
Total	20,256,071

Source: Based on data from European Commission Joint Research Centre, 2003.

have been directly under the influence of agricultural activities more or less intensively. Table 4.4 details the repartition of this stock by soil ecoregions.

The *Revised 1996 Guidelines for National Greenhouse Gas Inventories* (IPCC/UNEP/OECD/IEA, 1997: Reference Manual Chapter 5.3.7 and Workbook Chapter 5.6) reported default factors to be used in the calculation of the variations in soil C stocks as a function of changes in LULUCF activities. The impact factors (IF) represent changes in native C stock associated with conversion of the native vegetation to agricultural use, integrating the management practices (tillage intensity and input level). The IPCC methodology (IPCC/UNEP/OECD/IEA, 1997) proposed the following calculation to determine IF: IF = base factor × tillage factor × input factor, where the base factor represents changes in SOM associated with conversion of the native vegetation for a particular land use. Tillage and input factors account

FIGURE 4.2. Land-use categories derived from the Global Land Cover map for the year 2000 (*Source:* Created by the authors from data from European Commission Joint Research Centre, 2003; Lambert Equal-Area Azimuthal projection.) (See also color gallery.)

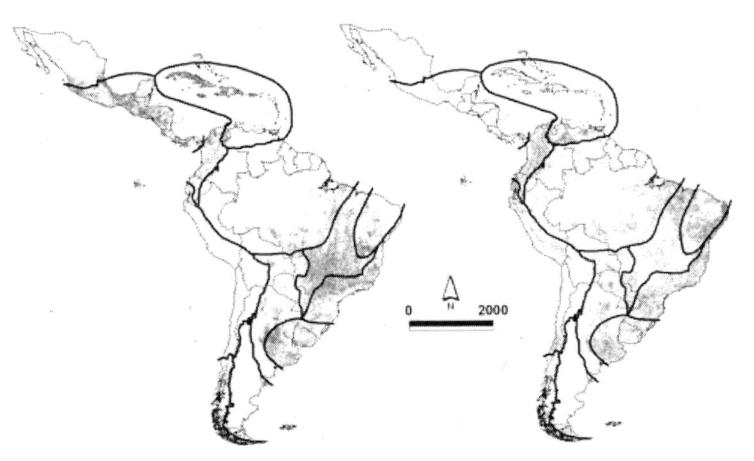

FIGURE 4.3. Localization of the categories "cultivated and managed areas" (left) and "mosaic cropland/native vegetation" (right); geographic projection.

TABLE 4.4. Inventory of the original soil carbon pools under native vegetation for the soil ecoregions of Latin America that have been impacted by land-use activities.

	Agricultural impacted areas			Mosaic of native and agricultural impacted area		
Soil ecoregion	1,000 km2	% of SER	C pool (Tg C)	1,000 km2	% of SER	C pool (Tg C)
Amazonian	196.3	2.9	881	277.4	4.1	1,367
Andean Equatorial	98.5	15.6	603	198.9	31.5	1,290
Brazilian Atlantic	363.0	32.0	1,740	345.2	30.5	1,879
Caribbean	141.9	21.6	759	129.6	19.8	658
Central American	415.5	36.5	2,378	4.2	0.4	32
Central Andes	14.2	0.8	78	120.7	6.7	685
Central Brazilian	697.8	37.1	2,963	320.0	17.0	1,291
East Brazilian	155.7	15.4	631	348.5	34.4	1,262
Eastern Pampas	235.1	28.5	1,624	205.4	24.9	1,338
Mexican	55.8	4.1	316	0	0.0	0
Preandean	148.8	8.4	768	154.9	8.8	799
South Argentina	1.5	0.2	9	6.5	0.7	7
South Chilean	18.7	5.4	213	12.8	3.7	119
Total	2,542.8	12.5	12,963	2,124.1	10.5	10,727

for effects of various management levels and practices that could be associated with this land use.

For the purpose of this first attempt to estimate the C stocks that may have been lost by agricultural activities, it was proposed that 30 percent of original C stock were lost for the "Cultivated and managed areas" and 10 percent for the "Mosaic cropland/native vegetation." It must be highlighted that these values are conservative in regard to IF components published by the IPCC methodology. The base factor is 0.6 in tropical conditions for long-term cultivation.

The first estimate of each soil ecoregion (Table 4.5) indicates that some soil ecoregions suffered important C loses and are therefore potential future sinks if the tendency can be reversed. The Central Brazilian region contributes to nearly one-fifth of the total. On the other hand, the South Argentina region is nearly at the same C level, and the potential to enhance the soil content is probably limited. More accurate values will be obtained when better IF values are known for each ecoregion.

TABLE 4.5. Estimated C pools lost by agricultural activities.

Soil ecoregion	Total C stock under native condition (Tg C)	Agricultural impacted area (Tg C)	Mosaic of native and agricultural area (Tg C)
Amazonian	33,711	264	137
Andean Equatorial	4,136	181	129
Brazilian Atlantic	6,267	522	188
Caribbean	3,370	228	66
Central American	7,140	713	3
Central Andes	7,822	23	69
Central Brazilian	7,986	889	129
East Brazilian	3,806	190	126
Eastern Pampas	5,419	487	134
Mexican	5,793	95	0
Preandean	8,125	230	80
South Argentina	3,185	3	1
South Chilean	3,743	64	12
Total	100,503	3,889	1,073

CONCLUSION

With an estimate of about 5,000 Tg C lost by agricultural activities, soil appears as a main component in the regional C pool and fluxes. Agriculture can be not only a source but also a sink for atmospheric CO_2. Adoption of recommended management practices can result in important C sequestration by cultivated lands.

REFERENCES

Batjes, N.H. 1996. Total carbon and nitrogen in the soils of the world. *European Journal of Soil Science,* 47, 151-163.

Bernoux, M., Carvalho, M.C.S, Volkoff, B., and Cerri, C.C. 2001. CO_2 emission from mineral soils following land-cover change in Brazil. *Global Change Biology,* 7, 779-787.

Bernoux, M., Carvalho, M.C.S, Volkoff, B., and Cerri, C.C. 2002. Brazil's soil carbon stocks. *Soil Science Society of America Journal,* 66(3), 888-896.

Bouwman, A.F. 1990. Global distribution of the major soils and land cover types. In A.F. Bouwman (ed.), *Soils and the greenhouse effect: Proceedings of the International Conference on Soils and the Greenhouse effect* (pp. 31-59). John Wiley and Sons, New York.

Brown, S. and Lugo, A.E. 1982. The storage and production of organic matter in tropical forests and their role in the global carbon cycle. *Biotropica,* 4, 161-187.

Dixon, R.K., Brown, S., Houghton, R.A., Solomon, A.M., Trexler, M.C., and Wisniewski, J. 1994. Carbon pools and flux of global forest ecosystems. *Science,* 263, 1985-1900.

European Commission Joint Research Centre. 2003. GLC2000—The global land cover map for the year 2000. GLC2000 database, http://www.gvm.sai.jrc.it/glc2000/defaultGLC2000.htm.

FAO. 1995. Digital Soil Map of the World (DSMW). Version 3.5. CD-ROM

IPCC/UNEP/OECD/IEA. 1997. *Revised 1996 IPCC guidelines for national greenhouse gas inventories*: *Reporting instructions* (Volume 1); *Workbook* (Volume 2); *Reference manual* (Volume 3). Intergovernmental Panel on Climate Change, United Nations Environment Programme, Organization for Economic Co-Operation and Development, and International Energy Agency, Paris.

Jenness, J. 2004. Grid and Theme Projector, version 2 (grid_theme_prj.avx) extension for ArcView 3.x.

Post, W.M., Emmanuel, W.R., Zinke, P.J., and Stangenberger, A.G.1982. Soil carbon pools and world life zones. *Nature,* 298, 156-159.

Schroeder, P.E. and Winjum, J.K.1995. Assessing Brazil's carbon budget: 1. Biotic carbon pools. *Forest Ecology and Management,* 75, 77-86.

Venkatapen, C., Blanchart, E., Bernoux, M., and Burac, M. 2004. Déterminants des stocks de carbone dans les sols et spatialisation à l'échelle de la Martinique. *Les Cahiers du PRAM,* 4, 35-38.

Whittaker, R.H. and Likens, G.E. 1975. The biosphere and man. In H. Lieth and R.H. Whittaker (eds.), *Primary production of the biosphere* (pp. 305-323). Ecological Studies 14. Springer Verlag, Heidelberg, Germany.

PART II: SOIL CARBON SEQUESTRATION IN DIFFERENT BIOMES OF LATIN AMERICA

Chapter 5

Soil Carbon Sequestration in Western Mountain Ridges and Deserts of South America

D. Gómez
C. Balbontín
A. Monterroso
J. D. Etchevers

INTRODUCTION

Over the past decade climatic change due to anthropogenic factors has been the focal point of a number of scientific studies. Scientists anticipate that the global average temperature will increase between 1.4 and 5.8°C from 1990 to 2100 (IPCC, 2000). This is due to the emission of greenhouse-effect gases (GEI) such as carbon dioxide (CO_2), methane (CH_4), nitrous oxide (N_2O), tropospheric ozone (O_3), and chlorofluoride carbons (CFCs). The concentration of these GEI continues building up mainly as a result of human activities such as the use of fossil fuels, land-use change, and agricultural practices.

The global climatic change will affect the entire economy, but the agricultural sector (including crops, stockbreeding, forestry, and fish farming) is perhaps the most sensitive and vulnerable, as it is highly dependent on climatic resources (Sombroek and Gommes, 1996). Among other mechanisms, the development of clean technologies aimed at reducing the emission of such gases has been proposed, as well as their application for a long time period in soil and vegetation (carbon sequestration) in order to counteract the adverse effect caused by the GEI increase.

Interest in developing national carbon inventories has increased in recent times, and the current efforts of the Kyoto Agreement countries to make the carbon emission reduction goals obligatory may be the prelude of a more understandable carbon accounting system (IGBP, 1998). This will be better

Carbon Sequestration in Soils of Latin America

doi:10.1300/5755_05

accomplished at a national and regional level, through the appropriate combination of databases, monitoring networks, and historical knowledge, and making serious methodological proposals to achieve such an aim.

Various attempts have been made to measure the carbon stored in the soil and vegetation, based on the information contained in the national inventories on these resources using different global models, yet these inventories do not have the information needed to reduce the variability of the results. For such purpose, an appropriate description is required of the vegetation and soils of the areas chosen to do such estimates (Eswaran et al., 1993; Sombroek et al., 1993; Dixon et al., 1994; Moraes et al., 1995; Batjes, 1996; Turner et al., 1998; Bernoux et al., 2002).

For the countries included in Biome B, carbon accumulation and sequestration in soils are issues scarcely documented. No accurate estimates of the size of the carbon stocks present in the soil and the way these stocks evolve have been conducted in these regions of the world, nor has the effect of different land-use systems been studied. However, in the Biome B variety of vegetation, systems are found in which carbon can be sequestered. In the present chapter an attempt to answer how much carbon can be accumulated in the soil is presented in a similar way as Lal et al. (1999) did for U.S. cropland.

Geography, Economics, and Social Aspects of Biome B

Biome B is located in a land extension going from the north of Colombia up to the south of Patagonia, in Argentina. It comprises the highlands and the western side of the Andes and the South American Pacific Coastal Plain, as well as the eastern side of the Argentine Andes. The Andes is divided into three segments: the Northern Andes extending from the coast of Venezuela up to Cajamarca in Peru, the Central Andes, extending from Cajamarca, Peru, to one line connecting Antofagasta in Chile with Cajamarca in Argentina, and the Southern Andes, from the Antofagasta-Catamarca line up to a line connecting Valdivia in Chile with Neuquén in Argentina. The southern limit is arbitrary for research purposes, as by definition the Andes extend to the southern end of South America. The Northern Andes, comprising the mountains and adjacent lands of Venezuela, Colombia, and Ecuador, are characterized as equatorial, humid, with symmetrical vegetation zones in the east-west cross sections, and with minimal seasonal climatic variations. This is the Andean subregion recording the highest vegetal productivity due to its natural ecological system and greater potential for plant productivity. Because of the high precipitation and low temperatures, the height limit for practical agriculture ranges from 3,200 to 3,400 m above sea level. The

Central Andes includes the subregions of Peru, Bolivia, and the north of Chile and Argentina. Here there is an asymmetry between pluvial precipitation and temperature with the Pacific hillsides which are drier and more fresh than the eastern hillsides, with well-marked patterns of seasonal pluvial precipitation in the highlands, tending toward the south to turn into real aridity, starting along the coastline and culminating in the Atacama Desert and in the highlands of the central-southern area, especially in the Peruvian-Bolivian high plateau. The Southern Andes of Chile and Argentina are less compact than the Northern and Central Andes and lack highlands, though there are many high peaks. The Cordillera, sharply steep, gradually declines in height down to Chiloé Island by the Pacific coast, where it is divided into a series of isolated highlands, of an altitude of some 1,000 m. The weather is cool and dry in the north, with a dry season in the central area and very humid in the south. Therefore, the arid vegetation zones in the north gradually change to humid forests in the south. Figure 5.1a shows the distribution of Biome B throughout South America, with a total surface of 3,160,844 square kilometers, as estimated by the authors.

The Andes mountain chain is the longest of the earth and has a great variety of climatic, geomorphological, flora, and fauna elements, at both micro and macro scales. In response to the opportunities offered by the various ecological systems, human populations have acquired a rich knowledge of the environment and developed a complex variety of adaptation mechanisms to such extreme situations as the impact of tensions at great heights,

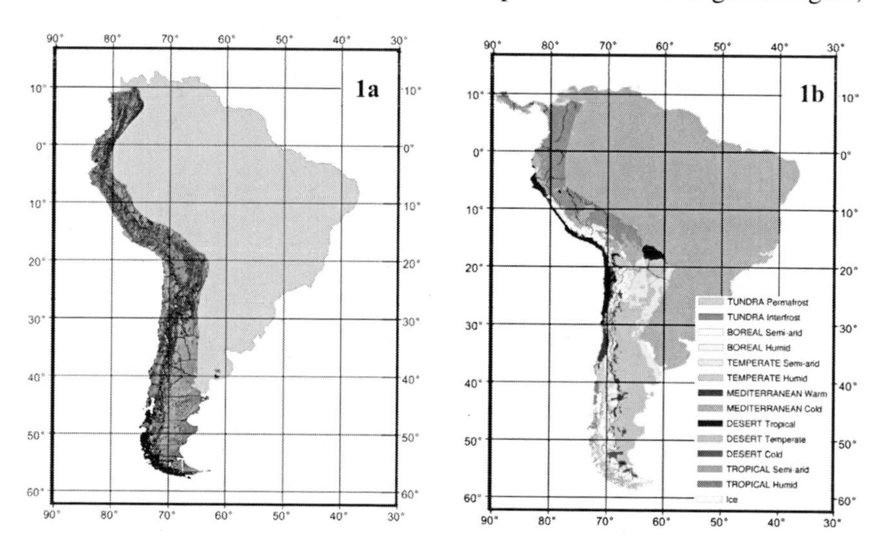

FIGURE 5.1. Distribution of Biome B and its ecological regions.

especially hypoxia and low temperatures, and human physiological adaptations, as well as cultural and socioeconomic development.

The main ecological regions within the study area are humid tropical, desert (tropical and temperate), humid, arid and semiarid temperate, and boreal at the southern end of the biome (Figure 5.1b).

Biome B comprises portions of six countries of the Andean axis: Colombia, Ecuador, Bolivia, Peru, Argentina, and Chile as a whole. These countries are characterized by a high population growth rate, which generates great pressure on natural resources. Table 5.1 shows the countries' population and selected economic indexes (adapted from WRI, 2004a,c).

The territorial surface of these regions comprises 3.16 million square kilometers, that is, 4.8 percent of the world terrestrial area. Nearly 50 percent of this area belongs to Argentina, 23.5 percent to Chile, 11.9 percent to Peru, and the remaining 14.6 percent is shared by Bolivia, Colombia, and Ecuador.

The GNI (gross national income) per capita converted to international dollars using purchasing power parity (PPP) in the region ranges from US$900 to US$4,250.

TABLE 5.1. Biome B: Population and economic indexes per country.

Country	Total land area[a] (10^3 km^2)	Land area in Biome B[b]	Population (10^3)			GDP[c]	GNI[d]
			1950	2002	2025		
Argentina	2736.69	1583.17	17.2	37.9	47.2	102.0	4220
Bolivia	1084.38	200.52	2.7	8.7	13.1	7.8	900
Chile	748.80	742.30	6.1	15.6	19.5	64.2	4250
Colombia	1038.70	175.06	12.6	43.5	56.2	80.9	1820
Ecuador	276.84	83.88	3.4	13.1	17.8	24.3	1490
Peru	1280.00	375.92	7.6	25.5	35.5	56.5	2020

Source: Adapted from WRI, 2004a,c.

[a]The total land area of the countries includes part of Biome D (Amazonia tropical wet forest).

[b]Estimated by the authors.

[c]GDP, billions of dollars, year 2002; gross domestic product, current dollars, is the sum of value added by all resident producers plus any product taxes (less subsidies) not included in the evaluation of output plus net receipts of primary income (compensation of employees and property income) from abroad. In other words, GDP measures the total income of all people who are citizens of a particular country, while GNI measures the total output of all persons living in that particular country's borders.

[d]GNI, dollars, year 2002, per capita; gross national income per capita, PPP, is gross domestic product converted to international dollars using purchasing power parity (PPP) rates, and divided by the population of the country that year. An international dollar has the same purchasing power in a given country as a U.S. dollar in the United States. In other words, it buys an equivalent amount of goods or services in that country.

GENERAL CHARACTERISTICS OF BIOME B BY COUNTRY

Colombia

The part of Colombia corresponding to Biome B can be divided into two large groups: the Andean region and the Pacific region. The Pacific Colombian region and the Andean region extend from the border with Panama up to the north of Ecuador, at an approximate 1,300 km longitude. This zone is the second richest and most important natural reserve of the planet and one of the most humid regions in the world. The proximity of the western cordillera with that bordering on the Pacific to the east and the combination of air currents in this equatorial zone makes it different from the rest of the country and the world. The high rainfall rate and temperatures in the Pacific region have contributed to create an environment of incredible biodiversity. Rains reach between 5,000 to 12,000 mm per year, and 75 percent of the area is still covered by tropical forest, estimated to amount to 8 million hectares, of which 50 percent has not been developed yet. This bioregion includes coastal ecosystems of mangrove, lakes and swamps, humid tropical forest, rain forest, and Andean Paramos.

In the Pacific region, soils are made up of highly weathered materials that have a low fertility potential. These soils are poor in nutrients due to intense leaching and at the same time acidic with high contents of interchangeable aluminum, in which the natural offer for biological cycles depends on the organic matter (IDEAM, 2001).

Ecuador

Ecuador has two zones within Biome B: the costal and the sierra. The coastal region is located between the Pacific Ocean and the Andes Mountains, and it is conformed of lowlands and mountains. The sierra includes two major chains of the Andes Mountains: the Cordillera Occidental and Cordillera Oriental and the intermontane basins in between.

Vegetation types are related to climatic conditions, and in particular to rainfall (Huber and Riina, 1997). The following types are identified: tropical rain forest found in the northern parts of the coast and on parts of the Andean piedmonts. The rain forest here is as rich as that of the Colombian rain forest. Its composition is influenced by altitude. At 1,000 to 2,000 m it is mixed with shrubs and ferns, whereas above 2,000 m a cloud forest is commonly encountered. Along the drier portions of the (southern) coast, a dry deciduous forest predominates with a savanna farther south, and scrubs composed mostly of *Mimosa* sp. that alternate with more open grassy types;

in the extreme southwestern part of the coast the savanna yields to desertic, xerophytic vegetation; in the Andes, vegetation depends on the altitudinal level and evolves from dry forest to grass "Paramo" or steppe as altitude increases, finally reaching the area of permanent snow.

Bolivia

The zone of Bolivia included in Biome B corresponds to the high plateau of the Andes, in which the main ecological zones are cold steppe, desert temperate cold shrubland, and temperate cold desert. It also includes a small part of the temperate humid and dry forest (PNCC, 2000).

One of the Bolivian regions belonging to Biome B is the Potosí Department composed of two regions, divided by the north-south Lipez Mountain chain. The western part, at an altitude of 4,000 m, is flat, very dry, and famous for its impressive salt lakes *(salares)*. The population density is very low as the natural resources are extremely limited. Agriculture consists mainly of potato and sorghum cultivation, but the extensive raising of llamas and alpacas plays an important role in household economies. The natural vegetation is limited to indigenous grasses and shrubs. On the eastern side of the chain of the Lipez Mountains the situation is very different, with altitudes varying between 2,500 and 3,900 m and various ecological zones. This allows for more diversified livestock activity and agriculture, as well as a diversity of tree species. In this environment of contrasts, natural resources are limited and deteriorating. Poor soils, shortage of water, high altitudes, and the corresponding climatic adversities (such as ice-winds and hail storms), combined with socioeconomic factors such as demographic pressure, decreasing agricultural production and productivity, and inadequate support services, are seriously affecting the sustainability of natural resource use and, hence, the living conditions of rural families.

Peru

Peru's geographical regions included in Biome B are the coastal zone with a mostly desert climate and the Andean area with a highly uneven topography in which most of the arable lands and natural grasses are on steep slopes with no possibility of being irrigated. Peruvians from ancient times solved this problem by constructing cultivation systems in terraces and irrigation channels for hillside soils, thus creating the famous platform systems, whose purpose was both to have a larger agricultural production than in hillside lands and to reduce the natural risks of production by providing lands with irrigation systems and diminishing erosion, in the context of increasing demographic pressure.

The main types of vegetation within the biome correspond to mangroves, dry forests, lakes, and coastal mountain vegetation, and at greater altitude in the sierra the forest patches of the Andean west slopes, the Paramo forests, the coastal forests of the sierra, the forests and plantations of *Polylepis, Buddleia* of the high sierra, and the plantations of *Pinus* and *Eucalyptus,* and in the higher zones the bushes, shrublands, and wetlands.

According to the soil classification proposed by the FAO, three large geopedological regions can be identified: the coastal desert (yermosolic region), the west side of the Andes (paramosolic region), and the high inter-Andean valleys and intermediate regions (kastañosolic region). The best agricultural lands are located in the coastal region, distributed along 52 alluvial low irrigation valleys. This has generated overirrigation of the area, causing salinization and poor drainage of the local lands. The sierra, made up of inter-Andean valleys distributed between 2,300 and 3,800 m.s.n.m, is the zone with the highest concentration of peasants and the center of dry farming. The main characteristic of the area is the lack of flat lands for agriculture, which together with the high demographic pressure has caused extensive erosion and desertification.

Chile

Nearly the entire territory of Chile is included in Biome B. Its extensive latitude distribution allows for the existence of various climates and, therefore, a great variety of vegetal communities. Climates range from the desert type in the northern end (the Atacama Desert), through the arid and semiarid in the center of the country, to end with the rainy and cold climates of the south. These climatic conditions, combined with the presence of two cordilleras (the Andes and the coast), limit the useful land extension for agricultural activities and favor erosive processes (FAO, 2003).

In the northern end of the country, in the high plateau area, is only crop agriculture that adjusts to the marginal conditions imposed by altitude, while in the valleys and foothills of the cordillera a low-yielding traditional form of agriculture is practiced, due to the effects of very dry weather and salty soils. Moving south, it is possible to find some vegetation in oasis or sites where there is a water source. In the northern desert zones there is not much agricultural development, and there are only valleys in which soils are permeable and the groundwater is closer to the surface but salty.

The central area, where there is a transition between these two extremes, presents a great deal of agricultural development favored by the natural bounty of water resources generated by the snow stored in the Andes. Unfortunately, this intensive land use has given way to the degradation of the soil, reducing its biological value.

Moving farther south, forest developments are found that compete for the agricultural use of the soil and, due to the high precipitation volumes, end up being predominant. It is precisely in the most southern areas of Chile where forest extensions are concentrated.

Finally, in the Patagonia and Tierra del Fuego zones the soil can be used only for natural fodder prairies because of the inclement climatic conditions. It is worth mentioning that in the Andes the vegetation line is at 1,500 m.s.n.m. in the north area, and at 500 m.s.n.m. in the south, severely limiting the development of communities at such heights.

Argentina

Argentina is in the southern area of Biome B and includes the zone called La Patagonia. Climatic conditions in this area are low precipitation (150 to 250 mm), a great day-night temperature fluctuation, strong winds, and few periods free from freezes. The climate is temperate-arid, and the few precipitations occur in the north throughout the year; in the south the Mediterranean weather conditions are present (winter rains), typical of Patagonia. Soils are mostly Aridisols, in accordance with the arid climate. Saline and stony phases are frequent.

High plateaus, terraces, sierras, and many depressions prevail in this zone. The main vegetal species are shrublike and Gramineae. In this context, where the ecosystem appears to be highly fragile, intensive use of natural resources has taken place, including overgrazing (mainly sheep), clearance, and farming, leading to rapid expansion of the desertification and degradation processes, the latter currently covering 80 percent of the country.

LAND USE AND LAND-USE CHANGE IN BIOME B

In the countries of Biome B most of the present farmland and pastureland come from the extensive change in land use that occurred after the conquest by the Spaniards. Also, the rate of land-use change has been higher as the population and animal mass have increased and new needs for farmland consequently emerged. Barren or sparsely vegetated areas cover 9 percent of the total biome and are related to very arid climates. Shrubland and mixed shrubland-grassland represent almost 35 percent of the total area and are related to arid and semiarid climates. Cropland and pasture cover 7 percent of the area. Evergreen forests associated with humid regions are located in the north coastal plain of the region and in the eastern side of the Andes from 10°N to 20°S latitude. Table 5.2 shows the land-use types in Biome B, and the map is presented in Figure 5.2.

TABLE 5.2 Land use in Biome B.

	Total land area covered (10^3 km^2)						Total
Land use	**Argentina**	**Bolivia**	**Chile**	**Colombia**	**Ecuador**	**Peru**	**land use**
Barren or sparsely vegetated	116.92	2.51	128.62	0.33	0.76	41.68	290.83
Cropland and pasture	21.57	1.98	28.33	18.22	10.09	8.18	88.36
Cropland/woodland mosaic	64.38	7.43	21.56	23.70	8.92	20.29	146.28
Evergreen broadleaf forest	7.46	35.99	3.12	78.82	26.36	119.75	271.50
Evergreen needleleaf forest	0.00	0.00	0.00	0.45	0.00	0.00	0.45
Grassland	219.88	21.15	140.21	10.09	4.73	50.53	446.59
Mixed forest	32.43	0.64	196.67	0.11	0.00	0.09	229.95
Mixed shrubland/grassland	245.65	0.00	6.50	0.00	0.00	0.00	252.24
Savanna	87.21	19.20	31.31	15.79	8.92	15.44	177.87
Shrubland	650.39	53.02	48.30	7.79	4.92	58.79	823.21
Snow or ice	4.61	0.00	36.92	0.00	0.00	0.00	41.53
Tundra	19.83	43.94	12.05	6.81	11.23	33.07	126.93
Urban and built-up land	0.27	0.00	0.85	0.20	0.20	0.08	1.60
Water bodies	27.87	2.75	87.77	3.95	1.10	7.80	131.24
Wooded wetland	0.00	0.48	0.00	0.07	0.00	0.09	0.63
Others	84.71	11.42	0.00	8.73	6.64	20.13	131.63
Total area	1583.17	200.52	742.30	175.06	83.88	375.92	3160.84

Source: Adapted from USGS, 1993.

Table 5.3 shows the changes in forest cover from ancient times and in recent ones according to the WRI (2004c). At the beginning of the twenty-first century the original forest cover has been reduced to almost 50 percent in most of the countries in Biome B. Changes in forest areas induced in the past ten years (1990-2000) range from a decrease of 12 percent in Ecuador to 1 percent in Chile.

Argentina has a higher area under cultivation, with 87 million hectares of the total land area of the country (32 percent) (see Table 5.4). Of the total area, Ecuador has 24 percent of its territory under cropland, Colombia has 17 percent, and the other countries have around 7 to 9 percent. Shrublands, savannas, and grasslands are the dominant land uses in Argentina and Bolivia, with more than 50 percent of the total land area. The forest area is dominant in Colombia and Peru, also with more than 50 percent of the total.

Table 5.5 shows the land-use change in a more disaggregated manner from 1961 to 2001. Data are reported by FAO (2003).

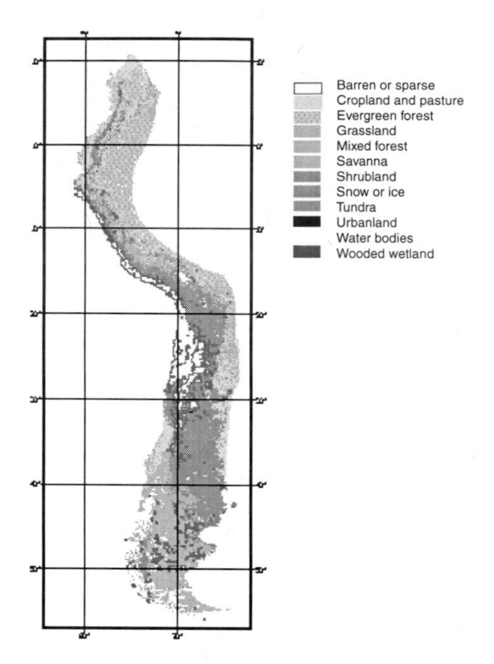

FIGURE 5.2. Land use in Biome B. (*Source:* Adapted from USGS, 1993.)

SOIL DEGRADATION IN BIOME B

Human activities have been the main cause of degradation of ecosystems. According to some estimates, 10 percent of the total surface land on the planet in 1990 had been transformed by anthropic activities, from the forest areas up to the pastures of arid zones, and more than 25 percent of the total land is at the risk of undergoing change (WRI, 1990). In Biome B different types of human-induced soil degradation (GLASOD) and their degrees of severity have been established (Oldeman et al., 1991). Table 5.6 includes information generated by the authors on the affected surface of the countries' portions making up this biome, with information from Oldeman et al. (1991).

Of the total Biome B surface, wind erosion is the main type of soil degradation, with almost 36 percent of the total area; water erosion is dominant in 30 percent of the area; chemical deterioration represents 4 percent; and physical deterioration only 1 percent. The area without human-induced soil degradation is about 11 percent, while 5 percent is wastelands. A map showing the soil degradation types and their severity is presented in Figure 5.3.

TABLE 5.3. Percentage of the original forest and change in forest area with respect to total land area[a] (10^3 km^2).

	Percentage of the total land		Change in forest area, 1990-2000 (%)[d]		
Country	**Original forest[b]**	**Forest area in 2000[c]**	**Total**	**Natural**	**Plantations**
Argentina	12	12	–8	–10	nd
Bolivia	54	48	–3	–3	4
Chile	40	21	–1	–7	5
Colombia	92	44	–4	–4	6
Ecuador	79	37	–12	–12	2
Peru	74	51	–4	–5	15

Source: Adapted from WRI, 2004b.
[a]Percentage refers to the total surface area of each country.
[b]Original forest as a percent of land area refers to the estimate of the percent of land that would have been covered by closed forest about 8,000 years ago assuming current climatic conditions, before large-scale disturbance by human society began.
[c]Forest area in 2000 as a percent of total land area is calculated by dividing total forest area by total land area.
[d]Change in forest area is the total percent change in both natural forests and plantations between 1990 and 2000.

Table 5.7 presents the causative factors indicating what type of physical human intervention has caused the soil to be degraded in Biome B. Overgrazing is the main causative factor of soil degradation, with 27 percent of the total land area; overgrazing and agriculture represent 15 percent of the total area; and deforestation represents about 16 percent. Agriculture alone contributes almost 6 percent of the causative factors, and deforestation, agriculture, and overgrazing in the same area represent almost 10 percent. The area without an explicit causative factor of degradation is around 25 percent of the total. In Figure 5.4 a map with the areas of causative factors of human-induced degradation and its severity is presented.

SOIL CARBON STORAGE IN BIOME B

In drawing the map on soil organic carbon density for Biome B, the database of soil subunits by SOTER (ISRIC, 1995) scale 1:5,000,000 was used and each polygon was labeled with the values reported by ISRIC (1991) of soil organic carbon density in kilograms of carbon per square meter at 1 m

TABLE 5.4. Land use in 2000.

Country	Total land covered (10^3 km^2)						Total land covered (%)					
	Forests[a]	Shrublands, savannas, and grasslands[a]	Cropland and crop/natural vegetation mosaic[c]	Urban and built-up areas	Sparse or barren; snow and ice	Wetlands and water bodies	Forests	Shrublands, savannas, and grasslands[a]	Cropland and crop/natural vegetation mosaic[c]	Urban and built-up areas	Sparse or barren; snow and ice	Wetlands and water bodies
Argentina	82.10	1587.28	875.74	2.74	164.20	27.37	3	58	32	0.1	6	1
Bolivia	412.06	563.88	97.59	0.10	10.84	10.84	38	52	9	0.0	1	1
Chile	187.20	217.15	67.39	0.75	247.10	29.95	25	29	9	0.1	33	4
Colombia	581.67	249.29	176.58	1.04	10.39	10.39	56	24	17	0.0	1	1
Ecuador	91.36	99.66	66.44	0.28	13.84	5.54	33	36	24	0.1	5	2
Peru	652.80	409.60	89.60	0.13	115.20	25.60	51	32	7	0.0	9	2

Source: Adapted from WRI, 2004b.

Note: Total land area and percentages refer to the total surface area of each country.

[a]Includes all areas dominated by evergreen or deciduous trees with a canopy cover greater than 60 percent and a height exceeding 2 m; both broadleaf and needle-leaf trees are included.

[b]Shrublands, savanna, and grasslands include lands dominated by woody vegetation less than 2 m tall and with a shrub canopy cover greater than 10 percent; the shrub foliage can be either evergreen or deciduous; this category also includes savannas and grasslands with herbaceous and other understory systems; land may have a tree or shrub cover of less than 60 percent.

[c]Croplands are lands covered by temporary crops followed by harvest and a bare soil period (e.g., single and multiple cropping systems); perennial woody crops are classified as forest or shrubland cover; cropland/natural vegetation mosaics are lands with a combination of croplands, forests, shrublands, and grasslands in which no one component comprises more than 60 percent of the landscape.

TABLE 5.5. Land use and land-use change.

	Area (10^3 of km^2) by year		
Land use (by country)	**1961**	**1981**	**2001**
Agricultural Area			
Argentina	1754.33	1733.00	1770.00
Bolivia	300.42	340.99	369.31
Chile	133.86	167.50	152.35
Colombia	399.70	453.08	460.49
Ecuador	47.10	67.59	80.75
Peru	301.47	306.95	313.10
Arable Land			
Argentina	272.77	290.00	337.00
Bolivia	12.94	19.80	29.00
Chile	36.40	37.33	19.82
Colombia	35.32	37.28	25.16
Ecuador	17.05	16.00	16.20
Peru	17.96	32.45	37.00
Permanent Crops			
Argentina	11.56	12.00	13.00
Bolivia	1.48	1.19	2.01
Chile	1.96	2.17	3.18
Colombia	14.38	14.80	17.33
Ecuador	8.05	9.42	13.65
Peru	1.60	3.30	5.10
Permanent Pasture			
Argentina	1470.00	1431.00	1420.00
Bolivia	286.00	320.00	338.30
Chile	95.50	128.00	129.35
Colombia	350.00	401.00	418.00
Ecuador	22.00	42.17	50.90
Peru	281.91	271.20	271.00
Forests and Woodland[a]			
Argentina	509.00	509.00	509.00
Bolivia	604.50	580.00	580.00
Chile	165.00	165.00	165.00
Colombia	590.00	550.00	530.00
Ecuador	181.50	155.00	156.00
Peru	848.50	848.50	848.00

TABLE 5.5 *(continued)*

	Area (10^3 of km^2) by year		
Land use (by country)	**1961**	**1981**	**2001**
All Other Land[a]			
Argentina	473.36	494.69	500.39
Bolivia	179.46	163.39	140.87
Chile	449.94	416.30	429.30
Colombia	49.00	35.62	60.05
Ecuador	48.24	54.25	39.55
Peru	130.03	124.55	120.36

Source: Adapted from FAO, 2003.
[a]Dates are 1961, 1981, and 1994.

depth to obtain the surface of the different levels of carbon content in the Biome B area, doing the estimation for each country's portion making up this biome.

Table 5.8 shows the area covered by the main soil units (ISRIC, 1995). The dominant soils in Biome B are the Leptosols, which account for about 20 percent of the total area; Cambisols, 15 percent; and Regosols, 13.5 percent. The first ones are surface soils and are mainly associated to the mountainous zone with steep slopes. Regosols and Cambisols are poorly developed soils and in general with low contents of organic matter. Soils severely affected by salinity account for 10 percent of the total. Figure 5.5 shows the map presenting the distribution of soil units.

Table 5.9 shows the surface areas by levels of the soil organic carbon density for each portion of the countries making up Biome B, highlighting that almost 73 percent of the total area has a carbon density in the soil between 4 to 10 kg per square meter at 1 m depth. In 16 percent of the area soil carbon density is reported to be between 10 and 20 kg C per square meter. Soil organic carbon contents above 20 kg per square meter represent 21 percent of the total area. We estimated that the carbon pool in soils of Biome B through 1 meter depth range from 25.9 to 34.2 Pg .

The distribution map of the soil organic carbon density in Biome B is shown in Figure 5.6, in which the highest carbon contents are in the area close to the south coast of Chile, the most northern region of Argentina, the highlands of Colombia and Ecuador, and the eastern side of the Peruvian Andes.

Within the strategies of carbon sequestration in Biome B, see Table 13.7 which shows data from two Colombian Andean hillsides regions for various land-use systems if converted from degraded land (Amezquita et al., 2004).

TABLE 5.6. Biome B: Types and severity of soil degradation.

Soil degradation types	Severity	Total area in Biome B (10^3 km^2)						Total type of degradation
		Argentina	Bolivia	Chile	Colombia	Ecuador	Peru	
Loss of nutrients and/or organic matter	Low	0.0	0.0	0.0	21.41	8.67	24.31	54.39
	Medium	0.0	0.0	0.0	0.0	0.0	0.47	0.47
	High	0.0	0.0	0	6.33	0.0	6.74	13.07
Salinization	Low	0.0	0.0	17.29	0.0	0.0	19.72	37.01
	Medium	0.0	0.0	0.0	0.0	0.0	20.92	20.92
Active dunes[a]	N/A	0.0	0.0	0.0	0.0	0.0	8.00	8.00
Wind erosion (terrain deformation)	Medium	998.27	1.14	55.54	0.0	0.0	0.0	1054.96
Wind erosion (loss of topsoil)	Low	2.43	16.27	59.19	0.0	0.0	1.79	79.68
	Medium	117.14	28.35	17.70	0.0	0.59	12.68	176.47
Ice caps[a]	N/A	0.41	0.0	12.98	0.0	0.0	0.0	13.39
Arid mountain regions[a]	N/A	0.0	0.01	108.38	0.0	0.0	0.65	109.04
Compactation, sealing, and crusting	Low	0.0	0.0	29.76	0.0	0.0	0.0	29.76
Terrain stabilized by human intervention[b]	N/A	0.0	1.11	0.0	0.0	0.0	7.22	8.33
Stable terrain under natural conditions[b]	N/A	136.58	15.53	140.15	6.66	5.37	48.78	353.06
Water erosion (terrain deformation / mass movement)	Low	0.0	0.0	0.0	16.92	0.10	12.53	29.55
	Medium	168.69	10.60	1.59	0.0	0.0	11.47	192.34
	High	0.0	6.29	0.0	0.0	0.0	0.0	6.29
Water erosion (loss of topsoil)	Low	0.33	29.23	0.0	38.76	35.46	58.08	161.86
	Medium	54.07	58.19	175.62	76.00	27.78	113.84	505.50
	High	0.0	13.80	25.02	0.01	0.0	4.89	43.72
Salt flats[a]	N/A	9.84	7.09	0.003	0.0	0.0	0.0	16.93
Mapped units without level	N/A	95.42	12.91	99.07	8.97	5.90	23.83	246.10
Total per country		1583.17	200.52	742.30	175.06	83.88	375.92	3160.84

[a]Wastelands
[b]Mapped units without human-induced soil degradation

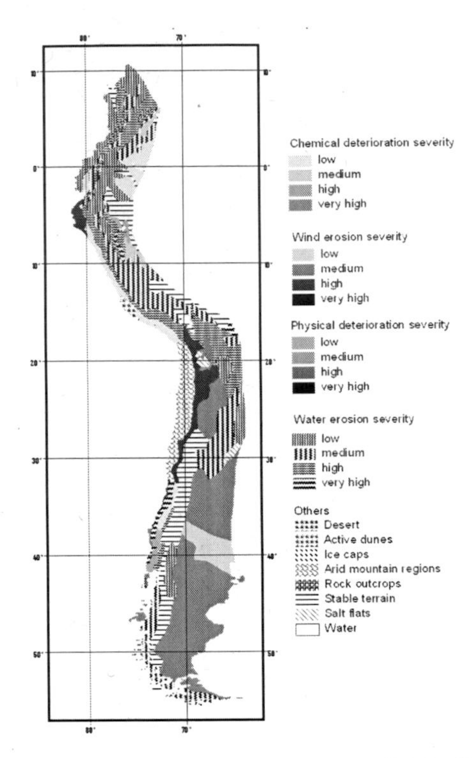

FIGURE 5.3. Human-induced soil degradation types (GLASOD) and their severity in Biome B. (*Source:* Created by the authors from available information in Oldeman et al., 1991.)

CONCLUSIONS

The soil organic carbon through 1 meter depth of Biome B is estimated by the authors to range from 25.9 to 34.2 Pg. Though there are some carbon sequestration rates for various land-use systems if converted from degraded land in the tropical ecosystems of Biome B, no information was obtained about the surfaces susceptible to change to estimate the soil C sequestration potential for this biome with the adoption of the recommended management practices and restoration of degraded soil (Lal et al., 2001).

For that reason, there is a need to determine the most viable land-use change to improve C sequestration in soils on the different ecological regions of the biome to determine the effect of different land-use systems in future carbon stocks.

TABLE 5.7. Causative factors of human-induced soil degradation and its severity in Biome B.

Causative factor	Severity	Land area by country (10^3 km^2)						Total for factor
		Argentina	Bolivia	Chile	Colombia	Ecuador	Peru	
Agricultural activities	Low	0.0	0.0	39.17	16.29	0.10	56.47	112.04
	Medium	0.0	13.42	16.39	7.43	2.85	20.94	61.03
Overexploitation of vegetation for domestic use	Medium	179.10	1.14	0.84	0.0	0.0	0.0	181.08
Deforestation	Low	2.43	9.04	48.70	51.64	41.31	34.87	187.99
	Medium	67.66	21.58	161.50	22.25	11.85	43.65	328.49
Deforestation and agriculture	Low	0.0	0.0	17.29	0.0	0.0	0.0	17.29
	Medium	65.04	0.13	0.0	0.0	0.0	0.0	65.17
Deforestation and overexploitation of vegetation for domestic use	Medium	82.55	0.0	0.0	0.0	0.0	0.0	82.55
Deforestation and overgrazing	Medium	0.0	0.0	0.0	0.9	11.72	0.14	11.86
Overgrazing	Low	0.33	36.46	1.08	9.16	2.82	25.09	74.94
	Medium	484.07	48.43	71.21	41.95	1.95	94.67	742.29
	High	0.0	11.61	25.02	0.0	0.0	0.01	36.65
Overgrazing and agriculture	Medium	459.74	0.0	0.50	4.37	0.0	0.0	464.61
Overgrazing and overexploitation of vegetation for domestic use	Medium	0.0	13.57	0.0	0.0	0.0	0.0	13.57
Total per country		1583.17	200.52	742.30	175.06	83.88	375.92	3160.84

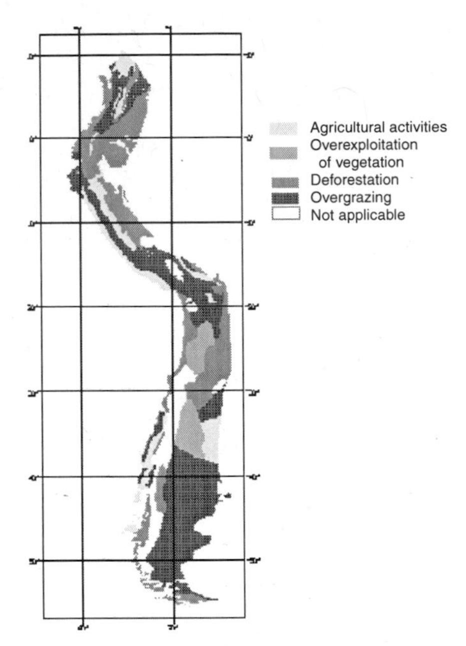

FIGURE 5.4. Causative factors of human-induced soil degradation in Biome B. (*Source:* Established by the authors from available information in Oldeman et al., 1991.)

TABLE 5.8. Soil units in Biome B.

	Total land area covered by country (10^3 of km^2)						
Soil units	Argentina	Bolivia	Chile	Colombia	Ecuador	Peru	Total
Acrisols	0.00	0.00	0.00	30.80	7.80	11.39	49.99
Alisols	0.00	0.00	12.49	0.00	0.00	0.00	12.49
Andosols	61.54	1.23	83.57	13.03	14.97	0.00	174.34
Arenosols	75.36	0.00	12.65	3.94	0.58	7.12	99.66
Chernozems	0.06	0.00	1.64	0.00	0.00	0.00	1.70
Calcisols	156.24	0.00	29.22	0.00	0.00	7.91	193.36
Cambisols	156.35	41.12	93.12	61.47	27.90	103.24	483.19
Fluvisols	60.26	0.03	4.63	0.96	1.76	3.35	70.99
Ferrasols	0.00	6.31	0.00	6.97	0.00	0.00	13.28
Gleysols	0.00	1.18	7.56	15.19	5.75	36.69	66.37
Gypsisols	17.63	0.00	0.00	0.00	0.00	0.00	17.63
Histosols	0.00	0.00	16.30	1.28	0.00	0.00	17.58
Katañozems	7.67	0.10	0.00	0.00	0.00	0.00	7.77
Leptosols	67.02	92.86	287.04	38.30	9.47	113.49	608.18
Luvisols	254.38	5.11	31.57	0.62	8.11	0.08	299.86
Nitisols	0.00	0.00	10.57	0.00	0.00	0.00	10.57

TABLE 5.8 *(continued)*

	Total land area covered by country (10^3 of km^2)						
Soil units	**Argentina**	**Bolivia**	**Chile**	**Colombia**	**Ecuador**	**Peru**	**Total**
Phaeozems	180.17	0.07	32.52	0.24	4.38	0.03	217.41
Planosols	0.00	0.05	0.70	2.26	0.61	27.71	31.34
Podsols	0.09	0.00	14.44	0.00	0.00	0.00	14.53
Regosols	275.24	43.97	50.05	0.00	2.00	56.73	427.99
Solonchaks	9.51	7.10	14.65	0.00	0.00	5.90	37.16
Solonetz	260.21	0.00	0.00	0.00	0.00	0.00	260.21
Vertisols	0.00	0.00	0.83	0.00	0.54	0.00	1.37
Ice	0.06	0.00	30.06	0.00	0.00	0.00	30.12
Water	1.39	1.39	8.68	0.00	0.00	2.30	13.76
Total area per country	1583.17	200.52	742.30	175.06	83.88	375.92	3160.84

Source: Adapted by the authors from available information in ISRIC, 1995.

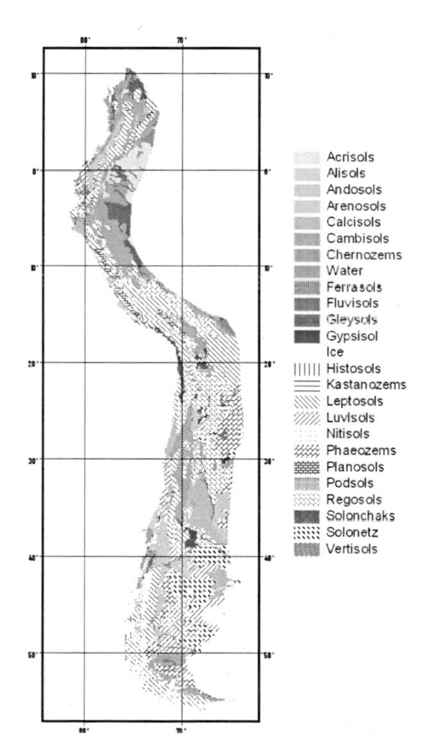

FIGURE 5.5. Soil units in Biome B. (*Source:* Established by the authors from available information in ISRIC, 1995.)

TABLE 5.9. Soil carbon density in Biome B.

Soil organic carbon density (kg C·m^{-2} to 1 m depth)	Total land area covered (10^3 km^2)						
	Argentina	Bolivia	Chile	Colombia	Ecuador	Peru	Total
<4	119.24	0	25.59	2.09	0	0	146.92
4-6	516.40	47.30	113.20	5.87	4.78	138.38	825.93
6-8	273.45	72.11	226.00	12.68	17.97	99.26	701.48
8-10	336.28	67.26	73.49	73.89	25.40	50.04	626.35
10-12	187.07	11.50	52.05	54.97	10.09	26.90	342.58
12-20	51.09	0.97	70.74	8.25	0.99	39.04	171.07
20-28	21.36	0.00	87.15	2.26	11.33	0.00	122.09
28-40	76.84	0.00	23.09	13.12	13.31	20.00	146.36
>40	0.00	0.00	32.25	1.92	0.00	0.00	34.18
Ice	0.06	0.00	30.06	0.00	0.00	0.00	30.12
Water	1.39	1.39	8.68	0.00	0.00	2.30	13.76
Total country	1583.17	200.52	742.30	175.06	83.88	375.92	3160.85

Source: Established by the authors from available information from ISRIC, 1991.

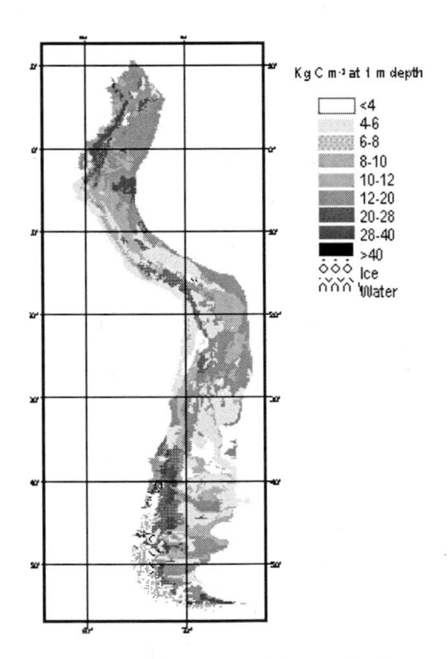

FIGURE 5.6. Soil organic carbon density in Biome B. (*Source:* Established by the authors from available information in ISRIC, 1991.)

BIBLIOGRAPHY

Amezquita, M.C., P. Buurman, M. Ibrahim, E. Murgueitio, and E. Amezquita. 2004. Carbon sequestration in pasture, agropastoral and silvo-pastoral systems in the American tropical forest ecosystems. Procedings I: World Agroforestry Congress, Orlando, FL, June 27-July 2.

Batjes, N.H. 1996. Total C and N in the soils of world. *Eur. J. Soil Sci.* 47:151-163.

Bernoux, M., Ma. Da C. Santana, B. Volkoff, and C.C. Cerri. 2002. Brazil's soil carbon stocks. *Soil Sci. Soc. Am. J.* 66:888-896.

CONAMA-CONAF-BIRF. 1999. Catastro y Evaluación de los Recursos vegetacionales nativos de Chile. Universidad Austral de Chile, Pontificia Universidad Católica de Chile Universidad Católica de Temuco. Santiago, Marzo de 1999.

CONDESAN. n.d. Consorcio para el Desarrollo Sostenible de la Ecorregión Andina (www.condesan.org; consulted May 25, 2004).

Dixon, R.K., S. Brown. R.A. Houghton, A.M. Solomon, M.C. Trexler, and J. Wisniewski. 1994. Carbon pools and flux of global forest ecosystems. *Science* 263:185-190.

Ecuador. 2001. Comunicación Nacional. República del Ecuador. Convención Marco de las Naciones Unidas. Cambio Climático. Quito: Comité Nacional sobre el Clima, Ministerio del Ambiente.

Eswaran, H., E. van Den Berg, and P. Reich. 1993. Organic carbon in soils of the world. *Soil Sci. Soc. Am. J.* 57:192-194.

FAO. 2001. Global forest resources assessment 2000: Main report. FAO Forestry Paper No. 140. Rome (www.fao.org/forestry/fo/fra/main/index.jsp; consulted May 25, 2004).

FAO. 2003. State of the world's forests: Main report. FAO Forestry Paper. Rome (www.fao.org/forestry/site/10168/en; consulted May 26, 2004).

Huber, O. and R. Riina (eds.). 1997. *Glosario Fitoecológico de las Américas,* Vol. 1: *América del Sur: países hispanoparlantes.* Caracas: UNESCO.

IDEAM. 2001. Adscrito al Ministerio de Ambiente, Vivienda y Desarrollo Territorial de Colombia. Sistema Nacional Ambiental (www.ideam.gov.co; consulted May 25, 2004).

IGBP Terrestrial Carbon Working Group. 1998. The terrestrial carbon cycle: Implications for the Kyoto Protocol. *Science* 280:1393-1394.

INRENA. 2002. *Perú Forestal en Números.* Lima, Peru.

Instituto de hidrología metereología y estudios ambientales. 2001. Colombia, primera comunicación nacional ante la convención marco de las naciones unidas sobre el cambio climático. Colombia.

IPCC (Intergovermental Panel on Climate Change). 2000. *Land use, land use change, and forestry: A special report of the IPCC.* Cambridge, UK: Published for the Intergovernmental Panel on Climate Change by Cambridge University Press.

ISRIC (International Soil Reference and Information Centre). 1991. Soil organic carbon density (www.isric.org; consulted May 20, 2004).

ISRIC (International Soil Reference and Information Centre). 1995. Soil and terrain data base (www.isric.org; consulted May 20, 2004).

Lal, R., J.M. Kimble, R.F. Follet, and C.V. Cole. 1999. *The potential of U.S. cropland to sequester carbon and mitigate the greenhouse effect.* Boca Raton, FL: Lewis Publishers.

Lal, R., J. Kimble, R.F. Follett, and B.A. Stewart (eds.). 2001. *Assessment methods for soil carbon.* Boca Raton, FL: CRC/Lewis Publishers.

León-Velarde, C. and F. Izquierdo. 1993. Producción y Utilización de los Pastizales de la Zona Altoandina. Compendio. Quito: Red de Pastizales Andinos-REPAAN.

Medina, G., and V. P. Mena. 2001. Los páramos de Ecuador. In V.P. Mena, G. Medina, and R. Hofstede (eds.), *Los Páramos del Ecuador. Particularidades, Problemas y Perspectivas.* Proyecto Páramo, Quito (www.paramo.org/; consulted May 20, 2004).

Moraes, J.L., C.C. Cerri, J.M. Melillo, D. Kicklighter, C. Neill, D.L. Skole, and P.A. Steudler. 1995. Soil carbon stocks of the Brazilian Amazon basin. *Soil Sci. Soc. Am. J.* 59:244-247.

Oldeman, L.R., R.T.A. Hakkeling, and W.G. Sombroek. 1991. World map of the status of human-induced soil degradation: Global assessment of soil degradation (GLASOD). Wageningen, the Netherlands: ISRIC and UNEP.

ONERN (Oficina Nacional de Evaluación de Recursos naturales). 1985. Los Recursos Naturales del Perú. Lima, Peru.

PNCC (Programa Nacional de Cambios Climáticos). 2000. Escenarios Climáticos, Estudio de Impactos y Opciones de Adaptación al Cambio Climático. MDSP–VMARNDF. La Paz, Bolivia.

Proyecto de Desarrollo Agropecuario Sostenido en el Altiplano (PRODASA). 1994. Informe anual. Proyecto colaborativo CIID-CIP-INIA.

Ramírez, P., F. Izquierdo, and O. Paladines. 1996. Producción y Utilización de Pastizales en Cinco Zonas Agroecológicas del Ecuador. Quito: MAG-GTZ-REPAAN.

Reinoso, J. 1997. Estrategias para conservación y desarrollo sostenible del altiplano. Centro de Investigaciones de RRNN y Medio Ambiente-CIRNMA, Peru.

Sombroek, W.G. and R. Gommes. 1996. The climate change-agriculture conundrum. In F. Bazzaz and W.G. Sombroek (eds.), *Global climate change and agricultural production: Direct and indirect effects of changing hydrological, pedological and plant physiological processes.* Chichester, UK: FAO and John Wiley.

Sombroek, W.G., F.O. Nachtergaele, and A. Hebel. 1993. Amounts, dynamics and sequestration of carbon in tropical and subtropical soils. *Ambio* 22:417-426.

Turner, D.P., J.K. Winjum, T.P. Kolchugina, T.S. Vinson, P.E. Scroeder, D.L. Phillips, and M.A. Cairns. 1998. Estimating the terrestrial carbon pools of the former Soviet Union, conterminous US and Brazil. *Climate Res.* 9:183-196.

USGS (United Status Geological Service). 1993. Global land cover characterization (www.edcdaac.usgs.gov/glcc/glcc.asp; consulted May 20, 2004).

WRI (World Resources Institute). 1990. *World resources 1990-1991: A guide to the global environment.* World Resources Institute in collaboration with the United Nations Environment Programme and the United Nations Development Programe. New York: Oxford University Press.

WRI (World Resources Institute). 2004a. *Earth trends: Economics, business, and the environment* (www.earthtrends.wri.org/country_profiles/index.cfm?theme=5; consulted May 21, 2004).
WRI. World Resources Institute. 2004b. *Earth trends: Forest, grasslands and drylands* (www.earthtrends.wri.org/country_profiles/index.cfm?theme=9; consulted May 18, 2004).
WRI (World Resources Institute). 2004c. *Earth trends: Population, health and human well-being* (www.earthtrends.wri.org/country_profiles/index.cfm?theme=9; consulted May 24, 2004).

Chapter 6

Carbon Sequestration in Soils of the Western Mountain Ridges and Deserts of Argentina

R. A. Rosell
J. A. Galantini

INTRODUCTION

High diversity of natural resources in South America is unique on the earth. There are mountains, steppes, dry areas, and more. South America has a surface of 18 million km^2 extending from latitude 12° N to 56° S. The region is 7,500 km in length and 5,000 km in width at its widest point. Altitude ranges from sea level to 7,000 masl.

The Andean mountains, like a long dorsal spine, are located in the west of the South American continent, crossing several countries such as Argentina, Chile, Peru, Ecuador, Colombia, and Venezuela from south to north. Due to the proximity of numerous volcanoes, large areas are dominated by Andosols derived of volcanic ash along mountains. Chile, for example, is rich in volcanic ash-derived Andosols.

The southern and southwestern areas of the mountain chain have a variable climate as a consequence of the differences in precipitation, vegetation, latitude, and altitude above sea level. Climate in the sequence (south to north) ranges from Antarctic, austral, temperate to subtropical, and tropical. These diverse climatic regions produce notable changes in the soil properties and are responsible for the existence of thousands of plant and animal species.

This chapter presents a general review of the conditions of change of natural resources related to the sequestration of organic carbon (OC) in those domains. The soil parent materials are variable. There are soils with active aluminum and smectite, chlorite, illite, vermiculite, regular, and/or interstratified minerals.

Carbon Sequestration in Soils of Latin America

doi:10.1300/5755_06

BIOME B OF ARGENTINA

Argentina's territory is 3,700 km long and 1,400 km wide. The country has a subtropical climate in the north and permanent frost in the south, and encompasses almost all climatic variability encountered on the earth. In general, however, a temperate, mild climate predominates in most of the country.

The extent and severity of soil erosion have increased during the past 50 years since the early 1950s (Table 6.1). Consequently, soil organic matter is severely affected by the intensity of erosion.

Ecotones

The subhumid, semiarid, and arid soils cover approximately two-thirds (75 percent) of the country. The arid lands (or "tierras secas") occur in the southern, southwest, and the western central areas of the country, occupying 51.5 percent of the total area. The country is divided into different ecoregions as follows.

Patagonian Region

It is a grass-shrub cold steppe south of the Colorado River, which extends from about 37° to 35°S. It has a land area of 780,000 km^2, or 28 percent of continental Argentina. The Patagonian "plateau" covers the provinces of Tierra del Fuego, Santa Cruz, Chubut, Neuquén, and Río Negro.

The Patagonian climate is mostly dry, cold, and windy. The Andean and sub-Andean regions have strong west-to-east precipitation gradients. In a strip 50 km wide, precipitation ranges from 600 to more than 3,000 mm in Chubut and from 400 to 2,000 mm in Neuquén. In the extra-Andean region or the continental "plateau," precipitation is concentrated in winter and de-

TABLE 6.1. Type and severity of soil erosion in Argentina (in 10^6 ha).

	Erosion type			Severity	
Year	Total	Wind	Water	Moderate	Severe
1956	34.2	16.0	18.2	27.1	7.1
1986	46.4	21.4	25.0	22.4	24.0
1990	58.0	28.0	30.0	27.9	31.0

Source: Adapted from Casas, 2001.

clines from 300 mm in the west to less than 150 mm in the east, increasing slowly toward the Atlantic Ocean.

Along the gradient of decreasing precipitation, beginning with the sub-Antarctic forest border, grass steppes give way to shrub-grass steppes, and eventually to deserts (del Valle et al., 1998). The region has a rich spectrum of vegetation types, from real desert to shrub or grass steppes. The distribution of different vegetation types includes 45 percent of shrub deserts, 30 percent of shrub-grass semideserts, 20 percent of grass steppes, and 5 percent of water surface and some humid regions (mallines) (Soriano, 1983; Rosell and Galantini, 1997; del Valle et al., 1999).

During the twentieth century, the Patagonian forest-steppe ecotone was adversely affected by pronounced overgrazing and also by natural and anthropic fires (Schlichter and Laclau, 2001; Aguiar and Sala, 1998; Golluscio et al., 1998). Approximately 70 percent of the whole Patagonia region has been affected by moderate to severe soil erosion, or desertification processes (Castro, 1985; del Valle, 1998). Ten million hectares (Mha) (approximately 15 percent) are affected by severe soil erosion. The degree of soil degradation in the region is summarized as follows (del Valle, 1998):

- overall erosion or desertification: 73.5 Mha (93 percent of total)
- moderate erosion: 27.8 Mha (35.4 percent)
- severe erosion: 25 Mha (31.8 percent)

The data in Figure 6.1 show percent of the area affected by desertification, and Table 6.2 presents the extent and severity of soil erosion in Patagonia.

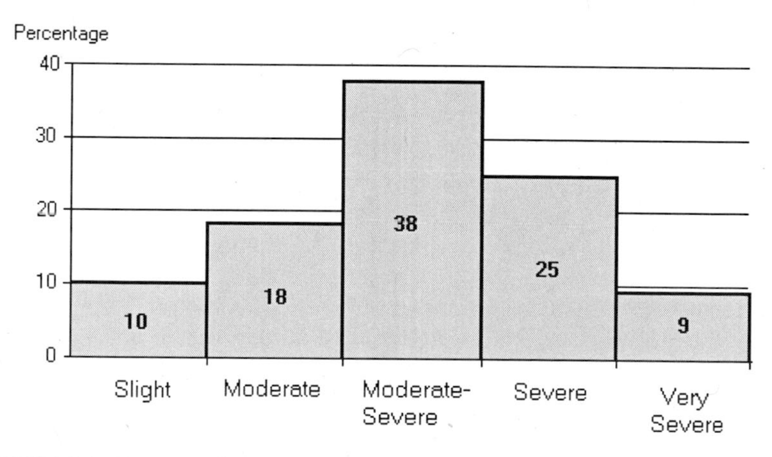

FIGURE 6.1. Degree of soil desertification in Patagonia, Argentina. (*Source:* Adapted from del Valle, 1998.)

TABLE 6.2. Extent and severity of soil erosion in the Patagonia region.

	Desertification degree (Mha)					
Zone	**Slight**	**Moderate**	**Moderate to severe**	**Severe**	**Very severe**	**Total**
North	3.92	5.32	10.32	7.51	1.21	28.28
Central	1.47	3.95	8.82	4.34	2.52	21.10
South	1.22	3.52	8.50	6.39	2.97	22.61
Southern Islands	0.69	0.72	0.15			1.56
Total	7.32	13.50	27.78	18.24	6.70	73.54

Source: Adapted from del Valle, 1998.

Soil biome transects. A study of soil along transects allows scientists to establish the gradual changes in soil properties along large areas and distances, reflecting the effects of environmental conditions on soil properties. This approach facilitates identification of changes establishing different values and tendencies (Doran and Smith, 1987). Two soil transects (A and B) were evaluated in Biome B (South) (Figure 6.2).

Table 6.3 presents some of the chemical parameters of the Ushuaia (south) to Río Grande (center) to Río Turbio (north) transect. In general, the soils are very acidic, especially in the surface horizons. There is a high litter and plant residue content in the upper few centimeters of soil. The base saturation is very low in Ushuaia as compared with other soils. Soil organic carbon (SOC) is very high in the upper horizons, thus indicating accumulation of detritus material and litter.

The SOC increases from east on the Atlantic Sea, in Península Valdez (Madryn), to the Andean Mountains in S.C. de Bariloche in the west. Table 6.4 shows the increase of SOC in several sites of the transect. The increase is slight in the eastern and central parts of the transect.

Detailed results reported by Rosell and Galantini (1997) indicated that SOC increases from 0.15 to 1.74 percent in the central "plateau." Increase in precipitation and SOC are pronounced toward the west and mountainous sites.

Monte

The Monte area extends from south to north in central and western Argentina. It is a rather continuous shrubland which covers 50 Mha. The

FIGURE 6.2. Ecoregions of Argentina and studied soils on the transects A and B. (*Source:* Adapted from Secretaría de Ambiente y Desarrollo Sustentable, 2004.)

northern part presents a typical landscape, with mountains, valleys, and slopes derived from La Pampa hills and large salty flats. The central part has an undulating topography with depressed loessic flats of fluvial, lacustrine, and quaternaric eolic origin. The southernmost part of the Monte merges with the northern Patagonia to form a wide, flat region (Guevara et al., 1997). The Monte has a dry climate, being warm in the north and gradually becoming cooler toward the south. Precipitation ranges from 100 mm per year in summer to 100 to 200 mm per year in winter. Aridity in the northern part is related to its position between the Andes to the west and the Pampean hills to the east. Both mountainous systems intercept moist winds from the pacific and the Atlantic, thus giving rise to the semiarid Pampean ecotone (Biome G).

TABLE 6.3. Some chemical parameters of soil transects from Ushuaia to Río Grande to Río Turbio transect (A in the map).

Horizon	Depth (m)	pH	OC (%)	CEC ($cmol \cdot kg^{-1}$)	S	Texture (%)		
						Clay	Silt	Sand
Ushuaia								
1	0.00-0.02	4.6	47.0	—	18.3	—	—	—
	0.02-0.06	4.8	2.5	—	1.5	—	—	—
	0.06-0.12	5.0	0.3	—	1.3	—	—	—
Ushuaia								
2	0.00-0.10	5.1	16.0	45	20.2	—	—	—
	0.10-0.20	5.0	2.9	18	3.6	24.0	47.4	28.6
	0.20-0.40	5.2	3.5	17	3.3	41.0	47.0	12.0
	0.40-0.60	5.2	1.6	12	3.8	21.0	45.2	31.8
Río Grande								
	0.00-0.22	3.7	4.3	53	—	33.2	43.7	23.1
	0.22-0.80	3.6	—	—	—	51.3	42.3	6.4
Río Turbio								
	0.00-0.10	5.1	6.9	—	47	—	—	—
	0.10-0.40	4.1	1.9	—	28	—	—	—
	0.40-0.60	3.8	0.9	—	46	—	—	—

Source: Adapted from Colmet Daage et al., 1991.
Note: SOC, soil organic carbon; CEC, cation exchange capacity; S, percentage of base saturation.

TABLE 6.4. Distribution of soil organic carbon concentration (percent) in a transect from east to west in the central Patagonian region (transect B in the map).

Site	MAP (mm)	MAT (°C)	SOC (%) by soil depth									
			0-10 cm	10-20 cm	20-30 cm	30-40 cm	40-50 cm	50-60 cm	60-70 cm	70-80 cm	80-90 cm	90-100 cm
1	225	12.4	0.86	0.20	0.20	0.20	0.16	0.16	0.15	0.15	0.15	0.13
2	400	16.0	1.1	1.1	1.1	1.0	1.0	1.0	1.0	0.8	0.8	0.8
3	585	16.0	1.6	1.6	1.2	1.2	1.2	1.2	1.1	1.0	1.0	1.0
4	690	14.0	13.1	7.6	7.6	5.0	3.6	3.6	3.0	2.8	1.3	1.3
5	1200		7.29	7.29	7.29	4.38	3.89	3.89	3.69	3.69	3.69	1.85
6	1600	12.0	25.0	6.2	5.1	5.1	5.1	4.1	4.1	4.1	2.6	2.6
7	2000		7.91	3.52	3.52	3.52	3.68	3.68	4.22	1.85	1.85	1.85

Source: Adapted from AACS Bariloche Congress, 1990; AACS Puerto Madryn Congress, 2002.
Note: MAP, mean annual precipitation, (mm); MAT, mean annual temperature (°C); Site distribution: 1, Península Valdez (Madryn); 2, INVAP, 46 km E Bariloche; 3, San Ramón, 30 km E Bariloche; 4, El Cóndor, 20 km E Bariloche; 5, 6, Llao Llao, 20 km W Bariloche; 7, West.

Natural resources of the Monte have been degraded by overgrazing and wood collection for more than 100 years. Overgrazing leads to the disappearance of useful perennial forage grasses followed by the invasion of unpalatable shrubs, weedy herbs, and annual grasses (Guevara et al., 1997).

Adoption of judicious land management and soil conservation measures in the Monte region are essential to maintaining its rather low productivity, especially in areas prone to wind erosion. One such approach includes cultivation of drought-resistant species, such as weeping lovegrass *(Eragrostis curvula),* which produce permanent pasture for many years after (Figure 6.3).

The most characteristic plant community dominating large areas of the Monte is the "jarillal" (Zygophyllaceae) with *Larrea divaricata, L. cuneifolia,* and *L. nitida* as the most abundant genus. The Monte vegetation has been degraded by overgrazing and wood collection for more than 100 years. Changes in plant species composition have also been caused by the replacement of cattle by goat. The change from natural to managed ecosystem accentuates soil degradation by increasing risks of wind erosion (Buschiazzo et al., 2004).

Caldenal

Physiography. This is one of the most interesting rangelands of the country from the point of view of potential productivity and management. It lies

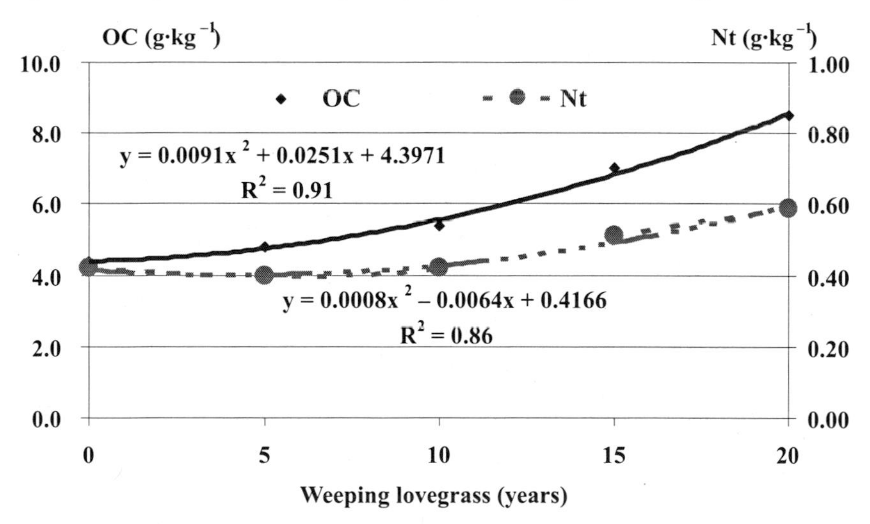

FIGURE 6.3. Effect of long-term cultivation of weeping lovegrass on SOC and Nt concentration in Ustipsamments of the western semiarid Pampas. (*Source:* Adapted from Demmi et al., 1986.)

in the ecotone between the dry Monte in the west and the cultivated humid Pampa grasslands in the east. The region is characterized by rainfall of 350 to 500 mm per year between autumn and spring, with annual evapotranspiration of 800 mm. The mean annual temperature is 15.3°C (max. 42.5 and min –12.8°C).

The cattle production industry is the most important activity within the Caldenal, where the grass steppe is used as the main food source for animals. The Caldenal is physiognomically represented by the presence of *Prosopis caldenia* (Calden), which is the typical and almost exclusive tree, giving its name to the region. It is endemic, deciduous, and may exceed 10 m in height. The recent human activity in the Caldenal has adversely affected both plant and animal productivity, and in many places the physical sustainability of the land itself.

Soil organic carbon in the Caldenal region. Crespo and Rosell (1990) reported the amount and quality of humified soil organic matter (SOM) of an isothermic climosequence (moisture sequence) in a 620 km transect sampled at seven noncultivated sites on or about 38°S latitude in the subhumid, semiarid, and aridic Pampean region of temperate (mean annual temperature of 15.2°C) Argentina. The area covers a wide variety of agricultural activities, including corn *(Zea mays),* soybeans *(Glycine max),* and potato *(Solanum tuberosum)* crops and grazed cattle on the east to wheat *(Triticum aestivum),* sunflower *(Heliantus annuus),* and sorghum (*Sorghum* sp.) in the center to small grains and range-fed cattle on the west. The sequence has a mean annual rainfall gradient of approximately 550 mm from Necochea, a province of Buenos Aires on the east with an Udic soil climate (916 mm precipitation), to Cuchillo-Co, a province of La Pampa in the west with an Aridic climate (365 mm precipitation).

The profile location, soil subgroup, mean annual precipitation, texture, pH, and SOC content and bulk density of the Caldenal soils are presented in Table 6.5. Concentrations and amounts of SOC and N decrease from east to west. The SOC has the highest concentration (2.56 percent) and amount (6.4 $kg \cdot m^{-2}$, 0 to 0.20 m) at Nueva Roma and the lowest corresponding values (0.38 percent and 1.0 $kg \cdot m^{-2}$) at Cuchillo-Co which has an Ustollic Paleorthid soil.

Analysis of the humus showed that humic and fulvic acids and humin decrease with increase in aridity from east to west. The majority of the data imply a decrease in the degree of humification from the humid to the arid regions. Furthermore, it was concluded that increased aridity reduces both the level and mobility of humus. These stable humus fractions create a nutrient reserve and contribute to maintaining soil aggregate stability by forming organic-metal-clay complexes and, ultimately, controlling "soil erosion" (Rosell and Galantini, 1997).

TABLE 6.5. Soil location, rainfall, texture, and pH from the semiarid and arid Pampean regions of temperate Argentina (Caldenal).

Name and location	Soil subgroup	Rainfall (mm per year)	Texture	pH	SOC %	SOC $kg{\cdot}m^{-2}$
Nueva Roma (38° 32' S, 62° 36' W)	Petrocalcic Argiudoll	490	clay loam loam	6.5	2.56	6.4
Chapalcó (38° 33' S, 63° 20' W)	Entic Haplustoll	400	sandy loam	6.5	1.84	4.6
Las Gaviotas (38° 44' S, 63° 40' W)	Ustic Torripsament	379	sandy	6.5	0.95	2.4
Cuchillo-Có (38° 28' S, 64° 53' W)	Ustollic Paleorthid	356	sandy loam	7.5	0.38	1.0

Source: Adapted from Crespo and Rosell, 1990.

Western Chaco

Western or dry Chaco is a large, low-elevation, washout plain covering 65 Mha developed on sediments derived from the eastern Andes. It covers the north of Argentina, west of Paraguay, and southeast of Bolivia. Rainfall increases gradually from west to east. It is the driest part of the so-called western or dry Chaco. The climate of western Chaco is humid and hot in summer (October to April), and mild and dry during winter. The absolute highest temperature can reach 48°C in summer and –8°C in winter.

The soils of western Chaco are coarse textured, with high base status, without Argillic horizon, and of dark surface color from the organic carbon content. The soil moisture balance is aridic or thermic in southern and northern parts of the region.

The native ecosystem has been altered due to overgrazing of the herbaceous grassland vegetation. Wood production is also an important activity. Agroforesty can be adopted in this area in order to restore western Chaco, which has proven successful. Grazing of livestock (goats, sheep, llamas *[Lama glama]*, cattle, etc.) is the principal economic activity of the region. Land cultivation is restricted to corn, quinoa *(Chenopodium quinoa)*, and soybean production, mostly by no-till farming techniques.

Puna

The Puna ecotone covers the arid Puna plateaus, high plains, and slopes of the Central Andes from latitude 8° to 27°S through Peru, Bolivia, Chile, and Argentina, at elevations between 3,200 and 4,500 masl. It constitutes one of the coldest rangelands of the world at these elevations. The climate of the Puna is very harsh: rainfall is under 200 mm per year, concentrated in the summer, with wide daily temperature amplitudes. The vegetation of Puna is very closely related to that of Patagonia. Many of its dominant genera are frequent in both regions. The poorly managed livestock grazing has adversely impacted the natural ecosystem. Many areas are markedly deteriorated by the loss of vegetation cover, extinction of desirable species, and accelerated soil erosion (Fernandez and Busso, 1999; Fernandez et al., 1989).

MANAGEMENT OF SOIL ORGANIC CARBON UNDER DIFFERENT ENVIRONMENTS

Arid and Semiarid Sandy Soils

Demmi et al. (1986) reported that weeping lovegrass (*Eragrostis curvula* Shrad Ness) pastures had a positive effect on several properties of Ustipsamments of the eastern part of San Luis province in the western semiarid Pampas. They observed significant increase in the SOC and total nitrogen (Nt) contents after more than ten years of having the perennial grass, which reseeded itself and was grazed regularly. Weeping lovegrass pastures increased SOC and Nt mainly in coarse (sandy)-textured soils, probably because of a larger root production than the native grass species in the same soils (Figure 6.3).

Management Impact on SOC Sequestration in Arid Chaco Soils

Cora and Bachmeier (2004) studied an arid Chaco soil under three different treatments: natural (Nat), forest without grazing (For), and forest with grazing (Fgr). They found that SOC and Nt in A horizon of the Nat situation were 1.33 and 0.125 percent, respectively. Forest soil increased SOC and Nt up to 2.72 and 0.273 percent, respectively. In contrast, forest with grazing system decreased both SOC and Nt to 0.64 and 0.082 percent, respectively. Forest systems without animal grazing showed an important potential to sequester SOC in the A horizon. Animal inclusion produced species replacement and soil compaction, negatively affecting the SOC sequestration capacity.

Forest Species Impact on SOC Sequestration

The native cypress *(Austrocedrus chilensis)* usually grows in pure stands or associated with radal *(Lomatia hirsute)* and ñire *(Nothofagus antarctica),* with remarkable tolerance to seasonal drought (Donoso, 1993; Veblen et al., 1996). Cattle breeding strongly affects the cypress forests because of soil compaction and browsing, or indirectly by induced fires of pasture (De Pietri, 1992).Wood exploitation has also contributed to forest degradation.

Establishing forest plantation has been promoted for many years through provisions of credit and tax exemption (Laclau, 1997). This policy resulted in approximately 70,000 ha of ponderosa pine [*Pinus ponderosa* (Dougl.) Laws] plantations, which are expanding at a rate of 5,000 ha per year. Conifers are planted mainly for lumber production but also for combined wood and beef production in forest-grazing systems (Schlichter and Laclau, 2001; Somlo et al., 1997).

Laclau (2003) estimated a sequestration rate of 2.6 $Mg{\cdot}ha^{-1}$ per year of biomass carbon, including dead foliage and litter, in natural pastures of the Rio Negro Province. He used this value as the baseline from which the forest land use would add carbon to the ecosystem. This study determined that pine plantation after pasture considerably increases the biomass carbon pool and could have neutral effects on soil carbon, at least in short rotations. The mean biomass carbon sequestration of 52.3 $Mg{\cdot}ha^{-1}$ estimated for a 14.1-year-old mean age of pine plantations represents a net addition of 3.5 $Mg{\cdot}ha^{-1}$ per year over constant biomass carbon storage of 2.6 $Mg{\cdot}ha^{-1}$, estimated for the pasture baseline. Cypress forests are important not only because of the biomass carbon actually sequestered (73.2 $Mg{\cdot}ha^{-1}$), but mainly for the larger soil carbon pool (116.5 $Mg{\cdot}ha^{-1}$), even under heavy disturbances of grazing, logging, or fire.

However, losses of carbon as a consequence of land-use change must be prevented, since both cypress forests and grassland store soil carbon to a large extent. After planting, special care must be taken to maintain or rapidly restore the labile soil carbon associated with rooting activity.

Biomass allocation to foliage and above- and below-ground woody tissues reflects not only the differential growth rates between ponderosa pine and cypress, but also different mechanisms of nutrient cycling and water uptake. Both forest species are important sequestration options. However, their efficiency varies with rainfall amount (Figure 6.4). The low rooting system relative to the aboveground biomass of cypress suggests stomatal

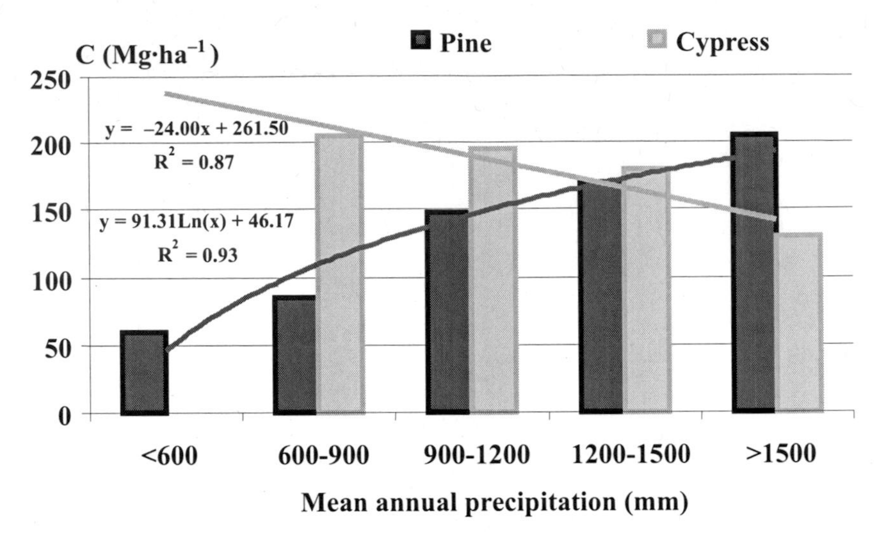

FIGURE 6.4. Organic carbon sequestration under pine and cypress for different precipitation regimes. (*Source:* Adapted from Laclau, 2003.)

control of water losses or another water-saving mechanism which differs from root activity under drought-stress conditions. The deeper and larger ponderosa pine roots make this species more efficient in terms of soil moisture uptake, and hence, in growth.

ESTIMATED POTENTIAL OF CARBON SEQUESTRATION

The soils of mountain ridges and desert lands are the poorest ones, due to the effect of natural and human intervention. The initial and parent materials undergo continuous changes due to extreme situations reached by the action of the environmental conditions. The Argentina Biome B covers about 200 Mha characterized by high variability in environmental conditions.

The limited data on the SOC pool show values between 28 to 116, 20 to 98, and 25 to 75 $Mg \cdot ha^{-1}$ (0 to 1 m depth) for the Patagonian, Central, and Chaco plains, respectively. Taking into account the mean value of SOC in these plains (excluding the mountain areas), the total SOC pool is estimated at 9,800 Tg of SOC.

The SOC content can improve by converting degraded lands to forest or permanent grasses. Increasing the SOC pool by 10 percent in the top 1 m depth can enhance the pool by 3.3, 2.5, and 3.0 $Mg \cdot ha^{-1}$ in the Patagonian, Central, and Chaco plains, respectively. These increases were variable, ranging from 2 to 8 Mg·ha^{-1} or from 0.2 to 0.4 Mg·ha^{-1} per year over a 10- to 20-year period (Demmi et al., 1986). Based on these data, the estimated sequestration potential of SOC in the 140 Mha of the biome plains is about 420±140 Tg of SOC.

In this arid and semiarid region, better management practices produce an important increase of the surface litter accumulation, due to the slow rate of decomposition under dry conditions. High levels of easily combustible materials represent a dangerous situation, which each year produces between 1 to 2 Mha of intentional and natural fires.

CONCLUSIONS

According to the variability of its climatic and edaphic conditions, the Biome B has several different ecoregions. The equilibrium level of SOC is moderated by temperature, precipitation, and soil parent materials.

There are small areas with high productivity and intensive management (irrigated soils), as well as larger areas with very low productivity and high risks of soil degradation.

Biome B is an arid region and is in unstable equilibrium. Maintaining and enhancing soil organic matter is important to avoiding soil desertification, maintaining the naturally fragile conditions, and improving the capacity to sequester carbon.

System production alternatives must be carefully analyzed for each area, taking into consideration climatic, soil, and sociological factors.

REFERENCES

Aguiar, M.R. and O.E. Sala. 1998. Interactions among grasses, shrubs and herbivores in Patagonian grass–shrub steppes. *Ecol. Aust.* 8, 201-210.

Asociación Argentina de la Ciencia del Suelo (AACS). 1990. *Guía de campo*. Argentine Congress of Soil Science, Bariloche, Rio Negro.

Asociación Argentina de la Ciencia del Suelo (AACS). 2002. *Guía de campo*. Argentine Congress of Soil Science, Puerto Madryn, Chubut.

Buschiazzo, D.E., H.M. Martinez, E. Fiorucci, and C. Guiotto. 2004. Mapas de erosión eólica potencial y actual de la región semiárida y subhúmeda pampeana Argentina. XIX Congreso Argentino de la Ciencia del Suelo, Paraná.

Casas, R. 2001. La conservación de los suelos y la sustentabilidad de los sistemas agrícolas. Vol. 4. Academia Nacional de Agronomía y Veterinaria, Buenos Aires.

Castro, J.M. 1985. Relevamiento de estados de erosión en la Precordillera Patagónica. *Revista INTA Presencia, S.C. de Bariloche* 2, 36-45.

Colmet Daage F., M.L. Lanciotti, and J. Irisarri. 1991. Suelos con aluminio activo y montmorillonita, clorita, illita, vermiculita, interestratificados, regulares o irregulars. Convenio Franco-Argentino: Grupo de estudio de suelos con Al activo.

Cora, A. and O. Bachmeier. 2004. Relación suelo-vegetación en sistemas sustentables del Chaco árido Argentino. XIX Congreso Argentino Ciencia del Suelo, Paraná.

Crespo, M.B. and R.A. Rosell. 1990. Change of properties of humic substances in an edaphic climosequence. *Agrochimica* 34(3), 193-200.

del Valle, H.F. 1998. Patagonian soils: A regional synthesis. *Ecol. Aust.* 8, 103-123.

del Valle, H.F., R.A. Rosell, and P.J. Bouza. 1999. Formation, distribution and physicochemical properties of plant litter in shrub patches of northeastern Patagonia. *Arid Soil Res. Rehab.* 13(2), 105-122.

Demmi, M.A., C.A. Puricelli, and R.A. Rosell. 1986. El efecto del pasto llorón en la recuperación de los suelos. INTA (Argentina) to San Luis Agric. Exp. St. Techn. Bull. 109.

De Pietri, D.E. 1992. Alien shrubs in a national park: Can they help the recovery of degraded forest? *Biol. Conserv.* 62, 127-130.

Donoso, C.D. 1993. *Bosques templados de Chile y Argentina. Variación: Estructura y Dinámica* (Temperate forests of Chile and Argentina: Variation, structure and dynamics). Editorial Universitaria, Chile.

Doran, J.W. and M.S. Smith. 1987. Organic matter management and utilization of soil and fertilizer nutrients. In R.F. Follet et al. (eds.), Soil fertility and organic matter as critical components of production systems. SSSA Spec. Publ. 19. SSSA and ASA, Madison, Wisconsin, pp. 53-72.

Fernandez, O.A., R.M. Bóo, and L.F. Sánchez. 1989. South America shrublands. In C.M. McKell (ed.), *The biology and utilization of shrubs.* Academic Press, San Diego, California, pp. 25-59.

Fernandez O.A. and C.A. Busso. 1999. Case studies of rangeland desertification. In A. Olafur and S. Archer (eds.), *Rala Report Nº 200.* Agricultral Research Institute, Iceland, pp. 10-60.

Golluscio, R.A., V.A. Deregibus, and J.M. Paruelo. 1998. Sustainability and range management in the Patagonian steppes. *Ecol. Aust.* 8, 265-284.

Guevara, J.C., J.B. Cavagnaro, O.R. Stevez, H.N. Le Hoerou, and C.R. Stassi. 1997. Productivity, management and development problems in the arid rangelands of the Central Mendoza plains (Argentina). *J. Arid Environ.* 35, 575-600.

Laclau, P. 1997. Los Ecosistemas Forestales y el Hombre en el Sur de Chile y Argentina (Forest ecosystems and man in southern Chile and Argentina). Boletín Técnico No. 34. Fundación Vida Silvestre Argentina, Buenos Aires.

Laclau, P. 2003. Root biomass and carbon storage of ponderosa pine in a northwest Patagonia plantation. *For. Ecol. Manage.* 173, 353-360.

Rosell, R.A. and J.A. Galantini. 1997. Soil organic carbon dynamics in native and cultivated ecosystems of South America. In R. Lal, J.M. Kimble, R.F. Follett, and B.A. Stewart (eds.), *Advances in soil science: Management of carbon sequestration in soil.* CRC Press, Boca Raton, Florida, pp. 11-33.

Schlichter, T. and P. Laclau. 2001. Valoración económica de la desertificación en Patagonia y algunas alternativas de desarrollo basadas en la conservación de los

Recursos Naturales (Economic valuation of the desertification in Patagonia and some development alternatives based on natural resources conservation). In *Situación Ambiental Argentina 2000.* Fundación Vida Silvestre Argentina, Buenos Aires, pp. 272-289.

Secretaría de Ambiente y Desarrollo Sustentable. 2004. Ecoregiones de la Argentina. Ministerio Salud y Ambiente. www.medioambiente.gov.ar/geoinformacion/ecorregiones.htm.

Somlo, R., G. Bonvissuto, T. Schlichter, P. Laclau, P. Peri, and M. Allogia. 1997. Silvopastoral use of Argentine Patagonian forest. In A.M. Gordon and S.M. Newman (eds.), *Temperate agroforestry systems.* CAB International, Wallingford, United Kingdsom, pp. 237-250.

Soriano A. 1983. Deserts and semidesert of Patagonia. In M.W. West (ed.), *Temperate deserts and semideserts.* Elsevier Publish. Co. Amsterdam, the Netherlands, pp. 440-453.

Veblen, T.T., T. Kitzberger, B.R. Burns, and A. Rebertus. 1996. Perturbaciones y dinámica de regeneración en bosques andinos del sur de Chile y Argentina (Regeneration disturbances and dynamic in the Andean forests of southern Chile and Argentina). In J.J. Armesto, C. Villagran, and M.K. Arroyo (eds.), *Ecología de los Bosques Nativos de Chile.* Editorial Universitaria, Chile, pp. 169-197.

Chapter 7

Soil Carbon Sequestration in Mexico and Central America (Biome A)

J. D. Etchevers
O. Masera
C. Balbontín
D. Gómez
A. Monterroso
R. Martínez
M. Acosta
M. Martínez
C. Ortiz

INTRODUCTION

Global Climate Change and Carbon Sequestration

The current change in the global climate is attributed to the unprecedented concentration in the atmosphere of greenhouse gases (GHGs). The main GHGs are CO_2, N_2O, CH_4 and CFCs. Agricultural activities, including land use and land-use change, constitute an important source of GHG, being responsible for 25 percent of CO_2, 50 percent of CH_4, and 70 percent of N_2O emitted by all human activities (Agriculture and Agri-Food Canada, 1998). At a global level, agricultural activities use approximately 35 percent of all the existing lands on the planet. If the accumulation of GHGs continues at the present rate, experts predict an increase in the average global temperature between 3 and 5°C by the end of the twenty-first century (IPCC, 1996, 2001). Because such increase may have serious consequences for humankind, mitigation measures must be adopted by all nations of the world. One of this mitigation measures is to increase carbon (C) stock in the major components of the ecosystems. Plants and soil are two major compo-

Carbon Sequestration in Soils of Latin America

doi:10.1300/5755_07

nents of the ecosystems with a high C sequestration potential which merit both basic and adaptive research.

The C sequestration by the ecosystems is considered an environmental service and, in addition, it may also have an added economic value. During the 1990s, C sequestration by forests and agriculture had been extensively studied worldwide. Some options, such as afforestation and reforestation, are already included within the Clean Development Mechanism (CDM) (UNFCCC, 2004). Currently, however, options where soil C sequestration is central, such as conservation tillage and other agricultural options, have not been approved within the CDM, partly because of the uncertainties regarding the rates, amount, cost, and verification issues associated with soil C measurements. Thus, there is a critical need for studies for addressing specifically the process and practices of soil C sequestration.

In Mexico and Central America, C sequestration in soils is an issue scarcely documented. No accurate estimates of the size of the C stocks present in the soil and the way these stocks evolve have been conducted in these regions of the world, nor has the effect of different land-use systems been studied. However, in the Biome A type of vegetation, there are ecosystems in which C can be sequestered. This chapter attempts to assess how much C can be sequestered in soil. It is prepared in a way similar to that presented by Lal et al. (1999) for the soils of U.S. croplands.

The C stock in the ecosystems can be roughly divided into above- and belowground stocks. The former stock accumulates C in the aboveground biomass and the litter, and the latter in the roots and the C (mainly organic) bound to inorganic soil compounds. The C in the aboveground component can be assessed by directly measuring the biomass using destructive measurements, indirect nondestructive methods (remote sensing the biomass by aerial photography or satellite images), or using simulation models. Appropriate C concentration coefficients are used to convert the mass of biomass into that of C. In contrast to the aboveground C stock estimation, soil C estimation is a more complex issue. Direct measurements of soil C in selected points and scaling them to larger areas (as soil units, regions, countries, and continents) have been employed for that purpose. However, because of the high spatial variability of soil C (Vergara et al., 2004) these data may be rather approximate.

Biome A (Mexico and Central America)

From a geographical, economic, and social point of view, Biome A constitutes an asymmetrical conglomerate of countries with a similar history and ethnic population but different degrees of economic development.

Mexico and Central America constitute Biome A. Table 7.1 shows population of Mexico and Central America along with selected economic indexes (adapted from WRI, 2004a,b,d). Important physiographic characteristics of the region are shown in Figure 7.1.

Mountain ranges of contrasting altitudes are predominant features of Biome A. Mountains in Mexico are much higher than the ranges that crisscross Central America. The territorial surface of these regions comprises 2.48 million Km^2, or 3.8 percent of the world land area. About 80 percent of the area of Biome A comprises Mexico. Eight countries of Central America share the remaining 20 percent of the area. Hillsides and mountains are more abundant than plateaus and plains. Climate ranges from desert to tropical humid forest (<200 mm to >2,000 mm rainfall per annum).

The average per capita GDP (gross domestic product converted to international dollars using purchasing power parity, PPP) of the region is

TABLE 7.1. Population and economic indexes of Mexico and Central American countries.

Country	Total land area (10^3 km^2)	Population (10^3) 1950	2002	2025	GNI[a]	GDP[b]
Belize	22.96	69	236	324	1258	5945
Costa Rica	51.11	862	4200	5929	30412	8193
El Salvador	21.04	1951	6520	8975	27699	4496
Guatemala	108.89	2969	11995	19624	42940	3821
Honduras	112.09	1380	6732	10106	15378	2454
Mexico	1958.20	27737	101842	130194	861284	8941
Nicaragua	130.00	1134	5347	8606	10565	2366
Panama	75.52	860	2942	3779	16234	6001
Total	2479.81	36962	139814	187537		

Source: Adapted from WRI, 2004a,b,d.
[a]GNI, Millions of dollars year 2000. Gross national income, or GNI, current dollars is the sum of value added by all resident producers plus any product taxes (less subsidies) not included in the evaluation of output plus net receipts of primary income (compensation of employees and property income) from abroad. In other words, GNI measures the total income of all people who are citizens of a particular country, while GDP (gross domestic product) measures the total output of all persons living in that particular country.
[b]GDP dollars year 2000 per capita. Gross domestic product per capita, PPP is gross domestic product converted to international dollars using purchasing power parity (PPP) rates, and divided by the population of the country that year. An international dollar has the same purchasing power in a given country as a U.S. dollar in the United States. In other words, it buys an equivalent amount of goods or services in that country.

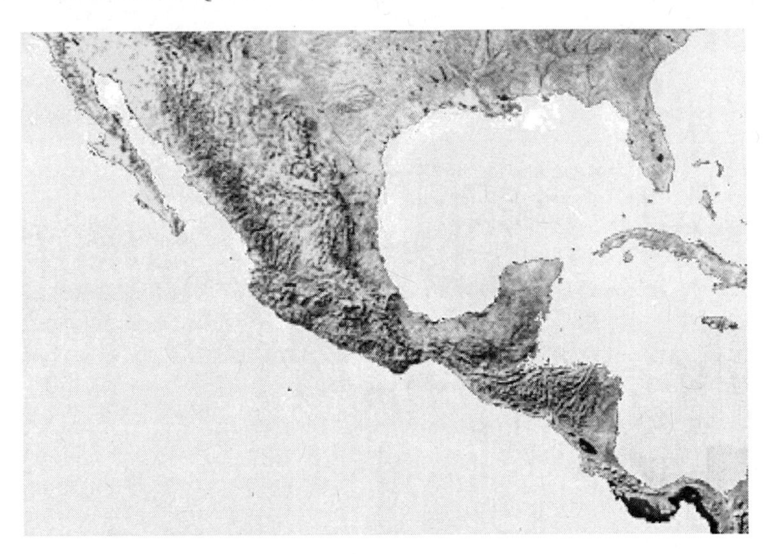

FIGURE 7.1. Land relief map of Mexico and Central America.

US$5,277, ranging from US$2,300 to US$8,200. Mexico is one of the most open economies in the world, but its natural resources are susceptible to severe degradation (WRI, 2004b).

Total population of the region is estimated at 140 million and is projected to increase by another 50 million by 2025. The population of Mexico was approximately 105 million in 2004 (average age 22 years old, birthrate of 2.5 children/female) and that of Central America was 35 million. Some Central American countries have high population density, with a strong adverse impact on the environment. The population of Biome A will increase to 187 million by the year 2025, a figure lower than that previously predicted (WRI, 2004d). No native forest is remaining in some areas of this region, and soil erosion in the hill slopes is extremely high.

Good farmland is scanty in the whole region, and whatever is available must sustain the growing population, which increased from 30 million to 140 million during the second half of the twentieth century. This high population growth has strong adverse impacts on natural resources of the region, particularly on the soil, water, and forests.

There are four distinct ecological macroregions in Mexico (Claverán, 2000): (1) arid and semiarid region (<500 mm annual rainfall) which covers approximately one-half of the country, (2) a dry tropical region (900 to 1,200 mm, with seasonal rainfall) which covers one-fourth of the total area, (3) the temperate hilly regions (600 to 900 mm) which cover 13 percent of the area, and (4) the humid tropics (>1,200 mm) which cover 8 percent of the area. Agriculture is practiced in all four regions. Only 16 percent of the arable land area is classified as prime farmland suitable for high-input agriculture. The rest of the arable land is characterized by steep slopes and that in marginal semiarid climates (INEGI, 1998; Tiscareño-López et al., 2000). Hillside agriculture is practiced along the sierras which crisscross the country, but mainly in the southern part where rainfall is abundant. Irrigated land, >6 million hectares (Mha), is concentrated in central and northern parts of Mexico (SEMARNAT, 2003). The lack of water is one of the most severe limiting factors for present and future agricultural expansion in this part of the country. Soil degradation and erosion are common features on most agricultural land (SEMARNAT-CP, 2001-2002).

There are about 3 million farmers in Mexico. Shortage of prime farmland is responsible for conversion of native forest and cultivation of steep slopes. Consequently, forested areas have been reduced by 30 percent in temperate and 75 percent in tropical regions of Mexico since 1960. Mexico has one of the highest rates of deforestation of the native forests (WRI, 1994). Similar conditions exist in Central America, particularly in countries with high population densities, such as El Salvador (WRI, 2002).

LAND USE AND LAND-USE CHANGE IN MEXICO AND CENTRAL AMERICA

The present farmland in Mexico and Central America is derived from widespread change in land use that occurred since the Spanish conquest. The rate of land-use change increased with increase in population and the attendant increase in demands. There was no need for pasture land prior to the European settlements, because the native population had neither domestic nor draft animals. Increase in cattle population led to increase in the area under pasture through conversion of native forests. Forest and woodlands covered most of the southern part of the region, especially the highlands. The north-central and the northwest part of the region are mostly deserts. Large plantations and animal husbandry have widely replaced the traditional agriculture.

The data in Table 7.2 show historic changes in the area under forest cover (WRI, 2004c). Little of the original forest cover remains in Biome A at the beginning of the twenty-first century. Changes in forest areas from 1990 to 2000 indicate a reduction of 37 percent in El Salvador (the most populous country) to a 7 percent decrease in Costa Rica. The latter has the most environmentally friendly legislation in the entire region.

The data in Table 7.3 show changes in land use between 1990 and 2000 (WRI, 1990-1991), and those in Tables 7.4 and 7.5 depict the current land

TABLE 7.2. Percentage of the original forest and change in forest area with respect to the total land area.

Parameter	Belize	Costa Rica	El Salvador	Guatemala	Honduras	Mexico	Nicaragua	Panama
Percentage of the total land								
Original forest[a]	92	98	99	99	100	56	100	97
Forest area (2000)[b]	59	39	6	26	48	28	25	38
Change in forest area (%)[c]								
Total, 1990-2000	–21	–7	–37	–16	–10	–10	–26	–15
Natural, 1990-2000	–21	–13	–11	–20	–11	–11	–27	–16
Plantations, 1990-2000	4	10	nd	nd	nd	nd	14	17

Source: Adapted from WRI, 2004c.

[a]Original forest as a percent of land area refers to the estimate of the percent of land that would have been covered by closed forest about 8,000 years ago assuming current climatic conditions, before large-scale disturbance by human society began.

[b]Forest area in 2000 as a percent of total land area is calculated by dividing total forest area (see above) by total land area.

[c]Change in forest area is the total percent change in both natural forests and plantations between 1990 and 2000.

TABLE 7.3. Total, cultivated, irrigated pasture, and forest lands (10^3 km^2) in Mexico and Central America during the 1980s.

Country	Total lands (1987)	Cultivated lands (1987)	Irrigated lands (1985-1987)	Pasture lands (1985-1987)	Forest lands (1985-1987)
Belize	23	1	0	1	10
Costa Rica	51	5	1	23	16
El Salvador	21	7	1	6	1
Guatemala	108	19	1	14	41
Honduras	112	18	1	25	36
Mexico	1909	247	52	745	446
Nicaragua	119	13	1	52	38
Panama	76	6	0	13	40
Total	2419	316	57	879	628

Source: Adapted from WRI, 1990-1991.

TABLE 7.4. Land use in Biome A in 2000 as total land area covered (10^3 km^2).

Land use	Belize	Costa Rica	El Salvador	Guate-mala	Honduras	Mexico	Nicaragua	Panama
Forests[a]	13.78	32.70	6.52	60.98	62.77	567.88	74.10	40.78
Shrublands, savanna, and grasslands[b]	1.38	3.57	0.21	8.71	7.85	1037.85	10.40	7.55
Cropland and crop/natural vegetation mosaic[c]	6.66	13.80	13.68	37.02	40.35	332.89	45.50	24.92
Urban and built-up areas	0	0	0.19	37.02	0.11	1.96	0	0.08
Sparse or barren vegetation; snow and ice	0	0	0	0	0	19.58	0	0
Wetlands and water bodies	1.14	1.02	0.42	1.09	2.24	19.58	10.4	2.27

Source: Adapted from WRI, 2004c.
[a]Forests include all areas dominated by evergreen or deciduous trees with a canopy cover greater than 60 percent and a height exceeding 2 meters. Both broadleaf and needleleaf trees are included.
[b]Shrublands, savanna, and grasslands include lands dominated by woody vegetation less than 2 meters tall and with shrub canopy cover greater than 10 percent. The shrub foliage can be either evergreen or deciduous. This category also includes savannas and grasslands with herbaceous and other understory systems. These lands may have a tree or shrub cover of less than 60 percent.
[c]Cropland and crop/natural vegetation mosaic. Croplands are lands covered with temporary crops followed by harvest and a bare soil period (e.g., single and multiple cropping systems). Perennial woody crops are classified as forest or shrub land cover. Cropland/natural vegetation mosaics are lands with a mosaic of croplands, forests, shrublands, and grasslands in which no component comprises more than 60 percent of the landscape.

use (WRI, 2004c). However, these data are not readily comparable because of differences in terminology and the classification systems used.

Of the total land area of Mexico, only 25 Mha (12 percent) were cultivated lands in the mid-1980s (see Table 7.2). A similar proportion of the total area under cropland existed for Central America. Area under pastures was 39 and 26 percent of the total land area in Mexico and Central America, respectively. Although the irrigated land area was not significant in Central America, it was about 20 percent of the total cultivated areas in Mexico. The area under forest was 23 and 36 percent of the total land area of Mexico and Central America, respectively. Mexico has extensive deserts not suited for forest growth. Climatic conditions in Central America are conducive to development of tropical forests in humid regions.

A large portion of the humid tropical forest occurs on plains, rolling hills, and along some hillsides. Large areas of of these forests have been converted into pastures. In some cases, pastures are abandoned and secondary forest vegetation allowed to regenerate. Some forests are being con-

TABLE 7.5. Land use in 2000 as percentage of total land area covered.

Land use	Belize	Costa Rica	El Salvador	Guate-mala	Honduras	Mexico	Nicaragua	Panama
Forests[a]	60	64	31	56	56	29	54	54
Shrublands, savanna, and grasslands[b]	6	7	1	8	7	53	3	10
Cropland and crop/natural vegetation mosaic[b]	29	27	65	34	36	17	35	33
Urban and built-up areas	0	0	0.9	0.2	0.1	0.1	0	0.1
Sparse or barren vegetation; snow and ice	0	0	0	0	0	1	0	0
Wetlands and water bodies	5	2	2	1	2	1	8	3

Source: Adapted from WRI, 2004c.
[a]Forests include all areas dominated by evergreen or deciduous trees with a canopy cover greater than 60 percent and a height exceeding 2 meters. Both broadleaf and needleleaf trees are included.
[b]Shrublands, savanna, and grasslands include lands dominated by woody vegetation less than 2 meters tall and with shrub canopy cover greater than 10 percent. The shrub foliage can be either evergreen or deciduous. This category also includes savannas and grasslands with herbaceous and other understory systems. These lands may have a tree or shrub cover of less than 60 percent.
[c]Cropland and crop/natural vegetation mosaic. Croplands are lands covered with temporary crops followed by harvest and a bare soil period (e.g., single and multiple cropping systems). Perennial woody crops are classified as forest or shrubland cover. Cropland/natural vegetation mosaics are lands with a mosaic of croplands, forests, shrublands, and grasslands in which no component comprises more than 60 percent of the landscape.

verted to industrial crops, such as henequen or coffee, which are also being abandoned because of poor market conditions. Areas under traditional slash and burn agriculture are also abandoned due to migration of young population to cities and aging of the remaining farm population.

Mexico is prone to intensive land-use changes with adverse consequences at both the local and global scales. The patterns of land-use change in Mexico are complex and vary by forest type and region. Deforestation has caused net C emissions of as much as 30 Mt C per year during the 1990s (Masera et al., 1997, 2003). The rate of deforestation was 413,000 ha per year between 1976 and 1993 (SEMARNAT, 2003). The rate of deforestation more than doubled between 1992 and 2000, reaching 1,057,000 ha per year. Deforestation rates are particularly high in the tropical regions, which are disappearing at a rate of 1.58 percent per year.

Land area under pastures increased at the rate of 4.07 percent per year. While temperate forests are disappearing at a rate of 0.79 percent mainly in the Sierra Madre Occidental and central Mexico, tropical forests are disappearing much faster, at the rate of 1.58 percent per year. The rate of deforestation is 2.5 percent per year in eastern and southern Mexico. The rate of deforestation of open semiarid forests is 0.48 percent per year. The states most affected by deforestation are Sinaloa, Hidalgo, Zacatecas, and Tamaulipas (SEMARNAT, 2003). Expanding livestock farming is responsible for more than half of total deforestation, and the remainder of deforestation is due to cultivation of crops and occurrence of forest fires. With the present rate of deforestation, the deforested area will double every 18 years. States in southern Mexico will be transformed at a high rate (SEMARNAT, 2003). The data in Table 7.6 show the vegetation land use and land-use change between 1976 and 2000 in Mexico.

The data in Table 7.7 show the synthesis of available information with regard to the land-use change in a more disaggregated manner. The information presented is based on the data on land use cartography published by INEGI (1980c), using aerial photographs taken in 1976, and a recent natural forestry inventory conducted by the National Autonomous University of Mexico (Instituto de Geografía-UNAM, 2002a,b) based on recent satellite images. Vegetation in both cases was divided into approximately 70 classes grouped according to climatic conditions listed in Table 7.7. Differences in total country area in 1976 and 2000 are due to more elaborate systems used in the 2002 inventory.

Land use and vegetation maps with information for 1976 and 2000, adapted by the authors from INEGI (1980c) and Instituto de Geografía-UNAM (2002a,b) are shown in Figure 7.2.

TABLE 7.6. Vegetation and land use and land-use change in Mexico between 1976 and 2000.

Vegetation	Area (km^2)			Area affected per year (km^2)		Rate of conversion (ha per year)	
and land use	**1976**	**1993**	**2000**	**1976-1993**	**1993-2000**	**1976-1993**	**1993-2000**
Forest	352.3	346.7	328.0	330.5	2,671.9	33,047	267,185
Tropical forest	378.6	343.6	307.4	2,055.80	5,177.9	205,576	517,785
Scrubland	605.3	576.5	557.4	1,694.60	2,723.1	169,461	272,314
Other vegetation	77.5	76.6	84.3	49.8	−1,102.9	4,976	0.0
Natural pastures	104.9	102.2	85.1	159.8	2,444.0	15,976	244,400
Induced pastures	150.3	175.7	232.4	−1,496.40	−8,097.4	0.0	0.0
Crops	263.4	302.1	327.6	−2,273.00	−3,642.9	0.0	0.0
Human settlements	2.3	11.2	12.4	−521.2	−173.6	0.0	0.0
Total	1,934.6	1,934.6	1,934.6	0	0.0	429,041	1,301,686
Subtotal forest						408,088	1,057,285

Source: Based on data from SEMARNAT, 2003.

TABLE 7.7. Land use in and land use change in Mexico between 1976 and 2000.

	10^3 km^2		Change	
Land use	**1976**	**2000**	**10^3 km^2**	**%**
Cropland	271.84	328.09	56.73	20.9
Natural forest areas	314.292	203.97	–110.31	-35.1
Secondary forest area	19.7	122.49	102.78	521.7
Water bodies	10.74	12.32	1.58	14.7
Natural shrublands	639.96	491.96	–148	–23.1
Secondary shrublands	29.17	101.35	72.17	247.4
Other vegetation types	60.51	85.51	24.98	41.0
Pastures/grassland	221.44	319.02	97.58	44.1
Plantations (forest)	0.1	0.26	0.15	150
Natural tropical forest	205.55	165.69	–39.86	–19.4
Secondary tropical forest	167.6	113.88	–53.72	–32.1
Urban areas	2.59	12.34	9.75	376.5
Total	1943.5	1957.34	5.673	

Source: Adapted by the authors from INEGI, 1980c, and Instituto de Geografía-UNAM, 2002a,b.

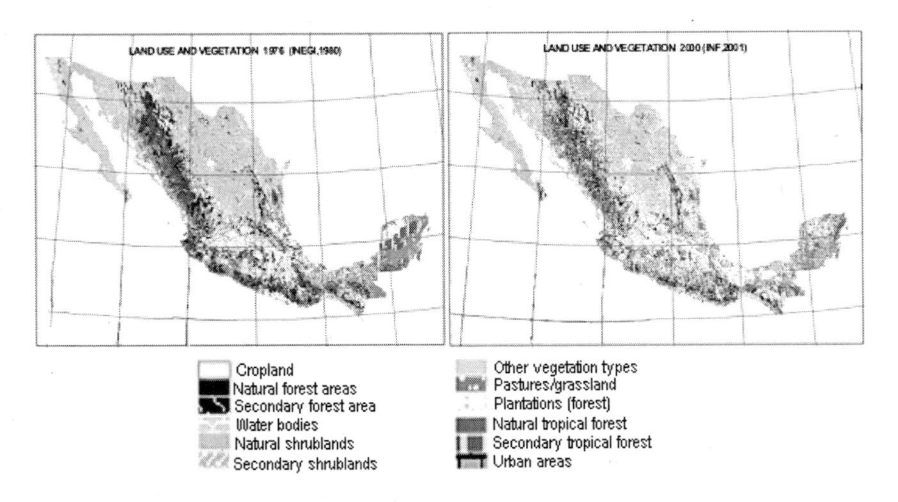

FIGURE 7.2. Land use and vegetation maps with data for 1976 and 2000. (*Source:* Redrawn from INEGI, 1980c, and Instituto de Geografia-UNAM, 2002a,b.)

SOIL DEGRADATION IN MEXICO AND CENTRAL AMERICA

Traditional agricultural cultivation systems have been responsible for exacerbating the soil degradation problems in Mexico and Central America. Table 7.8 shows the type of soil degradation and the fraction of the Mexican territory affected by different degradation processes during the 1990s (CONAZA, 1994). These estimates will be compared with recent data compiled in this chapter.

According to some estimates, as much as 1.6 Mha of land area was severely eroded, and and about 2.5 Mha were classified as highly erodable land in Mexico during 1980s. Accelerated soil erosion affected 80 percent of Mexico's land area (Maass and García-Oliva, 1990), and nearly 535 million Mg of soil were lost annually (SEMARNAP, 1997). More soil has been lost during the past 40 years than in the past four centuries (Maas and García-Oliva, 1990). Concurrent surface and gully erosion, exacerbated by deforestation and inappropriate cultivation of dry lands, have been observed on 65 to 85 percent of the land area in Mexico (Bocco and García-Oliva, 1992).

It has been estimated that at the beginning of the 1990s, 85 percent of the Mexican territory was affected by water erosion and 80 percent by biological degradation through the loss of soil organic matter (SOM), (CONAZA, 1994). These estimates were obtained by using a methodology of FAO-UNEP-UNESCO (1980). Figure 7.3, drawn with data from the CONAZA (1994) report, shows different degrees of soil erosion by water ranging from less than 10 $Mg \cdot ha^{-1}$ to 200 $Mg \cdot ha^{-1}$. Accelerated erosion also deleteriously affects agronomic productivity (Pagiola, 1999). The average soil loss in Mexico is estimated at approximately 2.8 $Mg \cdot ha^{-1}$ (Figueroa and Ventura,

TABLE 7.8. Estimates of soil degradation in Mexico.

Type of land degradation	Fraction of the total area affected (%)
Water erosion	85
Wind erosion	60
Lixiviation of bases	15
Physical degradation	20
Biological degradation	80
Salinity	20
Sodification	15

Source: Adapted from CONAZA, 1994.

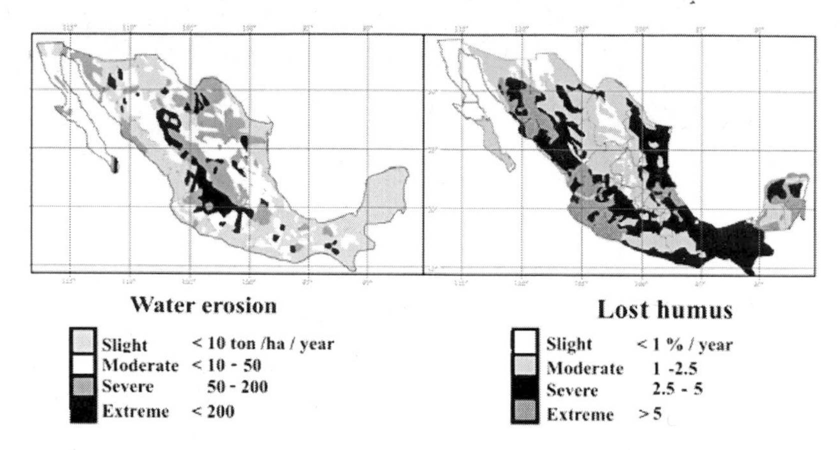

FIGURE 7.3. Areas prone to soil erosion by water and biological degradation (humus content) of soils from Mexico. (*Source:* Based on data from CONZA, 1994.)

1990). Water erosion is severe on slopes >10 percent and when protective management practices at the onset of the rainy season are not observed. Wind erosion is severe in arid and semiarid conditions. About 60 percent of the total land area was affected during the 1990s. Erosion rates of approximately 140 $Mg{\cdot}ha^{-1}$ of soil have been reported under specific conditions (Amante, 1989; Osuna, 1991), in comparison with only 40 $Mg{\cdot}ha^{-1}$ of soil loss by water erosion. Biological degradation is the second largest process of soil degradation in the country after water erosion, and represents the rate high of mineralization of SOM as determined by the methodology of FAO-UNEP-UNESCO (1980).

About 80 percent of the total land area of Mexico is affected by soil biological degradation, which is influenced by climate (Figure 7.3). Depletion of SOM is greater in areas closer to coastlines and less in semiarid and arid zones. Biological degradation is also caused by a reduction of the topsoil and excessive cropping. Water erosion and biological degradation can be reduced and even reverted by adopting recommended management, such as conservation tillage. Adoption of conservation tillage can increase soil C sequestration.

The data in Table 7.9 show the extent and severity of soil erosion by water and wind in Mexico.

Regions in Mexico prone to both water and wind erosion are presented in Figure 7.4. The available information was used to establish five classes of severity of water and wind erosion in Mexico (SEMARNAT-UACH, 2001-2002).

TABLE 7.9. Estimates of land area affected by water and wind erosion in Mexico.

	Surface affected and degree of erosion[a] by area and percentage of total country surface									
Erosion	None		Slight		Moderate		Severe		Very severe	
type	10^3 km^2	%	10^3 km^2	%	10^3 km^2	%	10^3 km^2	%	10^3 km^2	%
Water	1135.67	57.96	214.51	10.95	401.30	20.48	153.43	7.83	54.34	2.77
Wind	216.04	11.03	127.71	6.53	600.84	30.67	658.61	33.61	35.60	18.17

Source: Calculated from SEMARNAT-UACH, 2001-2002.
[a]For degree of erosion, None = 0 to 5 Mg·ha^{-1}; Slight = 5 to 10 Mg·ha^{-1}; Moderate = 10 to 50 Mg·ha^{-1}; Severe = 50 to 200 Mg·ha^{-1}; Very severe = >200 Mg·ha^{-1}.

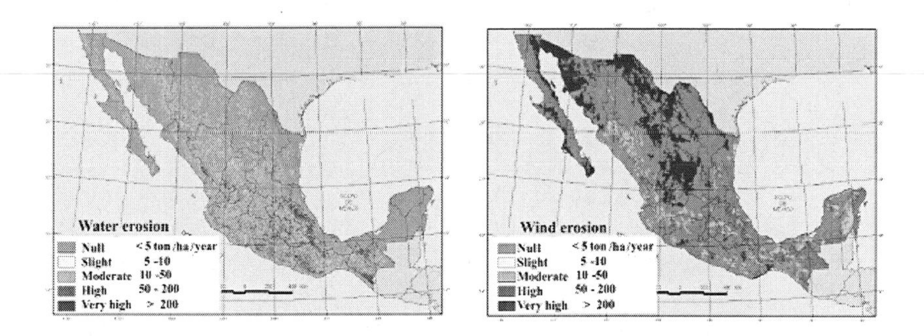

FIGURE 7.4. Severity classes of water and wind erosion in Mexico. (*Source:* Calculated and redrawn from SEMARNAT-UACH, 2001-2002.)

The data in Table 7.10 depict the dominant anthropogenic degradation processes in Mexico (SEMARNAT-CP, 2001). These calculations were based on the Physiographic Survey at the terrestrial systems level using a landscape approach at 1:250,000 scale (Ortiz and Cuanalo, 1984). The assessment was made at four levels: (1) national level, considering all terrestrial systems of the country; (2) state level, overlapping the state borders given in the space maps of the National Statistics Institute (INEGI, 1995); (3) hydrologic basin level, overlapping the hydrologic regions established by the Mexican Institute of Water Technology (IMTA); and (4) the ecological regions established by the SEMARNAT. These data were used to delineate homogeneous and observable units. The minimum mapping area was set at 1 cm^2. Each one of the resultant delimitations constituted the terrestrial systems (SEMARNAT-CP, 2001-2002).

Principal types of soil degradation were defined by using the ASSOD methodology (Van Lynden and Oldeman, 1997), which is a variation of the GLASOD methodology, proposed by Oldeman (1988). The latter was

TABLE 7.10. Predominant human-induced soil degradation processes in Mexico.

	Land area affected	
Process	**%**	**Mha**
Nutrient depletion	16.97	33.2
Water erosion	11.41	22.3
Pollution	0.64	1.3
Wind erosion	8.96	175.5
Soil structural decline	5.58	109.3
Salinity	0.57	1.1

Source: Based on data from SEMARNAT-CP, 2001-2002.

adopted by FAO worldwide and by the National Soil Inventory of the General Office of Soil Restoration and Conservation (SEMARNAT-CP, 2001-2002)

Land area affected by human-induced soil degradation in Mexico is estimated at 45.1 percent of the total area. In contrast, stable soils or those without apparent degradation cover 28.6 percent, and unused lands are estimated at 26.3 percent. The main degradation processes are chemical (18.4 percent), water erosion (11.4 percent), wind erosion (9.4 percent), and physical degradation (5.9 percent); soil degradation processes in Mexico are shown in Figure 7.5.

Predominant soil degradation processes are chemical degradation caused by a 17.0 percent decline in soil fertility; water erosion on 10.1 percent; wind erosion on 9.0 percent; and physical degradation by soil compaction on 4.1 percent. Of the 17 types of degradation processes seen, those not observed in Mexico include chemical degradation from acidification and physical degradation from the subsidence.

Agricultural activities and overgrazing are the main causes of soil degradation, accounting for 38.8 percent and 38.4 percent of the surface affected, respectively. Other activities affecting degradation include deforestation (16.5 percent), urbanization (3.5 percent), overexploitation of the vegetation (2.4 percent), and industrial activities (0.5 percent).

The data in Table 7.11 show the extent and severity of principal types of soil degradation in Central America in relation to the causative factors (ISRIC, 2004). The data in Table 7.12 show the extent and severity of soil degradation in Central America in relation to the principal processes (ISRIC, 2004).

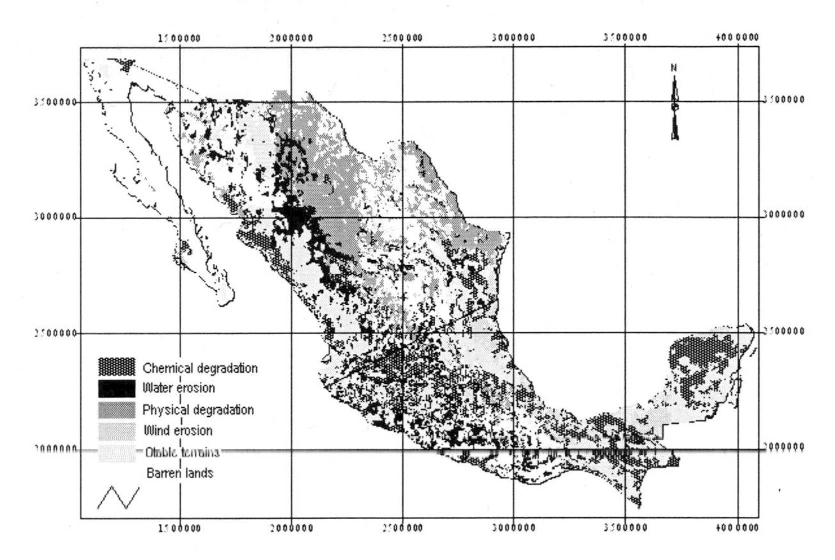

FIGURE 7.5. Area distribution of the human-induced soil degradation process in Mexico. (*Source:* Redrawn from SEMARNAT-CP, 2001-2002.)

TABLE 7.11. Extent and severity of soil degradation in Central America attributed to agricultural activities and deforestation.

		Land area by country (10^3 km^2)							
Causative factor	**Severity**	**Belize**	**Costa Rica**	**El Salvador**	**Guate-mala**	**Honduras**	**Nicar-agua**	**Panama**	**Total for factor**
Agricul-tural activi-ties	Low	<0.01	<0.01	0.46	7.42	<0.01	<0.01	<0.01	7.87
	Medium	<0.01	<0.01	<0.01	<0.01	2.03	5.10	<0.01	7.13
	High	<0.01	9.20	<0.01	0.17	17.03	78.17	<0.01	104.57
	V. high	<0.01	<0.01	16.27	19.48	38.47	0.36	<0.01	74.58
Deforesta-tion	Low	<0.01	<0.01	<0.01	1.21	<0.01	<0.01	7.23	8.44
	Medium	<0.01	8.78	<0.01	<0.01	<0.01	<0.01	0.98	9.76
	High	6.28	4.95	<0.01	59.21	41.11	15.21	40.21	166.96
	V. high	<0.01	25.40	<0.01	<0.01	<0.01	0.73	18.49	44.62
No label		15.89	3.28	3.96	22.02	14.22	29.48	7.78	96.64
Total by country		22.17	51.61	20.70	109.50	112.85	129.05	74.70	520.58

Source: Adapted from ISRIC, 2004.

TABLE 7.12. Extent and severity of soil degradation in Central America for principal degradation processes.

		Total area by country (10^3 km^2)							
Soil degradation types	**Severity**	**Belize**	**Costa Rica**	**El Salvador**	**Guatemala**	**Honduras**	**Nicaragua**	**Panama**	**Total types & severity**
Loss of nutrients and/or organic matter	High	6.08	9.20	<0.01	19.71	51.53	83.99	<0.01	170.52
Pollution	Low	<0.01	<0.01	0.46	7.42	<0.01	<0.01	<0.01	7.87
Wind erosion (loss of topsoil)	High	<0.01	<0.01	<0.01	0.17	<0.01	<0.01	<0.01	0.17
Stable terrain under natural conditions	N/A	13.70	<0.01	<0.01	17.69	11.87	17.16	2.41	62.82
Water erosion (terrain deformation)	Low	<0.01	<0.01	<0.01	1.21	<0.01	<0.01	7.23	8.44
Water erosion (loss of topsoil)	Medium	<0.01	8.78	<0.01	<0.01	2.03	5.10	0.98	16.89
	High	0.20	4.95	<0.01	39.50	6.60	9.38	40.21	100.84
	V. high	<0.01	25.40	16.27	19.48	38.47	1.09	18.49	119.21
No label		2.19	3.28	3.96	4.34	2.35	12.32	5.37	33.81
Total by country		22.17	51.61	20.70	109.50	112.85	129.05	74.70	520.58

Source: Adapted from ISRIC, 2004.

CARBON STOCKS IN SOILS OF MEXICO

A series of international symposia and workshops were organized during the 1990s (Kimble et al., 2002) in order to develop a scientific basis to understanding the relationship among C stocks, C sequestration, soils and climate change, and promoting the idea of soil as a C sink that can be managed and enhanced. The available information shows that the belowground C stocks in Mexican highlands are two to three times larger than the aboveground C stocks (Monreal et al., 2005; Acosta, 2003).

The adoption of recommended practices of management by farmers may reduce GHG emissions from the world's agricultural lands. Further, adoption of these practices may restore the quality of air, by being from net sources of GHGs into net sinks through the capture of atmospheric C and N and their sequestration in SOM (Agriculture and Agri-Food Canada, 1998). The C constitutes nearly 58 percent of the SOM, and it improves physical, chemical, and biological characteristics of soil and enhances agronomic yields (Herrick and Wander, 1998).

The project "Evaluation of Human-Induced Soil Degradation in Mexico, at 1:250,000 scale" (SEMARNAT-CP, 2001-2002) measured soil organic carbon (SOC) in the surface layer (0 to 20 cm) of soils in Mexico for 4,583 samples obtained from locations shown in Figure 7.6.

Soil C concentration was measured by using the Mexican Official Norm PRY-NOM-021-RECNAT-2000 (SEMARNAT, 2000) and the manual of ISRIC Procedures for Soil Analysis (van Reeuwijk, 1998). Maps were prepared depicting soil C concentration with the raster format for cartographic purposes. Figure 7.7 shows the map of C concentration in 0 to 20 cm layer for predominant soils of Mexico.

Later on, and with the purpose of obtaining estimates of the depth distribution of C, models of the distribution of C in subsoil were developed, based on field sampling data (Acosta and Etchevers, 2002). Such modeling was similar to that developed by Batjes and Sombroek (1997) but adapted to the specific Mexican soil conditions. The C variation model, including the depth level registered, is presented in Figure 7.8.

Corresponding soil units were assigned to each sampling point (surface 0 to 20 cm carbon values) based on the map of dominant soils in Mexico. This

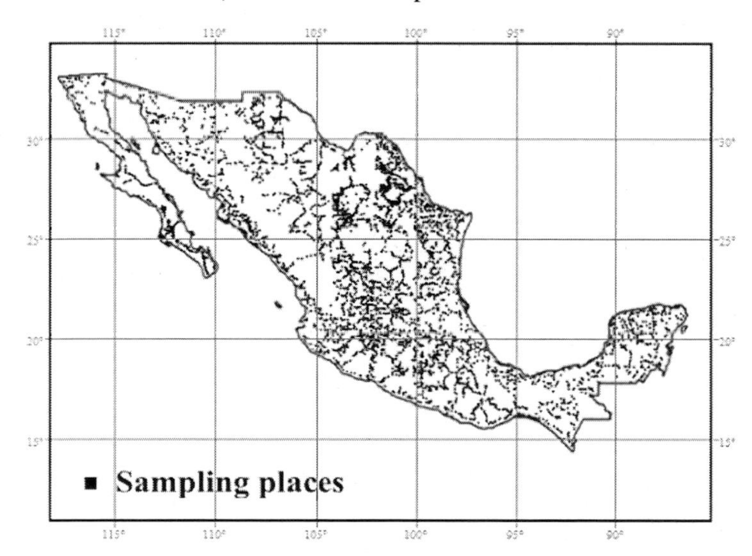

FIGURE 7.6. Distribution of the sampling locations to measure carbon concentration in 0 to 20 cm layer for predominant soils of Mexico.

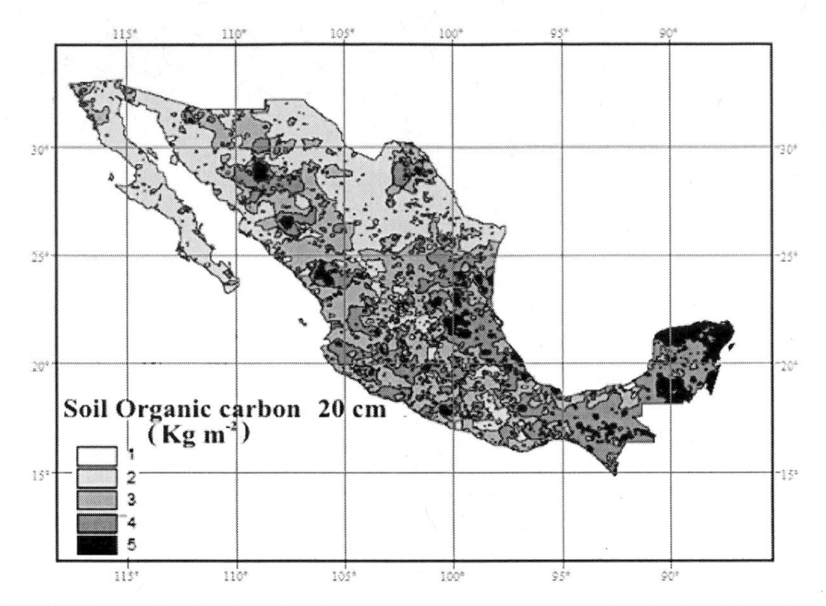

FIGURE 7.7. Distribution of soil organic carbon concentration in 0 to 20 cm layer for dominant soils of Mexico.

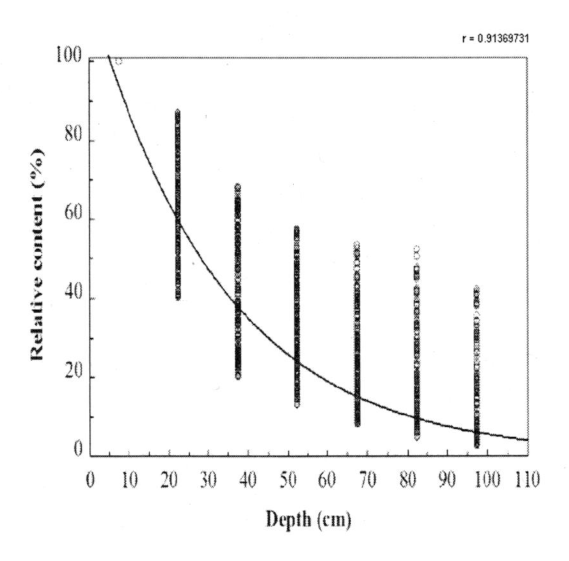

FIGURE 7.8. Model to estimate the depth distribution of C in the soil profile. (*Source:* Adapted from Acosta and Etchevers, 2002.)

map was a generalization of the pedological map, scale 1:1,000,000, of INEGI (1980b) updated to incorporate the 1988 version of the FAO/UNESCO/ISRIC classification system (Figure 7.9). Finally, the average soil depth for each soil unit was estimated based on the soil phases map (Figure 7.10).

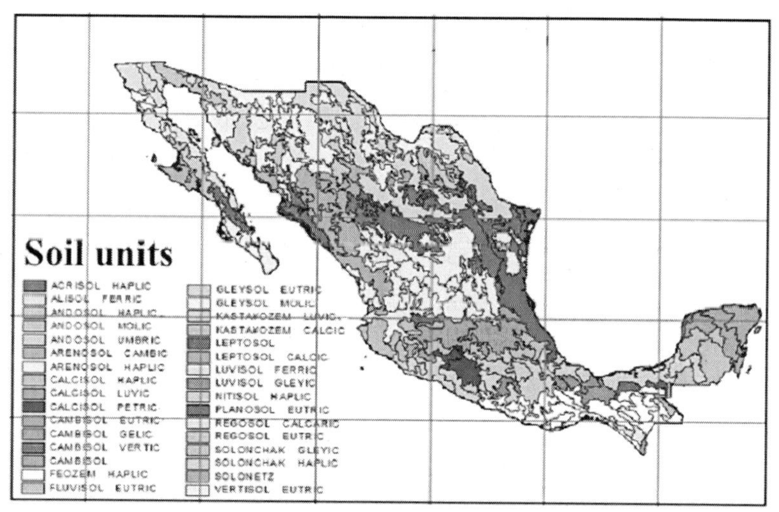

FIGURE 7.9. Soil units map of Mexico. (*Source:* Based on data from SEMARNAP, 1998.)

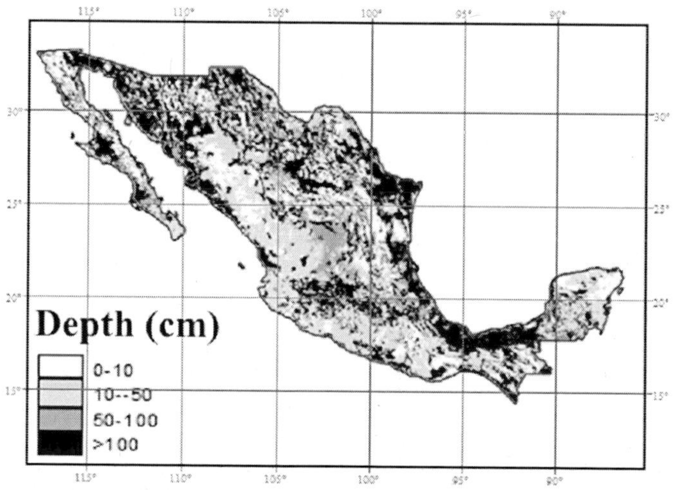

FIGURE 7.10. Average depth of soils in Mexico.

The model for C distribution in the soil profile (see Figure 7.8) was applied to calculate the amount of C accumulated in each soil unit. This calculation considered the particular soil depth estimated for each soil unit level (Figure 7.11). The data in Figure 7.12 show the distribution of soil C stocks in regions corresponding to the western ridges and the desert regions of

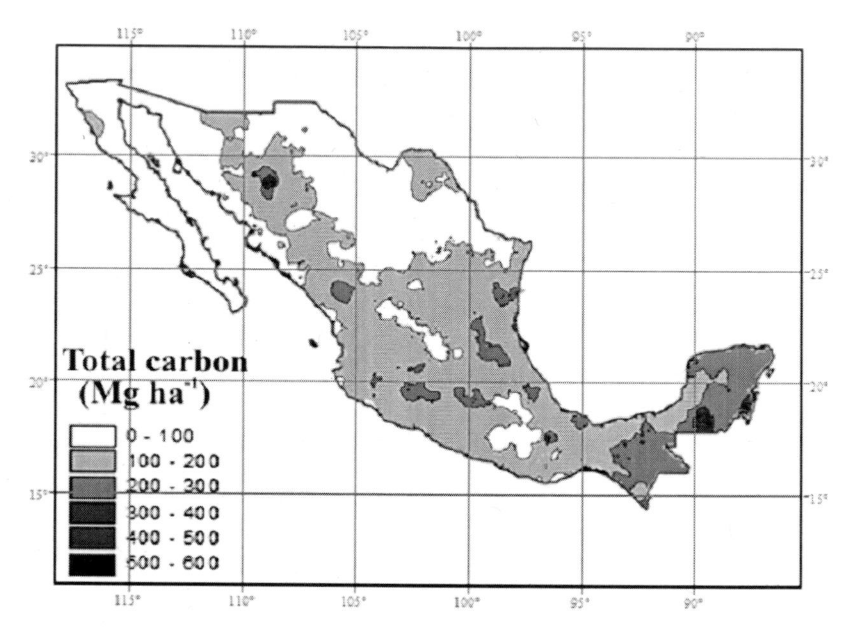

FIGURE 7.11. Total soil carbon (Mg·ha−1) stocks in soils in Mexico.

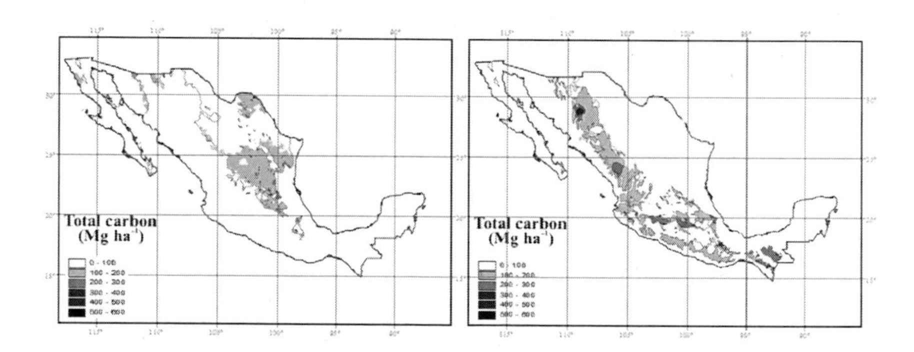

FIGURE 7.12. Total C stocks (Mg·ha−1) in soils of the desert and the western ridge regions of Mexico.

Mexico. The data in Table 7.13 show the organic C density and stocks for different soil units and different depths for principal soils of Mexico.

POTENTIAL SOIL CARBON SEQUESTRATION AND LAND-USE CHANGE

Research information with respect to rates of C sequestration in soils of Mexico is scanty. Thus far, most studies have focused on estimating C stocks in soils under different land-use classes. This section presents a preliminary estimate of the C sequestration potential in Mexico for some specific scenarios. Estimates of the rate soil C sequestration are obtained from published international literature and expert judgment. The future scenario is also a preliminary exercise which needs to be refined as more data become available.

The data in Table 7.14 show area under different land uses in Mexico, and those in Table 7.15 illustrate the potential C sequestration rate in relation to specific management options for agriculture and forestry sectors. Finally, Table 7.16 presents the total soil C sequestration potential in Mexico by 2020.

By the year 2020, approximately 8.8 Mha of marginal soils will be converted to restorative land use for C sequestration options in Mexico. The rate of soil C sequestration ranges between 0.1 $Mg \cdot ha^{-1}$ per year and 3 $Mg \cdot ha^{-1}$ per year. The estimated future C sequestration in the country ranges between 3.8 and 16.7 million Mg per year (Tg per year).

CONCLUSIONS

The present agricultural land in Biome A of Mexico and Central America has been created by extensive land-use change since the era of European settlement. A large proportion of this land is prone to accelerated erosion and depletion of the antecedent C stocks. A part of this land can be used to sequester C and to offset emissions. By the year 2020, about 8.8 Mha of land can be used for a range of soil C sequestration options in Mexico. Among recommended management practices by farmers are conservation tillage, irrigation plus conservation tillage, agroforestry, and restoration plantations Rate of soil C sequestration with adoption of recommended practices ranges between 0.1 $Mg \cdot ha^{-1}$ per year and 3 $Mg \cdot ha^{-1}$ per year. With these rates, the potential of soil C sequestration in the country could reach between 3.8 and 16.7 million $Mg \cdot ha^{-1}$ per year.

TABLE 7.13. Soil organic C density and stocks in principal soil units of Mexico.

Unit of soil	Subunit	Average depth (cm)	Surface carbon (Kg·m^{-2})	Carbon in effective depth (Mg·ha^{-1}) Mean	Carbon in effective depth (Mg·ha^{-1}) Mean + SD
Leptosol	Calcic	20.00	13.27	163.81	265.05
Gleysol	mollic	30.00	36.42	536.73	536.73
Leptosol		30.00	9.04	117.41	187.22
Regosol	Eutric	50.31	7.55	115.42	193.30
Luvisol	Ferric	52.50	9.14	142.77	228.55
Cambisol	Gellic	53.33	9.92	154.09	269.58
Regosol	calcaric	61.11	10.37	156.43	231.09
Acrisol	haplic	65.00	10.31	163.11	266.56
Kastañozem	Luvic	65.00	7.12	113.16	203.37
Luvisol	gleyic	68.57	13.95	215.28	349.99
Feozem	haplic	70.00	7.18	115.16	193.48
Planosol	Eutric	75.00	6.77	112.59	195.93
Alisol	Ferric	82.50	14.90	241.05	383.62
Arenosol	haplic	82.50	3.02	49.60	92.43
Solonchak	haplic	84.17	8.23	94.91	140.42
Cambisol	Eutric	84.44	9.81	159.78	266.03
Calcisol	Petric	91.67	3.69	63.25	112.69
Vertisol	Eutric	92.22	7.95	132.19	204.71
Calcisol	haplic	94.64	4.03	68.53	115.66
Andosol	haplic	100.00	14.04	238.95	408.13
Andosol	mollic	100.00	12.20	207.68	355.68
Andosol	umbric	100.00	15.85	269.70	451.45
Calcisol	Luvic	100.00	5.39	91.72	139.27
Cambisol	Vertic	100.00	10.43	177.54	226.67
Cambisol		100.00	10.09	171.71	239.65
Fluvisol	Eutric	100.00	1.10	18.76	28.54
Gleysol	Eutric	100.00	10.74	182.73	294.45
Kastañozem	Calcic	100.00	6.08	103.54	156.65
Nitisol	haplic	100.00	19.79	336.81	425.38
Solonchak	gleyic	100.00	11.04	187.89	187.89

Source: Adapted by the authors with data from SEMARNAT-CP, 2001-2002; SEMARNAP, 1998.

TABLE 7.14. Area under current land uses in Mexico.

	Area	
Land use	**%**	**Mha**
Total area	100.00	195.734
Cropland	15.29	29.93
Traditional[a]	12.09	23.66
Traditional on slopes[b]	18.90	5.66
Traditional in plains[c]	81.10	24.27
Irrigated[d]	3.20	6.26
Pastures and scrubland[e]	53.00	103.74
Improved pastures[f]	4.08	8
Restoration plantations	0.25	0.5

Source: Based on data from Inventario Nacional Forestal 2000 (Instituto de Geografa-UNAM, 2002a,b) and FAO (2000).
[a]Total cultivatable – irrigated lands.
[b]18.9 percent of cropland areas (USGS/EROS Data Center) include slopes >10 percent. Cropland areas include dry croplands and pastures, irrigated croplands, and areas of mixed cropland/grassland and cropland/woodland.
[c]Agriculture minus slope areas.
[d]4.08 percent of country area.
[e]1.3×10^6 ha with suspended irrigation.
[f]Pastures in tropical humid and subhumid areas as well as temperate areas.

TABLE 7.15. Future scenarios of soil carbon sequestration in Mexico.

Option	Percentage of area to be added to the option in the next 20 years	Area (10^6 ha)	Soil carbon increment ($Mg \cdot ha^{-1}$)
Traditional agriculture on slopes[a]			
Agroforestry	10.00	0.57	0.2-0.4
Fallow land to become secondary forests	10.00	0.57	0.3-1.0
Plains areas[b]			
Conservation tillage	20.00	4.85	0.2-0.3
Agroforestry	7.98	0.5	0.2-0.4
Irrigated land			
Irrigation plus cons. tillage	16.60	1.04	0.5-1.0
Improved pastures[c]	12.77	0.8	0.15-0.6
Commercial plantations	13.89	0.87	0.5-3.0
Restoration plantations	79.81	0.5	0.1-3.0

Source: Based on data from Inventario Nacional Forestal 2000 (Instituto de Geografia-UNAM, 2002a,b), FAO (2000), and Masera et al. (2000).
[a]18.9 percent of cropland areas (USGS/EROS Data Center) include slopes >10 percent. Cropland areas include dry croplands and pastures, irrigated croplands, and areas of mixed cropland/grassland and cropland/woodland.
[b]Agriculture minus slope areas.
[c]4.08 percent of country areas.

TABLE 7.16. Future scenarios of soil carbon sequestration in Mexico, and total C sequestration potential by 2020.

Option	Area by option (10^6 ha)	Unit carbon sequestration ($Mg \cdot ha^{-1}$ per year)	Total carbon sequstration (10^6 Mg per year)
Traditional agriculture on slopes[a]			
Agroforestry	0.57	0.5-1.8	0.28-1.01
Fallow land to become secondary forests	0.57	0.3-1.0	0.17-0.57
Plains areas[b]			
Conservation tillage	4.85	0.4-1.2	1.94-5.82
Agroforestry	0.5	0.5-1.8	0.25-0.9
Irrigated land			
Irrigation plus cons. tillage	1.04	0.4-1.2	0.42-1.25
Improved pastures[c]	0.8	0.2-0.6	0.16-0.48
Commercial plantations	0.87	0.1-4.6	0.09-4.00
Restoration plantations	5.0	0.1-2	0.5-10
Total	8.3		3.8-16.7

Source: Based on data from Inventario Nacional Forestal 2000 (Instituto de Geografia-UNAM, 2002a,b) and FAO (2000).

[a]18.9 percent of cropland areas (USGS/EROS Data Center) include slopes >10 percent. Cropland areas include dry croplands and pastures, irrigated croplands, and areas of mixed cropland/grassland and cropland/woodland.

[b]Agriculture minus slope areas.

[c]4.08 percent of country areas.

BIBLIOGRAPHY

Acosta, M. 2003. Diseño y aplicación de un método para medir los almacenes de carbono en sistemas con vegetación forestal y agrícolas de ladera en México. Tesis de Doctorado. Colegio de Postgraduados, Instituto de Recursos Naturales Programa Forestal, Montecillo, México.

Acosta, M. and J.D. Etchevers. 2002. Distribución de carbono orgánico en el perfil del suelo, en diferentes sistemas de bosques de la Sierra Norte de Oaxaca. In *Congreso Nacional de la Ciencia del Suelo* [CD-ROM]. Sociedad Mexicana de la Ciencia del Suelo, Torreón, Coahuila, México.

Agriculture and Agri-Food Canada. 1998. *The health of our air: Toward sustainable agriculture in Canada.* Compiled and edited by H. Janzen, R.L. Desjardins, J. Asselin, and B. Grace. Research Branch. Publication 1981/E.

Amante O.A. 1989. Variabilidad espacial y temporal de la erosión hídrica: Estudio de caso. Tesis de Maestría en Ciencia. Colegio de Postgraduados, Montecillo, México.

Batjes, N.H. and W.G. Sombroek. 1997. Possibilities for carbon sequestration in tropical and subtropical soils. *Global Change Biology* 3: 161-173.

Bocco, G. and F. García-Oliva. 1992. Researching gully erosion in México. *J. Soil Water Conserv.* 47: 365-367.

Claverán A.R. 2000. Conservation tillage in Mexico and Latin America: An overview. In *Mem. Simp. Intern. de Labranza de Conservación,* 24 al 27 de Enero 2000, Sinaloa, México. (CD-ROM).

CONAZA-Comisión Nacional de Zonas Áridas. 1994. Plan de acción para combatir la desertificación en México. Com. Zonas Áridas, Secr. de Desarrollo Social. México, D.F.

FAO. 2000. Two essays on climate change and agriculture: A developing country perspective. FAO 145. Economic and social development paper. FAO, Rome, Italy.

FAO-UNEP-UNESCO. 1980. *Metodología provisional para la evaluación de la degradación de los suelos.* FAO, Roma, Italy.

Feigl, B.J., J. Melillo, and C.C. Cerri. 1995. Changes in the origin and quality of soil organic matter after pasture introduction in Rodonia (Brazil). *Plant and Soil* 175: 21-29.

Figueroa B.S. and E. Ventura R. Jr. 1990. *Instructivo para la evaluación del proyecto: Efecto de la labranza en la estructura del suelo y su relación con el crecimiento, desarrollo y rendimiento de los cultivos.* Instituto Nacional de Investigaciones Forestales, Agrícolas y Pecuarias, Salinas, San Luis Potosí, México.

García-Oliva, F. and O.R. Masera. 2004. Assesment and measurement issues related to soil carbon sequestration in land-use, land-use change, and forestry (LULUCF) projects under Kyoto Protocol. *Climatic Change* 65: 347-364.

Herrick, J.E. and M.M. Wander. 1998. Relationships between soil organic carbon and soil quality in cropped and rangeland soils: The importance of distribution, composition, and soil biological activity. In R. Lal, J.M. Kimble, R.F. Follett, and B.A. Stewart (eds.), *Soil processes and the carbon cycle.* CRC Press, Boca Raton, FL, pp. 405-425.

INEGI-Instituto Nacional de Estadística, Geografía e Informática. 1980a. *Cartas de temperatura y precipitación media anual de la República Mexicana.* Instituto Nacional de Estadística, Geografía e Informática, México.

INEGI-Instituto Nacional de Estadística, Geografía e Informática. 1980b. *Cartas edafológicas de México, escala 1:250,000.* Instituto Nacional de Estadística, Geografía e Informática, México.

INEGI-Instituto Nacional de Estadística, Geografía e Informática. 1980c. *Cartografía de uso del suelo y vegetación Serie I, 1976.* Instituto Nacional de Estadística, Geografía e Informática, México.

INEGI-Instituto Nacional de Estadística, Geografía e Informática. 1995. *Espaciomapas escala 1:250,000.* INEGI, México.

INEGI-Instituto Nacional de Estadística, Geografía e Informática. 1998. *Estadísticas del medio ambiente: México 1997.* INEGI, SEMARNAP.

Instituto de Geografía. UNAM. 2002a. Análisis del cambio de uso del suelo. México, D.F.

Instituto de Geografía-UNAM. 2002b. Análisis del uso de suelo. México.

IPCC-Intergovernmental Panel on Climate Change. 1996. Climate change 1995. In R. Watson, M. Zinyowera, and R. Moss (eds.), *Impact, adaptations and mitiga-*

tion of climate change: Scientific-technical analyses. Cambridge University Press, Cambridge, UK.

IPCC-Intergovernmental Panel on Climate Change. 2001. Technical summary. Climate change 2001: Mitigation. A Report of Working Group III of the Intergovernmental Panel on Climate Change Available online at www.ipcc.ch/pub/wg3TARtechsum.pdf; consulted July 31, 2001.

ISRIC-International Soil Reference and Information Centre. 2004. Global assessment of soil degradation (GLASOD). Available online at www.isric.org; consulted May 20, 2004.

Kimble, J. M., L.R. Everett, R. F. Follet, and R. Lal. 2002. Carbon sequestration and the integration of science, farming and policy. In J.M. Kimble, R. Lal, and R.F. Follett (eds.), *Agricultural practices and policies for carbon sequestration in soil*. CRC/Lewis Publishers, Boca Raton, FL, pp. 3-11.

Lal, R., J.M. Kimble, R.F. Follet, and C.V. Cole. 1999. *The potential of U.S. cropland to sequester carbon and mitigate the greenhouse effect*. Lewis Publishers, Boca Raton, FL.

Maass, J.M.M. and F. García-Oliva. 1990. La conservación de suelos en zonas tropicales: el caso de México. *Ciencia y Desarrollo* 90: 21-36.

Masera, O.R., A.D. Cerón, and J. A. Ordóñez. 2001. Forestry mitigation options for Mexico: Finding synergies between national sustainable development priorities and global concerns. *Mitigation and Adaptation Strategies for Climate Change* (Special Issue on Land Use Change and Forestry Carbon Mitigation Potential and Cost Effectiveness of Mitigations Options in Developing Countries) 6(3-4): 291-312.

Masera, O.R., R. Martínez, T. Hernández, A. Guzmán, and A. Ordóñez. 2000. *Inventario Nacional de Gases de Efecto Invernadero 1994-96, Parte 6: Cambio de Uso del Suelo y Bosques*. Instituto Nacional de Ecología, SEMARNAP, México D.F.

Masera, O.R. and R.D. Martínez-Bravo. 2003. Diagnóstico de captura de carbono en el Paseo Ecológico Jaguaroundi. In *Conservación, Reforestación, Captura de carbono y Paseo Ecológico Jaguaroundi, Reporte*. CIECO-PUMA-PEMEX PPQ, México.

Masera, O.R., M.J. Ordoñez, and R. Dirzo, 1997. Carbon emissions from Mexican forests: Current situation and long-term scenarios. *Climatic Change* 35: 265-295.

Monreal, C.M., J.D. Etchevers, M. Acosta, C. Hidalgo, J. Padilla, R. M. López, L. Jiménez, and A. Velázquez. 2005. A method for measuring and monitoring above- and below-ground C stocks in hillside landscapes. *Canadian Journal of Soil Science* 85: 523-530.

Neill, C., J. Melillo, P.A. Steudler, C.C. Cerri, J.F.L. de Moraes, M.C. Piccolo, and M. Brito. 1997. Soils carbon and nitrogen stocks following forest clearing for pasture in the southwestern Brazilian Amazon. *Ecological Applications* 7: 1216-1225.

Oldeman, L.R. (ed.). 1988. Guidelines for general assessment of the status of human-induced soil degradation. ISRIC Working Paper and Preprint 88/4.

Ordóñez, J.A.B., B.H.J. de Jong, F. García-Oliva, F.L. Aviña, J.V. Pérez, G. Guerrero, R. Martínez, and O. Masera. 2004. Carbon content in vegetation, litter, and soil in ten different land-use land-cover classes in central highlands of Michoacan, Mexico. In Press.

Ortiz, S.C.A. and H.E. Cuanalo de la C. 1984. *Metodología del Levantamiento Fisiográfico: Un Sistema de Clasificación de Tierras*. Colegio de Postgraduados. Chapingo, México.

Osuna, C.E.S. 1991. Efecto de la cobertura vegetal en el proceso erosivo. Memorias Seminario. In *Conservación de Agua y Suelo (Manejo integral de cuencas)*. SARH-CNA-IMTA, México, D.F., pp. 148-166.

Pagiola, S. 1999. The global, environmental benefits of land degradation control on agricultural land: Global overlays program. Environment Paper: Number 16. World Bank, Washington, DC.

SAGARPA-INIFAP-MIAC. 2000. Simposium Internacional de Labranza de Conservación. Sinaloa, México 24 al 27 de Enero de 2000. SAGARPA-INIFAP-MIAC, México, D.F. (CD-ROM).

SEMARNAP. 1997. Secretaría de Medio Ambiente y Recursos Naturales y Pesca. Available online at www.semarnap.gob.mx; consulted January 24, 2003.

SEMARNAP–Secretaría de Medio Ambiente y Recursos Naturales y Pesca. 1998. Mapa de suelos dominantes de la República Mexicana. (Primera aproximación 1996). Subsecretaría de Recursos Naturales, México, D.F.

SEMARNAT–Secretaría de Medio Ambiente y Recursos Naturales. 2000. Norma Oficial Mexicana PROY-NOM-021-RECNAT-2000, que establece las especificaciones de fertilidad, salinidad y clasificación: Estudios, Muestreo y Análisis. Diario Oficial de la Federación del 17 de Octubre del 2000. México, D.F.

SEMARNAT. Secretaría de Medio Ambiente y Recursos Naturales. 2003. *Informe de la situación del medio ambiente en México: Compendio de estadísticas ambientales, 2002*. SEMARNAT, México, D.F.

SEMARNAT-CP, Secretaría de Medio Ambiente y Recursos Naturales y Colegio de Postgraduados 2001-2002. Memoria Nacional. Evaluación de la Degradación de los Suelos Causada por el Hombre en la República Mexicana, a escala 1:250,000. México, D.F.

SEMARNAT-UACH. Secretaría de Medio Ambiente y Recursos Naturales y Universidad Autónoma Chapingo. 2001-2002. Evaluación de la pérdida de suelo por erosión hídrica y eólica en la República Mexicana, escala 1:1,000,000. México, D.F.

Tiscareño-López, M., A.D. Báez-González, M. Velázquez-Valle, R. Claverán-Alonso, K. N. Potter, and J. J. Stone. 2000. Case study: Conservation tillage to save Patzcuaro Lake Watershed. In *Mem. Simp. Inter. de Labranza de Conservación, 24 al 27/01/2000*. Sinaloa, México. (CD-ROM).

UNFCCC. 2004. Clean development mechanism (CDM). Available online at cdm.unfccc.int/; consulted May 15, 2004.

Van Lynden, G. W. J. and L. R. Oldeman. 1997. *The assessment of the human-induced soil degradation in South and Southeast Asia*. International Soil Reference and Information Centre, Wageningen, the Netherlands.

Van Reeuwijk, L. P. 1998. Guidelines for quality management in soil and plant laboratories. Soils Bulletin 74. FAO, Rome.

Velázquez, A., J.F. Mas, J.R. Díaz-Gallegos, R. Mayorga-Saucedo, P.C. Alcántara, R. Castro, T. Fernández, G. Bocco, E. Ezcurra, and J.L. Palacio. 2002. Patrones y tasas de cambio de uso de suelo en México. INE-Semarnat. *Gaceta Ecológica* 62: 21-42.

Vergara, S.M.A., J.D. Etchevers, and M. Vargas H. 2004. Variabilidad del carbono orgánico en suelos de ladera del sureste de México. *Terra* 22: 359-368.

WRI. World Resources Institute. 1990-1991. *World resources, 1990-91*. Oxford University Press, New York.

WRI. World Resources Institute. 1994. *World resources, 1994-95—A guide to the global environment*. Oxford University Press, New York.

WRI. World Resource Institute. 2002. *World resources, 1998-1999: Regions at a glance, Central America*. Available online at www.wri.org; consulted August 2002.

WRI. World Resources Institute. 2004a. Earth trends: Agriculture and food. Available online at www.earthtrends.wri.org/pdf_library/country_profiles/Pop_cou_591.pdf; consulted May 18, 2004.

WRI. World Resources Institute. 2004b. Earth trends: Economics, business, and the environment. Available online at earthtrends.wri.org/pdf_library/country_profiles/Eco_cou_484.pdf; consulted May 18, 2004.

WRI. World Resources Institute. 2004c. Earth trends: Forest, grasslands and Drylands. Available online at earthtrends.wri.org/country_profiles/index.cfm?theme=9; consulted May 18, 2004.

WRI. World Resources Institute. 2004d. Earth trends: Population, health and human well being. Available online at earthtrends.wri.org/pdf_library/country_profiles/Pop_cou_591.pdf; consulted May 18, 2004.

Chapter 8

Potential of Soil Carbon Sequestration in Costa Rica

Alfredo Alvarado

INTRODUCTION

Costa Rica has extraordinary soil variability in a very limited area, greatly enhancing vegetation diversity and thus possibilities for any kind of agricultural operations (Bertsch et al., 2000). The reason for this diversity is the highly variable parent material, a heterogeneous relief, and the action of a greatly variable climate and biota. Most of the soils are young, reflecting the relatively recent formation of the Costa Rican landmass. Some shallow volcanic soils in mountainous regions are less than 400 years old, formed from volcanic ashes deposited by recent eruptions, and others have developed A horizons from ashes deposited in 1963-1965. The age of the organic soils in the alluvial plains has been reported to be less than 3,400 years old. These are remarkably young soils by world standards, and their youth has important implications for their potential for human use and carbon sequestration.

The parent material of Costa Rica's soils is dominated by Quaternary alluvial sediments and igneous rocks, with some additional influence of Holocene volcanic ashes and limited regions of older (Cretaceous and Tertiary) sedimentary formations that include limestone. In general, the terrain is very irregular, so the main economic activities are carried out in the flatter narrow coastal plains and the intermountain Central Valley. Major crops produced for local consumption and exports are coffee, bananas, sugar cane, vegetables, fruits (pineapples, oranges, mangos, cantaloupes, etc.), and grasslands. The most significant change in land cover is due to deforestation to convert the land into pastures.

Precipitation, evapotranspiration, and temperature variations, associated principally with changes in altitude, have produced 13 life zones with dif-

Carbon Sequestration in Soils of Latin America

doi:10.1300/5755_08

ferent biotic conditions. These different climatic conditions have been more determinant of the soil types found in Costa Rica. Variations in precipitation patterns associated with the intertropical convergence zone, hurricanes, and El Niño events dominate regional differences in climate. The northern and central Pacific regions receive less than 1.8 m per year of rain and have a distinctive dry season which varies between three and six months. The dry season lasts less than three months in the rest of the country, and the rainfall varies from 1.8 to over 4.8 m per year. In some locations, precipitation can increase up to as much as 7 or 8 m per year at relatively high elevations, although it drops to lower values at the Paramo, which is above 3,200 m elevation. On climatic grounds, it has been established that only one-third of Central America shows a good distribution pattern (two to three dry months of less than 50 mm of rainfall each), while one-third of the area, located mainly on the Pacific side is affected by dry periods longer than four to five months per year and are highly unreliable, and one-third, mostly in the Caribbean side, is permanently wet (less than one dry month per year).

It is possible to find in Costa Rica all 12 major soil orders recognized by soil taxonomists except desert soils (Aridisols) and frost soils (Gelisols). Six of these ten have major agricultural relevance: Inceltisols (38.6 percent), Ultisols (21 percent), Andisols (14.4 percent), Entisols (12.4 percent), Alfisols (9.6 percent), and Vertisols (1.6 percent); the values in parentheses represent the relative area covered by each soil order. General characteristics and managements practices recommended to crop these soils have been presented by Bertsch et al. (2000).

Several studies have attempted to describe major soils and land uses in Central America, but many of them have been questioned on methodological grounds and data reliability (Celis and Alvarado, 1994). For instance, one study uses only climatic regions and actual yield as parameters to classify land-use potential; another study uses a seven-class classification system that has not been internationally accepted yet. Most studies use the eight-class system from the USDA, which is based on philosophical theses such as capital availability, labor endowment, annual row crops, normal distribution of topographic land classes, and others that do not correspond to the conditions of Central America. In terms of soil classification, the FAO/UNESCO (1976) study is the only comprehensive study available for the region. Based on findings of Higgins et al. (1984) and Beets (1990), the major soil-limiting constraints for plant growth (hence, SOM accumulation) in Central America are drought (22 percent), soil depth (20 percent), mineral deficiencies (11 percent), poor drainage (8 percent), sandy texture (3 percent), and presence of 2:1 clays (2 percent); 34 percent of the regions present minor soil constraints.

Carbon sequestration in Costa Rica is a recent research topic. The accumulation of C in soils has been estimated routinely in soil survey projects for years and only in recent times in environmental projects. The data used in this chapter seek to document the status of soil organic matter (SOM) as affected by their genesis and environmental factors. Also, a summary of relevant information on C accumulation and turnover in the soils of the country is presented.

At present various projects on carbon sequestration are under execution in Costa Rica at the Organization for Tropical Studies (www.fiu.edu/~carbono/), Centro Agronómico Tropical de Investigación y Enseñanza (imhibrahim@catie.ac.cr), and Tropical Science Center (vwatson@cct.or.cr). The findings of these projects can be consulted at the addresses mentioned.

SOIL ORGANIC MATTER AS AFFECTED BY LIFE ZONE

The effect of natural communities on the accumulation of C in the soil was recognized long ago by different authors (Heal et al., 1997), and more recently in Australia (Khanna et al., 2000) and Africa (Berhard and Loumeto, 2002). Data presented in Table 8.1 show that in Costa Rica, SOM content increases from the tropical (warm) to the montane (cool) belt, both in the soil and the subsoil. Also, as precipitation increases among vegetation communities, soil organic matter increases, although the effect of temperature is more acute than that of moisture.

Organic matter content in the soil is a primary function of temperature and vegetative protection from surface dehydration. In forested areas in tropical cool climates, the forces of decomposition are less active than in

TABLE 8.1. Soil organic matter content as related to life zones in Costa Rica.

	SOM %[a]				
Life zone	**Dry**	**Moist**	**Wet**	**Rain forest**	**Average**
Tropical	3.3/1.0 (7)	3.3/0.9 (7)	4.3/1.3 (13)		3.6/1.1 (27)
Premontane		6.3/2.8 (2)	6.9/2.20(6)	6.6/2.0(5)	6.6/2.3 (13)
L.Montane		4.9/1.6 (1)	19.1/4.9 (1)	20.4/6.7 (3)	14.8/4.4 (5)
Montane				19.8/ Rock (1)	19.8/ Rock (1)
Average	3.3/1.0 (7)	4.8/1.8 (10)	10.1/2.8 (20)	15.62/ 4.31 (9)	

Source: Adapted from Holdridge et al., 1971.
[a]SOM (0-0.30 m) / SOM (0.31-1.00 m) (Number of samples).

warmer tropical environments, and organic matter content is usually higher in the former. Under cool weather conditions organic matter residues are waxy and hard to decompose, enhancing their accumulation and favoring the tendency to form shallow organic soils (case of the montane rain forest where the O horizon lies over rock). Drainage conditions have some effect, since poorly drained areas tend to encourage soil organic matter buildup (Holdridge et al., 1971). The relationships among forest soil organic carbon (SOC) and biotic and abiotic variables are different for low-elevation (<120 m) and high-elevation (120 to 800 m) sites, and elevation explains much of this variability in Costa Rica, since it is related to temperature and changes in clay mineralogy composition in Andisols (Powers and Schlesinger, 2002a).

SOIL ORGANIC MATTER AS AFFECTED BY TYPE OF SOIL

Small areas of peat deposits (Histosols) have been found in Costa Rica as (1) thin blanket deposits about 1 m thick in the highlands of the Talamanca Range, (2) peat layers interbedded with alluvium layers around the Nicaragua Lake, and (3) thick layers of peat at the Parismina River Basin. The accumulation of organic materials is the result of tectonism and volcanism (Cohen et al., 1986), including the burial of large amounts of vegetation by volcanic debris, and low temperature in the highlands. Apart from the catastrophic events, above- and belowground factors that decrease or increase carbon pools in the soil are presented in Exhibit 8.1 (Lal and Kimble, 2000; Buurman et al., 2004).

The SOC content of 167 soil samples representing the different environments of Central America was analyzed by Díaz-Romeu et al. (1970) to include 32 soils from Guatemala, 25 from El Salvador, 28 from Honduras, 32 from Nicaragua, and 59 from Costa Rica. The majority belong to the Andosols (56), Litosols (21), Fluvisols (20), and Cambisols (32) groups. The average SOC content was 2.96 percent (range 0.4 to 12.0 percent); 57 percent of the samples had values higher than 5 percent (Figure 8.1), and significant correlations were found with rainfall, life zone (*sensu* Holdridge et al., 1971), and soil pH. Andosols and Cambisols showed the highest SOC levels, organic C diminishing considerably with soil depth.

A more recent study conducted in Costa Rica by Cabalceta (1993) to evaluate different methods for extracting available nutrients also included SOM determinations as a part of the soils characterization. In this study, 25 soils of the four major soil orders of the country were sampled, and results are presented in Table 8.2. The author reported of a sequence in SOM content of the

EXHIBIT 8.1.
Major Factors Affecting Carbon Pools in the Soil

Aboveground

Decrease carbon pools

- Maximize soil temperature, aeration, and denitrification
- Mechanized deforestation, burning
- Ploughing, row cropping, monoculture
- Low-input agriculture
- Accelerated erosion
- Nutrient depletion and soil degradation

Increase carbon pools

- Minimize soil temperature, aeration, and denitrification
- Manual land clearing, growing cover crops
- No-till, mixed cropping
- Science-based agriculture
- Erosion control measures
- Soil restoration and soil fertility enhancement

Belowground

Decrease carbon pools

- High nutrient status
- High-quality litter
- High oxygen availability

Increase carbon pools

- Low pH
- High Al saturation or allophone content
- Saturation with water
- Mineral-organic bonding
- Protection by occlusion in aggregates
- Low-quality litter

Source: After Lal and Kimble, 2000; Buurman et al., 2004.

A horizon of the soil in the order Vertisols $<$ Inceptisols $<$ Ultisols $<$ Andisols; it is noteworthy that the larger the average of SOM of a particular soil order, the larger the range of its SOC content. Although working with more soil orders than those encountered in Costa Rica, Velayutham et al. (2000) described a similar pattern of SOC accumulation in soil orders of India.

Under natural conditions, the sequestration of SOC in each soil order can be increased up to the maximum value of its range. However, unpublished results in organic farming systems show that in spite of the large amounts of compost or organic residues applied (10 to 30 $Mg{\cdot}ha^{-1}$), the total amount of

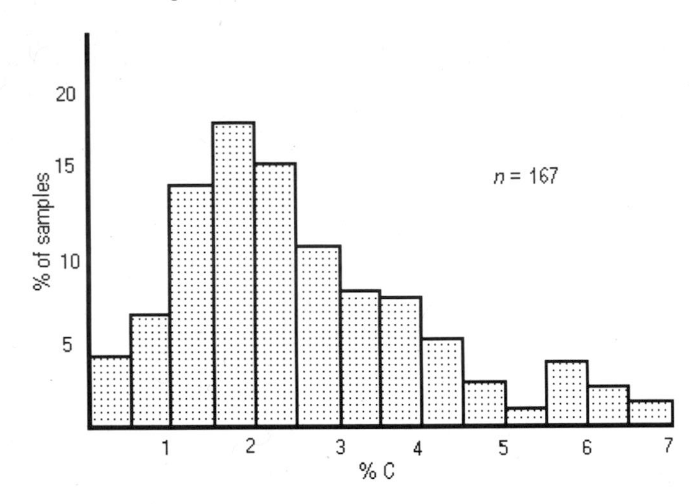

FIGURE 8.1. Frequency distribution of carbon content (A horizon) of some Central American soils. (*Source:* Díaz-Romeu et al., 1970. Used by permission.)

TABLE 8.2. Organic matter content in 100 soil samples of Vertisols, Inceptisols, Ultisols, and Andisols (25 A horizons of each order) from Costa Rica.

	SOM %		
Soil order	**Minimum**	**Maximum**	**Average**
Vertisols (n= 25)	1.6	5.9	3.5
Inceptisols (n= 25)	1.0	9.9	4.2
Ultisols (n= 25)	1.9	9.7	5.7
Andisols (n= 25)	4.8	24.0	10.9

Source: Adapted from Cabalceta, 1993.

C in soils rarely increases. According to Schlesinger (2000), only a small sink for SOC in soils may derive from the adoption of conservation tillage and the regrowth of native vegetation on abandoned agricultural land, but no net sink for SOC is likely to occur through application of manure to agricultural lands. These facts are relevant while considering the possibilities to sequester large amounts of SOC by different tropical soils.

TURNOVER OF SOIL ORGANIC MATTER ABOVEGROUND

According to Soto et al. (2002), several equations have being developed to describe aboveground turnover of residues:

$$Y = \Theta_0 e^{-kx} + \varepsilon \text{ (simple exponential)} \tag{8.1}$$

$$Y = \Theta_0 e^{-k1x} + (\Theta_1 - \Theta_0) e^{-k2x} + \varepsilon \text{ (double exponential)} \tag{8.2}$$

$$Y = \Theta_0 + (\Theta_1 - \Theta_0) e^{-kx} + \varepsilon \text{ (asymptotic)} \tag{8.3}$$

Sauerbeck and González (1977) study on the decomposition of ^{14}C-labeled wheat *(Triticum aestivum)* straw in 12 representative soils of Costa Rica showed that, under field conditions, after one year, 23 to 36 percent of the ^{14}C added in the wheat straw remained in the soil; however, four years later the residual ^{14}C ranged from 11 to 23 percent. The asymptotic model best fitted the turnover of the residues (Figure 8.2).

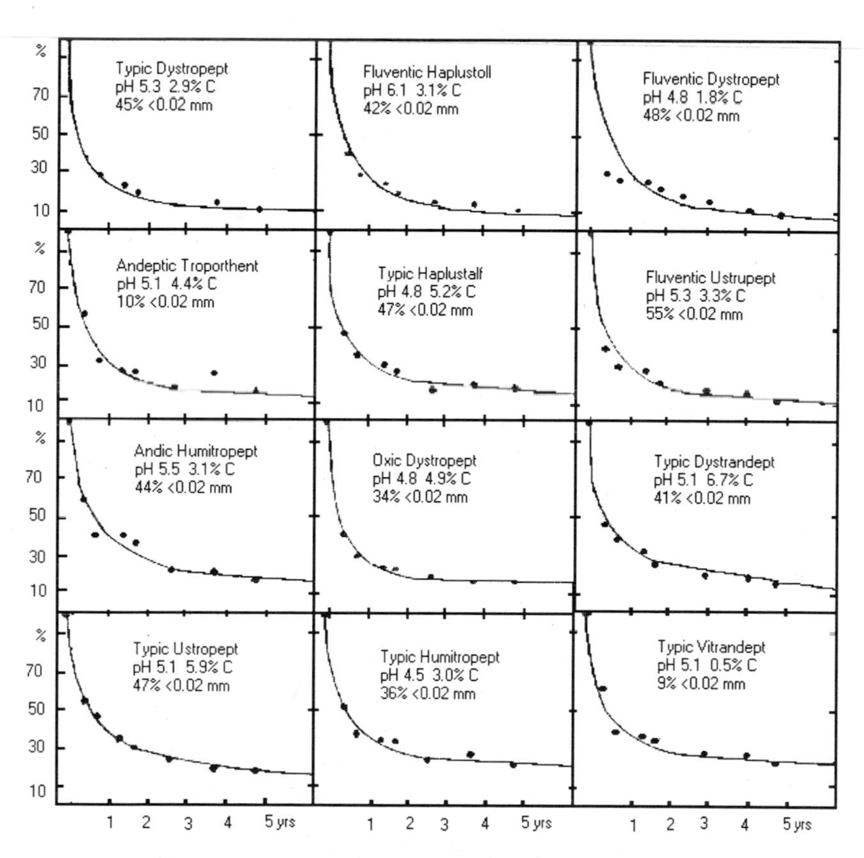

FIGURE 8.2. Decomposition of ^{14}C-labeled plant residues in different soils and climates of Costa Rica. (*Source:* Sauerbeck and González, 1977.)

Similar results were obtained while describing the turnover of banana (*Mussa* spp.) (Vargas and Flores, 1995) and peach palm *(Bactris gasipaes)* fresh residues under field conditions (Soto et al., 2002), using decomposition litterbags (Figure 8.3). In these studies, the fruits were harvested on a weekly basis; thus, the residues were applied after each harvest. A linear model that relates different stages of decomposed residues, such as the one proposed in Heal et al. (1997), is more appropriate. Other decomposition studies using rain forest species residues show similar results (Babbar and Ewell, 1989; Byard et al., 1996).

TURNOVER OF SOIL ORGANIC MATTER BELOWGROUND

According to Buurman et al. (2004), in all cases, the carbon pool in the soil depends on litter input and the rate of decay, as

Pool size = F_1 $(1 - k_f) / k_h$ in which

Pool size in mass·(surface unit)$^{-1}$
F_1 = litter deposition [mass·(surface unit)$^{-1}$ per year]

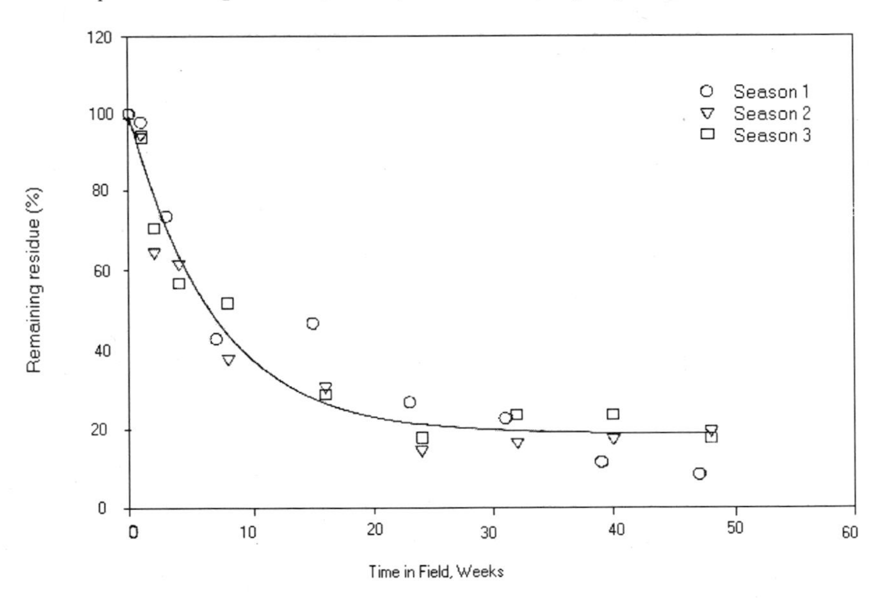

FIGURE 8.3. Percent of initial peach-palm foliage dry matter remaining in litterbags as a function of time placement in the field [$Y = 19.11 + (100 - 19.11)e^{-0.1472x}$] for each of the three seasons investigated. (*Source:* Soto et al., 2002. Used with permission.)

k_f = fraction of litter decomposed per year
k_h = fraction of humus decayed per year

This simple expression is valid for humus fractions with different turnover times or mean residence time: MRT = $1/k_h$.

Land management practices such as fire, grazing, tillage, and fertilizer application, among others, affect the distribution of SOC (Townsend et al., 2002). The accumulation of SOC underground is strongly related to (1) fine roots decomposing naturally (humification) into the soil, (2) self-pruning of roots during dry season in deciduous species (Ordóñez, 2003), (3) root chopping by ploughing the land (Veldkamp, 1994), (4) root decay after pruning the crop (coffee), and (5) decay of microorganism biomass.

A special case of C buildup in the soil is that of the so-called black carbon. This fraction results from burning the land for cropping and common savannah and forest fires that may convert up to 2 percent of the standing biomass into char/soot; char production upon burning of deciduous wood is higher. This fraction lasts for long periods of time in soils (Exhibit 8.2).

SOIL ORGANIC MATTER AND ECOSYSTEM MANAGEMENT

After deforestation, land-use changes had a profound influence on SOC and other soil properties (i.e., bulk density), since both residue addition and residue decomposition rates are affected (Ewell et al., 1981; Raich, 1983; Veldkamp, 1994; Johnson and Wedin, 1997; Guggenberger and Zech, 1999; Powers, 2001, 2004; Powers and Schlesinger, 2002a,b).

Slash and burning of an eight- to nine-year-old evergreen forest around Turrialba, Costa Rica, volatilized 31 percent of the initial amount of SOC,

EXHIBIT 8.2.
Age of Black Carbon Found in Soils of Some Parts of the World

Place	Age yr B.P.
North Queensland, Australia	27,000-12,000
Borneo	17,00-350
Amazon, Rio Negro	6,000-250
La Selva, Costa Rica	2,430
Chiripó, Costa Rica	1,180-1,100

Source: After Horn and Sanford, 1992; Buurman et al., 2004.

22 percent of N, and 49 percent of S. Soil CO_2 evolution was greater from beneath 11-week-old slash (3.6 $g \cdot m^{-2}$ daily of C) than from beneath the evergreen forest (2.5 $g \cdot m^{-2}$), probably because slashing conserved soil moisture better than did actively transpiring forest (Ewell et al., 1981).

According to Veldkamp (1994) deforestation followed by 25 years of pasture caused a net loss of 21.8 $Mg \cdot ha^{-1}$ of SOC for Eutric Hapludands and 1.5 $Mg \cdot ha^{-1}$ for Oxic Humitropets. In Andic Humitropepts the author found that the decomposition of tree roots caused an extra input of SOC during the first year after deforestation and a strong stabilization of SOC by forming Al-organic matter complexes.

To determine how the conversion in a tropical premontane wet forest and a nearby secondary forest affects SOC budget, major soil C storages, inputs, and CO_2 evolution from a tropical Inceptisol were measured by Raich (1983) over a six-month period. Total C storage in and on the mature forest soil comprised of 9,330 g $C \cdot m^{-2}$ in SOM, 1850 g $C \cdot m^{-2}$ in litter, and 340 g $C \cdot m^{-2}$ in small roots (diameters 5 mm); larger roots were not measured. Average daily inputs to the mature forest soil included 1.3 g $C \cdot m^{-2}$ in litterfall and 0.10 g dissolved organic carbon (DOC)$\cdot m^{-2}$ in precipitation (troughfall + stem flow). The evolution of CO_2 from the mature forest averaged 3.4 g $C \cdot m^{-2}$ per day, or 2.6 times the average rate of litterfall. Total C storage in and on the secondary growth soil was composed of 8600 g $C \cdot m^{-2}$ in SOM, 700 g $C \cdot m^{-2}$ in litter, and 157 g $C \cdot m^{-2}$ in small roots, or 2060 g $C \cdot m^{-2}$ less than in the mature forest. Average daily inputs to the mature forest soil included 0.7 g $C \cdot m^{-2}$ in litterfall and 0.12 g $DOC \cdot m^{-2}$ in precipitation (troughfall + stem flow). The evolution of CO_2 from the mature forest averaged 4.6 g $C \cdot m^{-2}$ per day, or 1.4 times the average rate in the mature forest. Measured inputs of C to the soil were considerably less than soil-CO_2 evolution rates in both sites. Johnson and Wedin (1997) reported a loss of SOM while comparing mature forest and grassland soils, an effect attributed to a larger rate of mineralization of residues due to higher temperatures in the latter ecosystem.

Particle-size separation of SOM, where particulate SOM (light fraction and sand-associated SOM) is separated from mineral-bond SOM (silt- and clay-associated SOM), revealed that under agricultural use of a soil formerly under primary forest a depletion of the particulate SOM occurred, whereas clay- and silt-bond SOM was less affected (Guggenberger and Zech, 1999). These authors also observed that abandonment of the pasture and regrowth of a secondary forest raised the SOC content in all separates to a precultivation level within 18 years, and sand-associated SOC was even higher as compared to that in the primary forest. The results suggest that land use primarily influences the balance across the light fraction and the size separates, with the particulate SOM pool being the most significant SOM component in the context of management impacts on these soils.

Powers (2001, 2004), examined changes in total SOC (0.30 cm depth) that accompanied different land-use transitions in northeastern Costa Rica, representing 12 conversions of primary forest to banana plantations, 15 conversions of pastures to cash crops, and four conversions of pastures to *Vochysia guatemalensis,* finding a decrease in SOC concentration and inventories (Mg C·ha^{-1}). In the conversion to bananas, the surface soil decreased SOC concentration and inventories by 37 and 16.5 percent, respectively. Similar results were obtained while converting pastures to crops, and soils under *V. guatemalensis* did not appear to increase SOC storage, at least over the first decade. Reduced productivity of C inputs to the soil is one mechanism that may explain losses of SOC pools with land-use changes in ecosystems that lose SOC rapidly due to cultivation, whereas the restoration rates are slow.

In a study to determine soil changes associated with the conversion of grasslands to 13-year-old, and 17-year-old secondary forests in tropical dry forest on Andic and Typic Haplusteps, Alfaro et al. (2001) found that plant cover did not affect SOM, pH, soil acidity, and K content; however, Ca and Mg contents were higher under the 13-year-old forest cover (nutrients associated to litter added to 7.03 percent) than the 17-year-old forest (nutrients associated to litter added to 4.51 percent).

CONSIDERATIONS ABOUT SOIL ORGANIC MATTER DETERMINATION

The efficiency of oxidation of organic carbon by the Walkley-Black method was studied by Bornemisza et al. (1979) by comparing determinations made using dry combustion equipment and gravimetric determination of the evolved CO_2. In this study, 52 surface soils (0 to 30 cm) were studied in groups of 10, each belonging to one of the following taxonomic suborders: Fluvents, Andepts, Tropepts, Udults, and Ustults. Samples from 17 Andept profiles were also analyzed to study changes in oxidation efficiency with depth. The results showed that 75 percent recovery of C was attained in the A horizons analyzed, and 86 percent of the studied samples deviated by <2 percent from this value. The practical application of this finding is that when organic matter is expressed as such (multiplying organic C × 1.724 and assuming a 78 percent C recovery), an overestimation of the amount of C in the subsoils occurs (Table 8.3).

The recovery of C in the deeper layers of Andisols is higher than that of the A horizons (Table 8.4), a fact related to the lower presence of plant resi-

TABLE 8.3. Organic carbon recovery in Andisols of Costa Rica.

Location and elevation	Horizon	Depth (m)	C organic, %		
			Dry combustion	Wet combustion	Recovery
Entic Dystrandept					
El Empalme (2,500 masl)	A1	0-0.15	5.3	4.0	74.9
	A2	0.15-0.30	10.1	8.7	86.4
	B1	0.30-0.45	5.7	4.8	83.7
	B2	0.45-0.95	17.6	16.8	95.3
	C	>0.95	5.3	4.4	82.7
Typic Dystrandept					
Agua Buena (1,200 masl)	A1	0-0.35	12.7	9.8	76.9
	2AB	0.35-0.60	8.2	7.2	87.4
	2B1	0.60-1.00	5.1	4.5	87.8
	2B2	1.00-1.45	3.4	3.3	97.9
	2B3	1.45-2.05	3.0	2.5	82.0
Typic Hydrandept					
Guayabo (900 masl)	A11	0-0.40	11.1	7.9	71.1
	A12	0.40-0.80	8.9	6.1	69.0
	AB	0.80-0.90	6.2	5.6	90.0
	2B2	0.90-1.30	2.6	2.4	92.2
	2C	>1.30	1.3	1.3	95.5

Source: Based on data from Bornemisza et al., 1979.

dues with depth and to an accumulation of fulvic acids in these layers (Alvarado, 1974). Van Dam et al. (1997) observed that decomposition rates decrease strongly with depth and that the diffusional transport alone was insufficient for the simulation of SOC movement into the soil; it had to be augmented by depth-dependent decomposition rates to explain the dynamics of SOC, delta ^{13}C, and delta ^{14}C. Cleveland et al. (2004) concluded that with regard to the presence of physicochemical reactions with soil surfaces, humic (hydrophobic) fractions of dissolved organic matter (DOM) become more abundant than nonhumic (hydrophilic) fractions over time. The latter fraction is the one that migrates into deeper layers of soil profiles. Both Cleveland et al. (2004) and Powers and Schlesinger (2002b) reported that neither the changes in delta ^{13}C isotopic fraction during DOM uptake by

TABLE 8.4. Soil organic fractions of two Typic Dystrandepts in a toposecuence of the Irazu Volcano.

Horizon	Depth (m)	SOM (%)	C-humic (Ch) (%)	C-fulvic (Cf) (%)	Ch/Cf
Profile 1					
A11	0-0.25	15.8	2.1	0.5	4.2
A12	0.25-0.50	8.8	1.3	0.3	4.3
A13	0.50-0.90	7.9	1.2	0.2	6.0
B1	0.90-1.30	2.0	0.1	0.2	0.5
B2	1.30-2.00	1.5	0.1	0.2	0.5
B3	2.00-2.50	0.9	0.1	0.1	1.0
Profile 2					
A11	0-0.24	17.4	2.8	1.1	2.6
12	0.24-0.37	10.8	2.0	0.7	2.9
A13	0.37-0.60	7.4	1.6	0.5	3.1
2A14	0.60-0.90	9.2	1.4	0.5	2.7
2B1	0.90-1.15	5.5	0.7	0.3	2.3
3C	1.15-1.25	0.2	0.1	0.1	1.0
4C	1.25-2.00	1.6	0.1	0.2	0.5

Source: Alvarado, 1974.

soil organisms nor the difference in composition of litter and roots explained the variation of delta ^{13}C values with soil depth.

One factor standing in the way of proving that silvopastoral systems contribute to mitigating the greenhouse effect is the generally inadequate calculation of SOC stocks. Comparisons between land-use systems must be based on soil depth (at least one meter) and the soil mass rather than on soil volume. The data in Table 8.5 show the relevance of bulk density (BD) to interpret the gravimetric percent of C. The relevance of BD to convert percent C to mass of $C \cdot m^{-3}$ has been pointed out by Forsythe and Díaz-Romeu (1969). This applies to all soil chemical analysis expressed as a gravimetric percent on an oven-dry basis.

SOIL ORGANIC MATTER STOCK IN COSTA RICA

Soil organic matter in Costa Rica is estimated by using two different databases. The first approach is done by using data from Holdridge et al. (1971), considering only the profiles in and the area of each life zone in the country (a total of 51 profile descriptions); the second approach is done by

TABLE 8.5. Carbon stocks based on standard depth section and on fixed soil mass.

Land use	Depth (m)	Thickness (dm)	BD ($kg \cdot dm^{-3}$)	C (%)	kg C (in 1 m^3)	Bulk weight ($kg \cdot m^{-2}$)	Equal weight (kg)	kg C in 860 kg
Native forest	0-0.10	1	0.64	6.56	4.20	64		4.20
	0.10-0.20	1	0.84	5.57	4.68	84		4.68
	0.20-0.40	2	0.94	4.49	8.44	188		8.44
	0.40-1.00	6	0.90	1.75	9.45	540		9.45
Total					26.77	876	876	26.77
Degraded pasture	0-0.10	1	0.89	4.71	4.19	89	89	4.19
	0.10-0.20	1	1.02	3.58	3.65	102	102	3.65
	0.20-0.40	2	1.01	2.22	4.48	202	202	4.48
	0.40-1.00	6	0.90	1.13	6.10	540	483	5.46
Total					18.43	933	876	17.79
Improved pasture	0-0.10	1	0.92	4.63	4.26	92	92	4.26
	0.10-0.20	1	1.15	3.27	3.76	115	115	3.76
	0.20-0.40	2	1.26	2.23	5.62	252	252	5.62
	0.40-1.00	6	1.16	1.11	7.73	696	417	4.63
Total					21.37	1155	876	18.27

Source: Based on data from Buurman et al., 2004.

collecting available profile descriptions by soil order (111 profiles in total) with reliable information and the consideration of area covered by each soil order in the country.

In both cases, it is assumed that (1) each life zone and soil order is homogeneous (no presence of other vegetation or soil impurities), (2) climatic changes during the past 50 years are "irrelevant," and (3) all aspects of soil management previously mentioned are combined in the soil order approach. In any case, consider 1 m^2 area of soil surface (10^4 cm^2) and multiply by bulk density ($g \cdot cm^{-3}$) × organic matter ($g \cdot g^{-1}$) × horizon thickness (cm) × 10^4 (m^2 in a hectare) to obtain g of OM per horizon per ha. The addition of all values to a depth of 0.3 or 1 m provides information for the two soil layers. The last computation was done to calculate SOM to 1 m depth per ha by regressing the values found for the two depths. Summarized data are presented in Tables 8.6 and 8.7.

On the basis of life zones (Table 8.6), a total of 1348.2 Tg of SOM was calculated for the country. As mentioned previously, SOM increases with decreasing ambient temperature and increasing the total amount of rainfall of life zones. However, the contribution of SOM of each life zone to SOM in

TABLE 8.6. Soil organic matter stock in Costa Rica calculated by life zones.

Life Zone (LZ)	Area (ha)	Mg SOM to a depth of (m)		SOM / LZ (Tg)	Sample no.	Regression equation	Regression coefficient
		x = 0.3	y = 1				
Dry forest	150328	82.7	142.9	21.5	7	$y = 1.2697x + 37.852$	0.955
Moist forest	1058193	119.9	192.0	203.2	7	$y = 1.4993x + 49.034$	0.5945
Wet forest	1083652	86.9	144.4	156.5	13	$y = 1.1008x + 4.7945$	0.9127
Premontane moist	556748	193.5	310.5	172.9	2	nd	nd
Premontane wet	1217681	191.6	321.1	391.0	6	$y = 2.6939x - 59.582$	0.9598
Premontane rain	445335	86.1	291.2	129.7	5	$y = 6.283x - 249.47$	0.9803
Lower montane rain	137621	309.6	586.2	80.7	3	$y = 1.7414x + 220.37$	0.9994
Montane moist	335478	188.2	342.0	114.7	1	nd	nd
Montane wet	1872	228.3	420.8	0.8	1	nd	nd
Montane rain	118678	244.3	651.6	77.3	6	$y = 1.4338x - 24.441$	0.9372
Costa Rica	5105586			1348.2	51		

TABLE 8.7. Soil organic matter in Costa Rica calculated by soil orders.

Soil order (SO)	Area (ha)	Mg SOM to a depth of (m)		SOM / SO (Tg)	Sample No.	Regression equation	Regression coefficient
		x = 0.3	y = 1				
Inceptisols	1975940	127.5	249.4	492.8	27	y = 2.1565 x – 17.388	0.8411
Ultisols	1068943	138.4	244.6	261.5	15	y = 2.0729 x – 42.378	0.7764
Andisols	749852	222.0	401.8	301.3	14	y = 1.8351 x – 5.6576	0.9133
Entisols	626940	116.7	286.1	179.4	8	y = 1.3424 x + 129.44	0.5315
Alfisols	487227	156.7	260.7	127.0	8	y = 1.4003 x + 41.212	0.8051
Vertisols	78446	111.2	223.8	17.6	20	y = 1.6993 x + 34.82	0.7232
Mollisols	68972	144.0	269.2	18.6	16	y = 1.8974 x – 4.1355	0.8006
Histosols	49266	362.6	967.3	47.7	3	y = 8.2793 x – 2035.1	0.9619
Costa Rica	5105586			1445.7	111		

the country depends on the area of each zone. The life zones that contributed the most to SOM in Costa Rica were premontane wet (391.0 Tg), tropical moist forest (203.2 Tg), and premontane moist (172.9 Tg), respectively. The correlation values between SOM at 0.3 and 1.0 m depth, where data allowed calculations, were reasonably good; these equations can be used to estimate SOM to a 1.0 m depth with lesser cost for analysis.

Soil organic matter value for the country calculated by soil orders came to a total of 1445.7 Tg (Table 8.7), larger than the value found by the life zone approach, but within reasonable assumptions. As mentioned earlier, SOM depends on the soil order, reflecting their genesis (Entisols < Ultisols = Inceptisols < Mollisols = Alfisols = Entisols < Andisols < Histosols). However, the soil order that contributed the most due to area covered were Inceptisols (492.8 Tg), Andisosls (301.3 Tg), and Ultisols (261.5 Tg), respectively. In this case, the correlations between SOM at 0.3 m versus 1.0 m depth for each order were very significant, with the exception of the Entisols order.

REFERENCES

Alfaro, E.A., A. Alvarado, and A. Chaverri. 2001. Cambios edáficos asociados a tres etapas sucesionales de bosque tropical seco en Guanacaste, Costa Rica. *Agron. Costar.* 25(1):7-20.

Alvarado, A. 1974. A volcanic ash soil toposequence in Costa Rica, Central America. Thesis M.Sc., North Carolina State University, Raleigh, North Carolina.

Babbar, L.I. and J.J. Ewell. 1989. Descomposición del follaje en diversos ecosistemas sucesionales tropicales. *Biotropica* 21(1):20-29.

Beets, W.C. 1990. *Raising and sustaining productivity of smallholder farming systems in the tropics.* Agbé Publishing, Alkmaar, Holland.

Berhard, F. and J.J. Loumeto. 2002. The litter systems in African forest-tree plantations. In M.V. Reddy (ed.), *Management of tropical plantation-forest and their soil-litter system.* Science Publisher, Inc., Enfield, New Hampshire, pp. 11-39.

Bertsch, F., A. Alvarado, C. Henríquez, and R. Mata. 2000. Properties, geographic distribution, and major soil orders of Costa Rica. In C.A.S. Hall (ed.), *Quantifying sustainable development.* Academic Press, San Diego, California, pp. 265-294.

Bornemisza, E., M. Constenla, A. Alvarado, E.J. Ortega, and A.J. Vázquez. 1979. Organic carbon determination by the Walkley-Black and dry combustion methods in surface soils and Andept profiles from Costa Rica. *Soil Sc. Soc. Am. Proc.* 43(1):78-83.

Buurman, P., M. Ibrahim, and M.C. Amézquita. 2004. Mitigation of greenhouse gas emissions by tropical silvopastoral systems: Optimism and facts. In L. Mannetje, M. Ramírez, C. Ibrahim, J. Sandoval, M. Ojeda, and J. Ku (eds), *The importance of silvopastoral systems for providing ecosystems services and rural livehoods*. Mérida, México, pp. 61-72.

Byard, R., K.C. Lewis, and F. Montagnini. 1996. Leaf litter decomposition and mulch performance from mixed and monospecific plantations of native tree species in Costa Rica. *Agriculture, Ecosystems & Environment* 58:145-155.

Cabalceta, G. 1993. Niveles críticos de fósforo, azufre y correlación de soluciones extractoras en Ultisoles, Vertisoles, Inceptisoles y Andisoles de Costa Rica. Thesis Mg. Sc. Universidad de Costa Rica, San José, Costa Rica.

Celis, R. and A. Alvarado. 1994. Land taxation for sustainable development in Central America and the role of soil and social scientists. In Transactions of the 15th World Congress of Soil Science, Acapulco, México, Volume 9: Supplement, pp. 104-106.

Cleveland, C.C., J.C. Neff, A.R. Townsend, and E. Hood. 2004. Composition, dynamics, and fate of leached dissolved organic matter in terrestrial ecosystems: Results from a decomposition experiment. *Ecosystems* 7:275-285.

Cohen, A.D., R. Raymond, S. Mora, A. Alvarado, and L. Malavassi. 1986. Características geológicas de los depósitos de turba en Costa Rica (preliminary report). *Rev. Geol. Amer. Central* 4:47-67.

Díaz-Romeu, R., F. Balerdi, and H.W. Fassbender. 1970. Contenido de materia orgánica en suelos de América Central. *Turrialba* 20(2):185-192.

Ewell, J.J., C.W. Berish, B.J. Brown, N. Price, and J.W. Raich. 1981. Slash and burn impacts on a Costa Rican wet forest site. *Ecology* 62(3):816-829.

FAO/UNESCO. 1976. *Mapa mundial de suelos,* Vol. 3, *México y Centro América*. Place de la Fontenoy, París, France.

Forsythe, W. and R. Díaz-Romeu. 1969. La densidad aparente del suelo y la interpretación del análisis de laboratorio para el campo. *Turrialba* 19(1):128-131.

Guggenberger, G. and W. Zech. 1999. Soil organic matter composition under primary forest, pasture, and secondary succession, Region Huetar Norte, Costa Rica. *Forest Ecology and Management* 124(1):93-104.

Heal, O.W., J.M. Anderson, and M.J. Swaift. 1997. Plant litter quality and decomposition: An historical overview. In G. Cadish and K.E. Giller (eds.), *Driven by nature: Plant litter quality and decomposition*. CAB International. Oxon, United Kingdom, pp. 3-29.

Higgins, C.M., A.H. Kassam, L. Naiken, G. Fisher, and M.M. Shah. 1984. Capacidades potenciales de carga demográfica de las tierras del mundo en desarrollo. Informe Técnico. INT/75/ P13. FAO, Rome.

Holdridge, L.R., W.C. Grenke, W.H. Hatheway, T. Liang, and J.A. Tosi. 1971. *Forest environments in tropical life zones: A pilot study*. Pergamon Press, Great Britain.

Horn, S.P. and R.L. Sanford. 1992. Holocene fires in Costa Rica. *Biotropica* 24(3): 354-361.

Johnson, N.C. and D.A. Wedin. 1997. Soil carbon, nutrients, and mycorrizhae during conversion of dry tropical forest to grassland. *Ecogical Applications* 7(1):171-182.

Khanna, P.K., P. Snowdon, and J. Bauhus. 2000. Carbon pools in forest ecosystems of Australia and Oceania. In R. Lal, J.M. Kimble, and B.A. Stewart (eds.), *Global climate change and tropical ecosystems*. CRC Press, Boca Raton, Florida, pp. 51-70.

Lal R. and J.M. Kimble. 2000.Tropical ecosystems and the global C cycle. In R. Lal, J.M. Kimble, and B.A. Stewart (eds.), *Global climate change and tropical ecosystems*. CRC Press, Boca Raton, Florida, pp. 3-32.

Ordóñez, H. 2003. Fenología de la copa y de las raíces finas de *Simarouba glauca* y *Dalbergia retusa* (cocobolo) con riego en una plantación mixta de Guanacaste. Thesis Mag. Sci. Universidad de Costa Rica, San Pedro de Montes de Oca, Costa Rica.

Powers, J.S. 2001. Geographic variation in soil organic carbon dynamics following land-use change in Costa Rica. Thesis PhD. Duke University, Graduate School and Management of Biology, Durham, North Carolina.

Powers, J.S. 2004. Changes in soil carbon and nitrogen after contrasting land-use transitions in northeastern Costa Rica. *Ecosystems* 7(2):134-146.

Powers, J.S. and W.H. Schlensinger. 2002a. Geographic and vertical patterns of stable carbon isotopes in tropical rain forest soils of Costa Rica. *Geoderma* 109(1-2):140-160.

Powers, J.S. and W.H. Schlensinger. 2002b. Relationships among soil carbon distributions and biophysical factors at nested spatial scales in rain forest of northern Costa Rica. *Geoderma* 109(3-4):165-190.

Raich, J.W. 1983. Effects of forest conversion on the carbon budget of a tropical soil. *Biotropica* 15(3):177-184.

Sauerbeck, D.R and M.A. González. 1977. Field decomposition of carbon-14-labelled plant residues in various soils of the Federal Republic of Germany and Costa Rica. In IAEA and FAO (eds.), in cooperation with Agrochimica, Vol. 1, Proc. Symp. Braunschweig, pp. 159-170.

Schlesinger, W.H. 2000. Carbon sequestration in soils: Some cautions amidst optimism. *Agriculture, Ecosystems and Environment* 82:121-127.

Soto, G., P. Luna, M. Wagger, T.J. Smyth, and A. Alvarado. 2002. Descomposición de residuos de cosecha en plantaciones de palmito en Costa Rica. *Agron. Costar.* 26(2):43-52.

Townsend, A.R., G.P. Asier, C.C. Cleveland, M.E. Lefer, and M.C. Bustamante. 2002. Unexpected changes in soil phosphorus dynamics along pasture chronosequences in the humid tropics. *J. Geophysical Res.* 107(34):1-9.

Van Dam, D., E. Veldkamp, and N. van Breemen. 1997. Soil organic carbon dynamics: Variability with depth in forested and deforested soils under pasture in Costa Rica. *Biogeochemistry* 39(3):343-375.

Vargas, R. and C.L. Flores. 1995. Retribución nutricional de los residuos de hojas, venas de hojas, pseudotallo y pinzote de banano (*Mussa* AAA) en fincas de diferentes edades de cultivo. *Revista CORBANA* 20(44):33-47.

Velayutham, M., D.K. Pal, and T. Bhattacharyya. 2000. Organic carbon stock in soils of India. In R. Lal, J.M. Kimble and B.A. Stewart (eds.), *Global climate change and tropical ecosystems*. CRC Press, Boca Raton, Florida, pp. 71-95.

Veldkamp, E. 1994. Organic carbon turnover in three tropical soils under pasture after deforestation. *Soil Sc. Soc. Am. J.* 58(1):175-180.

Chapter 9

Above- and Belowground Carbon Sequestration Under Various Land-Use Systems and Soil Types in Costa Rica

J. J. Jiménez
R. Lal

INTRODUCTION

The soil C pool is 2,400 Pg (1 Pg = 10^{15} g = 1 billion Mg), four and three times higher than the vegetation and atmospheric pools, respectively (Eswaran et al., 1993; Lal, 2003). The soil organic carbon (SOC) pool represents the largest reservoir in interaction with the atmosphere and is estimated at about 1,500 Pg C to 1 m depth and about 2,456 Pg C to 2 m depth. The soil inorganic C (SIC) pool comprises about 950 Pg and occurs in more stable forms. Vegetation (560 Pg) and the atmosphere (760 Pg) store considerably less C than soils do (Lal, 2004).

The development of agriculture has contributed to increasing CO_2 concentrations in the atmosphere, but now the combustion of fossil fuel by industry and transport sectors (6.5 Pg per year) represents the main source of CO_2 emission. Soil disturbance by plowing and other extractive practices leads to depletion of one-half to two-thirds of the SOC pool within the first five to ten years of deforestation in the tropics (Lal, 2003). At a global scale, it has been estimated that nearly 16 percent of the original SOC pool has been depleted (Rozanov et al., 1990). At present, while deforestation in many tropical areas leads to C emissions at the rate of about 1.5 Pg C per year, elsewhere around 1.8 to 2 Pg C per year is accumulating within the terrestrial ecosystems (IPCC, 2001). It has been estimated that 15 to 17 percent of the C emissions comes from SOC oxidation (Houghton and Hackler, 1994).

When natural ecosystems are converted into agricultural systems, the SOC is depleted by as much as 50 percent in temperate and 75 percent in

Carbon Sequestration in Soils of Latin America

doi:10.1300/5755_09

tropical regions (Lal, 2004). When undisturbed soil is converted to an agricultural use, the associated biological and physical processes result in a release of 20 to 50 percent of the SOC over a short period, with the amount varying by soil type, agricultural practices, and site-specific conditions (Rasmussen and Parton, 1994). This depletion of the C pool creates the potential to sequester C in soils. Because a substantial proportion of SOC originally present in soils has been released due to human land use, a large technical potential exists to sequester C in soils (Lal et al., 1998).

The CO_2 from the atmosphere can be "sequestered," accumulated, or stored in the soil as soil organic matter (SOM). The term *C sequestration* implies transferring CO_2 from the atmosphere into a long-lived pool. This means that SOC and SIC pools increase through certain land-use or recommended management practices (RMPs) (Lal, 2004). Adopting RMPs is an important strategy for storing C in soil under pastures, for example, where the highest rates of C accumulation may occur (Fisher et al., 1994). However, the rate of C sequestration may differ according to the composition of plant communities used in pastoral systems. The capacity for C sequestration in agricultural soils at a global scale has been estimated at 20 to 30 Pg C over the next 50 to 100 years (Paustian et al., 1997).

SOM has a very complex and heterogeneous composition, and is generally mixed or associated with mineral soil constituents. The traditional separation of SOM into fulvic and humic components does not separate fractions with different turnover rates (Balesdent, 1996). Physical separation methods allow separation of kinetically meaningful fractions (Feller, 1979; Balesdent, 1996), and among these fractions particulate OM (POM) or the light fraction (LF) is very sensitive to changes in land use (Cambardella, 1998).

The SOC pool in natural soils represents a dynamic balance between the input of dead plant material and loss from decomposition (mineralization), erosion, and leaching. The SOC is transformed biologically by the action of bacteria and is stabilized in each of the clay- or silt-sized organo-mineral complexes. Sand-associated SOM is important in short-term turnover, clay-bound SOM dominates the medium-term turnover, whereas silt-bound SOM participates in longer-term turnover. Fungi and bacteria govern most of the transformation and ensuing long-term transformations of organic C in soils (Hunt et al., 1989), although the burrowing and mixing activities of large soil invertebrates (termites, earthworms, and ants) may determine the activities of soil microorganisms in a set of hierarchical scales and functional domains (Lavelle and Spain, 2001).

Adoption of RMPs may increase C in soil by creating or expanding sinks (Fisher et al., 1994; Lal et al., 1998). This may be achieved through a variety of changes in land-use and management practices. Management practices that can increase soil C sequestration include land retirement (conversion to native vegetation or reversion to wetlands), afforestation, residue management, less-intensive tillage or minimum tillage, changes in crop rotations, conversion of cropland to pasture, especially in the case of deep-rooted grasses, and restoration of degraded soils.

The global potential of biological mitigation has been estimated at 100 Pg C (cumulative) by 2050, equivalent to about 10 to 20 percent of projected fuel emissions (IPCC, 2001). Lal et al. (1998; Lal, 2003) estimated that of the total 288 Tg C per year, 49 percent of C sequestration potential of the soils of the U.S. cropland can be achieved by adopting conservation tillage and residue management, 25 percent by changing cropping practices, 13 percent by land restoration, 7 percent through land-use change, and 6 percent by better water management.

Although some changes in land use and management practices can increase the SOC pool, the new equilibrium state depends on management and the biophysical conditions of the site. As the soil C pool increases, the rate of C sequestration decreases, and the soil's potential to become a future emission source increases because subsequent alteration of the management regime (for example, reversion to conventional tillage after the use of reduced tillage) can lead to mineralization of SOC and emission of CO_2.

In relation to the three main mitigation strategies identified by the Intergovernmental Panel on Climate Change (IPCC) to reduce emission of CO_2 into the atmosphere, this report assesses the potential of soil C sequestration under different land-use systems for principal soil types and agroecological zones of Costa Rica.

COSTA RICA: OVERVIEW

Costa Rica (CR) has a total land area of 5.1 million ha (Mha), and half of the land (2.9 Mha) is allocated to agricultural and livestock production.

Land use	Area (10^3 ha)
Arable land	225
Permanent crops	300
Permanent pastures	2,340

Total population of CR was 4.3 million in 2005, of which 61 percent is urban. By 2010, the urban population will increase by 18 percent, whereas rural population will remain stable (FAO, 2005). Population density is 78.7 hab per km^2; 31 percent is under 15 years old, and only 5 percent is over age 65. Life expectancy at birth is 76 and 70 years for females and males, respectively (WRI, 2002). The increase in the urban population may have some immediate consequences such as an increase in deforestation rate. Urban poverty is around 20 percent, and the number of people living on less than $1 a day is about 13 percent of the total population (WRI, 2002).

The economy of Costa Rica is based on tourism and agriculture (coffee, bananas, and other items such as sugar, meat, dairy products, and cocoa). The agricultural sector contributes about 30 percent to the gross domestic product (GDP). Despite a significant decrease in 2000, it provides 28 percent of jobs and contributes 60 percent of exports. According to official statistics, in the 1990s the forestry sector contributed 5 percent to the total agricultural product and about 1 percent to the GDP (WRI, 2002).

National parks cover 454,000 hectares (Tha), and the total protected area in the country is 23.4 percent (WRI, 2002). The protected area is divided into various management categories. Primary forest outside these areas covers about 180 Tha, 60 percent of which is concentrated in three regions—the northern, Baja Talamanca, and Osa Peninsula zones—and the remaining 40 percent is distributed in small areas. Two biosphere reserves make 729 Tha, so the total area protected is about 1.2 Mha. The total forest area is 1.97 Mha, with 1.79 M ha occupied by natural forest and the rest by plantations (178 Tha). During the 1990s there was a 7 percent reduction in the total forest area, and a 13 percent reduction in natural forest in contrast to a 10 percent increase in plantations. Originally, the forest area covered 98 percent of the total land area some 8,000 years ago, and the forest area in 2000 represented 39 percent of the total land area. Regarding ecosystem types and following the classification used by the World Resources Institute (2002, EarthTrends report), forests (not only the humid rain forest) cover some 64 percent of total land area (Exhibit 9.1), of which 45 percent is the protected forest. The Forest Stewardship Council (FSC) has been formulated to certify its forests and the products derived from the sustainable use of forest resources (Exhibit 9.1).

Planted forests covered about 158 Tha in 1999, and the forest industry is increasingly dependent on raw materials from these plantations. Establishment of plantations has not increased further, due to low prices for timber. In 1946, CATIE initiated experimental plantations to test various species. Plantations with exotic species have been emphasized, with more than 250 species introduced to date. The most predominant species is the alloctonous *Gmelina arborea,* accounting for 35.7 percent of the plantation area. However, some native species can also be productive with similar growth rates. The "Dirección General Forestal" describes the potential of many native tree species for commercial purposes (Motagnini and Sancho, 1990).

Since 1979, the government has provided several incentives to encourage plantation activities, although funds are insufficient to meet the de-

EXHIBIT 9.1. The Land in Costa Rica

Type of Ecosystem and Percentage of Total Land Area Covered

Forest	64%
Shrubland, savanna, grasslands	7%
Cropland, crops/natural vegetation mosaic	27%
Wetland and water bodies	2%

Extension of Forests Certified Through the Forest Stewardship Council

Forest	9,100 ha
Plantations	36,700 ha
Mixed forest	40,200 ha

Source: Adapted from WRI, 2002.

mand. In 1989, the total established area reached 9.2 Tha. However, this is not enough to offset the annual rate of deforestation which ranges from 50 to 60 Tha. The two primary incentive programmes are the "Certificado de Abonos Forestales" (CAF) and the "Fondo de Desarrollo Forestal" (FDF). The former was established in 1986. It reimburses landowners for the entire cost of the first five years after planting. The latter was established in 1989 for small holders. It pays them the cost of the first five years of plantation activities, through a local farmer organization.

POTENTIAL TERRESTRIAL CARBON SEQUESTRATION IN COSTA RICA

The potential soil C sink capacity of managed ecosystems in the world is estimated at 55 to 78 Pg (Lal, 2004), and the attainable capacity, defined by factors that limit the input C to soil systems, such as climate, for example (Ingram and Fernandes, 2001), is only 50 to 66 percent of the potential. Soil C sequestration is a cost-effective and an environment-friendly strategy (Lal, 2004). Regarding tropical sites, the total biological potential of annual C accumulation in aboveground biomass is about 8.7 Pg C per year according to different land management practices (Table 9.1).

TABLE 9.1. Carbon accumulation in aboveground biomass in tropical biomes.

Land management practice	Tropical biomes (Pg C per year)
Forestation	1.3
Agroforestry	2.1
Land rehabilitation	0.1
Conservation agriculture	2.4
End deforestation and desertification	2.8
Total	8.7

Source: Adapted from Dixon and Turner, 1991.

The SOC sequestration occurs through those management practices which add high amounts of biomass to the soil, cause minimal soil disturbance, adopt soil and water conservation strategies, improve soil structure, enhance activity of soil macrofauna such as earthworms and other soil engineers (Jones et al., 1994), and strengthen nutrient cycling (Lal, 2004). The potential for SOC sequestration in tropical countries is given in Table 9.2. In Brazil, for example, the total potential of SOC sequestration is estimated as 1.7 to 2.0 Pg C over a 25- to 50-year period (Lal, 2002). In addition to the 50 Tg C per year sequestered in soils, emission of additional 60 Tg C per year may be prevented by effective erosion control measures (Lal, 2002).

The rate of increase in the SOC pool through RMPs follows a sigmoid curve (Lal, 2004) attaining the maximum capacity of sequestration between 5 and 20 years after adoption of RMPs (Lal, 2004). Observed rates depend on several factors, including soil type, texture, and climate, and range from 0 to 150 kg $C \cdot ha^{-1}$ per year in the warm and dry regions and 150 to 500 kg $C \cdot ha^{-1}$ per year for humid climates (Lal, 2004). The continuous use of RMPs would sustain these rates for 20 to 50 years or until the soil sink capacity is filled (Sauerbeck, 2001).

Soils and ecosystems of CR have sink capacity to sequester C. However, the information dealing with the potential of SOC sequestration in CR is scarce. There have been some studies on the aboveground biomass accumulation in tree plantations by using young native trees and a first approximation to its contribution to the SOC pool (Fisher, 1995; Montagnini, 2000; Montagnini and Sancho, 1990; Montagnini and Porras, 1998; Stanley and Montagnini, 1999). Some information is also available about cultivation of maize (*Zea mays* L.) in association with the tree *Erythrina poeppigiana* (Kass et al., 1997).

The average SOC density under in tropical forest is about 8.3 $Kg \cdot m^{-2}$ to 1 m depth. When tropical forest is converted to agriculture, SOC losses range from 15 to 40 percent in a two-to-three-year period to 1 m depth

TABLE 9.2. The potential for soil C sequestration (in Pg C by 2050) in tropical countries.

Source	Plantations	Agroforestry	Forest regrowth
Trexler and Haugen (1994)	2.0-5.0	0.7-1.6	9.0-23.0
Brown et al. (1996)	16.4	6.3	11.5-28.7

Source: Adapted from Bloomfield and Pearson, 2000.

(Ingram and Fernandes, 2001). In an experiment conducted in native forest of CR, the SOC pool ranged from 191 to 285 Mg C·ha^{-1}. Recent estimates suggest a net release of C from the tropics due to deforestation in the range of 0.42 and 1.60 Pg C per year, of which 0.1 and 0.3 Pg C per year are attributed to decreases in SOM content (Veldkamp, 1993). Veldkamp quantified the changes in SOC (using pulse labeling with ^{14}C) after the conversion of tropical rain forest to pasture on two different soil types, Andisol and Inceptisol. He found that SOC decreased very slowly after conversion (Veldkamp, 1994), probably because of higher root biomass production under improved pastures and turnover rates (Lugo and Brown, 1993). Veldkamp (1993) reported that the release of CO_2 from low-productive grasses such as *Axonopus* ranged from 31.5 to 60.5 Mg C·ha^{-1} in the first 20 years (1.6 to 3.0 Mg C·ha^{-1} per year) after forest clearing. The emissions are reduced considerably when *Brachiaria* grasses were used (*B. dictyoneura*). Dominique (unpubl. M. Sc. Thesis, 1994; op. cit. Kass et al., 1997) conducted research on an Andic Eutropet soil (with high organic matter content) of CR that had been cleared from 20-year-old secondary forest. The SOC losses ranged from 1.8 to 4.4 Mg·ha^{-1} per year (Table 9.3).

There have been several studies regarding soil changes following reforestation of abandoned pastures, and most researchers expected to find significant increases in SOC pools (Fisher and Binkley, 2000). But the SOC pool increased slowly. In Puerto Rico, for example, the overall C sequestration was 0.1 Mg C·ha^{-1} (Lugo et al., 1986), and in CR, Reiners et al. (1994) found that pastures had 16 Mg C·ha^{-1} in the 0 to 10 cm depth compared to 15 and 21 Mg C·ha^{-1} in the 5-to-10 and 10-to-15-year-old regrowth forest (annual increase rate of 1.5 Mg C·ha^{-1}).

Fisher (1995) studied SOC dynamics under 11 native tree species established in a degraded pasture. He observed that soil properties changed significantly after three years and SOC concentration decreased beneath the pasture control (fallow) and three tree species. However, data must be carefully interpreted because bulk density values were not used to obtain a reliable estimate of the SOC pool. In fact, there was only an increase in the SOC pool under two tree species, i.e., *Pinus tecunumanii* and *Gmelina*

TABLE 9.3. Maize yields (Mg·ha^{-1} per year) and SOC losses (Mg·ha^{-1} per year) in 16 maize harvests in monoculture and in tree association at different tree spacing.

Tree spacing	Maize yield	Soil C loss
6 × 1 m	3.6	1.9
6 × 2 m	4.0	1.8
6 × 3 m	4.0	2.8
6 × 4 m	4.4	2.9
Fertilized—no trees	5.0	3.5
Unfertilized—no trees	2.8	4.4

arborea, and the general trend was that systems were losing between 0.01 and 0.03 Mg C·ha^{-1} per year (in the control fallow there were also losses), probably due to soil erosion. Although this study provides an approximation to SOC concentration under forest plantation management, there is a strong need to conduct a study on the actual dynamics of the SOC pool under different soil types and land-use systems.

CARBON SEQUESTRATION IN ABOVEGROUND BIOMASS

The production of aboveground biomass in plantations depends on the tree species and whether it is fast growing or slow growing. Total tree biomass production is estimated between 5.2 Mg·ha^{-1} per year to 10.3 Mg·ha^{-1} per year in pure tree stands, and 8.9 Mg·ha^{-1} per year in mixed plantations (Stanley and Montagnini, 1999). Montagnini (2000) observed that the aboveground C accumulation in native tree plantations ranged from 6.2 to 15.5 Mg·ha^{-1} per year, which is in the normal range reported in other studies (Lugo et al., 1988). Under tree plantations there is also an increase in SOM (Montagnini, 2000) even within 2.5 years, from 4.8 percent under fallow (pasture) to 5.3 to 6.6 percent under plantation. Similar trends were observed for 15 to 30 and 30 to 60 cm depths. The highest value was obtained under *Vochysia ferruginea,* with 20.3 ± 1.1, 7.2 ± 1.1, and 3.6 ± 0.4 Mg·ha^{-1} per year, in stem, branches, and foliage, respectively. The total production was 31.1 ± 2.4 Mg·ha^{-1} per year (Stanley and Montagnini, 1999), a value that is close to the one obtained in the forest (7.6 percent).

Montagnini and Porras (1998) suggested that there is a real option to use tree plantations as C sinks in CR by a judicious combination of fast- and slow-growing trees in a mosaic of pure and mixed plantations. In a study conducted with three different mixtures of tree plantations they observed an increase in C pool at a rate of 10.8 to 13.0 Mg·ha^{-1} per year, and the average annual C sequestration was highest in *Jacaranda copaia*. However, they used only stem biomass, since most leaves and a large portion of the branches turn over every year (a short-term C storage). The stem biomass was converted to total C content by assuming that it contains 50 percent C (Brown and Lugo, 1982) (Table 9.4). Calculations are based on the data obtained at an early stage of the plantations, so aboveground C sequestration can be overestimated, since most of the C uptake occurs in the youngest age classes (0 to 10 years) (Brown et al., 1986). Rotation is also a key factor in plantations' ability to remove C from the atmosphere (Schroeder, 1992), i.e., 12 to 15 years for fast-growing species and 20 years for slow-growing species.

TABLE 9.4. Total aboveground biomass (Mg·ha^{-1} per year) and mean annual C sequestration (Mg C·ha^{-1} per year) in native tree species[a] from agroforestry systems in Costa Rica.

Species	Total aboveground biomass	Annual C sequestration
Eucalyptus deglupta	13.1	6.5
Gmelina arborea	12.9	6.4
Jacaranda copaia	15.5	7.7
Vochysia guatemalensis	9.1	4.5
Calophyllum brasiliense	6.3	3.1
Stryphnodendron microstachyum	0.8	0.4
(J+V+C+S)	10.8	5.4
Terminalia amazonia	10.8	5.4
Dipteryx panamensis	9.7	4.8
Virola koschnyi	8.6	4.3
Albizia guachapele	3.4	1.7
(T+D+V+A)	13.0	6.5
Pinus elegans	11.1	5.5
Hieronyma alchornoides	12.0	6.0
Vochysia ferruginea	10.4	5.2
Genipa Americana	6.1	3.0
(P+H+V+G)	10.3	5.1

Note: Data have been estimated by assuming a 50 percent of C in total aboveground biomass (data collected from Lugo et al. [1988] for the first two species and Montagnini and Porras [1998] for the rest). Capital letters within brackets indicate the mixture system.

[a]Except the first two

There is a lack of measurements at intermediate and mature ages of different species (in plantations) to estimate accurately the C storage. In addition, one important aspect is that aboveground biomass is about 90 percent of the total tree biomass, and undergrowth represents between 2 to 4 percent (Montagnini and Sancho, 1994). Therefore, estimates of C sequestration are mainly based on the aboveground parts. Roots decompose more slowly than leaves, so they can be considered as a long-term C storage. In tropical plantations, total biomass in tree roots is lower than normally observed in natural forests (Vogt et al., 1997). There is a need to assess root biomass to obtain accurate estimations of C sequestration belowground.

Tree plantations in wet tropical climates have the potential to sequester large amounts of C in tree biomass (Brown et al., 1986), but plantations do not always accumulate soil C (Lugo et al., 1986; Bashkin and Binkely, 1998). Some research has been conducted in agroforestry systems with *Erythrina poepiggiana* and *Gliricidia sepium* of 4-, 10-, and 19-year-old systems (Oelbermann et al., 2004, 2005). These authors found that in a 10-year-old alley cropping system (*E. poepiggiana*), 0.4 Mg $C{\cdot}ha^{-1}$ per year is accumulated in coarse roots, and 0.3 Mg $C{\cdot}ha^{-1}$ per year in tree trunks. Branches and leaves from pruned trees are used as mulch, contributing to 1.4 Mg $C{\cdot}ha^{-1}$ per year (Table 9.5). Crop residues (maize, beans) can contribute 3.0 Mg $C{\cdot}ha^{-1}$ per year, resulting in an annual increase of the SOC pool by 0.6 Mg $C{\cdot}ha^{-1}$ per year. Limited data exist on root biomass and C content for *E. poepiggiana* (Oelbermann et al., 2005). The ten-year-old trees allocated 16 percent of total tree C to the root system, whereas the younger trees allocated 28 percent to the root system, revealing that differences exist, not only between the age of plantation trees but also in the composition of tree species. Carbon inputs from tree prunings of *E. poepiggiana* and *G. sepium* in the 19-year-old systems are within the range reported in other studies (Beer, 1993). This author found that pruning biomass from a

TABLE 9.5. Tree C sequestration (Mg $C{\cdot}ha^{-1}$ per year) in above- and belowground components of 19-, 10- and 4-year old *E. poeppigiana*.

	Years		
Component	**19**	**10**	**4**
Trunk (stem)	0.5	0.4	0.3
Branches	2.1	0.6	0.4
Leaves	1.9	0.8	0.7
Roots	—	0.4[a]	0.9[b]

Source: Adapted from Oelbermann et al., 2005.
[a]0.24, 0.1, and 0.11 $Mg{\cdot}ha^{-1}$ in 0-20, 20-40, and 40-60 cm depth, respectively
[b]0.22, 0.5, and 0.18 $Mg{\cdot}ha^{-1}$ in 0-20, 20-40, and 40-60 cm depth, respectively

ten-year old coffee shade tree system with *E. poeppigiana* in CR contributed 2.9 Mg C·ha^{-1} per year. In another study, the productivity of *G. sepium* in a 10 year-old alley cropping systems tree pruning biomass was 1.9 Mg C·ha^{-1} per year (Fassbender et al., 1991).

Oelbermann et al. (2005) also reported C inputs from maize and bean residues. They found no significant differences between the sole crop and the alley crop in the 19-year-old system (Table 9.6). C inputs from maize residues in the other two alley cropping systems (10- and 4-year-old) were significantly higher in the *E. poeppigiana* alley crop and the sole crop when compared to *G. sepium* alley crop. The opposite was observed regarding the bean residues. Roots are an important source of SOM, since only aboveground inputs can explain part of the soil C stocks and their variations (Fisher et al., 1994). It seems that there is a clear predominance of root-derived C in the stored soil C of croplands and grasslands (Balesdent and Balabane, 1996).

The SOC pool to 40 cm ranged from 110 to 160 Mg C·ha^{-1} in 4-, 10-, and 19-year old alley cropping systems (Figure 1 in Oelbermann et al., 2004). Fassbender (1998) reported increased levels of SOC in a *Theobroma cacao* L. and *E. poeppigiana* shade tree systems, from 115 to 140 Mg C·ha^{-1} over a 9-year period (2.8 Mg C·ha^{-1} per year). Thus, agroforestry systems are a solution, since 1 ha of agroforestry production avoids 5 ha of forest from being cleared; besides, converting tropical forests into agroforestry systems results in smaller losses of C than converting croplands to pastures (if agroforestry is established after slash and burn agriculture 35 percent of the original forest C stock can be regained (Sanchez-Azofeifa, 2000). However, Tornquist et al. (1999) observed no significant differences between the soil C pool under agroforestry and pastures, i.e., 50 and 62.6 Mg C·ha^{-1} (0 to 15 cm depth). The SOC pool in pastures was quite similar to that observed in

TABLE 9.6. Annual C inputs (Mg·ha^{-1} per year) from maize and beans in 19-year-old *E. poeppigiana* and *G. sepium* alley and sole crops.

	C allocation	
System	**Shoots**	**Roots**
Maize (M) sole crop	0.9-1.1	0.1
M + *E. poeppigiana* (19)	0.6-1.2	0.1-0.2
M + *G. sepium* (19)	0.7-0.9	0.1
Bean (B) sole crop	0.2-0.3	0.06
B + *E. poeppigiana* (19)	0.4	0.06
B + *G. sepium* (19)	0.3-0.5	0.07

Source: Adapted from Oelbermann et al., 2005.

TABLE 9.7. Summary of C sequestration rates (Mg C·ha^{-1} per year) above- and belowground (to 20 cm depth) in different land uses in Costa Rica.

Site	Soil type	Species	C seq. rate	Reference
Aboveground				
		Coffee shade tree *(E. poepiggiana)*	2.9	Beer (1993)
		G. sepium	1.9	Fassbender et al. (1991)
	Eutric Cambisol (FAO)	*E. poepiggiana*	1.4	Oelbermann et al. (2005)
	Eutric Cambisol	*E. poepiggiana*	1.8	Oelbermann et al. (2005)
	Eutric Cambisol	*E. poepiggiana*	4.5	Oelbermann et al. (2005)
Belowground				
	Typic tropohumults (USDA)	Native plantation	0.02	Fisher (1995)
	Eutric Cambisol	T. cacao + *E. poepiggiana*	2.8[a]	Fassbender (1998)
		E. poepiggiana	2.3-5.2	Nygren (1995)
		G. sepium	1.7	Kanninen (2002)
	Eutric Cambisol	*E. poepiggiana*	1.8[b]	Oelbermann et al. (2005)
	Eutric Cambisol	*E. poepiggiana*	2.0[c]	Oelbermann et al. (2005)

[a]To 45 cm depth.
[b]0.4 allocated in roots
[c]0.9 allocated in roots

agroforestry systems, and it ranged from 20.2 to 26.3 Mg C·ha^{-1} in the 0 to 5 cm depth, and 29.8 to 36.3 in the 5 to 15 cm layer. A summary of C sequestration rates above- and belowground is indicated in Table 9.7.

POTENTIAL OF CARBON SEQUESTRATION IN COSTA RICA

A summary of the potential of C sequestration in soils under agroforestry or sole crops shown in Table 9.8 is compiled from the data by Oelbermann et al. (2005) who reported SOC contents in different alley-cropping systems compared to sole crops of maize and beans. These data show that not only

TABLE 9.8. SOC pool (Mg C·ha^{-1}) under different agroforestry systems of varying ages.

Agroforestry system	SOC	SOC in crop (maize, bean)	Reference
E. poeppigiana (10 yr)	45.6	41.1	Oelbermann (2002)
E. poeppigiana (19 yr)	69.6	57.8	Oelbermann (2002)
G. sepium (10 yr)	71.8	41.1	Oelbermann (2002)
G. sepium (19 yr)	70.4	57.8	Oelbermann (2002)
E. poeppigiana (10 yr)	66.7	55.4	Mazzarino et al. (1993)
G. sepium (10 yr)	66.0	55.4	Mazzarino et al. (1993)
E. poeppigiana (6 yr)	37.0	27.2	Haggar (1990)
G. sepium (6 yr)	29.6	27.2	Haggar (1990)

Note: Soil type was classified as Eutric Cambisol (FAO).

can the changes on the SOC pool with age be analyzed but the potential of soil C sequestration can also be assessed. This potential ranges from 0.4 to 2.0 Mg C·ha^{-1} per year, depending on the agroforestry system, the composition of species, their growth rates, and the age of the plantation. This variability in the results may indicate that other factors are involved in the dynamics of SOC sequestration which have not yet been addressed, i.e., the amount of C in different soil aggregates, the contribution of large soil organisms in the dynamics of C in soil, etc. This emphasizes the need for conducting a detailed study on the C dynamics in soils of CR under different land-use systems compared to the original C pool in forest soils and also by taking into account the spatial assessment of the soil C sequestration.

A simple extrapolation of the possible potential of soil C sequestration in of CR is shown in Table 9.9. Besides, according to Oelbermann et al. (2004), if agroforestry practices are adopted and established in the 867 Tha of degraded lands, including some pastures, the potential of soil C sequestration in those areas is about 1.0 Tg C per year.

Finally, the calculations show a total potential of 42.2 to 56.5 Tg C for the next 50 years. This is a gross estimation, and a detailed assessment of soil C sequestration in different land uses is needed to provide an accurate measure. The cumulative CO_2 emissions in CR for the period 1990-2000 was 13.6 Tg C, where nearly 90 percent of it come from liquid fuels used in transportation, and the rest from manufacturing cement (CDIAC). The Kyoto Protocol was ratified by CR in 2002. The CO_2 emissions by deforestation, biomass burning, and land-use change are less than 1 Tg CO_2 per year (WRI, 2002). Through a judicious land use and by adopting soil and

TABLE 9.9. Potential of soil C sequestration in different land uses in Costa Rica.

Land use	Area (T ha)	C seq. rate (Mg C·ha^{-1} per year)	Potential (Tg C per year)	Total potential over 50 years (Tg C)
Cropland	300	1.8	0.54	27.0
Plantations	178	0.4-2.0	0.07-0.36	3.5-17.8
Natural[a]	64	0.02	0.13	6.5
Pasture	2,340	0.1	0.23	11.7
Total			0.84-1.13	42.2-56.5

[a]Regarding plantations, about 36 percent of total plantation area is occupied by *Gmelina arborea*. Using the average soil C rate of 0.02 Mg C·ha^{-1} per year, the potential is 0.13 Tg C per year.

water conservation measures CR has the potential to offset CO_2 emissions by 50 Tg C in the next 50 years.

The SOC pool under natural forest in CR (1.79 Mha) is 0.43 Pg C, from an original pool of 1.19 Pg C (98 percent of total land area covered by forest). Since 50 to 75 percent of SOC is lost when converting natural system to agricultural land the potential amount of C that can be lost from soils if current land area is cleared ranges from 0.22 to 0.33 Pg C. The potential C sink capacity equal to the historic loss of the SOC is 0.76 Pg C, of which the attainable capacity is 66 percent of the loss, 0.50 Pg C over a 50-year period, or about 0.01 Pg C per year. This maximum may never be reached.

CONCLUSIONS

There have been some attempts to quantify the amount of C that can be sequestered in terrestrial ecosystems of CR. Most of them have focused on agroforestry systems and aboveground C biomass accumulation. The quantification of SOC pool and the rate of C sequestration have been assessed in a few studies. This review outlines opportunities and knowledge gaps in SOC sequestration in CR. There is a need to conduct a study on the evaluation of the SOC pool under different land-use systems and soil types in different agroecological zones of CR. A detailed watershed study such as the watershed (catchment) near EARTH University facilities (acronym for "Escuela de Agricultura en la Región del Trópico Húmedo"), must be based on intensive soil sampling at different spatial scales.

The total potential of soil C sequestration in CR for the next 50 years is around 60 Tg C. Some land uses are more effective in terrestrial C sequestration than others. For example, in agroforestry systems tree prunings and in pastures high root biomass production and less CO_2 emissions, enhance SOC pool. Adopting soil and water conservation measures in croplands, and rational management of pastures will decrease the need to clear the natural forest, so that the current SOC pool of 0.46 Pg C can be maintained in the soil. This will also provide a good example to other countries in the region for the development of further initiatives together with international programs and other global efforts.

These are very preliminary estimates of the potential of soil C sequestration. We have not addressed the rate of soil loss due to erosion processes in different agroecosystems, but it is certainly a process that merits further analysis through a combination of soil erosion and surface runoff experiments. Strong rainy events may be a source of litter removal and transportation to other areas.

Aboveground C accumulation enters into the SOC pool, but none of the studies reviewed have mentioned the role of soil organisms in the incorporation of this material into the soil. This is a process where soil macro-, meso-, and microfauna have an important role. Litter is incorporated to the soil and will be decomposed by the action fungi and bacteria, so that there will be losses of CO_2 due to heterotrophic respiration and decomposition.

Finally, these are estimates from a biological point of view only. We have not considered the role of farmers and their decisions on the rate of C sequestration through their actions in land uses. The role of farmers and land users in this process together with the socioeconomic aspects of their livelihoods will certainly influence the rates of C sequestration and attention must be paid.

REFERENCES

Balesdent, J. 1996. The significance of organic separates to carbon dynamics and its modeling in some cultivated soils. *Eur. J. Soil Sci.* 47: 485-493.

Balesdent, J., and Balabane, M. 1996. Major contribution of roots to soil carbon storage inferred from maize cultivated soil. *Soil Biol. Biochem.* 28(9): 1261-1263.

Bashkin, M. A., and Binkley, D. 1998. Changes in soil carbon following afforestation in Hawaii. *Ecology* 79: 828-833.

Beer, J. 1993. *Cordia alliodora* and *Erythrina poeppigiana* spacing effects on the amount of *E. poeppigiana* pollarding residues in a coffee plantation. In Westley, S. B., and Powell, M. H. (Eds.), Erythrina *in the new and old worlds.* Nitrogen Fixing Tree Research Reports Special Issue, Paia, Hawaii, pp. 102-120.

Bloomfield, J., and Pearson, H.L. 2000. Land use, land use change, forestry, and agricultural activities in the clean development mechanism: Estimates of greenhouse gas offset potential. *Mitig. Adapt. Strat. for Glo. Chagen* 5: 9-24.

Brown, S., and Lugo, A. E. 1982. The storage and production of organic matter in tropical forests and their role in the global carbon cycle. *Biotropica* 14: 161-187.

Brown, S., Lugo, A. E., and Chapman, J. 1986. Biomass of tropical tree plantations and its implications for the global carbon budget. *Can. J. For. Res.* 16(2): 390-394.

Cambardella, C. A. 1998. Experimental verification of simulated soil organic matter pools. In Lal, R. Kimble, J. M., Follet, R. F., and Stewart, B. A. (Eds.), *Soil processes and the carbon cycle.* Advances in Soils Science Series, CRC Press, Boca Raton, Florida., pp. 519-525.

Dixon, R. K., and Turner, D. P. 1991. The global carbon cycle and climate change: Responses and feedback from belowground systems. *Environ. Pollut.* 73: 245-261.

Eswaran, H., Van den Berg, E., and Reich, P. 1993. Organic carbon in soils of the world. *Soil Scil Soc. Am. J.* 57: 192-194.

FAO. 2005. FAOSTAT 2005. [CD-ROM]. FAO, Rome.

Fassbender, H. W. 1998. Long-term studies of soil fertility in cacao-shade trees agroforestry systems: Results of 15 years of organic matter and nutrients research in Costa Rica. In Schulte, A., and Ruhiyat, D. (Eds.), *Soils of tropical forest ecosystems: Characteristics, ecology, and management.* Springer Verlag, Berlin and New York, pp. 150-158.

Fassbender, H. W., Beer, J., Heuveldop, J., Imbach, A., Enriquez, G., and Bonnemann, A. 1991. Ten year balances of organic matter and nutrients in agroforestry systems at CATIE, Costa Rica. *For. Ecol. Manag.* 45: 173-183.

Feller, C. 1979. Une méthode de fractionnement granulométrique de la matière organique des sols: Application aux sols tropicaux à texture grossière, très pauvres en humus. *Cahiers ORSTOM*, série Pédologie 17: 339-346.

Fisher, M. J., Rao, I., Ayarza, M.A., Lascano, C. E., Sanz, J. E., Thomas, R. J., and Vera, R. R. 1994. Carbon storage by introduced deep-rooted grasses in the South American savannas. *Nature* 371: 236-238.

Fisher, R. F. 1995. Amelioration of degraded rain forest soils by plantations of native trees. *Soil Sci. Soc. Am. J.* 59: 544-549.

Fisher, R. F. and Binkely, D. (Eds.) 2000. *Ecology and management of forest soils.* John Wiley, New York.

Haggar, J.P. 1990. Nitrogen and phosphorous dynamics of systems integrating trees and annual crops in the tropics. PhD thesis, University of Cambridge, Cambridge, United Kingdom.

Houghton, R. A., and Hackler, J. L. 1994. The net flux of carbon from deforestation and degradation in South and Southeast Asia. In Dale, V. (Ed.), *Effects of land use change on atmospheric CO_2 concentrations: South and Southeast Asia as case study.* Springer, New York, pp. 301-327.

Hunt, H. W., Elliott, E. T., and Walter, D. E. 1989. Inferring trophic transfers from pulse-dynamics in detrital food webs. *Plant Soil* 115: 247-259.

Ingram, J.S.I., and Fernandes, E. C. M. 2001. Managing carbon sequestration in soils: Concepts and terminology. *Agr. Ecosyst. Environ.* 87: 111-117.

IPCC. 2001. *Climate change 2001: The scientific basis.* Contribution of Working Group I to the Third Assessment Report of the Intergovernmental Panel on Climate Change [Houghton, J.T.,Y. Ding, D.J. Griggs, M. Noguer, P.J. van der Linden, X. Dai, K. Maskell, and C.A. Johnson (Eds.)]. Cambridge University Press, Cambridge, United Kingdom.

Jones, C. G., Lawton, J. H., and Shachak, M. 1994. Organisms as ecosystem engineers. *Oikos* 69: 373-386.

Kanninen, M. 2000. Secuestro de carbono en bosques: el papel de los bosques en el ciclo global del carbono. Internal report. CATIE, Turrialba, Costa Rica.

Kass, D. C. L., Sylvester-Bradley, R., and Nygren P. 1997. The role of nitrogen fixation and nutrient supply in some agroforestry systems of the Americas. *Soil Biol. Biochem.* 29(5/6): 775-785.

Lal, R. 2002. Soil carbon sequestration for sustaining agricultural production and improving the environment with particular reference to Brazil. In Proceedings on the International Technical Workshop "Biological Management of Soil Ecosystems for Sustainable Agriculture," FAO-EMBRAPA, Londrina, Brazil, June 2002.

Lal, R. 2003. Soil erosion and the global carbon budget. *Environment International* 29: 437-450.

Lal, R. 2004. Soil carbon sequestration impacts on global climate change and food security. *Science* 304: 1623-1627.

Lal, R., Kimble, J., and Follett, R. F. 1998. Pedospheric processes and the carbon cycle. In Lal, R. et al. (Eds.), *Soil processes and the carbon cycle.* Adv Soil Sci. CRC Press, New York, pp. 1-7.

Lavelle, P., and Spain, A. 2001. *Soil ecology.* Kluwer Academic Press, Dordrecht, the Netherlands.

Lugo, A. E., and Brown, S. 1993. Management of tropical soils as sinks or sources of atmospheric carbon. *Plant Soil* 149: 27-41.

Lugo, A. E., Brown, S. and Chapman, C. G. 1988. An analytical review of production rates and stemwood biomass of tropical forest plantations. *For. Ecol. Manag.* 23: 179-200.

Lugo, A. E., Sanchez, A. J., and Brown, S. 1986. Land use and organic carbon content of some subtropical soils. *Plant Soil* 96: 185-196.

Mazzarino, M. J., Scott, L., and Jímenez, M. 1993. Dynamics of soil total C and N, microbial biomass and water-soluble C in tropical agroecosystems. *Soil Biol. Biochem.* 25: 205-214.

Montagnini, F. 1994. Agricultural systems in the La Selva region. In McDade, L. A. et al. (Eds.), *La Selva: Ecology and natural history of a neotropical rain forest.* University of Chicago Press, Chicago, pp. 307-316.

Montagnini, F. 2000. Accumulation in above-ground biomass and soil storage of mineral nutrients in pure and mixed plantation in a humid tropical lowland. *For. Ecol. Manage.* 134: 257-270.

Montagnini, F., Jordan, C. F., and Machado, R. M. 2000. Nutrient cycling and nutrient use efficiency in agroforestry systems. In Ashton, M. S., and Montagnini, F. (Eds.), *The silvicultural basis for agroforestry systems.* CRC Press, Boca Raton, Florida, pp. 131-169.

Montagnini, F., and Porras, C. 1998. Evaluating the role of plantations as carbon sinks: An example of an integrative approach from the humid tropics. *Environ. Manag.* 22(3): 459-470.

Montagnini, F., and Sancho, F. 1990. Impacts of native trees on tropical soils: A study in the Atlantic lowlands of Costa Rica. *Ambio* 19(8): 386-390.

Montagnini, F., and Sancho, F. 1994. Aboveground biomass and nutrients in young plantations of four indigenous tree species: Implications for site nutrient conservation. *J. Sust. For.* 1: 115-139.

Nygren, P. 1995. Carbon and nitrogen dynamics in *Erythrina poepiggiana* (Leguminosae: Phasseoleae) trees managed by periodic prunings. PhD thesis, Department of Forest Ecology, University of Helsinki, Finland.

Oelbermann, M. 2002. Linking carbon inputs to sustainable agriculture in the Canadian and Costa Rican agro forestry systems. PhD Thesis, Department of Land Resource Science, University of Guelph. Guelph, Ontario, Canada.

Oelbermann, M., Voroney, R., and Gordon, A. M. 2004. Carbon sequestration in tropical and temperate agroforestry systems: A review with examples from Costa Rica and Southern Canada. *Agr. Ecosys. Environ.* 104: 359-377.

Oelbermann, M., Voroney, R. P., Kass, D. C. L., and Schlonvoigt, A. M. 2005. Above- and belowground carbon inputs in 19-, 10- and 4- year-old Costa Rican alley cropping systems. *Agr. Ecosyst. Environ.* 105: 163-172.

Paustian, K., Andren, O., Janzen, H. H., Lal, R., Smith, P., Tian, G., Tiessen, H., van Noordwijk, M., and Woomer, P. L. 1997. Agricultural soils as a sink to mitigate CO_2 emissions. *Soil Use Manage.* 13(suppl. 4): 230-244.

Rasmussen, P.E., and Parton, W. J. 1994. Long-term effects of residue management in wheat/fallow: I. Inputs, yield, and soil organic matter. *Soil Sci. Soc. Am. J.* 58: 523-530.

Reiners, W. A., Bouwman, A. F., Parsons, W. F. J., and Keller, M. 1994. Tropical rain forest conversion to pasture: Changes in vegetation and soil properties. *Ecol. Appl.* 4: 363-377.

Rozanov, B. G., Targulian, V., and Orlov, D. S. 1990. Soils. In Turner, B. L. II, Clark, W. C., Kates, R. W., Richards, J. F., Mathews, J. T., and Meyer, W. B. (Eds.), *The earth as transformed by human action: Global and regional changes in the bio sphere over the past 30 years.* Cambridge University Press with Clark University, Cambridge, pp. 203-214.

Sanchez-Azofeifa, G. A. 2000. Land use and cover change in Costa Rica: A geographic perspective. In Hall, C.A.S (Ed.), *Quantifying sustainable development: The future of tropical economies.* Academic Press, San Diego, pp. 477-505.

Sauerbeck, D. R. 2001. CO_2 emissions and C sequestration by agriculture—perspectives and limitations. *Nutr. Cycling Agroecosyst.* 60: 253-266.

Schroeder, P. 1992. Carbon storage potential of short rotation tropical tree plantations. *For. Ecol. Manag.* 50: 31-41.

Stanley, W. G., and Montagnini, F. 1999. Biomass and nutrient accumulation in pure and mixed plantation of indigenous tree species grown on poor soils in the humid tropics of Costa Rica. *For. Ecol. Manage.* 113: 91-103.

Tornquist, C. G., Hons, F. M., Feagley, S. E., and Haggar, J. 1999. Agroforestry system effects on soil characteristics of the Sarapiquí region of Costa Rica. *Agr. Ecosyst. Environ.* 73: 19-28.

Trexler, M. C., and Haugen, C. 1994. *Keeping it green: Tropical forestry opportunities for mitigating climate change.* World Resource Institute, Washington, DC.

Veldkamp, E. 1993. Soil organic carbon dynamics in pastures established after deforestation in the humid tropics of Costa Rica. PhD thesis, Wageningen University, Wageningen, the Netherlands.

Veldkamp, E. 1994. Organic carbon turnover in three tropical soils under pasture after deforestation. *Soil Sci. Soc. Am. J.* 58: 175-180.

Vogt, K., Asbjornsen, H., Ercelawn, A., Montagnini, F., and Valdes, M. 1997. Roots and mycorrhizas in plantation ecosystems. In Nambiar, E. K. S., and Brown, A. G. (Eds.), *Better management of soil, water and nutrients in tropical plantations.* Monograph 43. Australian Centre for International Agricultural Research (ACIAR), Canberra, Australia, pp. 247-296.

World Resources Institute (WRI). 2002. *Earthtrends report.* World Resources Institute, Washington, DC.

Chapter 10

Soil Organic Carbon Sequestration in the Caribbean

C. Feller
C. Clermont-Dauphin
C. Venkatapen
A. Albrecht
D. Arrouays
M. Bernoux
E. Blanchart
Y. M. Cabidoche
C. E. P. Cerri
T. Chevallier
M. C. Larré-Larrouy

INTRODUCTION

Soil organic matter (SOM) provides services that can be described as "soil fertility" functions from the farmer's viewpoint, and " environmental" functions as they are perceived by society (Feller et al., 2001). This chapter focuses on the environmental functions of SOM for soil organic carbon (SOC) sequestration for the Caribbean region (Biome C).

In Chapter 2, Cerri et al. reported that soil carbon sequestration (SoilCseq) must not be confused with the classical notion of soil C storage. SoilCseq is not only storage of carbon (C) in the soil but also the balance of all greenhouse gases (GHGs = CO_2, CH_4, N_2O) fluxes at the soil-plant-atmosphere interface. This balance is considered for a given area and a given span of time, and expressed on an equivalent C- CO_2 (eqC) basis tak-

This work was partly supported by funding from the Inter-institutional French Project (Orstom-Cnrs-Cirad-Inra) "Biological functioning of tropical soils and sustainable land management" and from the French Ministry of Environment (GESSOL Program).

Carbon Sequestration in Soils of Latin America

doi:10.1300/5755_10

ing into account the global warming potential (GWP) of each GHG involved in the balance with reference to CO_2 (Bernoux, Feller, et al., 2005a). This chapter on Biome C does not address consideration the non-CO_2 gases. It is concerned more with soil C storage than SoilCseq. In accord with other chapters, the term "soilCseq" rather than soil C storage is also used in this chapter. The case studies presented in this chapter are those of the French West Indies (e.g., Martinique and Guadeloupe) and Haiti. Indeed, soils of these islands are representative of Biome C.

POPULATION AND LAND USE

Population

The data in Table 10.1 show the change in population between 1961 and 2002. Total population increased at the rate of about 8 to 9 million inhabitants every 20 years due mainly to increase in urban population. The rural population remained relatively constant.

General Land Use for Biome C

The data in Table 10.2 show that agricultural area increased over 40 years from 9.8 to 13.1 Mha and represented about 57.3 percent of the total land area in 2002. Permanent pastures are the dominant land use, but the main change occurred in the permanent crops increasing from 0.93 Mha in 1961 to 2.2 Mha in 2002.

Detailed Land Uses in 2002 for "Arable Land" and "Permanent Crops" for Biome C

Agriculture is characterized by a very large diversity of cultivated plants, including 30 crops. Area planted to sugarcane is the largest, with 27.2 per-

TABLE 10.1. Population of Biome C.

	Population (million inhabitants)		
Area	**1961**	**1980**	**2002**
Rural	12.2	13.4	13.9
Urban	8.3	15.8	23.8
Total	20.4	29.3	37.7

Source: Based on data from FAO, 2004.
Note: Rural population did not increase during the 40-year period.

TABLE 10.2. Land use in Biome C.

Land use	Mha 1961	1980	2002
Total area	23.5	23.5	23.5
Land area (3+4+5)	22.9	22.9	22.9
Agricultural area (31+32+33)	9.8	12.4	13.1
Arable land	3.7	4.9	4.9
Permanent crops	0.9	1.7	2.2
Permanent pasture	5.1	5.9	6.0
Forest and woodland	4.0	4.4	nd
All other land (mountains, urban, mines, etc.)	9.0	6.1	nd
Waterbodies (1-2)	0.6	0.6	0.6

Source: Based on data from FAO, 2004.
Note: Note the importance of permanent pastures but also the increase of permanent crops over the past 20 years.

cent of the total area, and 14 crops represent 82 percent of the total area. Sugarcane, maize, rice, coffee, and banana (including plantain) are the main crops (Table 10.3).

In Martinique and Guadeloupe, banana and sugarcane are the principal crops (Bernoux et al., 2004). In Haiti, however, the cereal-legumes are the dominant agroecosystems.

Martinique's and Haiti's Agricultures

Martinique

Martinique is part of the French West Indies (14°N, 60°W) and an overseas department of France. Its surface area is about 1,100 km^2, and its population is about 433,000 inhabitants (in 2005). The GDP per capita is US$14,400 (in 2003), which is one of the highest in the Caribbean region. Agriculture accounts for 6 percent of GDP, while the industrial sector accounts for 11 percent and services for 83 percent. The economy is mainly based on sugarcane, bananas, tourism, and light industry. Rum and bananas are exported to Europe. Martinique is characterized by a strong decrease in the cropped area between 1973 and 2000 (37 percent decline) and in the number of farms (from 15,284 in 1989 to 8,039 in 2000) (François et al., 2004).

At the end of the nineteenth century, Martinique's agriculture was predominantly based on sugarcane. Today, bananas and sugarcane occupy 30

TABLE 10.3. Area under "arable land" and "permanent crops" for Biome C.

	Area harvested in 2002	
Crop	**ha**	**%**
Sugar cane	1,292,474	27.23
Maize	409,611	8.63
Rice, paddy	401,630	8.46
Coffee, green	297,353	6.26
Cassava	196,110	4.13
Beans, dry	196,047	4.13
Bananas	166,987	3.52
Cocoa beans	162,222	3.42
Plantains	160,430	3.38
Coconuts	139,236	2.93
Sorghum	133,037	2.80
Sweet potatoes	131,599	2.77
Vegetables, fresh	105,521	2.22
Mangoes	102,348	2.16
Partial sum in % sum		82.05
Pumpkins, squash, gourds	76,823	1.62
Oranges	73,774	1.55
Yams	56,026	1.18
Cow peas, dry	50,470	1.06
Tobacco leaves	49,709	1.05
Tomatoes	48,685	1.03
Partial sum in % sum		89.54
Fruit, fresh	43,402	0.91
Groundnuts in shell	41,929	0.88
Grapefruit and pomelos	31,710	0.67
Roots and tubers	28,360	0,60
Cucumbers and gherkins	27,713	0.58
Yautia (cocoyam)	25,506	0.54
Pigeon peas	23,120	0.49
Avocados	21,558	0.45
Partial sum in % sum		94.67
Others (n = 44)		5.33
Total sum		100.00

Source: Based on data from FAO, 2004.

percent and 12 percent of the usable agricultural area, respectively. Bananas represent the main part of agricultural production (58.4 percent). Vegetable crops represent 12 percent of the usable agricultural area and 28 percent of the total production. A large part of the agricultural area is under extensive pastures even if intensive pastures (irrigated and fertilized) have recently been introduced in the southeast of the islands on Vertisols. Most banana farms (43 percent) are <3 ha (they represent 3 percent of production), 4 percent of farms are between 20 and 50 ha (30 percent of production), and 2 percent are >50 ha (50 percent of production). The number of sugarcane farms decreased from 1,100 in 1980 to 300 in 2000.

Up to now, agricultural production has been intensive, based on high inputs (mechanical tillage, fertilizers, and pesticides), especially for bananas, and food crops and vegetables. With increasing environmental impacts (especially pollution by chlordecone), however, there is a tendency to reduce tillage and fertilizer and pesticide inputs and to use organic residues. For instance, conventional banana systems are replaced with sustainable systems: planting of "in vitro" plants (which highly decreases pesticide inputs), rotation with sugarcane or pineapple. Sugarcane is now harvested without burning to trash. There are still large problems associated with intensive vegetable crops and there is now a strong conviction of practicing organic farming.

Bananas are mainly grown on Nitisols (clayey brown tropical soils) developed on volcanic ash with rainfall of 2,000 to 3,000 mm per year. Sugarcane is grown in drier areas on Ferralsols and Vertisols (1,500 to 2,500 mm per year). Vegetable crops are grown all over Martinique, but especially in the south on Vertisols (driest area, with rainfall of 1,500 mm per year).

Haiti

With a surface area of about 27,750 km^2 (18-20°N, 71.5-73.5°W), and a population of 8 million inhabitants, this country's economy is mainly based on agriculture involving food staples. Yet the country is prone to food deficits. The per capita GDP is about US$460 (www.worldbank.org), which is lower than that of many sub-Saharan African countries, and also much lower than the average for Latin America and the Caribbean.

The land-use changes during the twentieth century (Bellande, 2004, unpublished data) were very important with a substantial decrease in areas occupied by natural ecosystems (from 50,000 to less than 1,000 ha) and under shifting cultivation (from 800,000 ha to 15,000 ha), a decrease (from 450,000 to 2000 ha) in "tree crops" (coffee- fruit- banana- root crop intercropping) and a large increase (from 75,000 to 1,050,000 ha) in cropland and urban land (from 2,000 to 50,000 ha).

An average peasant farmer comprises a family of about five, land holdings of about 2 ha, and one or two hand tools (hoe, machete) for cropping. About 80 percent of these farms are located in the uplands, under humid seasonal climates (7 to 9.5 months humid), where highly weathered and leached soils (Fe-Al oxides, Ferralsols) are dominant (Cabidoche, 1994). Others are located in the lowlands, under wet-dry climate (4.5 to 7 months humid) and on Vertisols, Mollisols, and shallow soils as dominant soil types. Only 11 percent of the lowland agricultural land is irrigated.

In the uplands, climatic conditions allow cropping throughout the year. The bean-maize intercrop (BMI) for instance, very common in these regions, can be grown either twice a year on the same field or once a year, in rotation with a market gardening crop or in rotation with a short fallow of about eight months. This fallow, during which only some annual wild species have time to grow and be grazed by the few animals, may have an important impact, not only on C sequestration, but more particularly, on the decrease of *Fusarium* inoculum for the following bean crop (Clermont-Dauphin et al., 2003). The use of fertilizers and pesticides is often limited to cash crops. For example, on the limestone Plateau des Rochelois, (altitude of 900 m) the 60-day cabbage crop (*Brassica oleracea* D.C) introduced into the rotation along with the BMI receives on average 128-35-133 $kg{\cdot}ha^{-1}$ N-P-K in the form of urea, triple superphosphate, and potassium sulphate, respectively.

In the lowlands, cereal-legume intercrops are the dominant system. Main species on irrigated land are bean, maize, and rice. In contrast, sorghum, cowpea, pigeon pea *(Cajanus cajan),* and cassava *(Manihot utilissima)* are dominant crops in the nonirrigated land. There is no fallowing, and each farmland is cropped every year. Mineral fertilizer is mainly used on irrigated lands.

Whatever the climatic region, most of the crop residues are returned to the soils of each farm. These residues are partly grazed by the cattle tethered on the fields after harvest and partly decomposed as a source of organic matter. The woody parts are piled up and burned on the fields before replanting. With grain yields ranging from 0.3 to 1 $Mg{\cdot}ha^{-1}$ for bean, and 0.5 to 1 $Mg{\cdot}ha^{-1}$ for maize and sorghum, respectively, these Haitian annual cropping systems leave on the fields between 0.5 to 2 $Mg{\cdot}ha^{-1}$ of residues each year. There is no off-farm input of crop residues or organic matter.

The forest land meets about 71 percent of the energy needs of Haiti, which corresponds to about three to four times higher than the annual forestry or agroforestry production estimated at about 1.6 million m^3 (BDPA/SCET-AGRI/World Bank, 1990). Beyond the reduction of the forest area, one can underline that in many situations it has been to a certain extent replaced with small clusters of different types of mixed tree crop systems with a range of tree densities. These do not appear as large continuous areas of perennial cover but nevertheless integrate different degrees of tree cover. The wooded area in Haiti is not a remainder of a secondary or tertiary forest but mainly a combination of perennial and annual crops established by peasant farmers on the deeper soils under humid conditions.

Soil degradation in Haiti is mainly caused by water erosion on steep lands cropped annually. The rate of soil loss is 46 $Mg \cdot ha^{-1}$ per year (Table 10.4). The area affected by erosion is about 900,000 ha (BDPA/SCET-AGRI/World Bank, 1990).

SOILS OF BIOME C: TYPES, DISTRIBUTION, CONSTRAINTS, AND DEGRADATION

The main soil types according to the FAO classification and the soil map of the Caribbean biome are presented on Figure 10.1. Eight dominant soil types are Andosols, Vertisols, Gleysols, Nitisols, Ferralsols, Lixisols, Alisols, and Luvisols. The available soil map at 1:5,000,000 scale is rather crude. As an example, only three soil types are mapped on the FAO soil map for Martinique and Guadeloupe: Ferralsols, Alisols, and Luvisols, whereas the dominant soils are Andosols, Nitisols, Ferralsols, and Vertisols (Colmet Daage and Lagache, 1965).

Are the French West Indies (Lesser Antilles) and Haiti Representative of the Major Soils of the Caribbean Biome?

Guadeloupe and Martinique are representative of the different Lesser Antilles islands (Figure 10.2) in relation to soils developed on recent volcanic material. Soils are distributed along topo-climato-chronosequences as

TABLE 10.4. Soil losses (1,000 Mg per year) in Haiti.

	Region				
Slopes (%)	**North**	**Transversal**	**West**	**South**	**Total**
0 to 20	105	209	113	109	536
20 to 50	1,591	5,128	1,975	2,881	11,575
> 50	2,962	8,343	10,155	3,068	24,528
Total	4,658	13,680	12,243	6,058	36,639

Source: After BDPA/SCET-AGRI/World Bank, 1990.

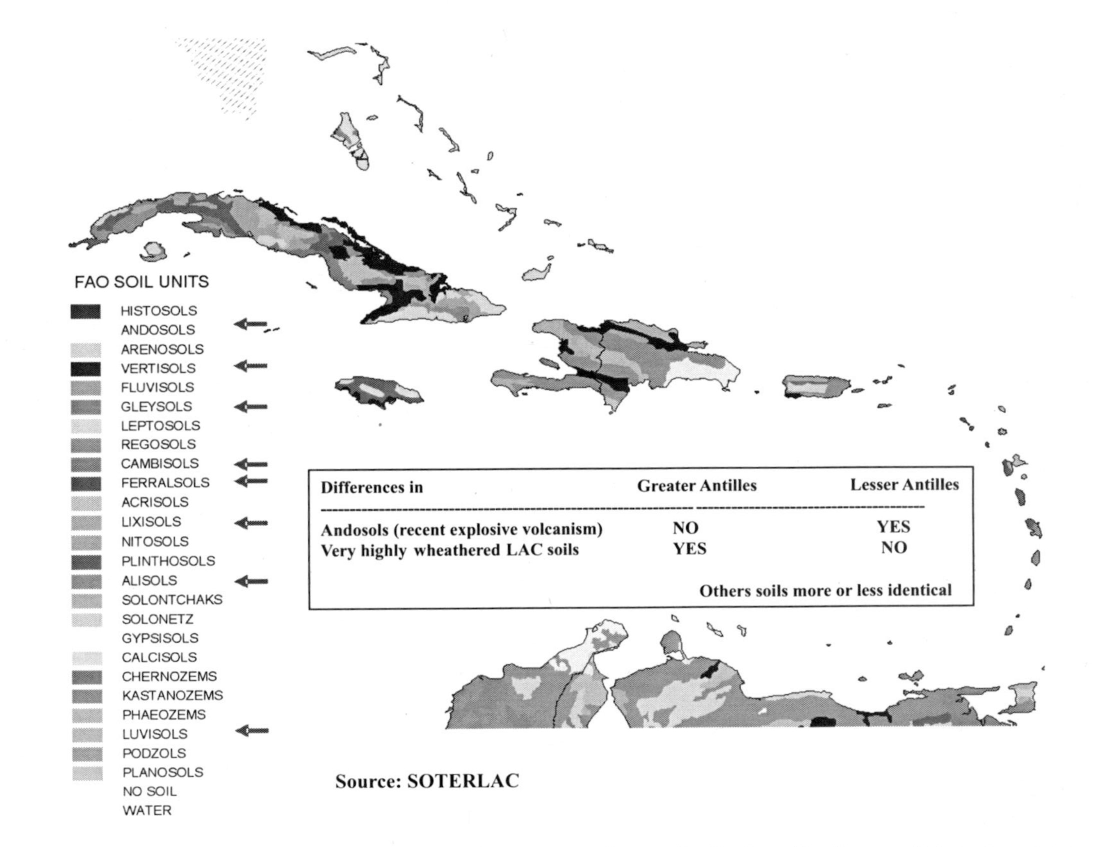

FIGURE 10.1. Main soil types for the Caribbean biome according to FAO classification and Soterlac database. Arrows correspond to the dominant soil types. (See also color gallery.)

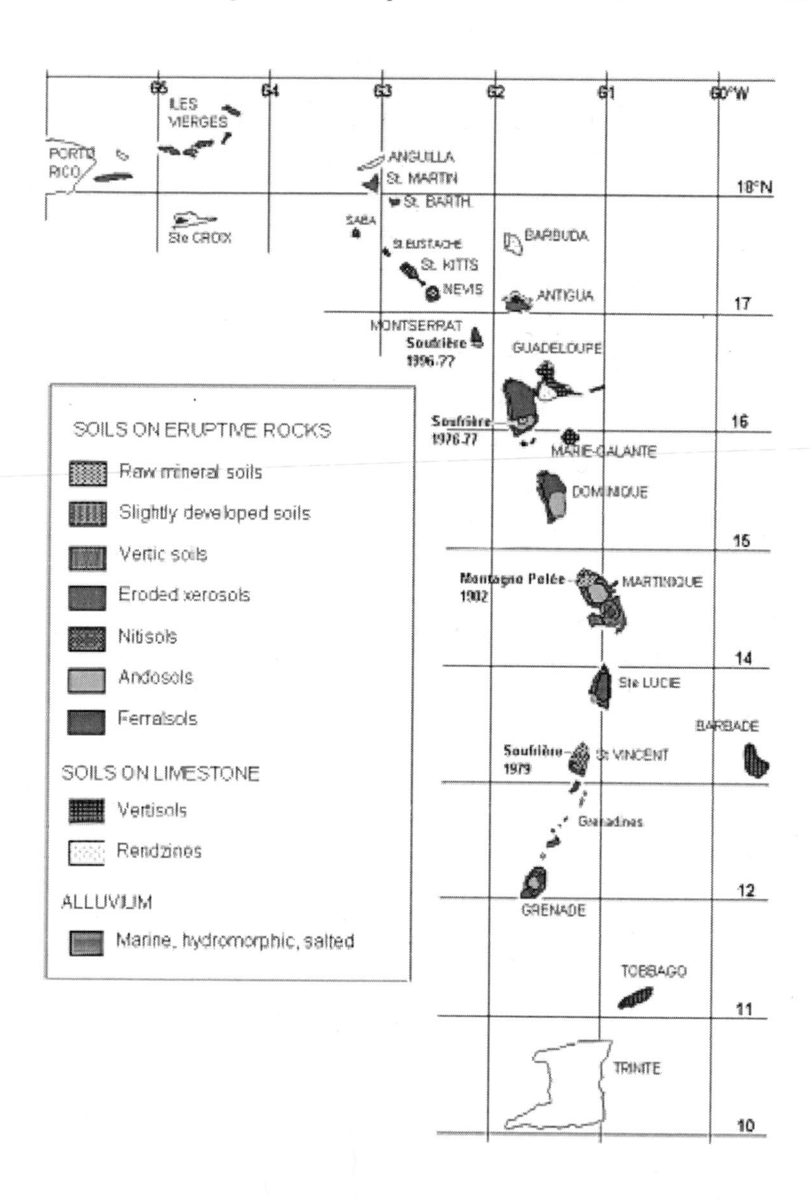

FIGURE 10.2. Martinique and Guadeloupe, representative of the diversity of soil types in the Lesser Antilles (Cabidoche, unpublished data, 1997) according to soil French classification (CPCS, 1967). (See also color gallery.)

- Young soils on pumices on very recent volcanic material,
- Andosols, Nitisols, and Ferralsols, or
- Andosols, Nitisols, Vertisols, according to the rainfall gradient on older volcanic deposits.

Each island does not have all soil types (i.e., the Vertisols of Barbados do not exist in Ste. Lucia and St. Vincent). However, Martinique and Guadeloupe have all the main soil types of the Lesser Antilles. It is one of the reasons for focusing on these two French West Indies islands for that part of the Caribbean biome.

For the Greater Antilles, the difference from the Lesser Antilles is in the absence of active volcanoes, and thus of Andosols. Even when the parent material is of volcanic origin as basalts in Haiti, St. Domingo, and Jamaica, soils developed on such material have Vertic characteristics. The highlands are developed under conditions of high rainfall and comprise soils derived from sedimentary materials, as hard limestone rather than marly limestone. These conditions give rise to soils of low-activity clays (LAC) and sometimes bauxitic soils as is the case in Haiti and Jamaica. One peculiar case is that of Cuba, where some elevated "diapirs" comprised of peridotites weathered into Ferritic soils (with pure goethite) occur because of the absence of aluminum in the parent material.

The available soil data for Martinique and Guadeloupe, and some complementary data from Haiti's Vertisols and LAC and bauxitic soils, comprises information on major soils of the Caribbean biome, except those of Trinidad, which are identical to Venezuelan soils.

Determinants of Soil Distribution in the Lesser Antilles

All islands of the internal arc of the Lesser Antilles formed from an explosive andesitic and acid-basaltic subduction volcanism. Parent rocks of soils are generally comprised of andesitic volcanic pyroclasts (pumices and ashes), even if the underlying rock is coral-reef limestone like in Barbados or Guadeloupe Grande-Terre. All minerals of the andesite can be weathered by water, and developed soils are all made up of fine secondary minerals (clays). When rainfall increases, silica and bases are leached during soil weathering, resulting clays are poor in silica, and soil becomes acidic. In the Lesser Antilles, different types of soils rich in secondary minerals occur, depending on the nature of these minerals, which depends on rainfall and soil age (Table 10.5) (Cabidoche et al., 2004 and unpublished data).

TABLE 10.5. Soil distribution in the islands of the Lesser Antilles in relation to rainfall, parent material, and age of the soils.

		Volcanic rocks		Sedimentary rocks	
		Basalt or Andesite		Limestone, marl	Marine alluviums
Rainfall (mm/year)	Silica and bases	10^3-10^4 years old	10^5-10^6 years old	10^5-10^7 years old	10^5-10^7 years old
1,000	Accumulation; alkaline pH	Vertisol (Smectite Mg+Na)[a]	Vertisol (Smectite Mg+Na)	Calcic Xerosol (Smectite Ca, $CaCO_3$)	Solontchak (Smectite Na)
1,500 (=ETP)	Maintenance; neutral pH			Vertisol (Smectite Ca)	Vertisol (Smectite Mg+Na)
2,000	Exportation; slightly acid pH	Nitisol (Halloysite)	Oxisol (Halloysite, Kaolinite)	Oxisol (Kaolinite, Al and Fe Oxyhydroxydes)	Not Found
> 2,500	Strong to very strong exportation; acidic pH	Andisol (Allophane)	Oxisol (Halloysite, Al and Fe Oxyhydroxydes)	Oxisol (Kaolinite, Al and Fe Oxyhydroxydes)	Not Found

Source: Adapted from Cabidoche, 1994; Cabidoche et al., 2004.
[a]Soil types are FAO classification; predominant secondary minerals are given in parentheses.

In all the islands of the Lesser Antilles, soil properties, in rows on the hillside, vary over short distances. Martinique represents soil properties and constraints observed in the Lesser Antilles (Figure 10.3).

Soils of the Lesser Antilles are not prone to the same fragility observed elsewhere in the tropics and especially in the Greater Antilles. In the latter islands, some soils are less clayey and more sensitive to crusting, erosion, and salinization. The clayey Ferralsols are strongly weathered, but some wet plateaus predominantly contain Ferralsols rich in Al or Fe oxyhydroxides: bauxite in Haiti and Jamaica, and Fersols in Cuba. The Vertisols predominate in areas characterized by a marked dry season.

Soil Constraints and Degradation

The Lesser Antilles

The general clayey nature of the soils of Lesser Antilles leads to high soil organic matter (SOM) content. The high SOM content and the associated important pool of nutrients allow a sustainable cultivation without any external inputs and only under conditions of long fallows and permanent

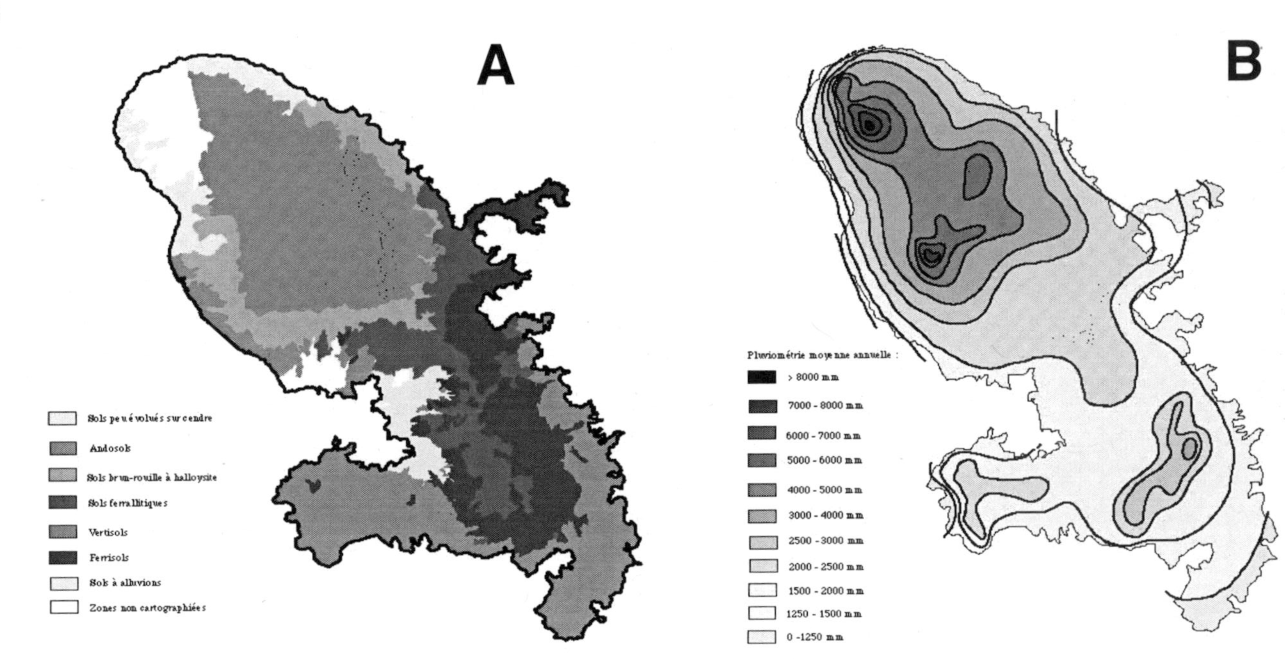

FIGURE 10.3. Simplified soil maps of Martinique (A) and rainfall distribution (B). Andosols in green; Nitisols in orange; Ferralsols in red and purple; Vertisols in blue. (*Source:* Created by authors from data of IRD-BOST after CNRS-IGN, 1976). (See also color gallery.)

cover, which slow down mineralization (Creole gardens, agroforestry). Nevertheless, the development of monoculture and the increase in yield require liming and application of potassium and phosphorus. Use of deep and frequent soil tillage completely alters the importance and functioning of SOM (Table II).

The main constraints of soils of Martinique, apparent after some decades of intensive agriculture (Cabidoche et al., 2004), are summarized in Table 10.6. Soils with Vertic properties exhibit a low level of P and K deficiency but poor physical properties and a low available water capacity. Andosols and Andic soils (young soils developed on volcanic ashes) exhibit K deficiency and Ferralsols P deficiency. All soils are prone to erosion, especially those with Vertic properties.

Haiti

In Haiti the soil degradation is mainly due to erosion by water, especially on annually cultivated land. About 36,639 Mt of soil is lost every year by erosion (BDPA/SCET-AGRI/World Bank, 1990). The magnitude of erosion depends on the slope (Table 10.4).

Cuba

The data on soil degradation for Cuba are shown in Table 10.7 (FAO, 2001). Low SOM contents represent the main soil constraint with negative impact on nutrient depletion, and with attendant degradation of physical properties along with high risks of poor drainage and accelerated water erosion.

TABLE 10.6. Soil-related constraints and degradation for soils of Martinique.

Soil type	Deficiency in P	Deficiency in K	Aluminium toxicity	CEC decrease	Erosion	Low water availability	Stones	Soil strength
Young soils on pumices and ashes (Arenosols)	+	++			++		++	
Andosols	+	++	+		+		+	
Nitisols	+	+	+		+	+	++	+
Vertic soils	+	+			++	++	++	++
Ferralsols	++	+	++	++	+			
Xerosols	+	+	++	+	+	++		+
Vertisols		+	+		+++	+++	+	++

TABLE 10.7. Extent of soil degradation in Cuba.

Process	Land area affected (1,000 ha)	% total area
Erosion by water	2,900	42.2
Erosion by wind	?	?
Elemental toxicity and salinization	?	?
Acidification:		
pH-KCl < 6.0	1,660	24.8
pH-KCl < 4.6	470	7.0
Nutrient depletion (low OM content)	4,600	69.9
Soil structural decline (compaction)	1,600	23.9
Others		
poor drainage	2,700	40.3
internal drainage	1,800	26.9
low water retention	2,500	37.3
stoniness and rockiness	800	11.9

Source: Based on data from FAO, 2001.

SOIL/CROP MANAGEMENT AND SOC STOCK

Edaphic and Agronomic Determinants of SOC Stocks in Martinique and Haiti

Martinique

In Martinique different determinants of SOC storage were studied for soils with different mineralogies: Andisols (with allophanes ALL), Ferralsols (with 1:1 clay type, low-activity clay [LAC] soils), and Vertisols (with 2:1 clay type, high-activity clay [HAC] soils). Results presented in Feller et al. (2001) are summarized next.

SOC stocks and texture. The SOC stocks strongly depend on soil texture, expressed on a clay+fine silt (0-20 μm) basis (Figure 10.4). The domain defined by the upper (natural vegetation) and lower (continuous cultivation with low level of organic restitution) limits represents, for a given texture, the maximum supplementary (or potential) SOC storage possible (Cpotential of Figure 10.4) when an RMP is adopted on a site corresponding to the lower line. Figure 10.4 shows that Cp varies with the texture, low values for coarse-textured soil and high for clayey soils.

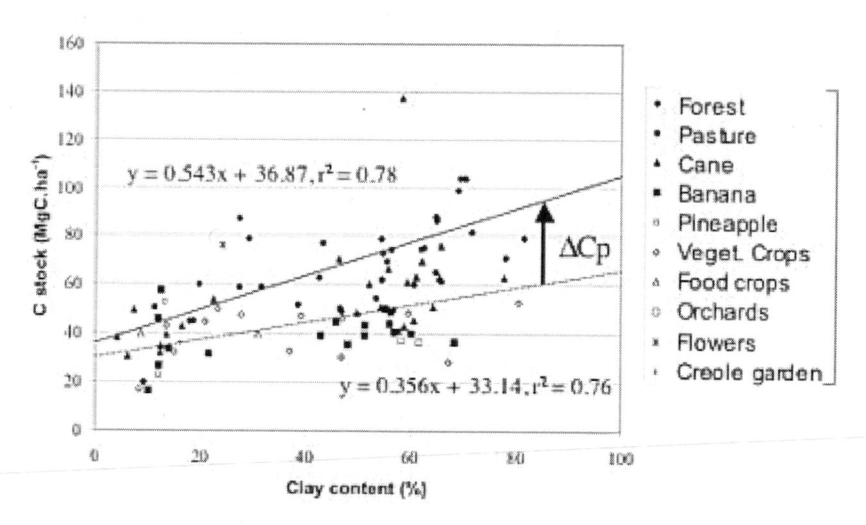

FIGURE 10.4. Variation of SOC stocks (0 to 30 cm) in relation to texture for different agrosystems in Martinique. The upper line represents natural vegetation + artificial meadows, and the dotted lower line all other crops.

SOC stocks and mineralogy. For a given texture, ALL soils (allophanes) always exhibit higher SOC contents and stocks than LAC and HAC soils. The SOC contents and stock do not differ among LAC and HAC (Feller et al., 1991, 2001; Venkatapen et al., 2004).

Feller et al. (2001) reported the following potential Cp for Martinique and/or Guadeloupe soils: 22.9, 23.3, and 30.8 t C·ha^{-1}, respectively, for ALL, LAC, and HAC soils. Hence, the higher SOC stocks in ALL soils do not imply a higher potential of SOC storage (Cp).

Aggregation and SOC protection against mineralization. The positive effect of SOM on aggregate stability is widely reported for tropical soils (Feller et al., 1996). It is also the case for Martinique, especially for LAC as compared to HAC soils. This results in an increasing protection of SOC against mineralization for high values of SOC content and aggregate stability. Chevallier et al. (2004) observed that about 40 percent of the SOC mineralization potential was in a protected form in aggregates of 0.2 to 2.0 mm.

Agronomic determinants. Some agronomic determinants were also studied by Feller et al. (2001). These are summarized here:

- *Plot history:* High rate of SOC sequestration in a Vertisol by fertilizer use and irrigated meadow was significantly higher (1.5 g C·kg^{-1} soil) for nondegraded soil than a degraded (poor in SOM) soil (1.0 g C $\cdot kg^{-1}$ soil).

- *Intensification of agricultural practices:* Fertilization and irrigation of pastures significantly increased SOC stock: 25.4 t $C \cdot ha^{-1}$ for low level to 53.5 t $C \cdot ha^{-1}$ for high rate.
- *Tillage:* Shallow tillage (10 cm) compared to deep tillage (30 cm) decreased SOC at the rate of 66 kg $C \cdot ha^{-1}$ during the first 15 months of cultivation (Blanchart et al., 2004).

Haiti

SOC stocks and texture. A positive linear relationship between clay content and SOC stock is also observed for soils of Haiti (Figure 10.5) at about the same level as for Martinique soils (Figure 10.4). According to soil texture, and for an identical cropping system, there is no significant difference

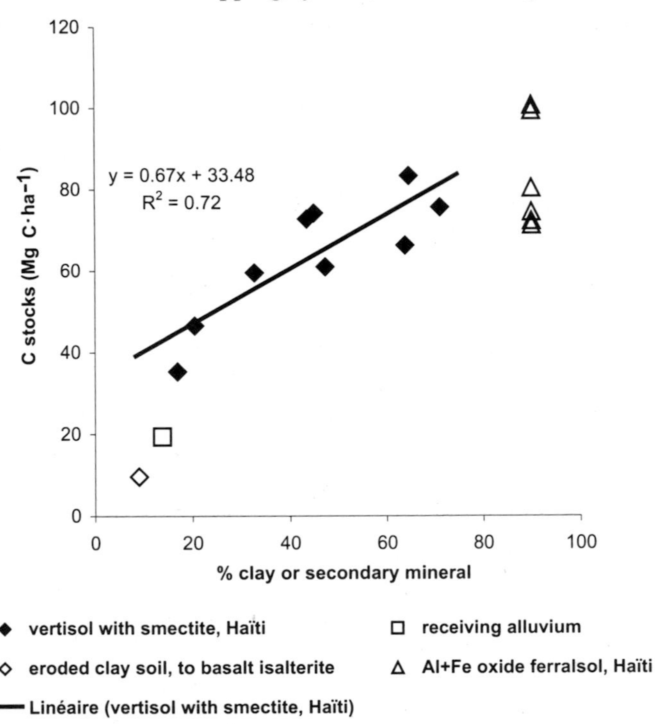

FIGURE 10.5. Variation of SOC stocks (0 to 30 cm) in relation to mineralogy and texture for several soils of Haiti (cereal + legume – short fallow cropping systems).

in SOC stocks of Vertisol or Ferralsol. Within two fields, one prone to soil erosion, the other receiving alluvial sediments, the SOC stock decreased below 30 Mg C·ha^{-1} for the eroded soil.

SOC stocks and pH. Figure 10.6 shows a positive relationship between SOC stocks and pH for the less weathered Ferralsols (medium deep calcic Ferralsol), whereas negative relationships are observed for the highly weathered Ferralsols (deep orthic and Al-Fe oxide). For medium deep calcic Ferralsols and Rendzinas, the positive relationships are due to the well-known effect of $CaCO_3$ on SOC storage. For the two other soil types, an increase in K and P was observed (data not shown) and it is increasing with pH. The results are lower yields (and hence lower SOC restoration) when pH is increasing.

SOC Stocks for the Agroecosystem As a Whole

The SOC stocks for the 0 to 30 cm depth were available only for Martinique and were determined according to soil texture (Figure 10.4) for several agroecosystems on LAC and HAC soils (252 studied sites). Andosols were not involved in these calculations.

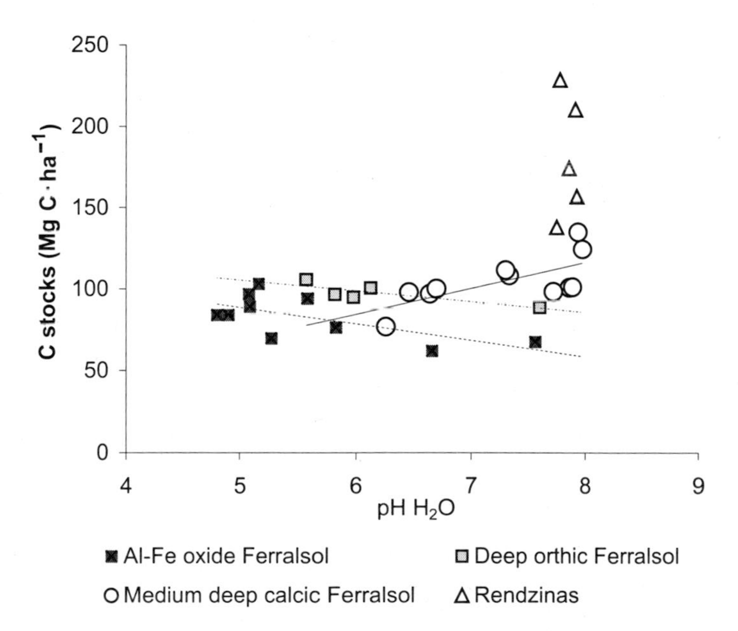

FIGURE 10.6. Variation of SOC stocks (0 to 30 cm) in relation to soil pH for upland soils of Haiti under annual low input cropping systems of staple food.

Venkatapen et al. (2004) reported that SOC stocks were not significantly different among LAC and HAC soils. Therefore, SOC stocks vary only with soil texture, land use, and management practices.

For each agroecosystem, the regression equation between SOC stock and clay+fine silt (c+fs) content (%) was established. The equation and statistical parameters are summarized in Table 10.8. The main RMP for existing agroecosystems in Martinique are sugarcane after banana and planted pastures after intensive (highly [hi] or medium [mi]) market gardening (MG) (Table 10.9). The annual rate of SOC sequestration (SOCseq rate) was low for the sugarcane-banana system but high to very high for the MG-pasture systems, ranging from 0.94 to 2.02 (Mg C·ha^{-1} per year) for coarse and fine-textured soils, respectively.

These data were used in section C to calculate (Table 10.10) the SOCseq rate for different RMPs and for two clay+silt contents: 20 percent (sandy) and 70 percent (clayey) at the Biome C scale.

Recommended Management Practices (RMPs) for Increasing SOC Stocks in the Caribbean Biome

Identifying RPMs comprises the principal goal of the research institutions in the Lesser Antilles and Haiti. It is assumed that these countries are representative of the Caribbean biome.

TABLE 10.8. Parameters of the regression equations between SOC stocks (0 to 30 cm layer) and clay+fine silt content (0 to 20 μm percent) for different agroecosystems in Martinique.

	Parameters				Soil texture (c+fs %)		SOC stocks (Mg C·ha^{-1})	
Land use	n	A	b	r^2	Coarse	Fine	Coarse	Fine
Forest	15	0.885	16.0	0.68	20	70	33.7	78.0
Meadow	58	0.681	17.9	0.43	20	70	31.5	65.5
Sugarcane	61	0.609	21.3	0.42	20	70	33.5	64.0
Banana	88	0.397	22.4	0.17	20	70	30.4	50.2
MG-hi	11	0.465	12.8	0.48	20	70	22.1	45.3
MG-mi	11	0.361	24.4	0.44	20	70	31.7	49.7
Creole garden	16	0.521	31.9	0.21	20	70	42.3	68.3

Note: Generally considered for a ten-year duration. Soil textures correspond to 20 percent (coarse) and 70 percent (fine) clay+fine silt contents (c+fs %). The regression equation is SOC Stocks (tC/ha) = a (c+fs %) + b.

TABLE 10.9. Calculations of SOC sequestration (Δ SOC) rate for some recommended management practices (RPM).

	Δ SOC stocks		rate (Mg C·ha^{-1}per year)	
RMP (10 yrs)	**Coarse**	**Fine**	**Coarse**	**Fine**
Sugarcane after banana	3.1	13.8	0.32	1.38
Meadow after MG-hi	9.4	20.2	0.94	2.02
Meadow after MG-mi	–0.2	15.8	–0.02	1.58

Note: Generally considered for a ten-year duration. Soil textures correspond to 20 percent (coarse) and 70 percent (fine) clay+fine silt contents (c+fs %). The regression equation is SOC Stocks (tC/ha) = a (c+fs %) + b.

TABLE 10.10. Ranges of the annual rates of soil carbon sequestration (SOCseq rate) with restorative land use and recommended management practices (RMP) in Biome C.

Alternatives of land use management (RMP)	Range of SOC sequestration rate according to soil texture (Mg C·ha^{-1} per year) Coarse	Fine
No burning of sugarcane[a]		
without no till	nd	0.20
with no till		1.0
Sugarcane after banana	0.32	1.38
5 years planted pastures after degraded soils under market gardening (vegetables & melons)	0.94	2.02
Improved pastures	2.0[a]	nd
No tillage with or without cover crops on cereals and pulses	nd, 0.4[b]	nd, 0.4[b]
or		
Agroforestry on cereals and pulses	nd, 0.5[c]	nd, 0.5[c]

Source: Based on values obtained by IRD in Martinique (Table 8), IRD/CENA in Brazil, and Watson et al., 2000.
Note: Generally considered for a 10-year duration. Coarse and fine textures correspond to 20 percent and 70 percent fine silt+clay contents, respectively, for the 0 to 30 cm layer.
[a]Based on values obtained by IRD/CENA in Brazil (Cerri, Bernoux, Feller, et al., 2004).
[b]Based on a literature review (mean value) by Cerri, Bernoux, Cerri, and Feller, 2004; Bernoux, Cerri, et al., 2005; Balesdent et al., 2005, for tropical and subtropical areas.
[c]Based on Watson et al. (2000) (mean value).
nd: not determined at Martinique for Biome C.

The SOCseq rate for "sugarcane after banana," "5 years planted pastures after degraded soils under market gardening (vegetables and melons)" and "improved pastures" were computed from equations in Table 10.8. For "no burning of sugarcane," data were adapted from recent CENA/IRD research in Brazil (Cerri, Bernoux, Feller, et al., 2004) and for "no tillage" systems from literature data for tropics and subtropics (Cerri, Bernoux, Cerri, and Feller, 2004; Bernoux, Cerri, et al., 2005; Balesdent et al., 2005).

The larger rates were observed for conversion to improved pasture after annual crops or after degraded pastures.

POTENTIAL FOR SOC SEQUESTRATION IN BIOME C

For each RMP the potential for SOC sequestration (potSOCseq) at the biome scale was calculated by the following equation:

potSOCseq = SOCseq rate × Surface area

and expressed in

Tg C per year (or 10^3 Mg C per year)

For each agroecosystem in Table 10.11, the considered total surface area (column 1) was extracted from Table 10.3.

The different percentages shown in column 2 were justified as follows:

For "no burning sugarcane": No burning generally implies mechanized harvest, but only for slopes lower than 12 percent. For steeper slopes, harvesting has to be manual and therefore with burning. Hence, only 50 percent of the total surface area was taken into consideration for that RMP.

For the "banana-sugarcane" rotation: 50 percent is probably the maximum proportion of land each year where sugarcane can be incorporated in the rotation. The areas concerned can be under either the present banana plots or sugarcane agrosystem. In the latter case, a small decrease in SOC sequestration is observed with the cultivation of banana in the sugarcane plots.

For "5 years planted pasture (P) after market gardening (MG)": The value of 25 percent was chosen for the following reasons: rotation MG-P was considered for only 50 percent of the present area under MG, and the duration of P (five years) is the half of the time span (ten years) used for the calculation of the SOCseq rate.

For "improved pastures": It is clear that land considered as permanent pastures corresponds probably more or less to abandoned plots, and it is presumptuous to calculate a potential with the total area. Thus, the value of 50 percent has been chosen arbitrarily.

TABLE 10.11. Annual potential of SOC sequestration (potSOCseq) for Biome C (data based on a ten-year scale basis).

Management	Total area extent of the present land use (1,000 ha)	Area under RPM (%)	Total area concerned (1,000 ha)	SOCseq rate (Mg C·yr–1)	Potential SOCseq (Tg C per year or 10^3 Mg C per year)
No burning of sugarcane					
without no till	1,292,474	50	646,237	0.20	0.129
with no till		50	646,237	1.00	0.646
Sugarcane after bananas+plantains					
coarse texture	327,417	50	163,709	0.32	0.052
fine texture	327,417	50	167,709	1.38	0.226
5 years planted pastures after degraded soils under market gardening (vegetables and melons)					
coarse texture	334,210	25	83,553	0.94	0.079
fine texture	334,210	25	83,553	2.02	0.169
Improved pastures (max.)	5,972,000	50	2,986.000	2.00	5.972
No tillage with or without cover crops on cereals and pulses	1,226,378	100	1,226. 378	0.40	0.491
or					
Agroforestry on cereals and pulses	1,226,378	50	613.189	0.50	0.307
Total 1 (coarse texture)					6.539
Total 2 (fine texture)					7.503

"No-till and cover plant systems": This system was assumed to apply only to cash crops. The example of Brazil shows that a complete transformation of such agrosystems is possible, and we accept the hypothesis that RMPs be adopted on all lands.

For "agroforestry": It is assumed that this alternative is more difficult to adopt in all situations, and that a significant part of the area (25 percent) will be covered by trees and not by the present crops. For these reasons, 50 percent was chosen.

Note the importance of the improved pasture management in the potential of soil C sequestration. Even no burning of sugarcane associated with no-tillage practice represents only 11 percent of the pasture potential.

Finally, the annual potSOCseq varies from 6.54 Tg C·yr^{-1} for coarse-textured soils to 7.50 Tg C per year for fine-textured soils, and the most important RMP at the biome C scale for SOC sequestration are concerned with improved pastures.

POLICY CONSIDERATIONS

The following policy issues are discussed for Haiti.

All RMPs proposed are not only efficient for SOC sequestration but also for enhancing sustainable plant and animal production.

Haitian farmers usually exist under highly precarious conditions. The cropland is highly eroded and, under low external input, does not produce enough food to meet the needs of the increasing population. It is not surprising, therefore, that in such a context, farmers are concerned more about survival than sustainability. Policy options to encourage adoption of restorative land use in Haiti must take into account four criteria:

1. *The need for an important investment in soil conservation:* The advanced state of erosion on large parts of the land may require, at least for the most degraded fields, drilling holes for tree plantations and implementing physical and not immediately rentable structures (terrace cultivation, drystone walls) in order to retain the soil where it still exists.
2. *Decreasing the population pressure on the most fragile soils:* This can be achieved by increasing the production of annual crops in the less eroded fields, by creating other sources of income for the rural population, by using agroforestry systems for timber and fuel production instead of petrol or charcoal from savannahs.
3. *Financial and technical assistance:* The precariousness in which most Haitian farms exist does not allow them to invest in soil restoration. Technical assistance must be provided for soil restoration and for the improvement of the productivity of annual crops.
4. *Economic incentives for restorative land use:* Any restorative land use must generate incomes for farmers. These land uses are not exclusively for erosion control or sequestering carbon. Economic incentives must be provided for farmers to maintain and increase area under forest cover and agroforestry systems. However, logistical support for marketing and safe export markets with good prices for farmers are conspicuously lacking.

CONCLUSIONS

Biome C is characterized by a very large diversity of ecological, agronomical, and socioeconomic conditions. Diverse conditions make quantification of SOC sequestration difficult at the biome scale. Important simplifications and assumptions must be made. The annual potential of SOC sequestration for Biome C ranges from 6.54 Tg C per year (coarse-texture soils) to 7.50 Tg C per year (fine-texture soils). The main potential of SOC sequestration is obviously for improved pastures (5.97 Tg C per year), and then for nonburned sugarcane (with no-till, 0.65 Tg C per year), no-tilled cereals and pulses (0.49 Tg C per year), agroforestry on cereals and pulses (0.31 Tg C per year), sugarcane after bananas or plantains (fine-texture soils, 0.23 Tg C per year), and pastures after market gardening crops (fine-texture soils, 0.17 Tg C per year). Thus, the main result for a substantial increase in SOC sequestration is a better management of pastures, either with development of improved pastures or by integration of artificial meadows in some rotations as market-gardening and/or food crop systems.

BIBLIOGRAPHY

Balesdent, J., Arrouays, D., Chenu, C., and Feller, C. 2005. Stockage et recyclage du carbone. In Girard, M.-C., et al. (Eds.), *Sols et Environnement.* Dunod, Paris, pp. 236-261.

BDPA/SCET-AGRI/World Bank. 1990. Gestion des Ressources naturelles en vue d'un développement durable en Haïti.

Bellande, A. and Paul, J.L. 1994. Les systèmes de culture d'altitude. In SACAD-FAMV (Eds.), *Paysans, Systèmes et Crise: Travaux sur l'agraire haïtien.* Pointe-à-Pitre, Vol. 3, pp. 307-356.

Bernoux, M., Blanchart, E., Venkatapen, C., Noronha, N.C., Burac, M., Colmet-Daage, F., and Scherer, C. 2004. Evolution de l'occupation des sols en Martinique. *Cahiers du PRAM* 4: 27-30.

Bernoux, M., Cerri, C., Volkoff, B., Carvalho, M.C.S., Feller, C., Cerri, C.E.P., Eschenbrenner, V., Piccolo, M.C., and Feigl, F., 2005. Gaz à effet de serre et stockage du carbone par les sols, inventaire au niveau du Brésil. *Cahiers Agriculture.* 14(1): 96-100.

Bernoux, M., Feller, C., Cerri, C.C., Eschenbrenner, V., and Cerri, C.E.P. 2005. Soil carbon sequestration. In Roose, E., Lal, R., Feller, C., Barthès, B., and Stewart, B.A. (Eds.), *Soil erosion and carbon dynamics.* Advances in Soil Science, CRC/Lewis Publishers, Boca Raton, FL, pp. 13-22.

Blanchart, E., Cabidoche, Y.M., Sierra, J., Venkatapen, C., Langlais, C., and Achard, R. 2004. Stocks de carbone dans les sols pour différents agrosystèmes des Petites Antilles. *Cahiers du PRAM* 4: 27-30.

Brochet, M. and de Reynal, V. 1977. *L'agriculture traditionnelle en Haïti: Fonctionnement des systèmes de culture et valorisation du milieu.* Port-au-Prince, Faculté d'Agronomie et de Médecine Vétérinaire.

Cabidoche, Y.M. 1994. Présentation du milieu physique. In SACAD-FAMV (Eds.), *Paysans, Systèmes et Crise: Travaux sur l'agraire haïtien.* Pointe-à-Pitre, Vol. 3, pp. 33-96.

Cabidoche, Y.M., Blanchart, E., Arrouays, D., Grolleaux, E., Lehmann, S., and Colmet-Daage, F. 2004. Les Petites Antilles: des climats variés, des sols de nature contrastées et de fertilités inégales sur des espaces restreints. *Cahiers du PRAM* 4: 21-25.

Cerri, C.C., Bernoux, M., Cerri, C.E.P., and Feller, C. 2004. Carbon cycling and sequestration opportunities in South America: The case of Brazil. *Soil Use and Management*. 20: 248-254.

Cerri, C.C., Bernoux, M., Feller, C., de Campos, D.C., de Luca, E.F., and Eschenbrenner, V. 2004. La canne a sucre au Brésil: agriculture, environnement et énergie. Canne a sucre et séquestration du carbone. C.R.Acad. Agric. France, séance du 17/03/2004. Available online at www.academie-agriculture.fr/publications/publications-html/notes_recherche.

Chevallier, T., Blanchart, E., Albrecht, A., and Feller, C. 2004. The physical protection of soil organic carbon in aggregates: A mechanism of carbon storage in a Vertisol under pasture and market gardening (Martinique, West Indies). *Agriculture, Ecosystems and Environment* 103: 375-387.

Clermont-Dauphin, C., Meynard, J.M., and Cabidoche, Y.M. 2003. Devising fertilizer recommendations for diverse cropping systems in a region: The case of bean maize intercropping in a tropical highland of Haiti. *Agronomie* 23: 673-681.

CNRS-IGN. 1976. *Atlas de la Martinique.*

Colmet Daage, F. and Lagache, P. 1965. Caractéristiques de quelques groupes de sols dérivés des roches volcaniques aux Antilles Françaises. *Cahiers ORSTOM,* Série Pédologie 3: 91-121.

CPCS. 1967. *Commission de Pédologie et Cartographie des Sols: Classification des sols.* Travaux CPCS, ENSA Grignon, France.

FAO. 1995a. Digital soil map of the world and derived soil properties (version 3.5). CD-ROM. FAO, Rome.

FAO. 1995b. Haïti: Analyse du secteur agricole et identification de projets. Rapport N° 75/95 TCP—HAI 23.

FAO. 2001. Land resources information systems in the Caribbean. Proceedings of a Subregional Workshop held in Bridgetown, Barbados, October 2-4, 2000. World Soil Resources Report 95. Available online at www.fao.org/DOCREP/004/Y1717e00htm.

FAO. 2004. FAOSTAT database query (and results). Available online at faostat.fao.org/faostat/form and faostat.fao.org/faostat/servelet.

Feller, C., Albrecht, A., Blanchart, E., Cabidoche, Y.M., Chevallier, T., Hartmann, C., Eschenbrenner, V., Larré-Larrouy, M.C., and Ndandou, J.F. 2001. Soil organic carbon sequestration in tropical areas: General considerations and analysis of some edaphic determinants for Lesser Antilles soils. *Nutrient Cycling in Agroecosystems* 61: 19-31.

Feller, C., Albrecht, A., and Tessier, D. 1996. Aggregation and organic carbon storage in kaolinitic and smectitic soils. In Carter, M.R. and Stewart, B.A. (Eds.), *Structure and organic matter storage in agricultural soils.* Advances in Soil Science, CRC Press, Boca Raton, FL, pp. 309-359.

Feller, C. and Beare, M.H. 1997. Physical control of soil organic matter dynamics in tropical land-use systems. *Geoderma,* 79: 49-67.

Feller, C., Fritsch, E., Poss, R., and Valentin, C. 1991. Effets de la texture sur le stockage et la dynamique des matières organiques dans quelques sols ferrugineux et ferrallitiques (Afrique de l'Ouest, en particulier). *Cahiers ORSTOM,* série Pédologie 26: 25-36.

François, M., Moreau, R., and Sylvander, B. 2004 *Organic farming in Martinique.* IRD Editions, Collection Expertise collégiale, Paris.

PRAM. 2004. *Les cahiers du PRAM* (Pôle de Recherche Agronomique de la Martinique). Numéro 4. CEMAGREF, CIRAD, INRA, IRD, MEDD, MNESR, Martinique, France.

SOTERLAC. 1997. Completion of a 1:5 million Soil and Terrain Digital Database (SOTER) for Latin America and the Caribbean, 1993-1997. Terminal Report. FAO, Rome.

Venkatapen, C., Blanchart, E., Bernoux, M., and Burac, M. 2004. Déterminants des stocks de carbone dans les sols et spatialisation à l'échelle de la Martinique. *Cahiers du PRAM* 4: 35-38.

Watson, R.T., Noble I.R., Bolin B., Ravindranath N.H., Verardo D.J., and Dokken D.J. (Eds.). 2000. *Land use, land-use change, and forestry.* Intergovermental Panel on Climate Change (IPCC), Special Report, Cambridge University Press, New York.

Chapter 11

Carbon Sequestration Potential of the Neotropical Savannas of Colombia and Venezuela

Marco A. Rondón
Dimas Acevedo
Rosa M. Hernandez
Yolanda Rubiano
Mariela Rivera
Edgar Amezquita
Mandius Romero
Lina Sarmiento
Miguel Ayarza
Edmundo Barrios
Idupulapati Rao

INTRODUCTION

Neotropical savannas of Latin America represent one of the last frontiers where agriculture could be expanded in the world. With an area of 269 million hectares (Mha), they account for nearly half of the total world's savannas. These savannas are located in Brazil (203 Mha), Venezuela (26.2 Mha), Colombia (23.6 Mha), Bolivia (13 Mha), and Guyana (4 Mha) (Rippstein et al., 2001). Savannas in Colombia and Venezuela are commonly known as *llanos* (the Spanish word for flatlands). Neotropical savannas are generally defined as continuous ecosystems dominated by perennial grasses with disperse woody species, where water availability follows distinct seasonal pat-

We wish to thank the support received during several years of research in Colombian Llanos from the Ministry of Agriculture and Rural Development MADR of Colombia.

Carbon Sequestration in Soils of Latin America

doi:10.1300/5755_11

terns, including a clear dry season. They are dominated by acidic soils with high aluminum toxicity and low nutrient availability, periodic burning, and moderate grazing pressure from herbivores (Walker, 1987; Huntley and Walker, 1982; Sarmiento, 1984; Bourlière, 1983). These factors not only regulate net primary productivity but also influence various ecological functional groups and their biodiversity within the ecosystem.

In contrast with the savannas in Brazil and Guyana, which have evolved over ancient Precambrian shields, the llanos from Colombia and Venezuela are quaternary plains linking the Andean and Caribbean piedmont to the west and north, with the borders of the Guyana shield to the south. The Llanos de Moxos or Beni in eastern Bolivia have clear commonalities with the llanos from the northern part of South America. They also link the Andes with the southern limit of the Brazilian shield (Sarmiento, 1990; Berroterán, 1988).

Although important physiographic, structural, functional, and ecologic differences exist within regions in these savannas, being a continuous ecosystem, the llanos of Colombia and Venezuela are comparatively more homogeneous in relation to climatic and edaphic conditions than the savannas in Brazil, Bolivia, or Guyana. On the other hand, the patterns of development and intensity of land use have been very contrasting in Colombia and Venezuela. Until approximately the mid-1950s, the llanos of both countries were exclusively used for low-intensity cattle ranching, with cattle feeding on native low-quality grasses. Stocking rates of less than 0.5 heads per hectare were common. In Venezuela, in the mid-1960s, government programs promoted the introduction of large-scale tree plantations of pines (*Pinus caribbea),* eucalyptus (*Eucalyptus deglupta*), and some commercial crops such as rice (Oryza *sativa*) and cotton (*Gossypium hirsutum*). With more than half a million hectares, the Venezuelan eastern llanos still host one of the largest continuous plantation of *Pinus caribbea.* In Colombia in the 1970s, research led by CIAT and ICA encouraged the intensification of livestock activities through the introduction of improved grasses (mainly *Brachiaria* species introduced from Africa) in association with forage legumes (*Arachis pintoi, Desmodium ovalifolium, Centrosema acutifolium*). More recently, rapid agricultural expansion is taking place in Colombia as well as livestock intensification in Venezuela. In spite of this, the dominant land use in the two countries is still extensive ranching on native pastures. The most important nonnative land use in the llanos is introduced grasses (5 Mha in Venezuela and 1 Mha in Colombia).

While in Brazil and Bolivia the main limitation is low soil fertility and high phosphorus fixation, in the llanos there are additional soil physical constraints such as low rainfall infiltration due to surface sealing and compaction in most of the Colombian llanos (Amézquita et al., 2002) or climatic constraints such as severe water deficits in the eastern parts of the Venezuelan llanos. There are also vast areas that are seasonally waterlogged land, thus limiting their use.

Soils in the llanos are considered fragile and very susceptible to degradation. Intensive tillage operations have resulted in serious loss of physical stability (Amézquita et al., 2002). Inadequate pasture management frequently results in rapid pasture degradation and associated decrease in soil quality. The recent development of appropriate agropastoral systems for the llanos enables, however, a sustainable use of the resources (Valencia et al., 2004). Long-term evaluations have shown that when properly managed, both crop and livestock systems could be sustainable activities in this ecosystem (Friesen et al., 1998; San José et al., 2003).

Pioneering research conducted in the elevated plateaus from Colombian savannas (Fisher et al., 1994, 1997) showed the high potential of improved pastures to accumulate carbon in the soil organic matter pool. Subsequent studies (Rondón, 2000) have shown that conversion of tropical savannas into pastures or even cropland with appropriate management could result in a net decrease in net fluxes of various greenhouse gases from the land into the atmosphere and could generate net carbon equivalent gains. These gains could potentially be traded in the emerging carbon markets.

Despite the apparent homogeneity of the llanos, the scenario of the changes in net carbon equivalent stocks when native land is converted into croplands or introduced pastures is very complex. Available information is very fragmented, and methodological homogeneity among countries, functional types, and physiographic units is lacking. Studies have been conducted using diverse approaches, ranging from ecology to agronomy and soil science, etc., which limits a proper extrapolation of available data, particularly for the Venezuelan savannas. Under these limitations, this chapter is aimed at estimating the current C stocks in the llanos, the maximum potential to sequester C in the soil, and the extent of sequestration that is realistic to expect with the likely development of the region in the coming decades. Estimated values should be viewed with some caution given the various assumptions in terms of homogeneity of functional subunits and carbon content in soils that we were forced to make to compensate for the lack of more precise data.

LANDSCAPE UNITS

The llanos are confined between the lower part of the Andes (3°N and 6°N) to the west, the Amazon forest margins to the south, the foothills of the coastal Caribbean range to the north, and the Orinoco River delta to the east

(Figure 11.1). Various important rivers cross the llanos from the Andes to the Atlantic. The Meta, Guaviare, Arauca, Apure, and Portuguesa are the most important, and they are also widely used for fluvial transportation of people and products (IGAC, 2002). Most of the rivers are tributaries of the Orinoco, which constitutes the main water stream in the llanos. The llanos cover 23.6 Mha in Colombia and 26.2 Mha in Venezuela, representing approximately 18 percent and 28 percent of the national territories, respectively.

Geology of the llanos is dominated by sedimentary deposits from the upper Tertiary and Quaternary. Central and eastern regions have been geologically elevated, favoring erosion and appearance of spots from the Tertiary on the surface. In the west, subsidence and accumulation of Quaternary deposits are dominant (González de Juana et al., 1980). The southernmost border shows undulated relief from the Pliocene and Pleistocene (Botero and Serrano, 1992).

For this study, areas for the various landscapes of the llanos as well as areas under different land uses were calculated based on satellite images from LANDSAT (Multispectral Scanner and Thematic Mapper). These images

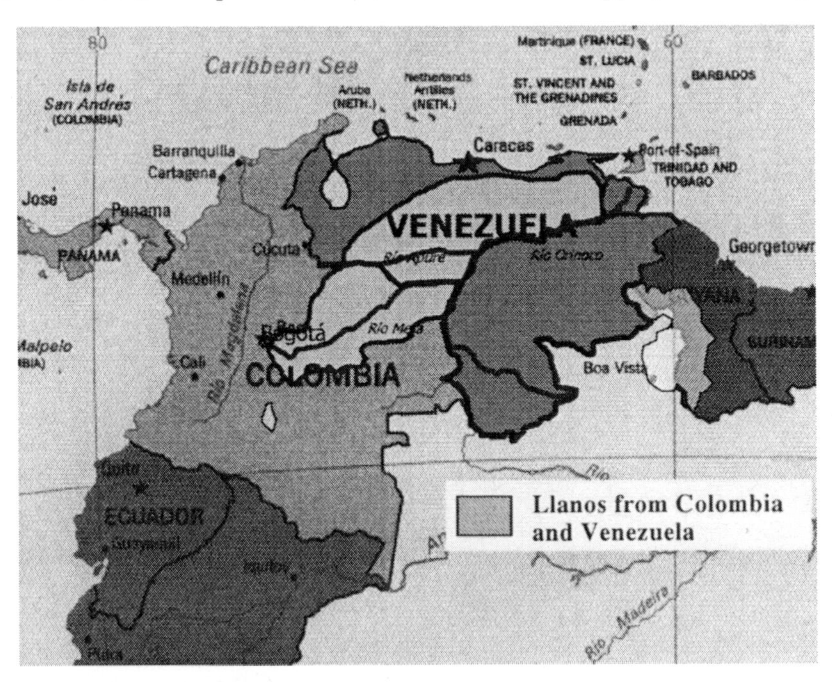

FIGURE 11.1. Neotropical savannas (llanos) from Columbia and Venezuela. (*Source:* Adapted from IGAC, 1999a,b.)

have been used previously to estimate land-use patterns in Colombia (IGAC, 2003) and Venezuela (Arias, 1992; Berroterrán, 1988; Schargel, 2003). The images were reprocessed for this study to group the landscape according to the classification described in this chapter. Four main landscape positions can be differentiated in the llanos: (1) alluvial plains, (2) eolic plains; (3) elevated or high plateaus; and (4) rolling hillsides (Berroterrán, 1988; Schargel, 2003). In Venezuela, alluvial plains dominate in the west and northeast. The eolic plains, including stabilized sand dunes, are located in the geographical center of the llanos. The elevated plateaus have an average elevation of 200 masl in Colombia and central Venezuela, but can reach around 400 masl in the eastern llanos of Venezuela (Sarmiento and Pinillos, 2001). In Colombia, alluvial and eolic plains are dominant between the Andes and the Meta River, while in the eastern side of the Meta River rolling terrain is characteristic (Goosen, 1964; IGAC, 1983).

In this chapter, we have simplified the landscape composition by combining eolic and alluvial plains. Figures 11.2 and 11.3 show the resulting landscape positions for the Colombian llanos and Venezuelan llanos, respectively: low-lying, poorly drained, seasonally flooded plains; well-drained lowlands; elevated flatland plateaus; and intermediate rolling hills. All of them include areas of evergreen gallery forest that spread along the main rivers and secondary water streams. Gallery forest accounts for 10 percent of total area in the llanos in Colombia (Rippstein et al., 2000) and around 8 percent in Venezuela (Schargel, 2003). Table 11.1 shows the area under the main landscape positions of the llanos from both countries.

CLIMATE

The llanos have a typical isothermal climate with a well-defined unimodal rainfall distribution. Rains are concentrated between April and October (Figure 11.4). The llanos are considered humid environments, but there is a decreasing west-to-east gradient in annual precipitation, with values ranging from 2,700 mm in the Andean piedmont to around 800 mm on the easternmost part of Venezuela on the border with the Orinoco delta (Sarmiento and Pinillos, 2001). There is also a gradient in the duration of the dry season, from one to two months in the southwest, to five to six months in the east. Evapotranspiration is high during the whole year (1,800 to 2,700 mm), with a peak of around 300 mm in March. Mean annual temperature is 26°C. Differences on the average daily temperature are very small between the rainy and dry seasons, but more accentuated differences of up to 12°C can be found between the minimum and maximum daily values. Solar radiation is high in the llanos, with values between 16.1 and 17.2 $MJ \cdot m^{-2}$ reported for

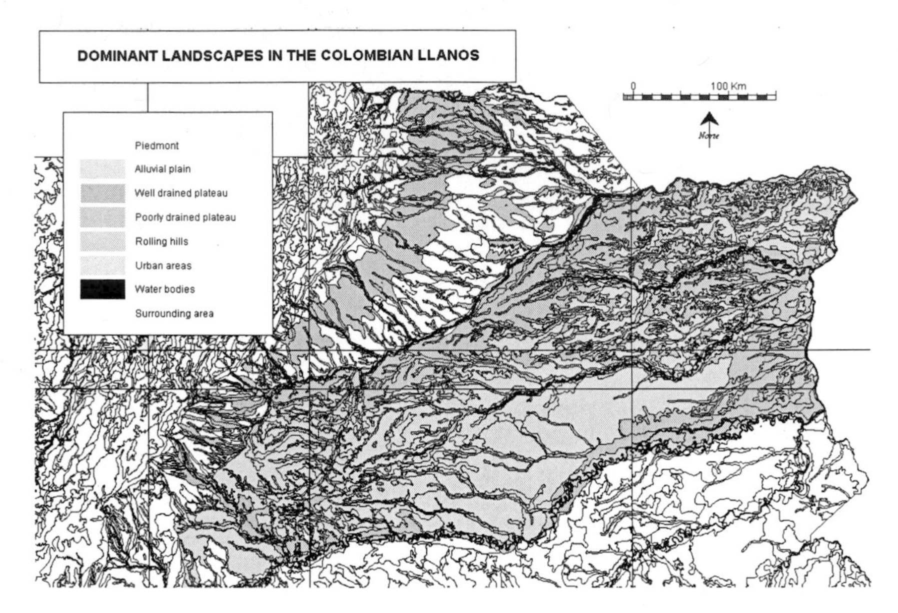

FIGURE 11.2. Dominant landscape positions in the Columbian llanos. (*Source:* Map prepared for this study using the soils map from IGAC [1983], and a landcover map for the Orinoco basin [IGAC, 1999a,b].)

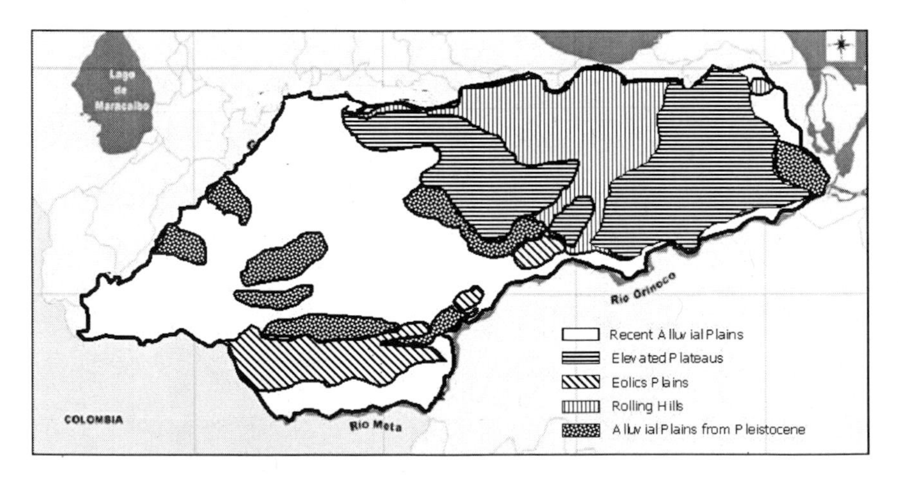

FIGURE 11.3. Dominant landscape positions in the Venezuelan llanos. (*Source:* Adapted from Schargel, 2003.)

TABLE 11.1. Areas under original main landscapes in the llanos (Mha).

Main landscape positions	Colombia[a]	Venezuela	Total per landscape
Elevated plateaus	8.27	6.48	14.75
Well-drained low plain savanna	2.85	6.42	9.27
Poorly drained lowland savanna	2.14	7.75	9.89
Rolling hills	6.98	3.15	10.13
Gallery and deciduous forest	2.65	1.52	4.17
Other (piedmont, urban, water bodies)	0.72	0.90	1.62
Total	23.6	26.2	49.8

Source: Based on data from Berroterán, 1988; Comerma and Luque, 1971; PINT, 1979, 1985, 1990; Schargel, 2003.
[a]Areas were calculated from a digital database on land use and land cover for the Orinoco basin of Colombia (IGAC, 1999a,b).

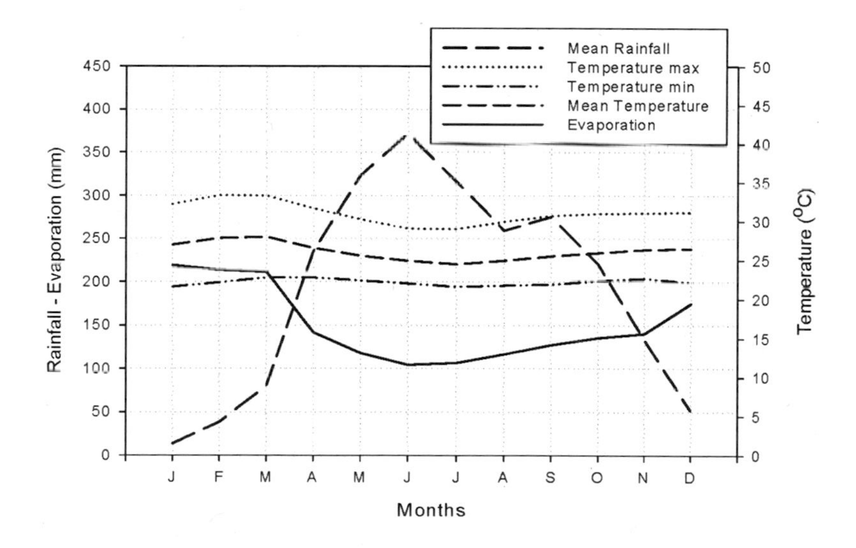

FIGURE 11.4. Climate data from Carimagua research station in the central llanos from Colombia. Rainfall data represent averages from a 20-year period. (*Source:* Adapted from Rippstein et al., 2001.)

the rainy and dry season, respectively, for the Carimagua research station in the middle of Colombian llanos (Hoyos et al., 2004).

This gradient in climatic conditions, associated with diverse landscape positions, results in a gradient in the species of the dominant vegetation, from dense gallery forest in the most humid areas to open savanna grasses and dispersed shrubs in the elevated plateaus, to semideciduous forest in the northern part of Venezuela. Native vegetation progressively dries with the advance of the dry season. Toward the end of it, natural or human-induced fires spread easily in the savanna. It is estimated that on average, native vegetation is burned every two years. Periodical burning plays a central role in maintaining the diversity and functions of the savanna ecosystem (Rippstein et al., 2001). Burning is also responsible in the llanos for important greenhouse gas emissions (CO_2, CH_4, and N_2O) to the atmosphere (Rondón, 2000). Though natural fires are still the dominant cause of biomass burning in the llanos, in recent decades, the use of controlled human-induced fires has increased. Controlled fires are promoted by ranchers to recycle the nutrients accumulated in the dry biomass and to promote the regrowth of more nutritious and palatable grasses for cattle, as well as to control cattle pests (e.g., ticks).

SOILS

Soils in neotropical savannas have been traditionally described as oligothrophic, with low to very low levels of organic matter and available nutrients. The diversity in the landscape, however, leads to a great diversity of soils. In the more recent alluvial plains, soils tend to be young Inceptisols, Entisols, and sparse spots of Vertisols, Alfisols, and even Mollisols. Dominant texture is clay-loam, and redox processes are common. The plains from the pleistocene have higher areas of sandy-loam to clay-loam Ultisols, low in organic matter (less that 1 percent). In the eolic plains, Ultisols are also dominant. Some soils located in the riverbanks or sand dunes have carbon content of <0.5 percent.

In the elevated plateaus, most of the soils are Oxisols with significant patches of Ultisols, though in the slopes of the hills separating one plateau from the next, Entisols are found, along with spots of Vertisols, Inceptisols, and Alfisols. Soils from the rolling hills are generally very superficial and dominated by Entisols in the higher slopes and Ultisols and Inceptisols in the lower slopes. Table 11.2 presents data for selected chemical characteristics of some of the most common soil types found in the llanos.

TABLE 11.2. Characteristics of some representative soil types found in the llanos.

Characteristic	Carimagua (N W, Colombian elevated plains)	Guarico (Central Llanos Venezuela)	Arauca (Eastern Lowlands Colombia)	Monagas (Elevated Plateaus Venezuela)
Soil type	Typic Haplustox	Ultisol	Humic Dystrudepts	Typic Kandiustult
pH	4.5	5.1	5.1	4.9
SOC (%)	2.0	1.3	1.6	0.5
CEC (cmol/kg)	4.2	4.6	8.0	1.9
Al saturation (%)	88	25	96	—
P (ppm)	3.9	—	23.3	2.0
K (cmol/kg)	0.08	0.40	0.12	traces
Ca (cmol/kg)	0.18	1.41	0.07	0.31
Mg (cmol/kg)	0.06	0.92	0.13	0.21
Bulk density (0-20 cm) ($Mg \cdot m^{-3}$)	1.38	1.46	1.27	1.55

Sources: Rao, 1998 (Carimagua); Hernández and López, 2002 (Guarico); Comerma and Chirinos, 1976 (Arauca); IGAC, 2003 (Monagas).

NATIVE VEGETATION

Seasonally dry climate, contrasting soil types, and varied landscape positions have generated a range of vegetation communities perfectly adapted to the environment. Ecologically, vegetation communities can be grouped into four main functional categories: seasonal savannas, hyperseasonal savannas, semiseasonal savannas, and gallery forest (Sarmiento, 1984). Seasonal savannas experience severe drought stress for four to six months annually. Vegetation is dominated by grasses with a C4 type of photosynthetic system such as *Trachypogon, Andropogon, Leptocoryphium, Axonopus, Paspalum, Panicum, Aristida, Echinolaena,* and *Tristachia.* Fire is common toward the end of the dry season, and burning usually removes most soil cover. Net primary productivity approaches $1 kg \cdot m^{-2}$ per year, with an estimated value of half of that from roots (Sarmiento, 1984). Hyperseasonal savannas are subject to shorter drought periods (two to three months) and a period of water excess favored by the location in the lower parts of the basin on poorly drained soils. Vegetation is mostly grasses (*Leersia, Paspalum, Sorghastrum,* and *Andropogon*), with little presence of woody species (*Copernicia tectorum* in Venezuela and *Caraipa llanorum* in Colombia). Sedge species are also important. Annual primary productivity is estimated at $1.2 kg \cdot m^{-2}$ per year, with some 60 percent from aerial biomass.

Semiseasonal savannas are characterized by water excess, including flooding conditions during eight to eleven months per year. Dominant grasses are *Hymenachne, Leersia, Oryza,* and *Panicum,* although plants from *Ciperaceae* and *Amaranthaceae* are also abundant in the wettest parts. In the lower parts that remain flooded most of the year, Palmacea such as *Mauritia flexuosa* are dominant. Fire plays a minor role in these areas. Soils have not been studied in detail, but they usually have higher levels of organic matter compared to other savanna subregions. In the gallery forest, plant cover is dominated by diverse evergreen or deciduous trees (*Duguetria riberensis, Nectandra pichurni, Chomelia polyanta, Copaifera officinalis, Covvoloba ontusifolia*) subject to periodic flooding resulting from river overflows.

LAND USE

Historically, the llanos have been mostly used for extensive cattle ranching with some subsistence agriculture concentrated in the most fertile soils along the river and occasional timber extraction from the gallery forest (Sarmiento, 2000). The main management practice in the llanos has been the burning of native herbaceous vegetation to promote the regrowth of more palatable grasses. This production system has been intensified in recent decades through the introduction of exotic grasses such as *Brachiaria decumbens, B. brizantha, B. humidicola, B. dictyoneura, Panicum maximum, Digitaria swazilandensis,* and *D. decumbens*, among others. New breeds of cows have also been introduced. Crop species such as sorghum (*Sorghum bicolor*), maize (*Zea mays*), beans (*Phaseolus vulgaris*), rice (*Oryza sativa*), sunflower (*Helianthus annuus)*, sugar cane (*Saccharum officinarum*), cotton (*Gossypium hirsutum)*, sesame (*Sesamum indicum)*, cassava (*Manihot esculenta*), and some fruits such as watermelon (*Citrullus lanatus)* and bananas (*Musa acuminate*) are gaining space, particularly in the Venezuelan llanos. Forest plantations of pinus (*Pinus caribbea*), eucalyptus (*Eucalyptus spp.*), and teak (*Tectona grandis*) are now also important.

Land-use intensity varies according to the landscape position in the llanos and shows different trends between the two countries. Most of the cropland and pastures are concentrated in the elevated plateaus and in parts of the alluvial plains. In Colombia, the main land use has been and continues to be traditional extensive cattle ranching. However, recent decades have seen an intensification of livestock systems. Currently there are nearly 1 Mha of introduced pastures. The conversion into cropland started during the 1970s as a result of new agropastoral systems being implemented in the llanos to cultivate mainly upland rice as a preliminary step to establish pastures (Sanz et al., 2004). More recently, during the 1980s, some areas were planted to oil palm, and in the 1990s, with the development of acid-soil-adapted maize varieties and hybrids, this crop is gaining importance, especially when planted in rotation with acid-soil-tolerant soybeans (Narro et al., 2004). Currently, 0.4 Mha are sown to crops and 0.1 Mha are planted in Colombian llanos with tree species such as pines, rubber (*Hevea brasiliensis*), and oil palm.

In the Venezuelan llanos, the main land use is also extensive cattle ranching, but the history and distribution of land use in the elevated plains is quite different. The plateaus of the western llanos, in the piedmont of the Andes, have more fertile soils and have experienced strong agricultural expansion. The main crops in this region are cotton, sorghum, upland rice, and peanuts. The livestock has evolved into dual-purpose (milk and beef), intensive, and semi-intensive systems. Although the soils in the plateaus in the eastern llanos of Venezuela are not very fertile, the subregion has seen important agricultural expansion, but livestock continues to be the dominant activity. Main crops are peanuts (*Arachis hypogaea*), sugar cane (*Saccharum officinarum)*, maize (*Zea mays*), sorghum (*Sorghum bicolor*), sesame (*Sesamum indicum)*, and watermelon (*Citrullus lanatus)*. In the plateaus of the central llanos, the main crops are cereals (sorghum and maize). There are around 1 Mha of croplands and at least 5 Mha under introduced pastures in the Venezuelan llanos. There are still important areas dedicated to extensive cattle ranching on large farms based on native vegetation. In the southern part of the Venezuelan llanos, a large-scale plantation of *Pinus caribbea* was established in 1969 and continues under production (Torres et al., 2003). Total area under tree plantations is 0.8 Mha.

The poorly drained areas from alluvial and eolic plains are located preferentially between the Meta and Apure Rivers, covering most of the Apure state in Venezuela and the Arauca and Casanare plains in Colombia. This seasonally flooded area accounts for 2.14 Mha in Colombia and 6.42 Mha in Venezuela (Sarmiento and Pinillos, 2001). Mean elevation ranges from 150 masl in Colombia to 75 masl in Venezuela. During three to six months of the year, depending on the altitude, the lowlands are flooded and cattle are forced to move somewhere else in search of pastures. At the end of the rainfall, the rapid grass regrowth again attracts the herds of cattle. This traditional livestock system has been used in the llanos for more than two centuries. In the Colombian llanos very little land from the poorly drained areas has been converted into other land uses. During the mid-1960s, the Venezuelan government promoted, and in most cases built, a system of low earth dykes, aimed to control flooding and to improve the economy based on livestock. The terrains encircled by dykes and river levees were called modules, as they operate both as hydrological and as grazing units. Some of these modules were planted later to improved grasses, and some patches of crops, but most of the area is still dominated by native vegetation. Due to edaphic constraints and water excess, agriculture is limited to the eolic plains and only traditional extensive livestock prevails.

The fourth main landscape of the llanos, the Lomerio, is used little for agriculture or pastures in Colombia, but receives greater use in Venezuela. Of the 7.75 Mha in Colombia, only some 0.01 Mha are used for crops, though some 0.05 Mha are now under pastures. The dominant use is extensive cattle raising. In Venezuela, this landscape coincides to most of the central llanos in the states of Cojedes and Guarico. This landscape occupies approximately 3.15 Mha. Extensive livestock on native vegetation (mostly of the genus *Trachypogon)* is the main land use. However, areas with lower slopes (less than 3 percent) have been converted to mechanized cropping (sorghum, maize), taking advantage of irrigation projects developed in the region. Forest plantations have replaced the patches of deciduous forest in Venezuela. Water is scarce in the region due to overuse of resources to cope with the high demand in the most populated areas that surround the region.

Table 11.3 summarizes current distribution by land use in Colombian and Venezuelan llanos. It is evident that the native savanna has preferentially been converted into introduced pastures in the llanos. As we mentioned before, the use of Venezuelan savannas is much more intensive than in Colombia, as a result of more access to infrastructure and clear government programs to develop the region. In Colombia, the lack of infrastructure in terms of roads, together with the remoteness of the region, contributed to relatively less intervention. The social conflict that the country has been facing for several decades has also played a role in limiting the interest of investors in the region.

TABLE 11.3. Distribution of the main land uses in the llanos (Mha).

Land use	**Colombia**[a]	**Venezuela**	**Llanos**
Native herbaceous vegetation	18.77	16.96	35.73
Improved pastures	0.98	5.00	5.98
Annual crops	0.39	1.00	1.39
Tree plantations	0.10	0.80	0.90
Gallery forest	2.65	1.52	4.17
Other (urban, water bodies)	0.72	0.90	1.62
Total llanos	23.6	26.2	49.8

Sources: Sarmiento, 1990; Schargel, 2003; PINT, 1990; Berroterán, 1988.
[a]Data were interpolated from a landcover map of Colombian llanos (IGAC, 2003).

CONSTRAINTS TO PRIMARY PRODUCTIVITY

Net standing biomass for the most prevalent open savanna type of vegetation is around 4.5 Mg C·ha^{-1} for the eastern Venezuelan llanos (San José et al., 1998) and 3.8 Mg C·ha^{-1} for the central part of the Colombian llanos (Rondón, 2000; Rippstein et al., 1996). This contrasts with the average 100 Mg C·ha^{-1} of a mature rainforest in the Amazon (Phillips et al., 1998). Apart from the limitations associated with a severe dry season, biomass productivity in the llanos is constrained by the low nutrient availability, high acidity, and very high aluminum saturation of the soils. As shown in Table 11.2, most soils are highly weathered Oxisols and Ultisols, with vast patches of Typic Haplustox Isohiperthermics (IGAC, 2003).

Fire is another important factor that limits productivity in the llanos. Some experiments to evaluate fire exclusion on native savanna vegetation have shown that secondary vegetation could accumulate much higher biomass (in the range of 10 Mg C·ha^{-1}) than the primary savanna on the uplands of the Colombian llanos (Garcia, unpublished data). Similarly, San José et al. (2003) indicate that when native savanna is protected from fire and grazing, a more dense type of vegetation can succeed even under the relatively drier conditions of the eastern llanos of Venezuela. Increases of up to 20 percent in aerial biomass have been observed in a period of 30 years by protecting the vegetation against fire in seasonal savanna from central Venezuela (Güerere, 1992). As was mentioned before, water deficits limit the photosynthetic activity during three to four months of the year. When these particular constraints are alleviated through liming, fertilization, and irrigation, very high net primary productivity can be obtained. Fisher and Thomas (2004) show that well-managed grass pastures can reach a net aerial primary productivity of 33.5 Mg·ha^{-1} per year. Biomass accumulation of 10 Mg·ha^{-1} per year have been reported for pine plantations in Venezuela (Hoyos, 1998). Solar radiation is not considered to be an important limiting factor for net primary productivity in the llanos.

CURRENT CARBON STOCKS IN SOILS

The Carbon Content of the Soils

A detailed survey of soils in the Colombian llanos reports carbon content in soils (0 to 30 cm depth) ranging from 1.2 to 3.5 percent (IGAC, 2003) in the well-drained upland savannas and from 1.0 to 4.0 percent in the poorly drained soils. Friesen et al. (1998) reported values of 2.3 percent for the 0 to 30 cm depth at the Carimagua Research Station, where the dominant soils have silty-loam texture representing around 70 percent of total soils in Colombian savannas. Soils with a more sandy texture show values of 0.7 percent. Values reported for Matazul near Puerto Lopez, representing the areas where most of the intensification is occurring in the Colombian llanos, are 2.3 percent for loam textures and 1.6 percent for sandy texture (Hoyos et al., 2004).

In the Venezuelan llanos, seasonal savannas account for 56 percent of the area. Soil carbon levels range from 0.8 to 3.5 percent, while values in the range of 2 to 2.5 percent are found in the high plateaus of the western llanos, where the more fertile soils are located. This contrasts with levels of 0.5 to 0.8 percent for the sandier Ultisols of the central llanos (Hernández and López, 2002; Schargel, 2003). In general, the eastern llanos in Venezuela have the lowest C content, with values as low as 0.6 percent. Soil carbon under gallery forest is usually higher, with values ranging between 3.4 and 5.5 percent (Schargel, 2003). Not surprisingly, much more soil data are available for agricultural soils than for areas under native vegetation, particularly on the poorly drained savannas. Values for the flooding lowlands in Venezuela range from 0.6 to 1.8 percent (Sarmiento and Pinillos, 2001), while C values between 1.1 and 1.6 percent have been reported for the Arauca drainage plain in Colombia (IGAC, 1991, 2003). Table 11.4 shows estimates of C content and stocks for different landscape positions and vegetation types for areas of the Venezuelan savannas. The western llanos of Venezuela store higher levels of carbon in soils, while the eastern llanos have consistently lower carbon stocks, due in part to their geologic origin associated with igneous acidic rocks from the Guyana shield.

A trend seems to exist in the C content in soils with the increase in distance from the Andes. The piedmont soils are clearly more fertile with higher organic C content, and for the elevated plateaus, the Colombian soils tend to have slightly higher C content than those of the equivalent landscape position in Venezuela. This could be the result of the contribution of eroded soil and nutrients coming from the Andes to lower areas which is progressively reduced eastward. However, more data are needed to support this hypothesis. As shown by previous data, within each landscape position, SOC (soil organic carbon) is relatively heterogeneous. This makes it difficult to extrapolate values to the overall landscape. Despite these limitations, an attempt is made herein to estimate a weighted average for C content in the top 30 cm of the soils for each landscape position in both countries, using an extensive literature review.

TABLE 11.4. Carbon content and stocks in the top 30 cm of soils from Venezuelan llanos.

Region	Landscape	C stock (Mg C·ha^{-1})	C content (%)
Young alluvial plains			
West:	Deciduous forests	77.0 ± 19.4	5.0 ± 1.3
	Hyperseasonal savannas (lowlands)	45.2 ± 17.3	3.2 ± 1.4
	Semiseasonal savanas (lowlands)	61.5 ± 12.0	5.2 ± 1.1
Alluvial plains from the Pleistocene			
West:	Hyperseasonal savannas (lowlands)	39.0 ± 8.5	2.6 ± 0.6
	Seasonal savanas	33.5 ± 6.4	2.2 ± 0.5
East:	Seasonal savanas	42	0.9
	Seasonal savanas	46	2.9
Eolic Plains			
West:	Poorly drained savanas	65 ± 7.1	4.2 ± 0.3
	Well-drained savanas	22 ± 4.2	1.4 ± 0.2
Elevated Plateaus			
West:	Seasonal savanas	24 ± 9.9	1.6 ± 0.7
East:	Seasonal savanas	17.7 ± 7.5	0.8 ± 0.5
Central:	Seasonal savanas	51.3 ± 22.5	3.5 ± 1.6
Rolling hills			
Central:	Seasonal savanas	31	3.3

Sources: Original data from Campos, 1999; Colmenares et al., 1974; Comerma and Chirinos, 1976; Gómez, 2004; Güerere, 1992; Hernández-Hernández et al., 2004; Hernández-Hernández and Dominguez, 2002; Hernández-Hernández and López-Hernández, 2002; Malavé, 1981; Pérez-Materán et al., 1980; PINT, 1979, 1985, 1990; Schargel, 1972, 1978; Westin, 1962; Zinck and Stagno, 1966.

Data on Soil Bulk Density Are Missing

Quantifying total C stocks in soils from the llanos depends on the availability of data on both SOC and soil bulk density (SBD). But there are considerably more data on SOC than the corresponding SBD. This missing information contributes to most of the uncertainty in the estimation of total C stocks, because SBD data are probably more heterogeneous than SOC in the llanos (Lozano et al., 2000).

SBD values have been reported for Colombian savannas for clay-loam and sandy soils (IGAC, 2003, 1991, 1983) as part of detailed soil surveys in the region. The SBD of soils at the Carimagua Research Station in the middle of Colombian llanos have been intensively measured (Amézquita et al., 2002), as well as the values for the clay-loam savannas near Puerto Lopez, where most agricultural intensification is taking place in Colombia (Hoyos et al., 2004). Values (in $Mg \cdot m^{-3}$) for high plateaus in Colombia range from 1.25 to 1.32 in the top 10 cm layer to 1.4 to 1.6 at 30 cm depth. Data on SBD are scarce for river floodplains. IGAC (2003) reported values ranging from 1.1 to 1.7 for the seasonally flooded lowlands in Arauca and Casanare. In Venezuela, SBD values between 1.46 and 1.7 are reported by Hernández and López (2002) for the rolling hills of the central llanos, and 1.4 for the Apure floodplains (Pinillos, 1999). Lobo et al. (2002) found values between 1.32 and 1.62 for the rolling hills of central Venezuela. Values for the gallery forest are also scarce. Rondón (2000) measured SBD values for the gallery forest in Colombian plateaus and reported an average value of 1.1 for the top 30 cm of the soil. In Venezuela, data from the soil survey (MARNR, 1986) indicated values ranging between 1.2 and 1.4 for native forest. For Venezuelan eastern plateaus, values range between 1.3 and 1.8 (San José and Montes, 1991; Sarmiento and Acevedo, 1991). For this analysis, data are considered down to 30 cm soil depth for two reasons: (1) the lack of available sufficient data for lower depths, and (2) the overarching assumption that it is in the 0 to 30 cm layer where most of the changes in SOC occur due to erosion, oxidative processes, or accretion of SOM from root and litter turnover.

Carbon Stocks

Using weighed averages for carbon content and SBD for the top 30 cm of the soil, total soil carbon stocks are calculated for the areas under native vegetation in various landscape positions, as well as for the areas under more intensive use: pastures, crops, and forest plantations. Data for Venezuela are the result of an extensive review of the literature in the country, while data for Colombia have been taken mainly from a detailed soil survey conducted by IGAC (2003) in the Colombian llanos, as well as data accumulated from 30 years of research from CIAT and CORPOICA. Table 11.5 shows integrated stocks calculated for the main landscape positions and land use in both countries to estimate global carbon storage in the top 30 cm of soils in the llanos.

Covering an area of nearly 50 Mha, the native savannas account for around 1 percent of the total area of 4,968 Mha of the humid tropics (Lal,

TABLE 11.5. Estimated carbon stocks for the main land-use systems in the llanos from Colombia and Venezuela.

	Area (Mha)		Estimated carbon stocks (0-30 cm depth)			
Landscape position/ land use	Colombia	Venezuela	Colombia (Mg C·ha^{-1})	Venezuela (Mg C·ha^{-1})	Colombia (Tg C)	Venezuela (Tg C)
Remaining natural systems						
Elevated plateaus	7.29	5.04	73.98	35.66	539.3	179.8
Well-drained lowlands	2.51	4.81	83.54	43.16	209.7	207.6
Poorly drained low plains	2.13	5.45	89.92	59.41	191.5	323.8
Rolling hills	6.84	1.66	48.07	40.00	328.8	66.4
Gallery and deciduous forest	2.65	1.52	138.60	75.00	367.3	114.0
Subtotal	21.42	18.48			1636.6	891.6
Modified systems						
Introduced pastures	0.98	5.00	98.20	78.00	96.2	390.0
Annual crops conventional tillage	0.39	0.83	70.49	38.10	27.5	31.6
Annual crops reduced tillage	0.001	0.17	72.90	43.20	0.07	7.34
Tree plantations	0. 1	0.80	73.39	27.00	7.34	21.6
Urban, water bodies, etc.	0.72	0.90			—	—
Subtotal	2.19	7.70			131.1	450.6
Total per country					1767.7	1342.2
Total for the llanos						3110

2002) and store in the top 30 cm of the soil an estimated 3.1 Pg of SOC. This is roughly 0.64 percent of the estimated 496 Pg of carbon stored by tropical soils down to 1 m depth (Lal, 2002). If we assume that a similar amount of C is stored by savanna soils between 30 and 100 cm depth, this results in estimates of total C stocks of around 1.3 percent for tropical savannas, indicating that soils from the neotropical savannas are similar to other tropical environments with regard to their ability to stock C in the soil profile (Houghton, 2003).

EFFECTS OF LAND-USE INTENSIFICATION

The llanos of Colombia and Venezuela have experienced different degrees of agricultural development or livestock intensification. In Colombia, essentially all agricultural lands have been opened on upland savannas in the vicinity of the main cities (Villavicencio, Puerto Lopez, Granada, Puerto Gaitan) and along the main connecting roads. There are currently some 0.01 Mha under plantation crops of oil palm and rubber, and 0.4 Mha dedicated to field crops (mainly maize, soybean, rice, and sorghum); most of the land-use intensification occurred on 0.98 Mha of improved pastures (*B. dictyoneura, B. decumbens*). Figure 11.5 shows that the total area under crops, pastures, and forestry plantations in Colombia is around 6 percent of the total area of the llanos. In Venezuela, agricultural lands account for 1 Mha, of which 0.8 Mha are on forest plantations, mainly of *Pinus caribbea,* and some 5 Mha are on improved pastures. The proportion of managed land is around 26 percent, indicative of the more intensive use of the llanos in Venezuela. Both countries still have vast areas of savannas that could be converted into more intensive land-use systems.

It is widely documented that agricultural intensification results in net changes in the amounts of carbon stored in soils (Bravo, 2000; Gomez, 2004). Conventional tillage management usually decreases the C from the topsoil by 20 to 40 percent over a 20-year period (Cihacek and Ulmer, 1995). There are limited data about C dynamics under cultivated lands in the llanos (Lozano, 1999; Hernández and López, 2002). Data from a long-term experiment conducted in the Carimagua Research Station in the mid-

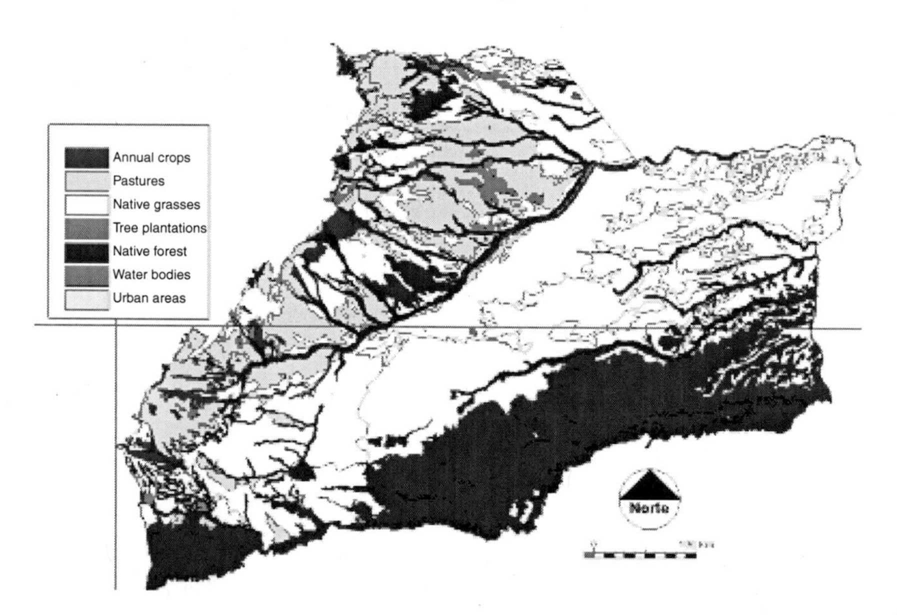

FIGURE 11.5. Current land-use distribution in the llanos from Colombia. (*Source:* Adapted from IGAC, 2003.)

dle of the Colombian llanos, to study the sustainability of different land-use options, including maize-based and rice-based systems, is presented herein. The experimental details were reported before (Friesen et al., 1998). Continuous cultivation of upland rice, maize, and rotation of these two crops with soybean or cowpea were evaluated during an eight-year period. Other systems included grass-legume pastures (*B. humidicola* and *A. pintoi; Panicum maximum and Phaseolus phaseoloides)* and native savanna. Each experimental plot measured 200 × 18 m (0.36 ha), and the experiment was replicated four times. The soil is a typic Haplustox (Table 11.1). At the time of establishment in 1993, a detailed soil sampling was done in the area. Soil C was determined to 40 cm depth with 5 cm increments to 20 cm, and then for every 10 cm depth. SBD was measured for the same soil increments. The plots were monitored annually until the end of the experiment in 2003. The data from the initial and final measurements for some of the treatments are summarized in Figure 11.6 for total carbon stocks as well as stocks in each soil layer increment for the maize-based systems.

After eight years of cultivation, there were no significant differences in the C stored in the top 40 cm of the soil among the savanna plots, the *P. maximum* plots, and the maize-based systems plots. The higher nitrogen inputs to these plots could probably explain this trend, which has been also reported previously for other soils (Halvorson and Reule, 1999). These results

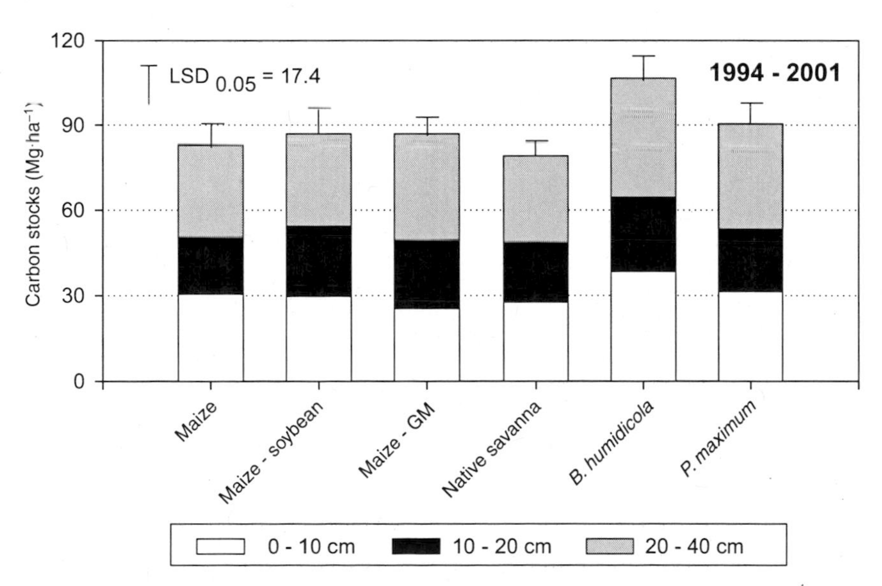

FIGURE 11.6. Soil carbon stock under a long-term experiment on clay-loam Oxisols from central Colombian llanos.

contrast with several studies both in temperate and tropical regions, indicating that cultivation usually a result in net losses of C (Reicosky et al., 2003; Freixo et al., 2002). However, long-term studies reported by San José et al. (2003) for the eastern Venezuelan llanos indicate also that after 30 years of continuous cultivation, there was a nonsignificant decrease in the net soil C stocks. For soils of the central llanos in Venezuela, intensive tillage resulted in net losses of up to 20 percent in 20 years (Hernández and López, 2002), while the use of no-till caused a net increase in total C stocks of 7 percent compared to native savanna (Hernández and López, 2002). The C accrual in no-till maize occurred at the rate of 0.8 Mg C per year in regions of the Venezuelan rolling hills (Hernández and López, 2002). Most of the new carbon goes to the soil maccroagregates and to the light fraction of soil organic matter. The data showed that there was no net loss in C from cropping in the Colombian llanos, compared with a weighted average loss of 2 Mg $C{\cdot}ha^{-1}$ estimated for Venezuelan croplands under conventional tillage systems. No-till cropping increased C content by 4 Mg $C{\cdot}ha^{-1}$ on average for the lands where this practice has been implemented in Venezuela. The area under no till is still incipient in Colombia. As shown in Figure 11.6, the conversion of native savannas on clay-loam soils into pastures of *B. humidicola* and *A. pintoi* resulted in net increases in soil C stocks of 25 Mg $C{\cdot}ha^{-1}$ over eight years. Very similar rates of SOC accumulation (3 Mg $C{\cdot}ha^{-1}$ per year) have been reported on pastures of *B. dictyoneura* associated with *C. macrocarpum* on a sandy soil from well-drained plains in Venezuela (Hernández et al., 2004). These results confirm previous findings (Rao, 1998; Trujillo, 2000) in clay-loam soils from central llanos in Colombia, indicating that introduced pastures can significantly enhance total C in soils. Though the magnitude of the accumulated C found in this study is smaller than previously reported in similar soils down to 2 m soil depth (Fisher et al., 1994), the C pool in this study was estimated to 1 m depth. However, our values are in agreement with a net average increase of around 18 Mg $C{\cdot}ha^{-1}$ in savannas converted to pasture (San José et al., 2003).

Figure 11.6 shows that most of the accumulation in grasses occurs in the top 10 cm of the soils. This may be the result of higher influence of litter added to surface layers and also of the high biomass of roots concentrated in that layer. Fisher and Thomas (2004) report that the rate at which litter decays at the soil surface has been grossly underestimated in the past. Rao (1998) evaluated different grasses in the same area and indicated that introduced grasses have up to 5.7 $Mg{\cdot}ha^{-1}$ of root biomass compared to 1.4 $Mg{\cdot}ha^{-1}$ for native savannas down to 80 cm depth. Rao also observed that up to 73 percent of the root biomass from the grasses is concentrated in the top 20 cm layer. Trujillo (2000) found in Colombian high plateaus that standing root biomass was about three times more in pastures of *B. dictyoneura* (8.6 Mg·ha^{-1}) than in native savanna (2.9 Mg·ha^{-1}).

The amount of total C that can be accumulated in soils when native savanna is converted into pastures with *Brachairia* species varies among locations in the llanos. Fisher et al. (1994) reported accumulations during a nine-year period in the top 1 m of soil of 25.6 Mg C·ha^{-1} for *B. humidicola* grass and of 70.4 Mg·ha^{-1} for associated pastures of *B. humidicola* and *A. pintoi* in the middle of the Colombian llanos. In contrast, San José et al. (2003) reported no net change in C stocks in the top 30 cm of the soil after 30 years of a pasture of *B. decumbens* in the state of Monagas in the eastern Venezuelan llanos. Estimates presented herein are based on values repeated in Figure 11.6 as representative of the high plateaus of Colombia, assuming that appropriate management, fertilization, and maintenance could prevent pasture degradation. It is uncertain, however, the span of time that pastures could maintain high productivity (Acevedo, 2003). For crops in the hillsides of Venezuela, values reported by Hernández and López (2002) were used which indicated a net decrease of 18 percent in SOC by conventional till versus a net gain of 7 percent relative to the native savanna by no-tillage systems. Data reported in the literature are used to estimate net change in C stock due to land-use change in other regions of the Venezuela llanos (Bravo, 2000; Campos, 1999; Larreal et al., 1975; Hernández and Domínguez, 2002; Hernández et al., 2004). Data from Campos (1999) report no net changes in C storage in the soil under a 30-year-old pine plantation in eastern Venezuela, despite high production of litter. Due to lack of data to calculate the C change under tree plantation in Colombia, no net change in soil C stocks under these plantations is assumed.

Using these data, net change in soil C stocks that have resulted from the conversion of native savannas into more intensive land use (i.e., agriculture, pasture, or forestry land) in the llanos have been estimated (Table 11.5). Given the relative small area converted into crops and the reported low or no losses of C due to cultivation in the region, agriculture and forestry are not seen to have caused large impacts on total stocks of C in soils from the llanos, accounting for a net loss of 1 Tg C. Pastures dominate the managed lands in the llanos and have had a clear impact in increasing the overall C stocks of the llanos to 115 Tg C . The available data show that tree plantations are neutral in terms of soil C stock change. Overall, a net increase in C stocks in the top 30 cm of soil from the llanos of 114 Tg C has been estimated as a result of land intensification in this ecosystem. With an estimated total C stock in the llanos prior to changes in land use of about 2,985 Tg C, the soil C accrual due to pastures represents an increase of about 4 percent over the baseline level.

CARBON SEQUESTRATION POTENTIAL

The previous analysis was based on the assumptions that 12 percent of current cropland is managed with no-till systems. If all current crops in the llanos can be transformed into no-till systems, this could result in a net additional increase of 5 Tg C in the soil C stocks. Intensification expected to occur in the region in the two coming decades foresees an additional 2 Mha converted to agriculture. Assuming that no-till may be able to increase soil C by 4 Mg C·ha^{-1} relative to savannas (Hernández and López, 2002), and that the new land is managed using conservation practices, this may result in a an additional 8 Tg C being sequestered in soils. Prospects of increases in C stocks associated with cropping seems, therefore, to be modest in the llanos.

It is expected that the area under forestry plantations will grow steadily at an average rate of 0.05 Mha per year. With currently available data there is no indication that net changes in soil C stocks will result from expansion of tree plantations. However, planting trees has the advantage that significant amounts of C can be accumulated in the tree biomass, as compared with other land-use systems. Recently, the Ministry of Agriculture of Colombia launched a program to promote the plantation of rubber, pines, and oil palm in vast regions of the central Colombian llanos. It is feasible that tree plantation under soils of higher clay and SOM content than the soils from the eastern Venezuelan llanos, where most of current forest plantations are located, may result in net increases in SOM of soil under these plantations.

An important opportunity for C sequestration in the region lies in the restoration of degraded pastures in the area and the establishment of productive new grass-legume pastures to intensify cattle raising. With current expansion, it is foreseen that some 5 Mha of new land will be converted into pastures in the next two decades, 2 Mha of which is in Colombia. This would in itself result in approximately 148 Tg C sequestered in soils. If intensification of livestock occurs in lieu of converting part of the savannas to pastures, other areas of savanna could be protected against fire through proper policies and incentives. This could accelerate the recolonization of woody vegetation with potentially high C gains in the biomass in the range of 1 Mg C·ha^{-1} per year. (San José et al., 2003) to 8 Mg C·ha^{-1} per year (Güerere, 1992), depending on the original vegetation cover. This possibility is, however, more remote, as it would require further management strategies to protect the vegetation against fire, at least during the initial years. Present analysis is based on the assumption that the area under gallery forest will continue to be unaffected. This is desirable for several reasons, ranging from preserving endemic biodiversity hosted in such areas to maintaining high levels of SOC. It is unlikely than any alternative land use may increase soil carbon levels beyond the levels of the gallery forest. Gallery forests are very important to moderate the hydrological cycles in the llanos; therefore governments must aim at their preservation.

The establishment of agroforestry systems has been very slow in the llanos, and high labor requirements for these activities would probably limit the areas that may be converted in the near future. Oil palm plantations are promising alternatives for expansion of agroforestry systems. Trees are, however, expected to play a central role in the future development of the llanos in both countries. The introduction of silvopastoral systems offers probably the most beneficial option to enhance C in the system, by combining accumulation in both soils and biomass.

The maximum potential of the llanos to sequester carbon in the soils would result from a total conversion of available land into improved pastures. This may theoretically generate a total increase of soil C stocks of about 1.02 Pg C. Given current trends in the development of the llanos, it may be expected that approximately 160 Tg C can be added to soil C stocks in the next 20 years.

UNCERTAINTIES IN THE SOCIAL AND INTERNATIONAL CONTEXTS

Although the estimates presented here have still a range of uncertainty due to limited availability of data for C content in more points of the landscape and principally due to poor data on SBD profiles, estimates presented within this study provide a reasonable gross estimate of the potential of the region. Nevertheless, much higher uncertainties appear in the social and political context, to foresee how much of the total potential to accumulate C in soils from the llanos may be realized in the coming decades.

Livestock intensification will be the main avenue for enhancing C stocks in soils, but whether the projected expansion will happen depends on numerous factors beyond the control of farmers, regional, and even national authorities. The influence that the common-market schemes which are being negotiated between the governments of Colombia and Venezuela and other countries in Latin America, as well as the ALCA, the free trade agreement with the United States, will have on agricultural expansion is difficult to predict. Most economists seem to agree that this will result in a deceleration of the agricultural expansion in the llanos (L. Rivas, personal communication). Nevertheless, current political trends in Venezuela strongly oppose the ALCA and propose an alternative economic model based on endogenous development, agrarian reform, strengthening of cooperatives, and the egalitarian access to credits, which could result in agriculture expansion and livestock intensification. However, this political process is new and it is too soon to predict its agricultural and environmental impacts.

In Colombia, the effect of the social conflict which has afflicted the country during past decades has already had a negative effect on agricultural expansion and slowed governmental plans to increase maize production in the region. This has moved investors to look into plantations of industrial crops such as oil palm or rubber trees, which are considered safer than annual crops or livestock. It is likely that at least this sector will continue to grow at the expected rate in Colombia. Indications of this were recently given by the government when launching a development plan for the central llanos of Colombia.

OPPORTUNITIES FOR CARBON TRADING IN THE LLANOS

Despite all uncertainties, the vast area of land that is suitable for intensive land use in the llanos makes the region a good candidate for carbon trading projects when suitable markets materialize through the CDM (Clean Development Mechanism) under the Kyoto Protocol or other mechanisms which are being promoted by the European community. Nevertheless, under current agreements, for the first period of the Kyoto Protocol (2008 to 2012), soils are excluded for trading within the CDM (UNFCC, 2004). This makes forest plantations, agroforestry, or silvopastoral systems the most economically acceptable options available now to trade C and the ones that most likely could start materializing in the region.

Currently, the economic margin of agricultural activities is modest in the llanos. Any additional income resulting from carbon trading would make investments in the llanos an attractive alternative for current farmers and for new investors. The monitoring of C trading projects in this area is easier and should be less expensive than in other more heterogeneous environments such as the hillsides. Establishment of forestry plantations, silvopastoral, or agroforestry systems in the region are the most attractive options, given that in addition to possible C accumulation in the soils, large amounts of C could be accumulated in the standing biomass of trees. For the pine plantations in Venezuela, as much as 50 Mg $C{\cdot}ha^{-1}$ can be accumulated in the biomass in a 10-year period (Hoyos, 1998). This is twice the equivalent amount accumulated in the soils under improved pastures. Combining trees and improved pastures in properly managed silvopastoral systems offers a win-win situation to sequester C in the llanos. Gains in soil C stocks favored by the pastures may be complemented by CO_2 capture in tree biomass. Nevertheless, tree plantations present important agronomic challenges, as evidenced by the large single-species plantations at Uverito, Venezuela, where plantation health and timber production are hampered by several pests and diseases difficult to control. The selection of the most appropriate combinations of pastures, forage legumes, and trees and its effect on total C balances will need further research.

This chapter has focused only on the carbon in soils. Nevertheless, it has been reported that in the llanos, net balances of greenhouse gases are strongly influenced by the emissions resulting from burning the native vegetation (Scharfe et al., 1990). As much as 41 percent of total methane and 35 percent of the nitrous oxide emissions in savannas from high plateaus in Colombia result from periodic burning (Rondón, 2000). Conversion of native vegetation into permanent crops, pastures, agroforestry, or silvopastoral systems eliminates annual emissions due to fire and may greatly contribute to net CO_2 equivalent gains in the system. Although some advances have been made in understanding the role of agriculture and livestock intensification on net balances of greenhouse gases in the region (Rondón, 2000), a proper accounting of the integrated balance of emissions associated with land-use intensification is still a major challenge that national and regional research institutions need to address to successfully implement any potential C trading scheme in the region.

CONCLUSIONS AND FUTURE PERSPECTIVES

Soils from the llanos have a large potential to sequester up to 1.02 Pg C, assuming that all the native land could be converted into well-managed and sustainable pastures. However, this option is not necessarily desirable. Neotropical savannas play key roles in planetary biogeochemical cycles that could seriously change if a few exotic species completely displaced the large biodiversity of this ecosystem. Savannas are a major repository of animal diversity as well, particularly birds, and even serve as a summer habitat for migratory birds coming from as far as Alaska. Savannas should be preserved and protected to ensure that coming generations of humans may enjoy their diverse environmental services and functions. On the other hand, the countries with savannas probably need and are willing to develop parts of these environments to improve the production of food, services, and income for a continuously increasing population.

Technologies are currently available to allow sustainable agriculture, livestock, and forestry in the region. Most of these technologies also sequester atmospheric carbon in the biomass or in the soils. The payments for equivalent carbon accumulation in the llanos that could result from implementing these methodologies may accentuate the adoption process. The decision to use these technologies lies in the hands of farmers and local and national governments. However, the final decision to what extent the llanos may exploit its inherent potential to sequester carbon in soils and biomass mostly lies with international policymakers and negotiators. Fortunately, for the sake of the soils, most trends are showing an interest in the region to move into sustainable agriculture, and this will result in more organic matter being sequestered into the soils of the llanos. This could have important benefits not only for the global environment through reductions in the net fluxes of greenhouse gases from the llanos to the atmosphere, but most importantly to the health status of the soils of the region. Healthier and more resilient soils are the best legacy that current farmers could leave for the future generations. It is a great challenge for researchers, institutions, farmers, and local and national authorities to find the mechanisms to enable that payments from environmental services resulting from land-use intensification in the llanos could be translated into sustainable development of the region. Another similarly daunting task will be to find the proper balance between development and conservation, so the neotropical savanna ecosystem may continue playing its important, though not properly understood role in the planet.

REFERENCES

Acevedo, D. 2003. Producción primaria y acumulación de nitrógeno en una pastura tropical bajo tratamientos de corte y fertilización. Tesis doctoral, Facultad de Ciencias, Universidad de Los Andes, Mérida, Venezuela.

Amézquita, E., D.K. Friesen, M. Rivera, I.M. Rao, E. Barrios, J.J. Jiménez, T. Decaëns, and R.J. Thomas. 2002. Sustainability of crop rotation and ley pasture systems on the acid-soil savannas of South America. 17th World Congress of Soil Science, Bangkok, Thailand, August 14-21, 2002.

Amézquita, E., D.K. Friesen, and J.I. Sanz. 2004. Sustainability indicators: Edapho-climatic parameters and diagnosis of the cultural profile. In E.P. Guimarães, J.I. Sanz, I.M. Rao, M.C. Amézquita, E. Amézquita, and R.J. Thomas (eds.), *Agropastoral systems for the tropical savannas of Latin America.* Publication 338. CIAT-EMBRAPA, Cali, Colombia, pp. 59-76.

Arias, L. 1992. Uso de la tierra. Del conuco a la agricultura industrial. Imagen Atlas de Venezuela, una visión espacial. Editorial Arte, Caracas, Venezuela.

Berroterán, J.L. 1988. Paisajes ecológicos de sabanas en Llanos Altos Centrales de Venezuela. *Ecotrópicos* 1(2): 92-107.

Botero, P.J. and D.H. Serrano. 1992. Estudio comparativo de Orinoquia-Amazonía (ORAM) Colombianas. *Revista CIAF* 13(1): 87-115.

Bourlière, F. (ed). 1983. *Ecosystems of the world* Vol. 13, *Tropical savannas.* Elsevier, Amsterdam.

Bravo, C. 2000. Efecto de la siembra directa y labranza convencional sobre las propiedades físicas y químicas del suelo y su influencia en el rendimiento del maíz en un Alfisol del Estado Guárico. Trabajo de ascenso para la categoría de agregado. IDECYT-UNESR.

Campos, A. 1999. Efecto de la siembra de *Pinus caribaea* L. en fracciones de materia orgánica de un suelo de sabana, Uverito-Estado Monagas. Tesis de Licenciatura, Facultad de Ciencias, Universidad Central de Venezuela, Caracas.

CA/OCEI (Censo Agrícola/Oficina Central de Estadística e Informática). 1961-1998. *Censo anual.* Caracas, Venezuela.

Cihace, L.J. and M.G. Ulmer. 1995. Estimated soil organic carbon losses from long-term crop fallow in the northern great plains of the USA. In R. Lal, J. Kimble, E. Levine, and B. Stewart (eds.), *Soil management and greenhouse effect.* Advances in Soil Science. Lewis Publishers, Boca Raton, FL, pp. 85-93.

Colmenares, E., A. Corzo, and P. Hernández. 1974. *Estudio de suelos gran visión Uribante-Arauca: Informe de Avance.* División de Edafología, Ministerio de Obras Públicas, San Cristóbal, Venezuela.

Comerma, J.A. and A. Chirinos. 1976. *Características de algunos suelos con y sin horizonte argílico en las Mesas Orientales de Venezuela.* Centro Nacional de Investigaciones Agropecuarias-CNIA, Maracay, Venezuela.

Comerma, J.A. and O. Luque. 1971. Los principales suelos y paisajes del estado Apure. *Agronomía Tropical* 21(5): 379-396.

Fisher, M.J., J.J. Jiménez, T. Decaëns, A.G. Moreno, J.P. Rossi, P. Lavelle, and R.J. Thomas. 1994. Dynamics and short term effects of earthworms in natural and managed savannas at the Eastern Plains of Colombia. In *Annual report 1994.* Working document 148. Centro Internacional de Agricultura Tropical (CIAT), Cali, Colombia, pp. 208-216.

Fisher, M.J. and R.J. Thomas. 2004. Implications of land use change to introduced pastures on carbon stocks in the central lowlands of tropical South America. *Environ. Devel. Sust.* 6: 111-131.

Fisher, M.J., R.J. Thomas, and I.M. Rao. 1997. Management of tropical pastures in acid soils savannas of South America for carbon sequestration in the soil. In R. Lal, J. Kimble, R. Follett, and B. Stewart (eds.), *Management of carbon sequestration in soil.* CRC Press, New York, pp. 405-420.

Freixo, A.A., P. Machado, C.M. Guimarães, C.A. Silva, and F.S. Fadigas. 2002. Carbon and nitrogen storage and organic fraction distribution of a Cerrado Latosol under different cultivation systems. *Revista Brasileira de Ciencia do Solo* 26(2): 425-434.

Friesen, D.K., R.J. Thomas, M. Rivera, N. Asakawa, and W. Bowen. 1998. Nitrogen dynamics under monocultures and crop rotations on a Colombian savannas Oxisol. 16th World Congress of Soil Sciences, Montpellier, France, August 20-26, 1998.

Gómez, I. 2004. Efecto del manejo agrícola sobre algunos parámetros bioquímicos y microbiológicos del suelo. IVIC.

González de Juana, C., J.M. Iturralde de Arozena, and X. Picard-Cadillat. 1980. *Geología de Venezuela y de sus cuencas petrolíferas.* Ediciones FONINVES, Caracas, Venezuela.

Goosen, D. 1964. Geomorfología de los Llanos Orientales. *Revista de la Academia Colombiana de Ciencias* 12(49): 129-139.

Güerere, I. 1992. Comparación de parámetros químicos, físicos y de la biomasa microbiana del suelo entre una sabana protegida del fuego y una sabana quemada anualmente. Trabajo especial de grado, Universidad Central de Venezuela.

Halvorson, A.D. and C.A. Reule. 1999. Long-term nitrogen fertilization benefits soil carbon sequestration. *Better Crops with Plant Food* 83(4): 16-22.

Hernández-Hernández, R.M. and C. Domínguez. 2002. Efecto de prácticas agrícolas usadas en distintas unidades de producción de maíz y sorgo sobre algunas propiedades bioquímicas de suelos del Estado Guárico. *Agrobiológica* 2: 10-18.

Hernández-Hernández, R.M. and D. López-Hernández. 2002. El tipo de labranza como agente modificador de la materia orgánica: un modelo para suelos de sabana de los Llanos Centrales Venezolanos. *INTERCIENCIA* 27(10): 529-536.

Hernández-Hernández, R.M., Z. Lozano, C. Bravo, B. Moreno, and L. Piñango. 2004. Alternativas para el mejoramiento de la productividad del sistema maíz-ganado en suelos del Estado Guárico. Proyecto FONACIT, Venezuela.

Houghton, R.A. 2003. Revised estimates of the annual net flux of carbon to the atmosphere from changes in land use and land management. *Tellus,* Ser. B, 55: 378-390.

Hoyos, J.F. 1998. Sembrando petróleo: Sabanas que se transforman en bosques. *Natura* 113: 41-45.

Hoyos Garcés, P., E. Amézquita Collazos, and D.L. Molina López. 2004. Estrategias para la construcción de capas arables productivas en dos suelos de la Altillanura Colombiana. Informe Final 2001-2003. Proyecto No. 201504060, CIAT-PRONATTA.

Huntley, B. and B.H. Walker (eds.). 1982. *Ecology of tropical savannas*. Springer. Berlin.

IGAC. 1983. Mapa de suelos de la República de Colombia. Escala 1:500000. Instituto Geográfico Augustín Codazzi, Bogotá, Colombia.

IGAC. 1991. Estudio semidetallado de suelos, sector Carimagua, Gaviotas, Departamento del Meta y Vichada. Instituto Geográfico Augustín Codazzi, Bogotá, Colombia.

IGAC. 1999a. Casanare, características geográficas. Instituto Geográfico Augustín Codazzi, Bogotá, Colombia.

IGAC. 1999b. *Paisajes Fisiograficos de la Orinoquia y Amazonia Colombiana, Analisis Geograficos,* Vol. 27-28, pp. 4-24. Instituto Geográfico Agustín Codazzi, Bogotá, Colombia.

IGAC. 2002. *Atlas de Colombia,* 5 edición. Instituto Geográfico Augustín Codazzi, Bogotá, Colombia.

IGAC. 2003. Cobertura y Uso actual de las Tierras en Colombia: La Orinoquia Colombiana. Instituto Geográfico Augustín Codazzi, Bogotá, Colombia. CD No. 2.

Lal, R. 2002. The potential of soils of the tropics to sequester carbon and mitigate the greenhouse effect. *Advances in Agronomy* 76: 1-30.

Larreal, M., L. Graterol, and S. Mazzei. 1975. Estudio de suelos semidetallado unidad agrícola de Turén. División de Edafología, Ministerio de Obras Públicas, Guanare.

Lobo, D., Z. Lozano, and I. Pla. 2002. Limitaciones físicas para la penetración de raíces de maíz (*Zea mays* L.) y sorgo (*Sorghum bicolor* L.) en cuatro suelos de Venezuela. *Revista Venesuelos* 4(1-2): 19-24.

Lozano, Z. 1999. Evaluación de propiedades físicas y químicas de dos suelos de los Llanos Occidentales con sistemas de labranza convencional y reducida. Trabajo de ascenso para optar a la categoría de asistente. Facultad de Agronomía. UCV-Maracay, Venezuela. 63 p.

Lozano, Z., D. Lobo, and I. Pla. 2000. Diagnóstico de limitaciones físicas en Inceptisoles de los Llanos Occidentales Venezolanos. *Biogra* 12(1): 15-24.

Malavé, V. 1981. El proceso de mineralización de fósforo orgánico en suelos de Uverito en relación con algunas variables bióticas y abióticas. Trabajo especial de grado, Universidad Central de Venezuela.

MARNR (Ministerio del Ambiente y de los Recursos Naturales Renovables). 1986. Inventario nacional de tierras de los Llanos Occidentales. División de Información e Investigación del Ambiente, Programa Inventario Nacional de Tierras, Caracas, Venezuela.

MARNR (Ministerio del Ambiente y de los Recursos Naturales Renovables). 1996. Inventario emisiones de gases de efecto invernadero Venezuela. Gráfica León, SRL, Caracas, Venezuela.

Narro, L., S. Pandey, A. León, J.C. Pérez, and F. Salazar. 2004. Maize varieties for acid soils. In E.P. Guimarães, J.I. Sanz, I.M. Rao, M.C. Amézquita, E. Amézquita, and R.J. Thomas (eds.), *Agropastoral systems for the tropical savannas of Latin America.* Publication 338. CIAT-EMBRAPA, Cali, Colombia, pp. 141-155.

Pérez-Materán, J.R., A. Corzo, J.M. Gómez, M. Larreal, and T. Valerio. 1980. Estudio de suelos preliminar del polígono de expropiación de módulos de Apure. MARNR. Serie de Informes Científicos Zona 3/1C/27, Barquisimeto, Venezuela.

Phillips, O.L., Y. Malhi, N. Higuchi, W.F. Laurence, P.V. Nunez, R.M. Vásquez, S.G. Laurence, L.V. Ferreira, M. Stern, S. Brown, and J. Grace. 1998. Changes in the carbon balance of tropical forests: Evidence from long-term plots. *Science* 282: 439-442.

Pinillos, M. 1999. Modelo hidrológico de simulación en los Llanos inundables del estado Apure. Tesis de Maestría, Postgrado de Ecología Tropical, Instituto de Ciencias Ambientales y Ecológicas, Facultad de Ciencias, Universidad de Los Andes, Mérida, Venezuela.

PINT. 1979. Inventario nacional de tierras Llanos Centro Occidentales. MARNR, Programa Inventario Nacional de Tierras. Serie de Informes Científico Zona 2/1C/22. Maracay, Venezuela.

PINT. 1985. Inventario nacional de tierras Llanos Occidentales. MARNR, Programa Inventario Nacional de Tierras. Serie de Informes Científico Zona 2/1C/63. Maracay, Venezuela.

PINT. 1990. Inventario nacional de tierras Guárico Central y Sur de Aragua. MARNR, Programa Inventario Nacional de Tierras. Serie de Informes Científicos Zona 2/1C/66. Maracay, Venezuela.

Rao, I.M. 1998. Root distribution and production in native and introduced pastures in the South American savannas. In J.E. Box Jr. (ed.), *Proc. 5th Symposium of the International Society of Root Research: Root demographics and their efficiencies in sustainable agriculture, grasslands and forest ecosystems.* Kluwer Academic Publishers, Dordrecht, the Netherlands, pp. 19-41.

Reicosky, D.C., L. García-Torres, J. Benites, A. Martínez-Vilela, and A. Holgado-Cabrera. 2003. Tillage-induced CO_2 emissions and carbon sequestration: Effect of secondary tillage and compaction. In L. García-Torres, J. Benites, and A. Martínez-Vilela (eds.), *Conservation agriculture: Environment, farmers experience, innovations, socio-economy, policy.* Kluwer Academic Publishers, Dordrecht, the Netherlands, pp. 291-300.

Rippstein, G., G. Allard, and J. Corbin. 2000. Fire management of natural grasslands and cattle productivity in the lower Eastern Plains of Colombia. *Revue-d'Elevage et de Medecine Veterinaire des Pays Tropicaux* 53(4): 337-347.

Rippstein, G., E. Amézquita, G. Escobar, and C. Grollier. 2001. Condiciones naturales de la sabana. In G. Rippstein, G. Escobar, and F. Motta (eds.), *Agroecología y Biodiversidad de las Sabanas en los Llanos Orinetales de Colombia.* CIAT, Colombia, pp. 1-21.

Rippstein, G., C. Lascano, and T. Decaëns. 1996. La production fourragère dans les savanes d'Amérique du Sud intertropicale. *Fourrages* 145: 33-52.

Rondón, M.A. 2000. Land use change and balances of greenhouse gases in Colombian tropical savannas. PhD dissertation, Cornell University, Ithaca, NY.

San José, J.J. and R.A. Montes. 1991. Regional interpretation of environmental gradients which influence Trachypogon savannas in the Orinoco Llanos. *Vegetation* 95: 21-32.

San José, J.J., R.A. Montes, and M.R. Farinas. 1998. Carbon stocks and fluxes in a temporal scaling from a savanna to a semi-deciduous forest. *Forest Ecology and Management* 105(1-3): 251-262.

San José J.J., R.A. Montes and C. Rocha. 2003. Neotropical savanna converted to food cropping and cattle feeding systems: Soil carbon and nitrogen changes over 30 years. *Forest Ecol. Managm.* 184(1-3): 17-32.

Sanz, J.I., R.S. Zeigler, S. Sarkarung, D.L. Molina, and M. Rivera. 2004. Improved rice/pasture systems for native savannas and degraded pastures in acid soils of Latina America. In E.P. Guimarães, J.I. Sanz, I.M. Rao, M.C. Amézquita, E. Amézquita, and R.J. Thomas (eds.), *Agropastoral systems for the tropical savannas of Latin America.* Publication 338. CIAT-EMBRAPA, Cali, Colombia, pp. 240-247.

Sarmiento, G. 1984. *The ecology of neotropical savannas.* Harvard University Press, Cambridge, MA.

Sarmiento, G. 1990. Ecología comparada de ecosistemas de sabanas en América del Sur. In G. Sarmiento (ed.), *Las sabanas Americanas: Aspectos de su Biogeografía, Ecología y Utilización.* CIELAT, Fundación Fondo Editorial Acta Científica Venezolana, Mérida, Venezuela, pp. 15-56.

Sarmiento, G. 2000. *La transformación de los ecosistemas en América Latina.* Laffont Ediciones Electrónicas, Argentina.

Sarmiento, G. and D. Acevedo. 1991. Dinámica del agua en el suelo, evaporación y transpiración en una pastura y un cultivo de maíz sobre un Alfisol en los Llanos Occidentales de Venezuela. *Ecotropicos* 4(1): 27-42.

Sarmiento, G. and M. Pinillos. 2001. Patterns and processes in a seasonally flooded tropical plain: The Apure Llanos Venezuela. *J. Biogeography* 28: 985-996.

Scharfe, D., W. Hao, L. Donoso, and P. Crutzen. 1990. Soil fluyes and atmospheric emissions from soils of the northern part of the Guyana shield, *Venezuela. J. Geophy. Res.* 95(D13): 22475-22480.

Schargel, R. 1972. Características y génesis de una cronosecuencia de suelos desarrollada sobre depósitos aluviales entre los Ríos Boconó y Masparro, Estado Barinas. *Agronomía Tropical* 22(4): 345-373.

Schargel, R. 1978. Características de algunos suelos con arcilla de baja actividad de los Llanos. MARNR, Serie de Informes Técnicos Zona 8/1T/24. Guanare, Venezuela.

Schargel, R. 2003. Geomorfología y suelos de los Llanos venezolanos. In J.M. Hétier and R. López Falcón (eds.), *Tierras Llaneras de Venezuela.* IRD-CIDIAT.

Torres Lezama, A., M. Díaz, W. Franco, and H. Ramírez Angulo. 2003. Mortalidad en plantaciones de *Pinus caribaea* var. *hondurensis* en el Oriente de Venezuela. Available online at iufro.boku.ac.at.

Trujillo, W. 2000. Accretion of organic carbon in the acid soils of the eastern plains of Colombia. PhD dissertation, The Ohio State University, Columbus, OH.

Trujillo, W., E. Amézquita, M.J. Fisher, and R. Lal. 1998. Soil organic carbon dynamics and land use in the Colombian savannas. In R. Lal, J. Kimble, R. Follett, and B. Stewart (eds.), *Soil processes and the carbon cycle.* CRC Press, New York, pp. 267-280.

UNFCC. 2004. The mechanisms under the Kyoto Protocol: Joint implementation, the clean development mechanism and emissions trading. United Nations Framework Convention on Climate Change. Available online at: unfccc.int/kyoto_ mechanisms/itms/1673.php.

Valencia, R.A., C.R. Salamanca, G.E. Navas R., J.E. Baquero P., A. Rincón, and H. Delgado. 2004. Evaluation of agropastoral systems in the Colombian Eastern Plains. In E.P. Guimarães, J.I. Sanz, I.M. Rao, M.C. Amézquita, E. Amézquita, and R.J. Thomas (eds.), *Agropastoral systems for the tropical savannas of Latin America.* Publication 338. CIAT-EMBRAPA, Cali, Colombia, pp. 291-306.

Walker, B. (ed.). 1987. *Determinants of tropical savannas.* IUBS Monograph Series No. 3. IRL Press.

Westin, F. 1962. *The major soils of Venezuela.* Project VEN/TE/LA, FAO, Rome.

Zinck, A. and P. Stagno. 1966. *Estudio edafológico de la zona Santo Domingo-Paguey, Estado Barinas.* División de Edafología, Ministerio de Obras Públicas, Guanare, Venezuela.

Chapter 12

Potential of Soil Carbon Sequestration in the Amazonian Tropical Rainforests

C. E. P. Cerri
C. C. Cerri
M. Bernoux
B. Volkoff
M. A. Rondón

INTRODUCTION

Land-use change usually causes a modification in land cover and an associated change in carbon stocks (Bolin and Sukumar, 2000). The change from one ecosystem to another could occur naturally or be the result of human activity. The latest process has occurred in the tropical rain forest, which is a diverse and complex system that occupies approximately 10 percent of the world's area and comprises about 40 to 50 percent of the earth's species (Meyers, 1981).

Tropical rain forests represent significant sources/sinks for trace gases, and the exchange of CO_2 between forest and the atmosphere is an important component of the global carbon cycle. Global warming damage from tropical deforestation alone is estimated to be $1.4 to 10.3 billion per year (Pearce and Brown, 1994). Fearnside (1996) estimates that the value of global warming damage in the Amazon region is approximately $1,200 to 8,600 per hectare.

Despite the small fraction (about 14 percent) of cleared forest in the Amazon, the total deforested surface is larger than the territory of France or the state of Texas in the United States. In addition to its vast surface, this

Research that led to this study was supported partly by the Global Environment Facility (project number GFL/2740-02-4381), Conselho Nacional de Desenvolvimento Científico e Tecnológico (CNPq), Fundação de Amparo a Pesquisa do Estado de São Paulo (Fapesp), and Institut de Recherche pour le Dévelopement (IRD-France).

Carbon Sequestration in Soils of Latin America

doi:10.1300/5755_12

ecoregion also presents adequate climate conditions for plant growth, expressed by high temperatures over the year and well-distributed precipitation. Therefore, the Amazonian tropical rain forest has, in principle, great potential for soil carbon sequestration, even considering its poor to moderate soil fertility.

The objective of this chapter is to estimate, using data from the literature, the effects of some land-use change scenarios on soil carbon sequestration for the Brazilian Amazon and then broadly extrapolate it to the entire area of the Amazon tropical rain forest.

DESCRIPTION OF THE STUDY AREA

The Amazonian tropical rain forest (Figure 12.1) is the world's largest region of continuous intact tropical forest, accounting for 17 percent of the world's forest area (Figure 12.2). The total area of lowland, humid Amazon forest in 1990 was 710 million ha (Phillips et al., 1998), 500 million ha of which was closed canopy forest (Houghton, 1997).

The Amazon basin covers an area of approximately 700 million ha (Pires and Prance, 1986) and occupies large portions of the national territories of Venezuela, Colombia, Peru, Guyanas, Bolivia, Ecuador, and Brazil (Figure

FIGURE 12.1. The Amazonian tropical rain forest and the Brazilian Amazon.

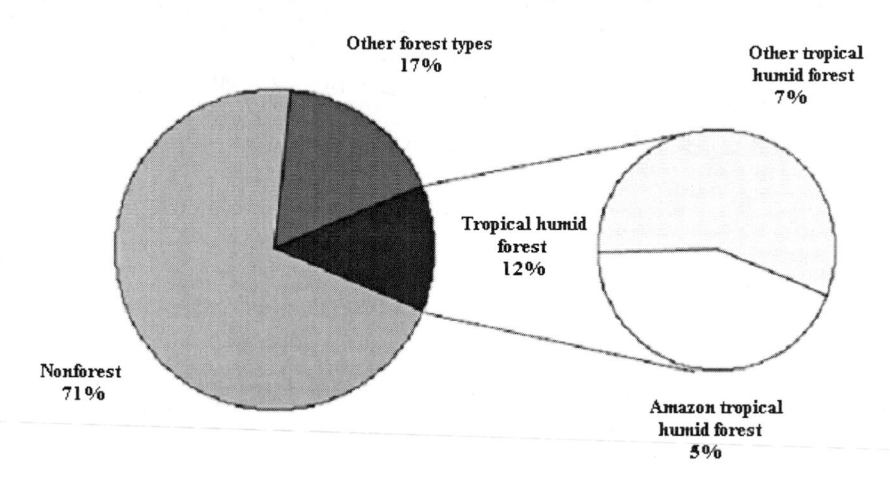

FIGURE 12.2. Global forest cover. (*Source:* Adapted from FAO, 2001, and Phillips et al., 1998.)

12.1). The basin is limited to the west by the Andean Mountains, to the north by the crystalline mass of Guyana and the savannas of Colombia and Venezuela, to the south by the plateau of Mato Grosso, and to the east by the Atlantic Ocean. The length of the basin is more than 3,000 km from west to east, and its width ranges from 300 km in the west to 800 km to the east. Its central part is almost entirely located within Brazilian territory, and forms the Brazilian Amazon area (Figure 12.1), which is mainly developed on sediments from the Pleistocene (Prance and Lovejoy, 1984). As a whole, the Amazon has a hot and humid climate, characterized by only small variations in the diurnal and monthly temperatures. However, because of its vast size and geomorphological heterogeneity, this region represents a large diversity of local climates, with different annual rain distribution and, sometimes, different temperature extremes (Marengo and Nobre, 2001).

This region has the highest rates of deforestation in the world (Skole and Tucker, 1993). For instance, deforestation has been higher in Latin America than in Asia or Africa not only in area (4.3 million ha per year) but also in percentage of forest cleared (0.64 percent per year) (Anderson, 1990). From the Latin American forests standing in 1850, 370 million ha, or 28 percent, had been cleared by 1985. Of that area, 44 percent was converted to pasture, 25 percent to cropland, 20 percent had been degraded, and 10 percent had changed to shifting cultivation (Houghton, 1991).

In the Brazilian Amazon, estimated deforestation rates ranged from 1.1 million to 2.9 million ha per year, and the total area cleared reached approximately 55 million ha, which is about 14 percent of its total area (INPE,

2004). According to Laurance et al. (2001a) this rapid pace of deforestation has several causes. First, nonindigenous populations in the Brazilian Amazon have increased tenfold since the 1960s, from about 2 million to 20 million people, as a result of immigration from other areas of Brazil and high rates of intrinsic growth. Second, industrial logging and mining are growing dramatically in importance, and road networks are expanding that sharply increase access to forests for ranchers and colonists. Third, the spatial patterns of forest loss are changing; past deforestation has been concentrated along the densely populated eastern and southern margins of the basin, but new highways, roads, logging projects, and colonization are now penetrating deep into the heart of the area. Finally, human-ignited wildfires are becoming an increasingly important cause of forest loss, especially in logged or fragmented areas (Laurance et al., 2001a).

Despite the magnitude of deforested area, the Brazilian Amazon still contains about 40 percent of the world's remaining tropical rain forest (see native vegetation types in Figure 12.3) and plays vital roles in maintaining biodiversity, regional hydrology and climate, and the terrestrial carbon storage (Laurance et al., 2001a). The forest accounts for about 10 percent of the

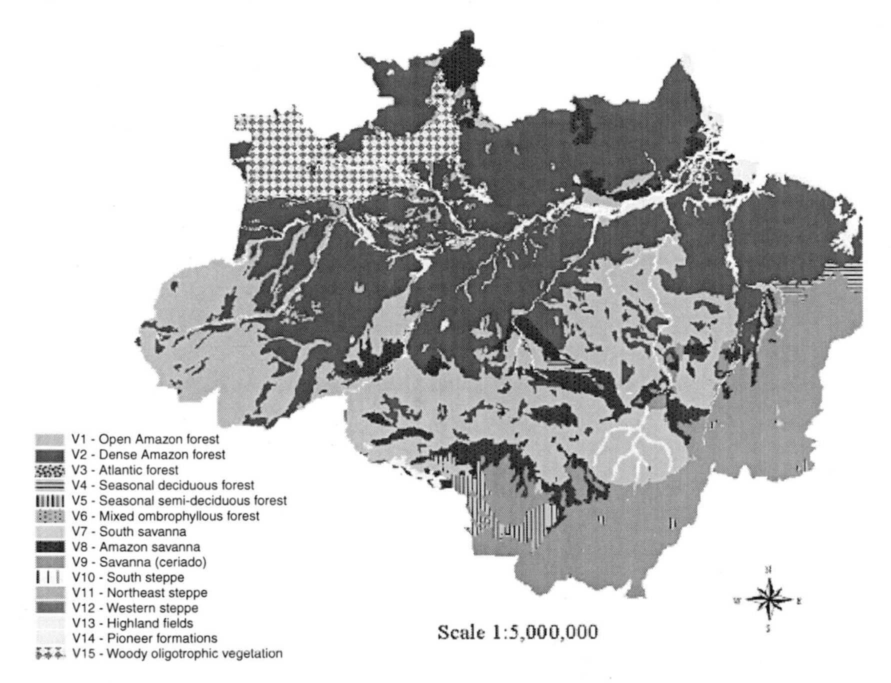

FIGURE 12.3. Native vegetation types in the Brazilian Amazon. (*Source:* Adapted from Bernoux et al., 2002.)

world's terrestrial primary productivity and for a similar fraction of the carbon stored in land ecosystems (Keller et al., 1997). Cattle pasture dominates this once-forested land in most of the basin (Pires and Prance, 1986; Skole and Tucker, 1993; Fearnside and Barbosa, 1998; Dias-Filho et al., 2001).

According to Jacomine and Camargo (1996) two main soil divisions of the Brazilian soil classification, Latossolos (Oxisols) and Podzólicos (mainly Ultisols), cover nearly 75 percent of the area in the Amazon basin (Figure 12.4). The remainder is distributed among 13 soil divisions, only two of which cover more than 5 percent of the Amazon area: Plintossolos (Inceptisols, Oxisols, and Alfisols) and Gleissolos (Entisols and Inceptisols), representing 7.4 percent and 5.3 percent, respectively (Figure 12.5).

The Brazilian Latossolos, Oxisols in the U.S. soil taxonomy, are old, deep, permeable, and well-drained soils. The clay mineral component is predominantly kaolinite, a low-activity clay, with varying amounts of iron and aluminum oxides (Cerri et al., 2000). Generally, the cation exchange capacity is only partially saturated with bases, and the exchangeable Al^{3+} is relatively high (Lal, 1987). The Podzólicos are mainly Ultisols of the U.S. soil taxonomy. They also contain low-activity clays and unsaturated cation exchange capacity. Eutrophic soils (Alfisols) are only occasionally found. Ultisols usually occupy younger geomorphic surfaces than Oxisols, with which they are often associated in the landscape. These soils are thick mineral soils with profiles often deeper than 2 m (Moraes et al., 1996). Moraes

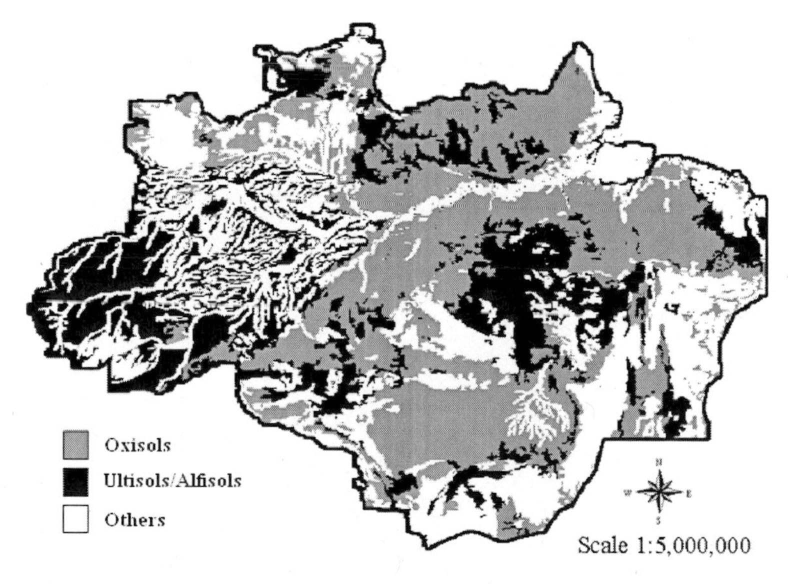

FIGURE 12.4. Simplified soil map of the Brazilian Amazon.

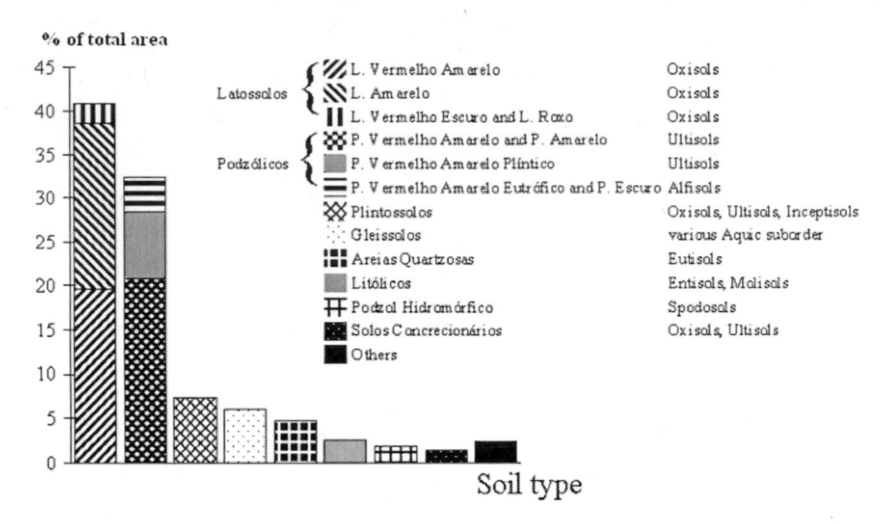

FIGURE 12.5. Relative distrubtion of main soil types in the Amazon. (*Source:* Adapted from Cerri et al., 2000.)

et al. (1995) estimated that approximately 47 Pg C are contained in the soils of the Brazilian Amazon to a depth of 2 m. Of that, 21 Pg occur in the top 20 cm, where changes to soil C stocks that follow land-use alterations are most rapid. Cerri et al. (2000) estimated a carbon stock of 41 Pg C for the 0 to 100 cm, with 23.4 Pg C (i.e., 57 percent) stored in the top 0.3 m.

Ecological characteristics of the Brazilian Amazon described by many researchers (see McClain et al., 2001) include high plant biomass and concentration of nutrients within the plant biomass (Bernoux et al., 2001), rapid rates of nutrient recycling (Cuevas, 2001), high annual rainfall with little seasonal variation in temperature and humidity (Marengo and Nobre, 2001), and a relatively closed system for nutrient and water cycling (Melack and Forsberg, 2001).

SCENARIOS OF SOIL CARBON SEQUESTRATION IN THE AMAZON

As mentioned previously, the objective of this chapter is to estimate, using data from the literature, the effects of some land-use change scenarios on soil carbon sequestration for the Brazilian Amazon and then broadly extrapolate it to the entire area of the Amazon tropical rain forest. In order to do so, we compiled available data from the literature on soil carbon accumulation rates for the following categories of land use and land conversion:

(1) primary forest (avoided deforestation), (2) from forest to well-managed pasture, (3) from degraded pasture to well-managed pasture, (4) from degraded pasture to secondary forest (abandonment) and existing secondary forests, and (5) from degraded pasture to agroforestry.

In this study, the definition of *primary forest* refers to native forest before it is cleared for other land use. *Well-managed pasture* is land used for productive cattle grazing, mainly under African grasses (often from the genus *Brachiaria* and *Panicum*), where weed invasion is controlled and adequate animal stock rate is respected. *Degraded pasture* is dominated by weeds and sheet erosion occurs. *Secondary forest* is the vegetation that develops naturally on abandoned land formerly used for other purposes, mainly pasture. *Agroforestry* is a dynamic management system that diversifies and sustains production for increased social, economic, and environmental benefits for land users at all levels, through the integration of trees in farmland and rangeland.

Primary Forest (Avoided Deforestation)

Undisturbed, mature forests were once thought to be in a steady state in terms of CO_2 flux, with CO_2 uptake by photosynthesis being balanced by CO_2 releases through respiration and biomass decay. However, there is strong evidence to suggest that intact forests are actually responding to the increasing levels of CO_2 in the atmosphere caused by anthropogenic emissions of carbon dioxide, the so-called CO_2 fertilization effect (ECCM, 2002). The extent to which CO_2 fertilization affects productivity of natural forests will depend on a number of factors, including the availability of water and nutrients. However, even a small increase in productivity (results from Norby et al. 1999 suggest a 0.3 percent increase in productivity per year) could produce an observable increase in carbon uptake by forests.

Eddy-covariance studies (monitoring of gas movements in air flows inside and immediately above the forest) allow direct measurement of changes in all ecosystem carbon stocks. However, eddy-covariance studies are limited in temporal and spatial scales, and there are claims that carbon uptake may be overestimated due to various limitations in the method. Several studies have been carried out in so far in the Amazon forests using not only eddy covariance but also forest inventory and carbon model approaches (ECCM, 2002). For instance, Grace et al. (1995), Fan et al. (1990), and Malhi et al. (1998) concluded that Amazon forests were accumulating 1.0, 2.2, and 5.9 Mg C·ha^{-1} per year, respectively. Higuchi et al. (1997) have estimated an uptake of 1.2 Mg C·ha^{-1} per year in 3 ha of forest growth measurements over the 1986-1996 period near Manaus in the central Amazon.

Malhi and Grace (2000) estimated that a 0.3 percent increase would translate to an annual uptake of 1 Mg C·ha^{-1} per year by the Amazon forests. Phillips et al. (1998) concluded that Amazon forests were accumulating 0.62±0.37 Mg C·ha^{-1} per year. A study by Chambers et al. (2001) in the central Amazon reported an accumulation rate of 0.5 Mg C·ha^{-1} per year. Tian et al. (1998) and Prentice and Lloyd (1998) both estimated that average uptake by Amazon forests was 0.3 Mg C·ha^{-1} per year. According to Lal and Kimble (1999), experiments in the Amazon have shown a C sequestration rate of 2 to 3 Mg·ha^{-1} per year. This sequestration rate may be attributed to the CO_2 fertilization effect, tree mortality due to catastrophic events, and effects of El Niño in exacerbating tree mortality. The authors mentioned that, apparently, the rate of new growth exceeds the decomposition and mineralization of dead biomass and soil organic carbon.

In summary, if we take the mean of all available estimates, this would suggest that carbon uptake is currently around 1 Mg C·ha^{-1} per year. This is the value we used in our estimate of potential soil C sequestration.

This value needs to be used with caution, since some of the studies show considerable year-to-year variation. This is particularly important in the modeling studies by Tian et al. (1998) and Prentice and Lloyd (1998), who conclude that in some years Amazon forests acted as a net source of carbon of 0.3 and 0.6 Mg C·ha^{-1} per year, respectively. These variations are primarily due to changes in rainfall and temperature associated with the El Niño weather phenomenon which causes drought in some area of the Amazon.

The future rate of carbon uptake by Amazon forests will therefore depend on the extent to which forests continue to respond to CO_2 fertilization. Some studies (e.g., Cox et al., 2000) suggest that climate change will cause significant drying in the Amazon, creating a carbon source as forests decline. There is also evidence to suggest that CO_2 fertilization will continue to result in carbon uptake for many years to come (Chambers et al., 2001).

Given the vast area of standing forest in the Brazilian Amazon, estimated to be around 344 million ha (Laurance et al., 2001b), and also considering an average uptake of 1 Mg C·ha^{-1} per year (mean of the values presented before), this results in a potential of C sequestration of about 344 Tg C per year (1Tg = 10^{12}g), which is a significant contribution to the global carbon balance.

It is important to stress that large spatial coverage is needed in order to draw better conclusions, as uptake at one site may be balanced by emissions at other sites. According to Fearnside (1999) time scale is undoubtedly also important: over the long term, "mature" forest cannot continue to grow in biomass, but imbalances over periods of years or decades are still important for understanding global carbon dynamics, including clarifications of the "missing sink." An uptake would increase the impact of deforestation by eliminating part of the sink. Although the amount of sink loss due to deforestation in a single year may appear modest compared to the emissions from forest biomass caused by the clearing, the fact that the sink represents an annual flux rather than a one-time emission means that it would have significant consequences over the long term if the sink can be assumed to have a duration of decades or more.

From Forest to Well-Managed Pasture

Cattle pasture represents the largest single use of cleared forest land in most of the Brazilian Amazon. Estimates show that 70 percent of the deforested land has been managed as pasture at one stage or another (Serrão and Toledo, 1990; Dias-Filho et al., 2001). According to Fearnside and Barbosa (1998), about 45 percent of the Brazilian Amazon is occupied by actively grazed cattle pasture which accounts for approximately 24.7 million ha. This is in agreement with other studies (Homma, 1994; Kitamura, 1994; Camarão and Souza Filho, 1999). Farmers were obviously motivated to convert lands cleared from forest into pasture because of the real, or at least perceived, increases in land value which could result from the conversion. Farmers not only maintained cattle as standing "bank accounts" and obtained cash from sales of animals and milk, but also built savings by investing time and resources in pasture, fencing, corrals, and ponds (Fujisaka et al., 1996).

Despite the enormous scale of pasture expansion in the Amazon, there is yet no clear understanding of the direction of the resulting changes in soil C stocks. Fearnside and Barbosa (1998) found that conversion of Amazon forest to pasture can produce a net soil C sink (well-managed pasture) or a net C source (overgrazed pasture), depending on management. According to Neill and Davidson (1999), pasture formation in the Amazon occurs on a variety of soils and in regions that differ in the amount and timing of precipitation. The sequence leading to pasture formation also differs. Some pastures are created by planting grasses directly into forest slash. Others are created after one or two years of annual cropping or after a cropping and fallow sequence. Grass species and practices of interplanting with legumes also differ. All of these factors can influence whether a pasture soil will accumulate or lose C. After establishment, pasture management by stocking rate, burning frequency, effectiveness of weed control, fertilizing, or disking may also affect soil C balance (Neill and Davidson, 1999).

Therefore, in some locations, C stocks in pastures are lower compared with the original forest (Luizão et al., 1992; Desjardins et al., 1994). In other locations, the pasture's grass productivity declines in older pastures, but soil C concentrations remain relatively constant (Falesi, 1976; Serrão et al., 1979; Buschbacher et al., 1988). In yet other locations, inputs of C from roots of pasture grasses cause increases in soil C stocks (Cerri et al., 1991; Bonde et al., 1992; Trumbore et al., 1995; Moraes et al., 1996; Neill et al., 1997; Bernoux et al., 1998; Cerri et al., 2003).

Neill and Davidson (1999) reported that a literature survey of studies in which soil C stocks were followed after deforestation for pasture in the Amazon showed that 19 of 29 pastures examined accumulated C in surface soils and 10 showed C loss. The authors also found a strong relationship between pasture grass species with the change in surface soil C stocks. Pasture planted with *Brachiaria humidicola* tended to lose C, and pastures planted with *Panicum maximum* and *Brachiaria brizantha* tended to gain C.

Moraes et al. (1996) found that total soil C contents to 30 cm in the 20-year-old well-managed pastures were 17 to 20 percent higher than in the original forest sites of the western part of the Amazon region. A comparison of C budgets for forest and pastures in the eastern Amazon was made by Trumbore et al. (1995). In a rehabilitated and fertilized pasture of *Brachiaria brizantha*, they estimated gains, relative to forest soil C stocks, of over 20 Mg $C{\cdot}ha^{-1}$ in the top 1 m of soil and a loss of about 0.5 Mg $C{\cdot}ha^{-1}$ in the 1 to 8 m soil depth interval during the first five years following pasture rehabilitation. More than 50 percent of the forest-derived C in surface soils of pastures on converted Amazon forest turns over in 10 to 30 years (Choné et al., 1991; Trumbore et al., 1995). Cerri et al. (1999) found an accumulation rate of 0.27 Mg $C{\cdot}ha^{-1}$ per year for the layer 0 to 30 cm depth. Neill et al. (1997) reported annual soil C accumulation rate, in the top 50 cm, in the range of 0.2 to 0.3 $Mg{\cdot}ha^{-1}$. Those results are within the range (0.2 to 3.9 Mg $C{\cdot}ha^{-1}$ per year) of the ones reported by Watson et al. (2000) for pastures in wet tropical areas of the world.

For the purpose of the present study, we considered a mean accumulation rate of 0.27 Mg $C{\cdot}ha^{-1}$ per year in the first 0 to 30 cm soil layer. If we assume that only one-fifth of the pasture areas are under well-managed systems (about 5 million ha), it is possible to broadly estimate the potential of soil C sequestration in already established well-managed pasture is approximately 1.35 Tg C per year.

Unlike annual crops, pasture grasses continuously maintain a cover of vegetation on the soil, reduce soil temperatures, and sometimes have high productivity and turnover rates that add organic matter, particularly from belowground, to the soil (Brown and Lugo, 1990). However, any lack of difference between forest and pasture soil contents of C and N does not diminish the fact that very significant aboveground stores for C (100 to 300 Mg $C{\cdot}ha^{-1}$) are lost when forests are converted to agricultural uses (Fearnside, 2000). Dias-Filho et al. (2001) reported that forest-to-pasture conversion released 100 to 200 Mg C·ha^{-1} from aboveground forest biomass to the atmosphere. There are also additional benefits of intact forests to ameliorating floods, conserving soils, maintaining stable regional climates, preserving biodiversity, and supporting indigenous communities and ecoturism industries (Laurance et al., 2001a). Therefore, any policy changes that reduce the rate of deforestation would have the greatest potential for reducing the net emission of greenhouse gases. Moreover, it is also greatly desired that those policies enhance the rate of carbon sequestered from the atmosphere to be incorporated into the soil.

From Degraded Pasture to Well-Managed Pasture

More than half of the cattle pasture areas in the Brazilian Amazon are degraded (Serrão et al., 1979; Dias-Filho et al., 2001), which represents approximately 13 million ha. Although during the first three to five years after establishment the productivity of Amazonian pastures is often good, after that period a rapid decline in productivity of the planted grasses associated with an increased presence of herbaceous and woody invaders is generally observed (Uhl et al., 1988; Serrão and Toledo, 1990). If left uncontrolled, these invader species slowly become dominant and lead to pasture degradation, a condition characterized by a complete dominance of the weedy community (Dias-Filho et al., 2001).

If the entire area under degraded pasture could become well managed, and considering that the mean accumulation rate of soil C sequestration for well-managed pastures is 0.27 Mg C·ha^{-1} per year in the first 0 to 30 cm soil layer, the potential of soil C sequestration of converting degraded to well-managed pasture in the Brazilian Amazon would be about 3.5 Tg C per year.

Many factors and processes should be taken into consideration in order to determine the direction and rate of change in soil organic carbon content when a soil management practice is changed. Post and Kwon (2000) indicated some that may be important for increasing carbon sequestration: (1) increasing the input rates of organic matter; (2) changing the decomposability of organic inputs that increase the light fraction organic carbon (in particular); (3) placing organic matter deeper in the soil either directly by increasing belowground inputs or indirectly by enhancing surface mixing by soil organisms; and (4) enhancing physical protection through either intraaggregate or organomineral complexes.

From Degraded Pasture to Secondary Forest (Abandonment) and Existing Secondary Forest

Secondary forest associated with agriculture in Amazon follows a clear pattern of development. During pasture use, burning and weeding delay succession, but once the field is abandoned, the forest begins to regenerate. Secondary vegetation establishes itself through four main processes: regeneration of remnant individuals, germination from the soil seedbank, sprouting from cut or crushed roots and stems, and seed dispersal and migration from other areas (Tucker et al., 1998). Variance in the speed of forest regrowth is evident across regions and along a soil fertility gradient in the Brazilian Amazon. The rate of forest succession is determined by several factors. Original floristic composition, neighboring vegetation, and soil fertility and texture may affect regrowth. In addition, farmers' land-use decisions, such as clearing size, clearing procedures, crops planted, frequency of use, and duration of use, influence tree establishment and direct the path of secondary succession. At the regional scale, soil fertility and land-use history emerge as the critical factors influencing the rate of forest regrowth (Tucker et al., 1998).

Secondary forests in the Amazon have shown relatively high rates of regeneration, both after slash-and-burn agriculture and after abandonment of degraded pasture. For instance, Brown and Lugo (1990) reported that abandoned agricultural lands reverted to forests accumulated carbon at rates proportional to initial forest biomass. Rates ranged from about 1.5 Mg C·ha^{-1} per year in forests initially with biomass <100 Mg C·ha^{-1} to about 5.5 Mg C·ha^{-1} per year for forests initially >190 Mg C·ha^{-1}. Woomer et al. (1999) found a value of 6.2±1.3 Mg C·ha^{-1} per year of carbon sequestration in secondary forest regrowth in the Brazilian Amazon. Watson et al. (2000) suggested a range of carbon accumulation of 3.1 to 4.6 Mg C·ha^{-1} per year for tropical regions with duration of 40 years. Schroth et al. (2002) determined on an infertile upland soil in central Amazon that secondary forest accumulated carbon in above- and belowground biomass and litter at a rate of about 4 Mg·ha^{-1} per year. The rate of accumulation in aboveground biomass reported by Nepstad et al. (2001) ranged from 2.5 to 5 Mg C·ha^{-1} per year for 20-year-old Amazon secondary forest. A study in the Central Amazon (Feldpausch et al., 2004), reported carbon accumulation of 128 Mg C·ha^{-1} for a 12-year-old secondary forest dominated by *Vismia* ssp. The secondary vegetation was regenerated on an abandoned severely degraded pasture near Manaus.

In order to estimate the potential of carbon sequestration when degraded pastures in the Brazilian Amazon are abandoned for secondary forest regrowth, we multiplied 13 million ha (degraded pasture area) by the 4 Mg

C·ha^{-1} per year (mean accumulation rate took from the figures presented previously) resulting in 52 Tg C per year. It needs to be mentioned that this potential of carbon sequestration accounts for carbon in soil + aboveground biomass. Another fundamental point is that secondary forests cover approximately one-third of the area of the Brazilian Amazon that has been cleared for agriculture (Houghton et al., 2000), and if left abandoned, it can also sequester carbon from the atmosphere to the soil and plant biomass. Thus, multiplying the same accumulation rate value (4 Mg C·ha^{-1} per year) by the 18 million ha already under secondary forest regrowth results in an additional 72 Tg C per year. Therefore, it is possible to end up with a broadly estimated figure of 124 Tg C per year of potential soil + biomass carbon sequestration in the Brazilian Amazon due to secondary forest regrowth.

Moreover, these forests that develop on such abandoned land counteract many of the deleterious impacts of forest conversion to agriculture and cattle pasture. They play an important role for the regional carbon budget, as it reassimilates part of the carbon that was released when cutting and burning the original forest vegetation. They reestablish hydrological functions performed by mature forests, and they reduce the flammability of agricultural landscapes. Secondary forests transfer nutrients from the soil to living biomass, thereby reducing the potential losses of nutrients from the land through leaching and erosion. They also allow the expansion of native plant and animal populations from mature forest remnants back into agricultural landscapes (Nepstad et al., 2001).

From Degraded Pasture to Agroforestry

Agroforestry is a possible option not only for carbon sequestration but also for the valorization of previously cleared forest land in the humid tropics (Fujisaka and White, 1998). Where this land-use system succeeds in maintaining soil fertility on a satisfactory level and increasing farmers' income, further clearing of primary forest and accompanying carbon emissions may be reduced. Furthermore, timber trees and tree crops in these systems accumulate carbon in their biomass and, if planted on degraded areas, possibly also in the soil, and they may provide firewood and charcoal as substitutes for fossil fuel. On the other hand, when agroforestry systems or tree crop plantations are established on previously cleared fallow or secondary forest land, carbon is released from the fallow vegetation, and the successional processes, which would have led to progressive further accumulation of carbon in biomass and litter, are interrupted. Instead, tree crops, timber trees, annual crops, and eventually cover crops are established, whose growth and development are influenced by management practices such as fertilizer application and suppression of spontaneous vegetation through weeding (Schroth et al., 2002).

Few studies are available on agroforestry carbon accumulation rates in the Amazon. We have calculated a mean value of 2.7 Mg C $\cdot ha^{-1}$ per year during a timescale of 25 to 30 years based on the research described as follows. Woomer et al. (1999), measuring total system carbon in chronosequences in Brazilian Amazon, Cameroon, Indonesia, and Peru, found that agroforests sequestered, in soils and vegetation, about 3.3 Mg $C \cdot ha^{-1}$ per year. Watson et al. (2000) mentioned a range of 0.5 to 1.8 Mg $C \cdot ha^{-1}$ per year of carbon accumulation for agroforest management in the tropics. McCaffery et al. (2000) found that agroforestry systems based on native fruits and palms, planted on a severely degraded pasture in the Central Amazon, accumulated up to 33 Mg $C \cdot ha^{-1}$ on the aboveground biomass after 12 years of management. Biomass from degraded pasture was 9 Mg $C \cdot ha^{-1}$, indicating a net C uptake of 2 Mg C ha^{-1} per year. For the same systems, Rondón et al. (2000) reported soil carbon stocks (to 1 m depth) under agroforestry systems of 120 Mg $C \cdot ha^{-1}$ as compared with the degraded pasture soil which stored 110 Mg $C \cdot ha^{-1}$. This resulted in soil carbon accumulation rates of 0.83 Mg $C \cdot ha^{-1}$ per year. Schroth et al. (2002) reported that monocultures accumulated carbon at lower rates, i.e., 1.0 $Mg \cdot ha^{-1}$ per year for citrus, 1.3 $Mg \cdot ha^{-1}$ per year for cupuaçu, and 2.5 $Mg \cdot ha^{-1}$ per year for rubber. From the viewpoint of the carbon balance, multistrata agroforestry was an interesting alternative to the monocultures. A fast-growing system in Central Amazon accumulated 3.8 and 3.0 $Mg \cdot ha^{-1}$ per year of carbon in the full (N, P, K and dolomitic lime) and the low-fertilization (only 30 percent of the fertilizer and lime of the former treatment) treatment, respectively. These relatively high rates compared to most monocultures were due to a relatively high tree density and to the association of the smaller tree crops, such as cupuaçu and peach palm for palmito, with larger and faster-growing trees, i.e., peach palm for fruit, rubber, and Brazil nut trees. Those authors also reported that in all of the investigated plantation systems, there was more than twice as much carbon in the soil organic matter than in the biomass and litter combined. Changes in the soil organic matter stocks could, therefore, be of crucial importance for the net carbon effect of land-use transformations. However, their study found no effects of vegetation types and plant species on the organic matter stocks of the soil profile to 2 m depth, although there were indications for effects on the carbon content of the topsoil. They give two possible explanations for this. First, the conversion of primary forest into different tree crop plantations could have affected the distribution of carbon in the soil, but not its total quantity. Such changes could arise through an altered distribution of root mass in the soil profile, or through differences in the abundance and activity of burrowing soil fauna between vegetation types and plant species. Alternatively, the total carbon stock in the soil to 2 m depth could only be less sensitive than the carbon content of the topsoil as a measure for soil organic matter loss over a relatively short time period (Schroth et al., 2002).

Another significant result of the Schroth et al. (2002) study is that the tree crops with low litter quality, cupuaçu and Brazil nut, were able to build up and maintain organic matter levels in the topsoil comparable to those in the primary forest, even when they were grown in association with tree and cover crops that produced easily decomposable litter. When integrated into multistrata agroforestry systems, such tree crops could act as an insurance against soil organic matter loss, in addition to their direct production role. According to the same authors, such systems should be established on sites with low standing biomass, such as degraded pastures or other degraded areas, whereas older and vigorously growing secondary forests should be conserved. However, Amazonian pastures are often characterized by topsoil compactions and soil erosion (Fearnside, 1985), and the growth and yields of tree crops would probably be reduced on such sites.

After discussing all these issues and their complexities, one can visualize how "simplistic" is our estimate of 35 Tg C per year for the potential of carbon sequestration in the conversion of degraded pasture to agroforestry. We are basically multiplying the 13 million ha of degraded pasture area by the mean soil + biomass carbon accumulation rate of 2.7 Mg C·ha^{-1} per year showed before. In this potential figure, we assume that the entire area under degraded pasture would be converted to agroforestry, which is unrealistic. Moreover, it is important to stress that soil organic carbon accumulation in agroforestry is potentially affected by species, as some species produce and accumulate more litter and roots than others and these differential rates of organic matter production influences soil organic carbon. The figures we present here, do not take into consideration these specificities and therefore, should be used with caution. Though the total area in the Amazon dedicated to agroforestry seems relatively small compared to other management systems, agroforests have been encouraged as an alternative land use for the region, and their area has been increasing steadily over the past decade.

FINAL CONSIDERATIONS

The potential of soil+biomass carbon sequestration for the Brazilian Amazon was estimated to be in the range of 421 to 470 Tg C per year (Table 12.1). Only soil carbon sequestration accounts for 126 to 141 Tg C per year and the rest (295 to 329 Tg C per year) is related to carbon sequestration in the plant biomass. This range takes into account the following categories of

TABLE 12.1. Potential of carbon sequestration in the Brazilian Amazon.

Land use	Area extent (million ha)	Mean rate of C sequestration ($Mg\ C \cdot ha^{-1}$ per year)	Total potential (Tg C per year)
Primary forest (avoided deforestation)	344	1.00[a]	344
Primary forest to well-managed pasture	5	0.27[b]	1.35
Degraded to well-managed pasture	13	0.27[b]	3.50
Degraded pasture to secondary forest	13	4.00[a]	52
Existing secondary forest	18	4.00[a]	72
Degraded pasture to agroforestry	13	2.70[a]	35

[a]Soil+biomass carbon sequestration
[b]Soil carbon sequestration

land use and land conversion: primary forest (avoided deforestation), from forest to well-managed pasture, from degraded pasture to well-managed pasture, from degraded pasture to secondary forest (abandonment) and existing secondary forest, and from degraded pasture to agroforestry. As expected, the largest contributor of carbon sequestration in the Brazilian Amazon is related to avoided deforestation, which represents about 75 to 80 percent of the total potential. Deforestation in the Amazon releases quantities of greenhouse gases that are significant both in terms of their present impact and in terms of the implied potential for long-term contribution to global warming from continued clearing vast area of remaining forest. Therefore, it is important to avoid deforestation by intensification of agriculture on existing land through rehabilitation of degraded pastures and adoption of recommended management practices.

Assuming that is possible to simply extrapolate the potential of soil carbon sequestration we found for the Brazilian Amazon to the entire area of the Amazonian tropical rain forest, we would end up with a range of 177 to 197 Tg C per year that could be sequestered in soils of the mentioned ecoregion. Assuming also that the current situation described in this study is going to be maintained in a timescale of, say, 30 years, the total soil carbon sequestration in the Amazonian tropical rain forest would be about 5.3 to 5.9 Pg C.

As stressed before, our broad estimates should be used with caution, since uncertainties are probably very large. Just to illustrate this issue, the largest contributor of carbon sequestration (avoided deforestation) was estimated multiplying a mean value taken from several data on annual carbon accumulation rates by another estimated value of the extent of area under primary forest. In these calculations we are probably propagating errors. Also, there are some uncertainties in the primary data obtained by other studies that we used here. For instance, current estimates of total C storage in Brazilian Amazon vary by more than a factor of 2, from 39 to 93 Pg C, largely as a result of uncertainty in the quantity and spatial distribution of forest biomass (Houghton, 2003). Houghton concluded that 60 percent of the uncertainty in their estimates of annual carbon flux from Brazilian Amazon resulted from varying estimates of forest biomass.

Clearly, there is a need for additional measurements across large expanses of the Amazonian tropical rain forest. Moreover, studies of specific management regimes maintained over the long term (many years or decades) are needed to improve our understanding of the consequences of land-use and management practices for soil C stocks and related soil C and nutrient cycling processes.

REFERENCES

Anderson, A. 1990. Deforestation in Amazônia: Dynamics, causes, and alternatives. In A.B. Anderson (ed.), *Alternatives to deforestation: Steps toward sustainable use of the Amazon rain forest.* Colombia University Press, New York, pp. 3-23.

Bernoux, M., D. Arrouays, C.C. Cerri, and H. Bourennane. 1998. Modeling vertical distribution of carbon in Oxisols of the western Brazilian Amazon. *Soil Science* 163: 941-951.

Bernoux, M., M.C.S. Carvalho, B. Volkoff, and C.C. Cerri. 2002. Brazil's soil carbon stocks. *Soil Sci. Soc. Am. J.* 66: 888-896.

Bernoux, M., P.M.A. Graça, C.C. Cerri, P.M. Fearnside, B.J. Feigl, and M.C. Piccolo. 2001. Carbon storage in biomass and soils. In M.E. McClain, R.L. Victoria, J.E. Richey (eds.), *The biogeochemistry of the Amazon basin.* Oxford University Press, New York, pp. 84-105.

Bolin, B. and R. Sukumar. 2000. Global perspective. In R.T. Watson, I.R. Noble, B. Bolin, N.H. Ravindranath, D.J. Verardo, and D.J. Dokken (eds.), *Land use, land use change, and forestry.* Cambridge University Press, Cambridge, UK, pp. 23-51.

Bonde, T.A., B.T. Christensen, and C.C. Cerri. 1992. Dynamics of soil organic matter as reflected by natural ^{13}C abundance in paticle size fractions of forested and cultivated Oxisols. *Soil Biol. Biochem.* 24: 275-277.

Brown, S. and A.E. Lugo. 1990. Tropical secondary forests. *J. Trop. Ecol.* 6: 1-32.

Buschbacher, R., C. Uhl, and A.S. Serrão. 1988. Abandoned pastures in eastern Amazonia: II. Nutrient stocks in the soil and vegetation. *J. Ecol.* 76: 682-699.

Camarão, A.P. and A.P.S. Souza Filho. 1999. *Pastagens nativas de Amazônia.* Embrapa. Amazônia Oriental, Belém, Brasil.

Cerri, C.C., M. Bernoux, D. Arrouays, B.J. Feigl, and M.C. Piccolo. 2000. Carbon stocks in soils of the Brazilian Amazon. In R. Lal, J. Kimble, R. Follet, and B.A. Stewart (eds.), *Global climate change and tropical ecosystems.* Advances in Soil Science. CRC Press, Boca Raton, FL, pp. 33-50.

Cerri, C.C., M. Bernoux, B.J. Feigl, and M.C. Piccolo. 1999. Carbon dynamics in forest and pasture soils of the Brazilian Amazon. Workshop on Tropical Soils, Academia Brasileira de Ciências, Rio de Janeiro.

Cerri, C.C., B. Volkoff, and F. Andreux. 1991. Nature and behavior of organic matter in soils under natural forest, and after deforestation, burning and cultivation, near Manaus. *For. Ecol. Manage.* 38: 247-257.

Cerri, C.E.P., K. Coleman, D.S. Jenkinson, M. Bernoux, R.L. Victoria, and C.C. Cerri. 2003. Soil carbon dynamics at Nova Vida Ranch, Amazon, Brazil. *Soil Sci. Soc. Am. J.* 67: 1879-1887.

Chambers, J.Q., N. Higuchi, E.S. Tribuzy, and S.E. Trumbore. 2001. Carbon sink for a century. *Nature* 410: 429.

Choné, T., F. Andreux, J.C. Correa, B. Volkoff, and C.C. Cerri. 1991. Changes in organic matter in an oxisol from the central Amazonian forest during eight years as pasture, determined by ^{13}C composition. In J. Berthelin (ed.), *Diversity of environmental biogeochemistry.* Elsevier, New York, pp. 307-405.

Cox, P.M., R.A. Betts, C.D. Jones, S.A. Spall, and I.J. Totterdell. 2000. Acceleration of global warming due to carbon-cycle feedbacks in a coupled climate model. *Nature* 408: 184-187.

Cuevas, E. 2001. Soil versus biological controls on nutrient cycling in terra firme forests. In M.E. McClain, R.L. Victoria, and J.E. Richey (eds.), *The biogeochemistry of the Amazon basin.* Oxford University Press, New York, pp. 53-67.

Desjardins, T., F. Andreux, B. Volkoff, and C.C. Cerri. 1994. Organic carbon and 13C contents in soils and soil size-fractions, and their changes due to deforestation and pasture installation in eastern Amazonia. *Geoderma* 61: 103-118.

Dias-Filho, M.B., E.A. Davidson, and C.J.R. Carvalho. 2001. Linking biogeochemical cycles to cattle pasture management and sustainability in the Amazon basin. In M.E. McClain, R.L. Victoria, and J.E. Richey (eds.), *The biogeochemistry of the Amazon basin.* Oxford University Press, New York, pp. 84-105.

ECCM. 2002. Cabon sinks in the Amazon—the evidence. Produced for the World Land Trust by The Edinburgh Centre for Carbon Management. October 2002. Available online at www.worldlandtrust.org/carbon/carbonfigures.pdf.

Falesi, I.C. 1976. Ecosistema de pastagem cultivada na Amazônia Brasileira. Centro de Pesquisa Agropecuária do Trópico Úmido. Empresa Brasileira de Pesquisa Agropecuária, Boletim Técnico.

Fan, S.M., S.C. Wofsy, P.S. Bakwin, and D.J. Jacob. 1990. Atmosphere-biosphere exchange of CO2 and O3 in the central Amazon forest. *Journal of Geophysical Research-Atmospheres* 95: 16851-16864.

FAO. 2001. Forest resource assessment 2000. Available online at www.fao.org/forestry.

Fearnside, P.M. 1985. Agriculture in Amazonia. In G.T. Prance and T.E. Lovejoy (eds.), *Key environments: Amazonia.* Pergamon, New York, pp. 393-418.

Fearnside, P.M. 1996. Human carrying capacity estimation in the Brazilian Amazon: Research requirements to provide a basis for sustainable development. In R. Lieberei, G. Retsdorff, and A.D. Machado (eds.), *Interdisciplinary research on the conservation and sustainable use of the Amazonian rain forest and its information requirements.* Report on a workshop held in Brasilia, November 20-22, 1995, pp. 274-291.

Fearnside, P.M. 1999. Greenhouse gas emissions from land-use change in Brazil's Amazon region. In R. Lal, J.M. Kimble and B.A. Stewart (eds.), *Global climate change and tropical ecosystems.* CRC Press, Boca Raton, FL, pp. 231-249.

Fearnside, P.M. 2000. Global warming and tropical land-use change: Greenhouse gas emissions from biomass burning, decompositions and soils in forest conversion, shifting cultivation and secondary vegetation. *Climate Change* 46: 115-158.

Fearnside, P.M and R. I. Barbosa. 1998. Soil carbon changes from conversion of forest to pasture in Brazilian Amazon. *Forest Ecology and Management* 108: 147-166.

Feldpausch, T., M. Rondón, E. Fernandes, S. Riha, and E. Wandelli. 2004. Carbon and nutrient accumulation in secondary forests regenerating on pastures in central Amazonia. *Ecological Applications* 14: S164-S176.

Fujisaka, S., W. Bell, N. Thomas, L. Hurtado, and E. Crawford. 1996. Slash-and-burn agriculture, conversion to pasture, and deforestation in two Brazilian Amazon colonies. *Agriculture, Ecosystems & Environment* 59: 115-130.

Fujisaka, S. and D. White. 1998. Pasture or permanent crops after slash-and-burn cultivation? Land-use choice in three Amazon colonies. *Agrofor. Syst.* 42: 45-59.

Grace, J., J. Lloyd, J. McIntyre, A.C. Miranda, P. Meir, H.S. Miranda, C. Nobre, J. Moncrieff, J. Massheder, Y. Malhi, I. Wright, and J. Gash. 1995. Carbon dioxide uptake by an undisturbed tropical rain forest in southwest Amazonia, 1992 to 1993. *Science* 270: 778-780.

Higuchi, N., J. dos Santos, R.J. Ribeiro, J.V. de Freitas, G. Vieira, A. Cöic, and L.J. Minette. 1997. Crescimento e incremento de uma floresta amazônica de terra-firme manejada experimentalmente. In N. Higuchi, J.B.S. Ferraz, L. Antony, F. Luizão, R. Luizão, Y. Biot, I. Hunter, J. Proctor, and S. Ross (eds.), *Bionte: Biomassa e nutrients florestais.* Relatório Final. Instituto Nacional de Pesquisa da Amazônia (INPA), Manaus, Brazil, pp. 87-132.

Homma, A.K.O. 1994. Amazônia: desenvolvimento econômico e questão ambiental. In E.F. Vilhena and L.C. Santos (eds.), *Agricultura e meio ambiente.* UFV-NEPEMA, Viçosa, pp. 25-37.

Houghton, R.A. 1991. Tropical deforestation and atmospheric carbon-dioxide. *Climatic Change* 19: 99-118.

Houghton, R.A. 1997. Terrestrial carbon storage: global lessons for Amazonian research. *Ciencia e Cultura Journal of the Brazilian Association for the Advancement of Science* 49: 58-72.

Houghton, R.A. 2003. Why are estimates of the terrestrial carbon balance so different? *Global Change Biology* 9: 500-509.

Houghton, R.A., D.L. Skole, C.A. Nobre, J.L. Hackler, K.T. Lawrence, and W.H. Chomentowski. 2000. Annual fluxes of carbon from deforestation and regrowth in the Brazilian Amazon. *Nature 403*: 301-304.

INPE. 2004. *Amazônia: desflorestamento 2002-2003*. Instituto Nacional de Pesquisas Espaciais, São José dos Campos, SP.

Jacomine, P.K.T. and M.N. Camargo. 1996. Classificação pedológica nacional em vigor. In V.H. Alvarez, L.E.F. Fontes, and M.P.F. Fontes (eds.), *Solos nos grandes domínios morfoclimáticos do Brasil e o desenvolvimento sustentado*. SBCS-UFV, Viçosa-MG, Brazil, pp. 675-689.

Keller, M., J.M. Melillo, and W. Z. deMello. 1997. Trace gas emissions from ecosystems of the Amazon basin. *Cienc. Cult. J. Braz. Assoc. Adv. Sci.* 49: 87-97.

Kitamura, P.C. 1994. *A Amazônia e o desenvolvimento sustentável*. Embrapa, Brasília.

Lal, R. 1987. *Tropical ecology and physical edaphology*. John Wiley and Sons, New York.

Lal, R. and J.M. Kimble. 1999. What do we know and what needs to be known and implemented for C sequestration in tropical ecosystems. In R. Lal, J.M. Kimble, and B.A. Stewart (eds.), *Global climate change and tropical ecosystems*. CRC Press, Boca Raton, FL, pp. 417-431.

Laurance, W.F., M.A. Cochrane, S. Bergen, P.M. Fearnside, P. Delamônica, C. Barber, S. D'Angelo, and T. Fernandes. 2001a. The future of the Brazilian Amazon. *Science* 291: 438-439.

Laurance, W.F., M.A. Cochrane, S. Bergen, P.M. Fearnside, P. Delamônica, C. Barber, S. D'Angelo, and T. Fernandes. 2001b. Re: The future of the Brazilian Amazon. *Science Online*. February 2001. Available online at www.sciencemag.org/cgi/eletters.

Luizão, R.C., T.A. Bonde, and Rosswall, T. 1992. Seasonal variation of soil microbial biomass: The effect of clearfelling a tropical rainforest and establishment of pasture in the central Amazon. *Soil Biol. Biochem.* 24: 805-813.

Malhi, Y. and J. Grace. 2000. Tropical forests and atmospheric carbon dioxide. *Trends in Ecology and Evolution* 15: 332-337.

Malhi, Y., A. Nobre, J. Grace, B. Kruijt, M. Pereira, A. Culf, and S. Scott. 1998. Carbon dioxide transfer over a Central Amazonian rain forest. *Journal of Geophysical Research* 103: 593-612.

Marengo, J.A. and C.A. Nobre. 2001. General characteristics and variability of climate in the Amazon Basin and its links to the global climate system. In M.E. McClain, R.L. Victoria, and J.E. Richey (eds.), *The biogeochemistry of the Amazon basin*. Oxford University Press, New York, pp. 17-41.

McCaffery, K., M. Rondón, J. Gallardo, S. Welsch, T. Feldpausch, E. Fernandes, S. Riha, and E. Wandelli. 2000. Carbon and nutrient stocks in agroforestry systems and secondary forest in the Central Amazon. Proeceedings of the Second Scientific Conference of the LBA Project, Atlanta, GA.

McClain, M.E., R.L. Victoria, and J.E. Richey. 2001. *The biogeochemistry of the Amazon basin*. Oxford University Press, New York.

Melack, J.M. and B.R. Forsberg. 2001. Biogeochemistry of Amazon floodplain lakes and associated wetlands. In M.E. McClain, R.L. Victoria, and J.E. Richey (eds.), *The biogeochemistry of the Amazon basin.* Oxford University Press, New York, pp. 235-274.

Melillo, J. M., R. A. Houghton, D. W. Kicklighter, and A.D. McGuire. 1996. Tropical deforestation and the global carbon budget. *Annu. Rev. Energy Environ.* 21: 293-310.

Meyers, N. 1981. Conversion rates in tropical moist forests. In F. Mergen (ed.), *Tropical forests utilization and conservation.* Yale University, New Haven, CT, pp. 48-66.

Moraes, J.F.L., C.C. Cerri, J.M. Melillo, D. Kicklighter, C. Neill, D. Skole, and P.A. Steudler. 1995. Soil carbon stocks of the Brazilian Amazon basin. *Soil Sci. Soc. Am. J.* 59: 244-247.

Moraes, J.F.L., B. Volkoff, C.C. Cerri, and M. Bernoux. 1996. Soil properties under Amazon forest change due to pasture installation in Rondônia, Brazil. *Geoderma* 70: 63-81.

Neill, C., C.C. Cerri, J. Melillo, B.J. Feigl, P.A. Steudler, J.F.L. Moraes, and M.C. Piccolo. 1997. Stocks and dynamics of soil carbon following deforestation for pasture in Rondonia. In R. Lal, J.M. Kimble, R.F. Follett, and B.A. Stewart (eds.), *Soil processes and the carbon cycle.* CRC Press, Boca Raton, FL, pp. 9-28.

Neill, C. and E.A. Davidson. 1999. Soil carbon accumulation for loss following deforestation for pasture in the Brazilian Amazon. In R. Lal, J.M. Kimble, and B.A. Stewart (eds.), *Global climate change and tropical ecosystems.* CRC Press, Boca Raton, FL, pp. 197-211.

Nepstad, D., P.R.S. Moutinho, and D. Markewitz. 2001. The recovery of biomass, nutrient stocks, and deep-soil functions in secondary forests. In M.E. McClain, R.L. Victoria, and J.E. Richey (eds.), *The biogeochemistry of the Amazon basin.* Oxford University Press, New York, pp. 139-155.

Norby, R.J., S.D. Wullschleger, C.A. Gunderson, D.W. Johnson, and R. Ceulemans. 1999. Tree responses to rising CO2 in field experiments: Implications for the future forest. *Plant Cell and Environment* 22: 683-714.

Pearce, D.W. and K. Brown. 1994. Saving the world's tropical forests. In K. Brown and D. W. Pearce (eds.), *The causes of tropical deforestation: The economic and statistical analysis of factors giving rise to the loss of the tropical forests.* University College London Press Limited, London, pp. 1-26.

Phillips, O.L., Y. Malhi, N. Higuchi, W.F. Laurence, P.V. Nunez, R.M. Vasquez, S.G. Laurence, L.V. Ferreira, M. Stern, S. Brown, and J. Grace. 1998. Changes in the carbon balance of tropical forests: Evidence from long-term plots. *Science* 282: 439-442.

Pires, J.M. and G. T. Prance. 1986. The vegetation types of the Brazilian Amazon. In G.T. Prance and T.M. Lovejoy (eds.), *Amazonia.* Pergamon Press, Oxford, pp. 109-115.

Post, W.M. and K.C. Kwon. 2000. Soil carbon sequestration and land-use change: Processes and potential. *Global Change Biology* 6: 317-327.

Prance, G.T. and T.M. Lovejoy. 1984. *Amazonia.* Pergamon Press, Oxford, UK.

Prentice, I.C. and J. Lloyd. 1998. C-quest in the Amazon basin. *Nature* 396: 619-620.

Rondón, M., E. Fernandes, R. Lima, and E. Wandelli. 2000. Carbon storage in soils from degraded pastures and agroforestry systems in Central Amazonia: The role of charcoal. Proceedings of the Second Scientific Meeting of the LBA Project, Atlanta, GA.

Schroth, G., S.A. D'Angelo, W. G. Teixeira, D. Haag, and R. Lieberei. 2002. Conversion of secondary forest into agroforestry and monoculture plantations in Amazonia: Consequences for biomass, litter and soil carbon stocks after 7 years. *Forest Ecology and Management* 163: 131-150.

Serrão, E.A.S., I.C. Falesi, J.B. da Veiga, and J.F.T. Neto. 1979. Productivity of cultivated pastures on low fertility soils of the Amazon Basin. In P.A. Sanchez and L.E. Tergas (eds.), *Pasture production in acid soils of the tropics*. Centro Internacional de Agricultura Tropical, Cali, Colombia, pp. 195-225.

Serrão, E.A.S. and J.M. Toledo. 1990. The search for sustainability in Amazonian pastures. In A.B. Anderson (ed.), *Alternatives to deforestation: Steps toward sustainable utilization of Amazon forests*. Columbia University Press, New York, pp. 195-214.

Skole, D. and C. Tucker. 1993. Tropical deforestation and habitat fragmentation in the Amazon satellite data from 1978 to 1988. *Science* 260: 1905-1910.

Tian, H., J.M. Mellilo, D.W. Kicklighter, A.D. McGuire, J.V. K. Helfrich III, B. Moore III, and C.J. Vorosmarty. 1998. Effect of interannual climate variability on carbon storage in Amazonian ecosystems. *Nature* 396: 664-667.

Trumbore, S.E., E.A. Davidson, P.B. Camargo, D.C. Nepstad, and L.A. Martinelli. 1995. Below-ground cycling of carbon in forest and pastures of eastern Amazonia. *Global Biogeochem. Cycles* 9: 515-528.

Tucker, J.M., E.S. Brondizio, and E.F. Moran. 1998. Rates of forest regrowth in eastern Amazonia: A comparison of Altamira and Bragantina regions, Para State, Brazil. *Interciencia* 23: 64-73.

Uhl, C., R. Buschbacher, and E.A.S. Serrão. 1988. Abandoned pastures in eastern Amazonia: I. Patterns of plant sucession. *Journal of Ecology* 76: 663-681.

Watson, R.T., I.R. Noble, B. Bolin, N.H. Ravindranath, D.J. Verardo, and D.J. Dokken. 2000. *Land use, land use change, and forestry: A special report of the IPCC*. Cambridge University Press, Cambridge, UK.

Woomer, P.L., C.A. Palm, J. Alegre, C. Castilla, D.G. Cordeiro, K. Hairiah, J. Kotto-Same, A. Moukam, A. Reise, V. Rodrigues, and M. van Noordwijk. 1999. Slash-and-burn effects on carbon stocks in the humid tropics. In R. Lal, J.M. Kimble and B.A. Stewart (eds.), *Global climate change and tropical ecosystems*. CRC Press, Boca Raton, FL, pp. 99-115.

FIGURE 4.2. Land-use categories derived from the Global Land Cover map for the year 2000 (*Source:* Created by the authors from data from European Commission Joint Research Centre, 2003; Lambert Equal-Area Azimuthal projection.)

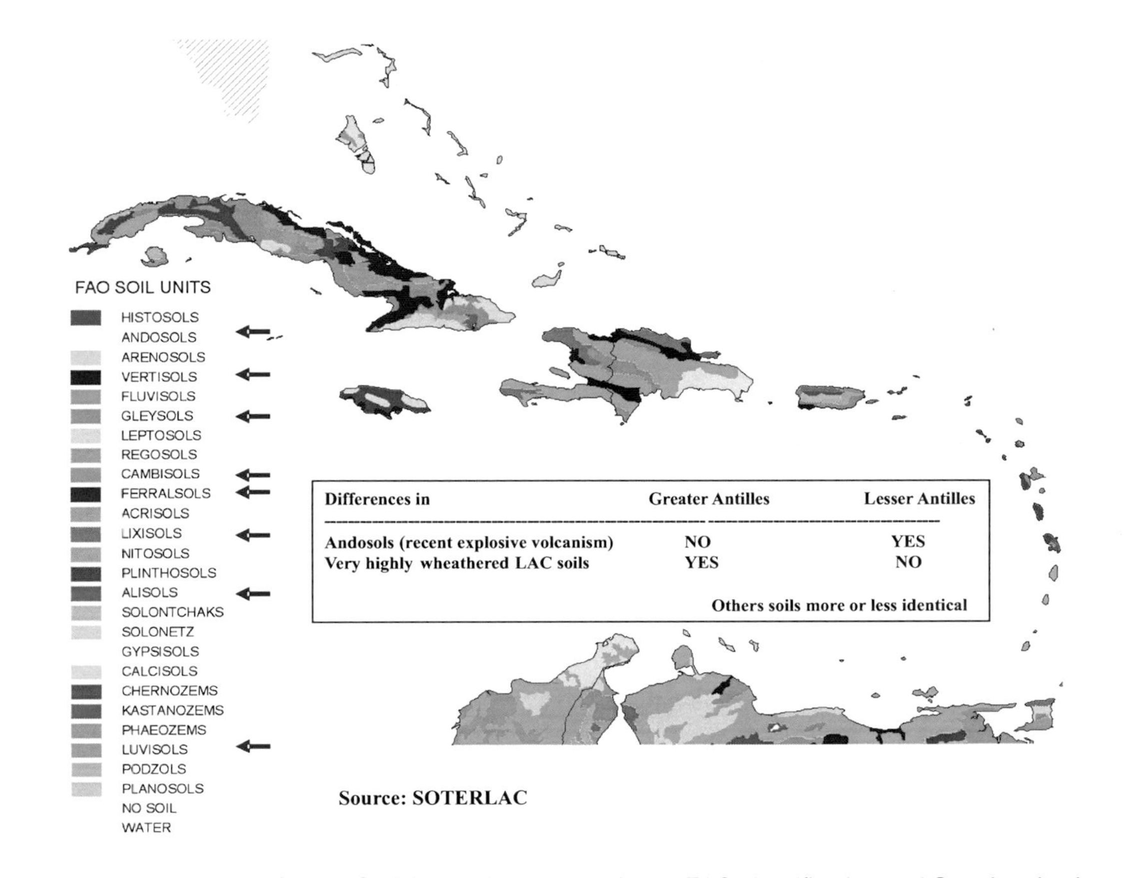

FIGURE 10.1. Main soil types for the Caribbean biome according to FAO classification and Soterlac database. Arrows correspond to the dominant soil types.

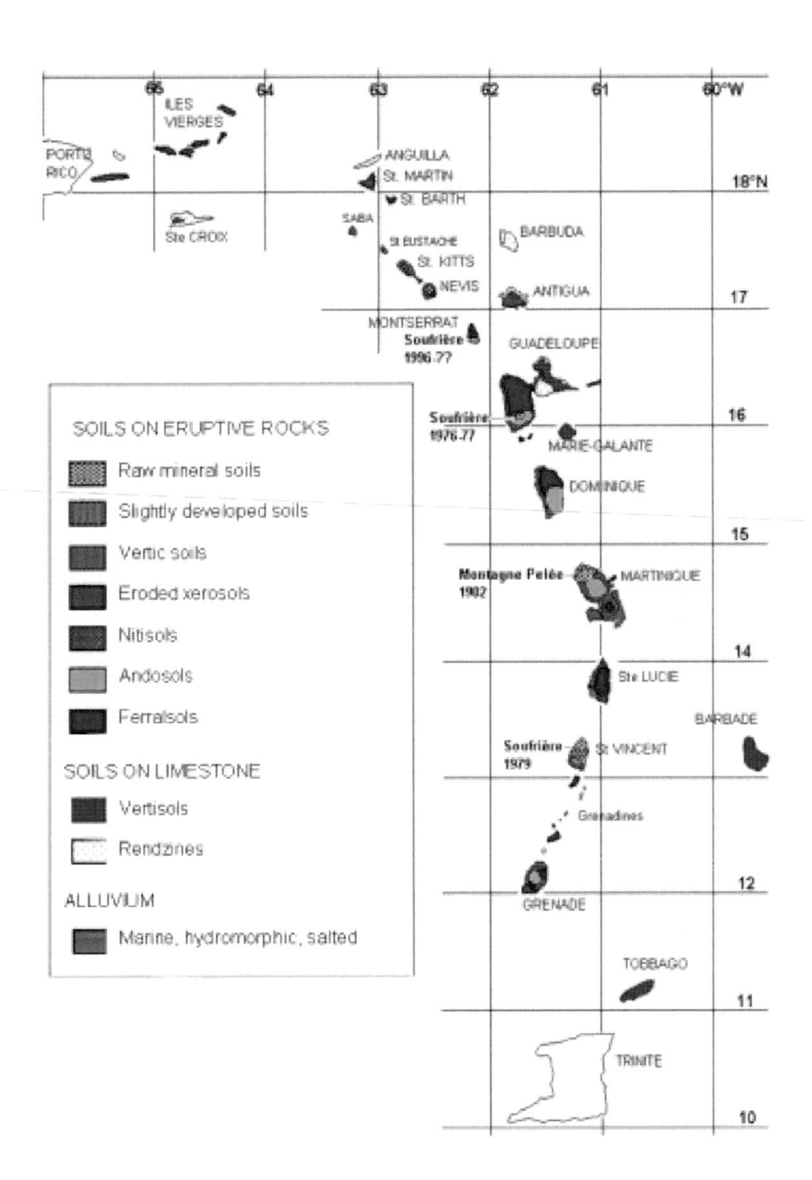

FIGURE 10.2. Martinique and Guadeloupe, representative of the diversity of soil types in the Lesser Antilles (Cabidoche, unpublished data, 1997) according to soil French classification (CPCS, 1967).

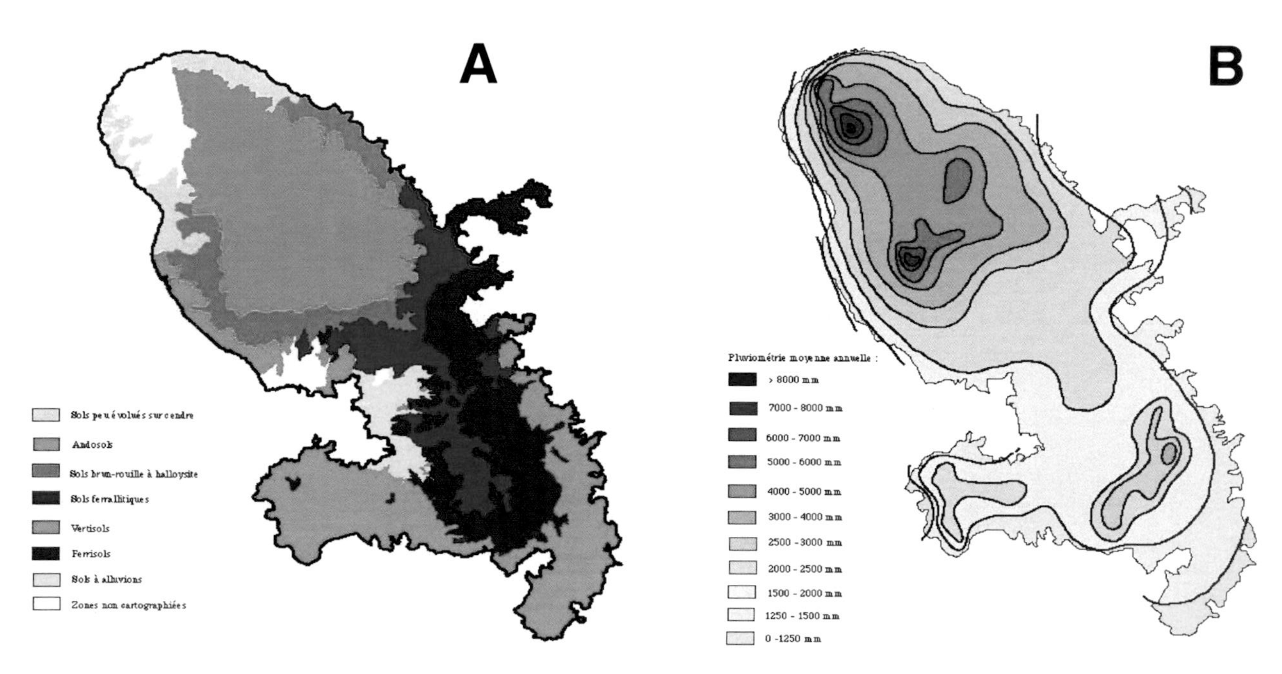

FIGURE 10.3. Simplified soil maps of Martinique (A) and rainfall distribution (B). Andosols in green; Nitisols in orange; Ferralsols in red and purple; Vertisols in blue. (*Source:* Created by authors from data of IRD-BOST after CNRS-IGN, 1976).

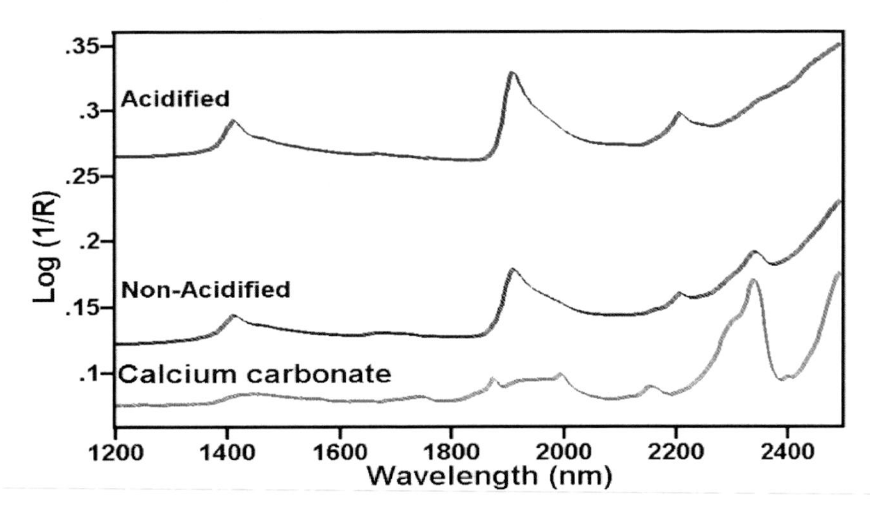

FIGURE 19.2. Near-infrared spectra of acidified and nonacidifed soil and calcium carbonate.

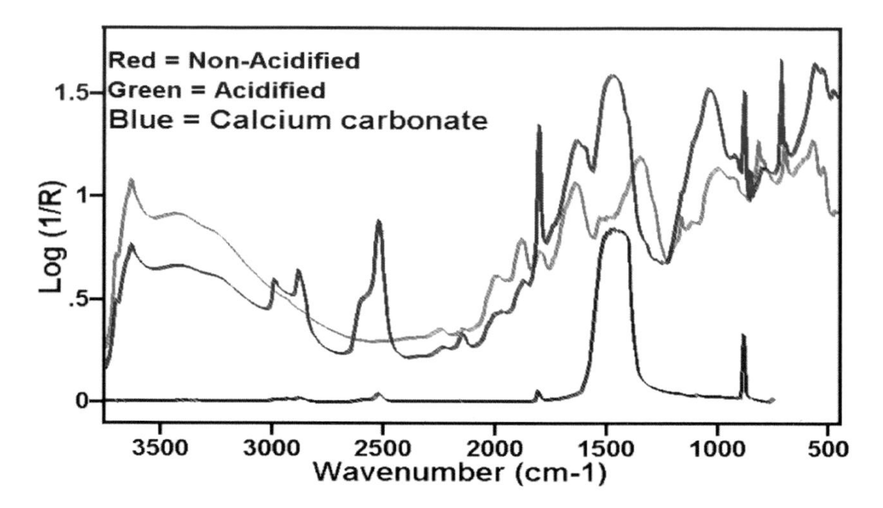

FIGURE 19.3. Mid-infrared spectra of acidified and nonacidified soil and calcium carbonate.

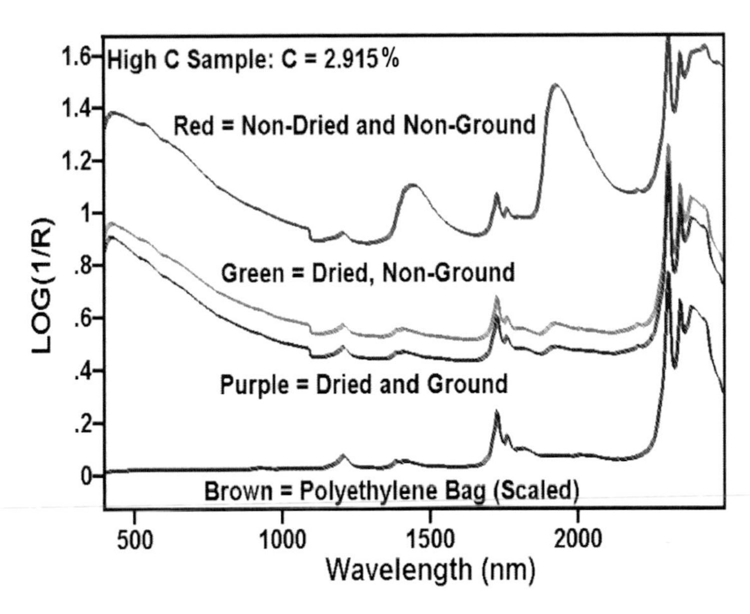

FIGURE 19.4. Near infrared spectra of soils taken under different soil conditions.

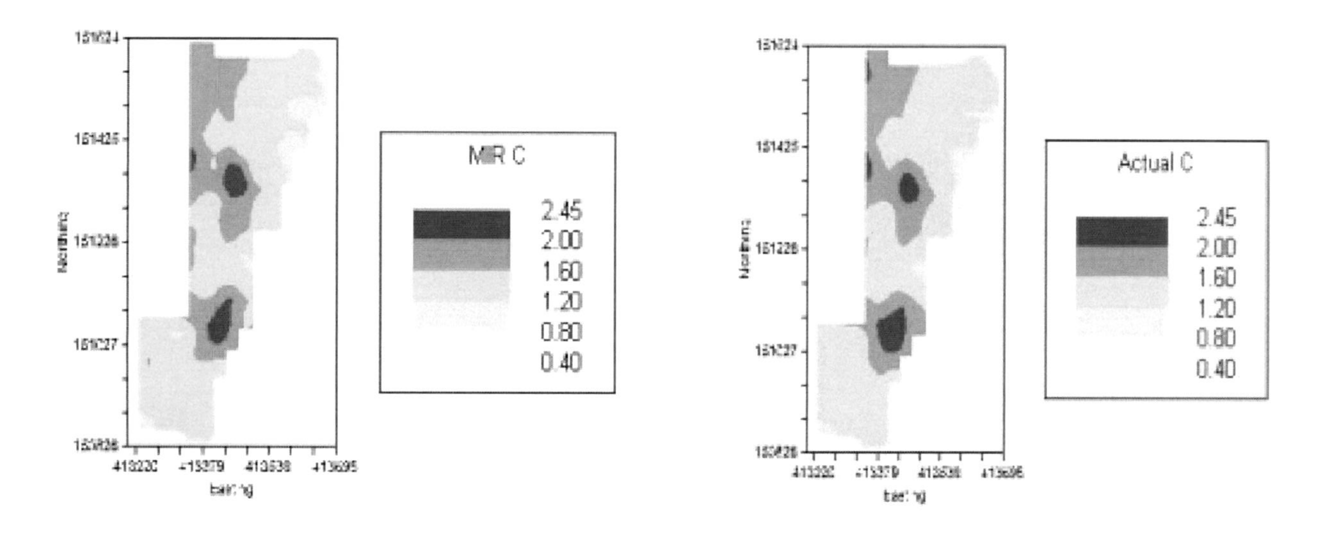

FIGURE 19.10. Results using a calibration developed using two-thirds of samples from a small watershed to develop a calibration to determine the remaining one-third.

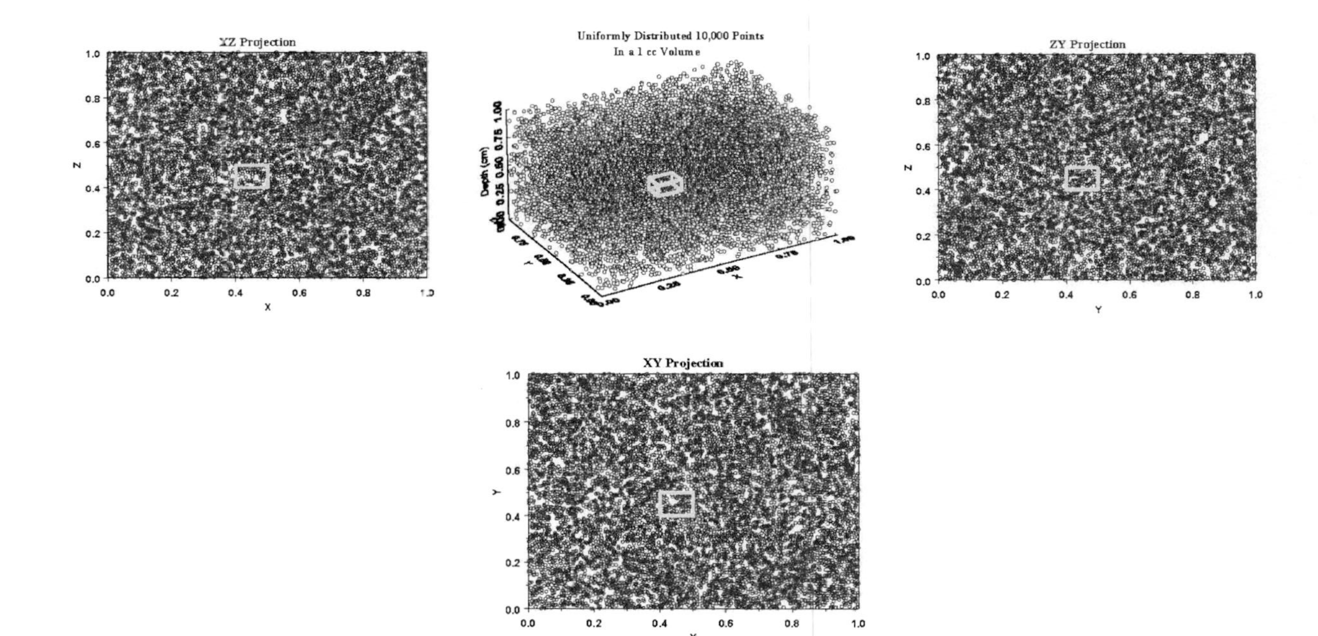

FIGURE 20.5. Ten-thousand points uniformly distributed in a unit volume (center) and projection on each plane. The sampling volume is marked in the center of each panel.

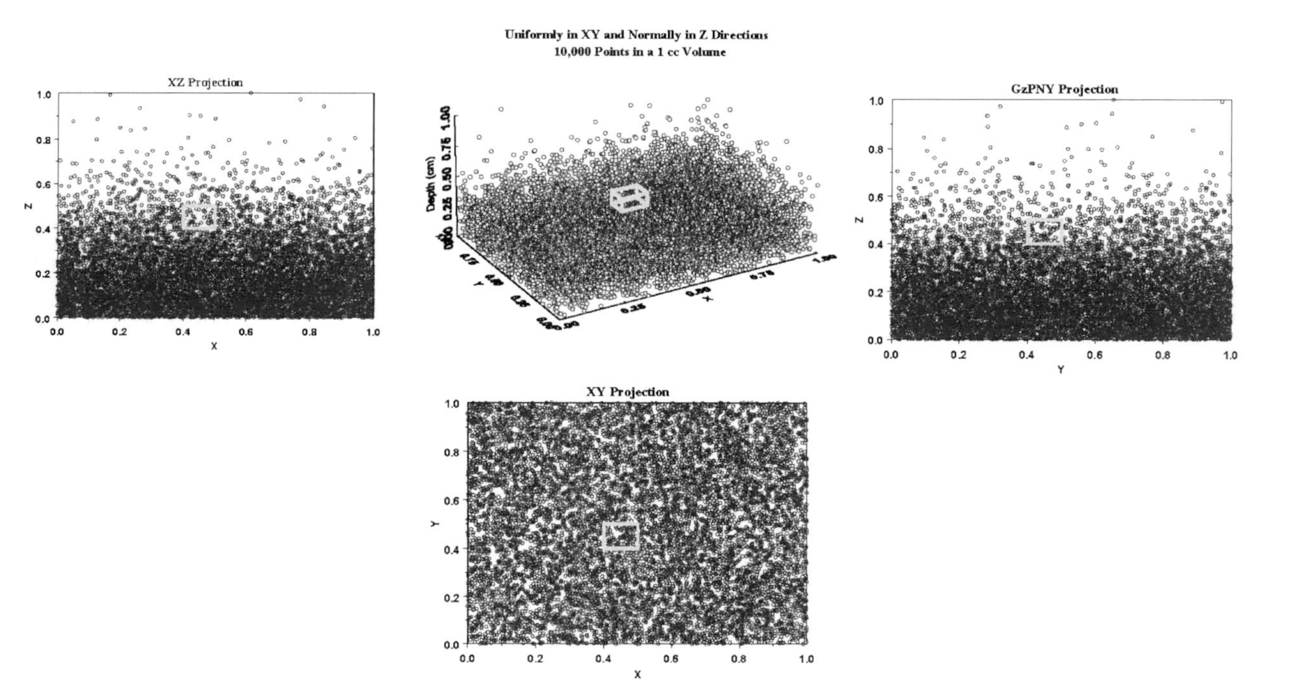

FIGURE 20.6. Ten-thousand points uniformly distributed in the lateral directions and normally distributed with depth in a unit volume (center) and projection on each plane. The sampling volume is marked in the center of each panel.

PLATE 22.1. A soil from Florida that has very little variability.

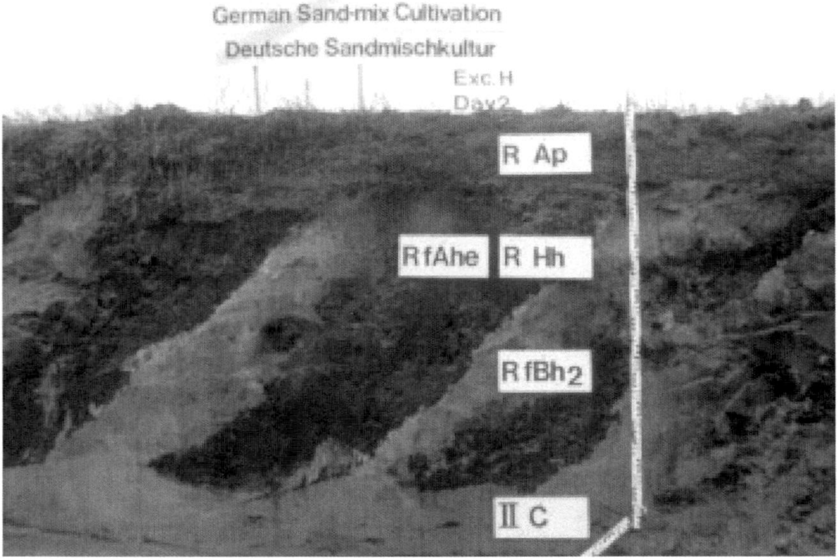

PLATE 22.2. Soil in Germany showing extreme variability but in a pattern that can be explained.

Chapter 13

Carbon Sequestration Potential of Pasture and Agro-Silvo-Pastoral Systems in Tropical Andean Hillsides

M. C. Amézquita
P. Buurman
E. Murgueitio
E. Amézquita

INTRODUCTION

Tropical America: Population, Land Resources, Agricultural and Livestock Production

Tropical America (TA) comprises Mexico, Central America, the Caribbean, and South America, excluding Argentina, Chile, and Uruguay. It covers 1,688 million ha representing 11 percent of the world continental area, and is a home to 432 million people representing 8 percent of the total world population. Forests cover 41 percent of its territory—a very high proportion compared with the world proportion in forests (28 percent)—and represent 22 percent of the world forest area. Consequently, renewable water resources in the region represent 22 percent of the world water resources. The per capita water resources in TA (35,405 m^3) are almost five times the corresponding world average of 7,176 m^3. TA's agricultural land area is 548 Mha corresponding to 32 percent of its territory and to 11 percent of the world agricultural land area. Of its 432 million inhabitants, 100 millions (23 percent) are farmers, who live on agricultural and livestock production activities, and represent 4 percent of the world population in agriculture. TA's agricultural land includes crop, pasture (both native and introduced), and agro-silvo-pastoral land uses (Table 13.1).

Carbon Sequestration in Soils of Latin America

doi:10.1300/5755_13

TABLE 13.1. Population, land resources, agricultural, and livestock production in 2000 in tropical America and the world.

Resources	TLA[a]	Temperate LA[b]	Total LA	World
Population				
Total, millions of people (world percent)	431.5 (7.5)	52.8 (0.9)	484.3 (8.4)	5757.8 (100.0)
In agric., millions of people (world percent)	100.3 (3.9)	2.5 (0.1)	102.8 (4.0)	2592.4 (100.0)
% population in agriculture in the region	23.24	4.73	21.85	45.02
Land resources and land use				
Total continental land,[c] Mha (world percent)	1688 (11.3)	371 (2.5)	2059 (13.7)	14991 (100.0)
Agricultural land, Mha (world percent)	548 (11.3)	201 (4.1)	749 (15.4)	4866 (100.0)
• Pastures, Mha (world percent)	424 (12.5)	169 (5.0)	593 (17.4)	3399 (100.0)
• % pasture within agric. land in the region	77.37	84.08	79.17	69.85
• Crops, Mha (world percent)	124 (8.5)	32 (2.2)	156 (10.6)	1467 (100.0)
• % crops within agric. land in the region	22.63	15.92	20.83	30.15
• Pastureland /cropland ratio in the region	3.42	5.28	3.80	2.32
Forest land, Mha (world percent)	938 (22.5)	68 (1.6)	1006 (24.1)	4172 (100.0)
% of forest land in the region	41.32	18.38	38.11	27.83
Water resources				
Total, km^3 (world percent)	9005 (22.0)	1586 (3.9)	10591 (25.8)	41002 (100.0)
Per capita, m^3 (region/world ratio)	35405 (4.9)	30479 (4.3)	34831 (4.9)	7176.8 (1.0)
Cattle inventory				
Total, million heads (world percent)	280 (21.2)	68 (5.2)	348 (26.4)	1320 (100.0)
Lactating cows, million heads (world percent)	41 (17.9)	4 (1.8)	45 (19.7)	229 (100.0)
% of lactating cows in the region	14.64	5.88	12.93	17.35
Meat and milk production				
Meat, million metric tons (world percent)	9 (16.7)	3 (5.6)	12 (22.2)	54 (100.0)
Milk, million metric tons (world percent)	42 (9.0)	12 (2.6)	54 (11.6)	466 (100.0)

Source: FAO, 2000; Rivas et al., 1998.
[a]Mexico, Central America, the Caribbean, and South America, excluding Argentina, Chile, and Uruguay.
[b]Argentina, Chile, and Uruguay.
[c]Original data expressed in millions km^2 (*Nieuwe Grote Wereldatlas,* 1977).

Pasture and agro-silvo-pastoral land represent 77 percent of TA's total agricultural land, mostly on poor, acidic soils. Cropland covers the remaining 23 percent, mostly located on better-quality soils. Thus, the pastures: crop land ratio of 3.4 is higher than the world ratio of 2.3. TA represents 21 percent of the world's cattle inventory and 18 percent of the world's lactating cow inventory. Pasture and agro-pastoral land, as well as meat and milk production, are concentrated in four countries: Brazil, Mexico, Colombia, and Venezuela. Together these countries hold 76 percent of the pasture and agro-silvo-pastoral land, 84 percent of total cattle inventory, and 85 percent and 83 percent, respectively, of meat and milk production of TA (Vera et al., 1993). The major tropical ecosystems where meat and milk are produced are the Savannah Ecosystem (250 Mha, of which 243 Mha are in the four countries), the Tropical Forest Ecosystem (about 44 Mha in those four countries (Vera et al., 1993), and the Andean Hillsides Ecosystem (96 Mha, in Peru, Bolivia, Ecuador, Colombia, and Venezuela).

Meat and milk in TA have important socioeconomic significance. Its consumption is high in most LA cities, representing, respectively, 12 to 26 percent and 7 to 13 percent of total family expenditure (de Rubinstein and Nores, 1980). The annual per capita meat consumption in TA ranges from 7 to 38 kg compared with 0.7 to 2.6 kg in Southeast Asia and 3.6 to 9.6 kg in tropical Africa (Valdés and Nores, 1979). The effect of improved and well-managed pasture systems on increase in productivity, socioeconomic benefit, export market competitiveness, and economic development of TA's countries has been amply documented (CIAT, 1976-1996, 1994, 1997-1999, 1999; de Rubinstein and Nores, 1980; Sanint et al., 1984; Toledo, 1985; Rivas et al., 1998; Vera et al., 1993).

Land-Use Change

Conversion of forests to crops and pastures has been the most important land-use change in TA during the second half of the twentieth century (Kaimowitz, 1996). The decline in forest area in Columbia has been accompanied by an increase in pasture area with no major change in cropped area (Table 13.2), as pastures are established once soils are too degraded for crop production. However, cattle production systems are not the only cause of deforestation (Kaimowitz, 1996). Deforestation is attributed to a variety of factors, including production and marketing interests of multinational companies, national policies, ease of access to the forests, high initial soil fertility, and favorable climatic conditions (Browder, 1988; Sader and Joyce, 1988; Veldkamp, 1993). After deforestation and crop and pasture establishment, large areas are abandoned due to decline in productivity caused by

TABLE 13.2. Land-use change in Colombia, 1950-2000 (area in Mha, and percent).

Land use	1950 area (%)	1970 area (%)	1978 area (%)	1987 area (%)	1995 area (%)	2000 area (%)
Crops	5.0 (4)	7.6 (7)	8.8 (8)	5.3 (5)	4.4 (4)	5.0 (5)
Pastures	14.6 (13)	17.5 (15)	20.5 (18)	40.1 (35)	35.5 (31)	45.0 (39)
Forest	94.6 (83)	89.1 (78)	84.9 (74)	68.7 (60)	74.2 (65)	64.2 (56)
Total	114.2 (100)	114.2 (100)	114.2 (100)	114.2 (100)	114.2 (100)	114.2 (100)

Sources: DANE, 1960-1970; Balcázar, 1994; IGAC-ICA, 1987; DANE-SISAC, 1995; in Ramírez and Ortiz, 1997.

mismanagement, leading to degradation of more than 60 percent of TA's pasture area (CIAT, 1999).

Carbon Sequestration and Land Use

Interest in carbon sequestration has arisen since the 1990s. The Kyoto Protocol of 1997 and subsequent agreements of the United Nations have considered reforestation and aforestation to be land-use systems suitable for economic incentives in developing countries through the Clean Development Mechanism (CDM) and trading carbon credits. Although environmental considerations may suggest partial reforestation of areas currently under pasture, thus potentially contributing to carbon sequestration, this would cause a serious threat to the economic welfare of farmers and to food availability for the population, especially for milk and meat. Therefore, the combination of agricultural production and environmental services (particularly carbon sequestration) through improved and well-managed pasture, agro-pastoral, and silvo-pastoral systems, appears to be a good alternative. The approval for European Union countries to contribute to their GHG emission reduction levels through carbon sequestration in grassland systems and the United States' motivation to provide farmers incentives for carbon sequestration in grasslands) makes this alternative particularly relevant for TA.

As a pilot project, we investigated the potential for carbon sequestration in a number of pasture, agro-pastoral, and silvo-pastoral systems in various subecosystems of Tropical America, including the Andean hillsides in two zones of the Colombian Andean region.

THE TROPICAL ANDEAN HILLSIDES ECOSYSTEM

General Characteristics

The Andean Mountain chain *(Cordillera de los Andes)* is the longest mountain complex on Earth. Located in the western part of South America, parallel to the Pacific coast, it has a total length of 7,200 km from Cabo de Hornos, Argentina, to the Caribbean Sea in Venezuela. It has a minimum width of 100 km in the southern part, a maximum width of 600 km in the central part of the continent between Peru and Bolivia, and a mean width of 241 km. Its mean altitude is 3,660 masl and the maximum altitude is 6,959 masl at the *Aconcagua* peak in Argentina (Encarta, 2003; Kindersley, 2004). Consequently, the region exhibits a wide range of climatic conditions, landscapes, hydrology, hydrological cycles, and biodiversity. In the central part of South America, the Andean Mountains divide into two chains that cross Peru and Ecuador (central and eastern mountains), and into three chains in Colombia (western, central, and eastern mountains) forming inter-Andean valleys of flat lands with high available water and soil fertility.

The Tropical Andean Hillsides occupy 96 Mha from southern Peru at 7°S to Sierra Nevada de Santa Martha in Colombia, at 12°N. This region excludes Brazil, which also has a large area corresponding to the hillside environment (Jones, 1993), the hillsides of Central America, and the foothills of the Andes in Brazil. These are extensive areas which are essentially identical (Amézquita et al., 1998).

The region is characterized by a very high population density and high anthropogenic activity. In Colombia, for example, 70 percent of the country's total population lives in the Andean Region, with 78 inhabitants per km^2 compared to 29 per km^2 in the rest of the country (DANE, 1996) comprising a rural population density of 21 inhabitants per km^2 compared to 8.4 inhabitants per km^2 in the rest of the country (Etter and van Wyngaarden, 2000). About 70 percent of the Andean Hillsides area in Colombia is strongly influenced by human activity (Instituto Alexander von Humboldt, 1998). A similar situation holds for Andean Hillside areas in Peru, Bolivia, Ecuador, and Venezuela (Amézquita et al., 1998). The region is also characterized by poverty in the small-farm sector, which often leads to the deforestation of steeply sloping lands which are highly vulnerable to degradation and only marginally suited to production of semisubsistence crops (Pachico et al., 1994).

Land Use and Land-Use Change

Due to population pressure, the Tropical Andean Hillsides, initially covered by forest, were intensively deforested. The conversion of forest to annual crops was done initially on highly productive lands but was followed later by annual crops of lower productivity, and in turn by establishment of pasture, agro-pastoral, and silvo-pastoral systems. The latter was due to poor management, exacerbated by soil degradation, and to the abandonment of the land. The data in Table 13.3 show a 50-year history of land-use change (1950s to 2000s) for some small farms in two regions of the Tropical Andean Hillsides in Colombia: Dovio, at 1,900 masl, and Dagua, at 1,350 masl (Amézquita, 2003).

The Tropical Andean Hillsides are presently characterised by land-use patterns of native forest, pasture systems, crops [intensive coffee production, fruit production, and various annual crops such as maize *(Zea mays),* beans (*Phoseolus* spp.), and cassava *(Manihoc esculenta)*] and natural regeneration areas (secondary forest and fallow land, "rastrojos") each with variable degree of degradation (Table 13.4). Pasture systems are important in the region. The data in Table 13.5 illustrates the predominance of pasture systems at the microwatershed level in the Andean Hillsides of Colombia, between 800 and 2,000 masl (Gómez, 2002). There are three types of cattle production systems in the region: milk production (at altitudes between 2,000 and 3,500 masl), dual-purpose cattle for milk and beef production (at altitudes between 1,000 and 2,000 masl), and beef production in the lower areas (between 200 and 1,000 masl) (Murgueitio, 2003). Dual-purpose cattle production is practiced irrespective of soil, altitude, or topographic conditions, in small farms (1 to 12 ha) of subsistence agriculture, medium-size farms (13 to 50 ha), and large commercial farms (50 to 500 ha) of agro-industrial production.

Extent of Soil Degradation

Of the total 92 Mha corresponding to the nonurban part of the Tropical Andean Hillsides, about 20 Mha (22 percent) are already highly degraded, and an estimated 50 Mha (55 percent) are prone to rapid degradation (Amézquita et al., 1998). Expert knowledge on land-use distribution and degradation levels (Table 13.4) is consistent with these statistics. More than 80 percent of the Tropical Andean Hillsides area exhibits different states of erosion. The ISRIC World Map on the Status of Human-Induced Soil Degradation and its corresponding report (Oldeman et al., 1990) indicate water erosion as the main limiting factor in the Tropical Andean Hillsides, with

TABLE 13.3. Land-use change between 1950 and 2002 in the tropical Andean Hillsides.

Present Land Use	Initial Land Use	1950s	1960s	1970-1977	1977-1986	1986-1988	1988	1988-2002
Area 1: Dovio (1,900 masl), Colombia[a]								
Degraded pasture	Forest	Sugar cane	Abandoned land	Fruit trees (Tomate de árbol)	Pasture (*Melinis minutiflora*)	King grass var Taiwan	Degraded King grass + trees + maize + pine-apple	Degraded King grass pasture
Improved pasture	Forest	Sugar cane	Coffee + Guamo	Fruit trees (Tomate de árbol)	Abandoned land	*Brachiaria decumbens* under grazing		
Mixed forage bank	Forest	Maize-beans-sweet potatoes		Fruit trees + maize	Star grass	5-species forage bank *Trichanthera gigantea*, *Morus* spp, *Erthrina edulis*, *Boehmeria nivea*, *Tithonia diversifolia*		
Forest	Forest	Forest (intervened)				Forest (nonintervened)		
Area 2: Dagua (1,350 masl), Colombia								
Degraded pasture	Forest	*Hyparrhenia ruffa* pasture under grazing				Degraded *Hyparrhenia rufa* pasture		
Improved pasture	Forest	Coffee	*H. rufa* pasture			*Brachiaria decumbens* under rotational grazing		
Mixed forage bank	Forest	Coffee	*H. rufa* pasture			4-species forage bank *Trichanthera gigantea*, *Morus* spp, *Erthrina fusca*, *Tithonia diversifolia*		
Forest	Forest	Forest (intervened)				Forest (regenerated)		

Source: Amézquita et al., 2004.
[a]Research areas, tropical Andean Hillsides, Colombia. Carbon Sequestration Project, the Netherlands Cooperation C0-010402.

TABLE 13.4. Approximate land-use distribution and degradation stage in tropical Andean Hillsides.[a]

			Degradation stage					
			Severe		Moderate		Nondegraded	
Land use	Area (%)	Area (Mha)	%	Mha	%	Mha	%	Mha
Pasture systems	50	46.0	20	9.2	70	32.2	10	4.6
Crops	25	23.0	25	5.8	60	13.8	15	3.5
Native forest (nonintervened)	13	12.0	0	0.0	0	0.0	100	12.0
Secondary forest (intervened)	7	6.4	30	1.9	60	3.8	10	0.6
Fallow land ("rastrojo")	5	4.6	80	3.7	20	09	0	0.0
Total	100	92.0	22.4	20.6	55.0	50.7	22.6	20.7

[a]Excludes urban areas and other uses (mines, recreation areas, natural parks, and lakes).

TABLE 13.5. Land-use distribution at microwatershed level in Andean Hillsides (800-2,000 masl), Colombia.

	Microwatershed		
Land use	(1) Alto Dagua—La Cumbre—Restrepo	(2) Alto Garrapatas	(3) Pance-Meléndez—Cali—Aguacatal
Pastures	53.5 (62%)	53.1 (47%)	18.3 (61%)
Crops	18.5 (22%)	26.5 (23 %)	1.5 (5 %)
Forest & nat. regen.	12.9 (15%)	34.4 (30%)	8.3 (28%)
Other uses[a]	1.2 (1%)	0.18 (0.1%)	1.9 (6 %)
Total area studied (thousand ha)	86.1 (100%)	114.2 (100%t)	30.0 (100%)

Source: Gómez, 2002; modified from CVC, 1998.
[a]Other uses: mines, recreation areas, natural parks, lakes, urban areas.

loss of topsoil, terrain deformation, and mass movement. Moderate to strong chemical deterioration caused by loss of nutrients and organic matter is also present. Wind erosion does not occur. General soil degradation severity in the region ranges from low to very high.

CARBON SEQUESTRATION RESEARCH IN THE TROPICAL ANDEAN HILLSIDES

The present chapter reports results of a three-year research study on the evaluation of soil carbon stocks (SCS) carried out as part of the research agenda of the international project "Research Network for the Evaluation of

Carbon Sequestration Capacity of Pasture, Agro-pastoral and Silvopastoral Systems in the American Tropical Forest Ecosystem," sponsored by the Netherlands Cooperation as Activity CO-010402, and implemented by CIPAV (Cali, Colombia), Universidad de la Amazonia (Florencia, Colombia), CIAT (Cali, Colombia), CATIE (Turrialba, Costa Rica), and Wageningen University (the Netherlands). The project aims at identifying pasture, agro-pastoral, and silvo-pastoral systems that present an attractive economic alternative to the farmer and show high levels of carbon sequestration and carbon stocks, comparing them with two reference states: degraded land (negative control) and native forest (positive control). It also aims at providing recommendations to policymakers at local, national, and regional levels, about appropriate pasture systems, considering their socioeconomic benefit and provision of environmental services, particularly carbon sequestration. The project works on three contrasting subecosystems of the American Tropics: Tropical Andean Hillsides, Colombia; Subhumid and Humid Tropical Forest, Costa Rica; and Humid Tropical Forest, Amazonia and Colombia, with a possible addition from 2005 onward of the Savannah subecosystem with research sites in Colombia.

This chapter reports methodology and preliminary research findings on the Tropical Andean Hillsides, from two experimental areas whose generalization domain corresponds to specific environmental characteristics of 800 to 2,000 masl, acidic soils (pH between 5.2 and 6.2), mean temperature between 14.0 and 23.5°C during the growing season, mean rainfall (1,500 to 1,900 mm per year) exceeding 60 percent of potential evapotranspiration during six to nine months annually, and moderate to steep slopes (15 to 83 percent depending on land-use system). Regions with these characteristics cover 30 Mha in Peru, Bolivia, Ecuador, Colombia, and Venezuela (Pachico et al., 1994).

Methodology

Soil carbon stocks were evaluated in long-established pasture and silvopastoral systems (14 to 16 years), on small farms belonging to the "Network of small farmers of the Andean Hillsides of dual-purpose cattle under cut and carrying" used by research purposes by CIPAV (6 small farms of 2 to 12 ha per farm, located in areas representative of the above-mentioned conditions).

A soil-sampling design controlling factors affecting SCS (site conditions, slope gradient, land use, and soil depth) was used. Soil carbon contents were measured for four soil depths (0-10, 10-20, 20-40, 40-100 cm) using 2 space replications per land-use system and 12 sampling points per land-use system per space replication. All soil samples taken were composite samples and analyzed for bulk density, total C, oxidizable C, total N, P, CEC, pH, and soil texture for each soil pit and depth. Total oxidizable and stable C (the latter expressed as the difference between total C and oxidizable C) were corrected by bulk density and expressed as Mg C·ha^{-1} in 10 cm soil layers, and for 0-40, 40-100, and 0-100 cm depth.

For statistical comparisons of SCS among land-use systems, calculations based on fixed soil mass according to Ellert et al. (2002), but without subdivision in layers as modified by Buurman et al. (2004) were carried out for total and stable C, using ANOVA models consistent with the sampling design. Following Buurman et al. (2004), the minimum soil mass per sampling point to 1 m depth was used as a reference for each experimental area. Although the fixed soil mass method is more accurate, many authors conveniently use fixed-depth SCS estimates instead. We therefore present both estimates for these research sites. A perfect correlation between fixed soil mass and fixed soil depth estimates is expected exclusively when bulk densities do not show a large variation with depth. The two calculation methods—fixed soil depth and fixed soil mass—were compared in terms of absolute SCS estimates and statistical significance of land-use system comparisons.

Soil Carbon Stock Data

The data in Table 13.6 show statistical comparisons of SCS among land-use systems using fixed soil mass and fixed soil depth estimates. When SCS estimates were adjusted to fixed soil mass, corresponding rankings among land-use systems remained the same, but absolute SCS estimates and statistical comparisons among pairs of systems changed, indicating an overestimate of stocks when using fixed-depth calculations.

Fixed soil mass-based SCS estimates presented in Table 13.6 show the following. For area 1 (Dovio)—with higher altitude, more steep slopes and higher soil fertility—native forest (234 Mg·ha^{-1} per 1 m-equivalent) had statistically higher SCS than improved *B. decumbens* pasture, degraded pasture, and mixed forage bank for cut and carrying (162, 156, and 138 Mg·ha^{-1} per 1 m-equivalent, respectively). On area 2 (Dagua)—with lower altitude, less steep slopes, and lower soil fertility—lower levels of SCS were measured for all systems. Native forest (186 Mg·ha^{-1} per 1 m-equivalent) had statistically higher SCS than secondary forest, natural regeneration of a degraded pasture, and improved *B. decumbens* pasture (152, 147, and 142 Mg·ha^{-1} per 1 m-equivalent, respectively), and these in turn were statistically higher than those of degraded soil and mixed forage bank for

TABLE 13.6. Soil carbon stocks (Mg C·ha^{-1}) for land-use system, estimated based on fixed soil mass (Method 1) and fixed soil depth (Method 2), tropical Andean Hillsides, Colombia.[a]

	Total C (Mg·ha^{-1})		Oxidizable C (Mg·ha^{-1})		Stable C (Mg·ha^{-1})	
System	Meth 1	Meth 2	Meth 1	Meth 2	Meth 1	Meth 2
Area 1: Dovio (1,900 masl)						
Native forest	234 a	262 a	169 a	184 a	67 a	79 a
Improved *Brachiaria decumbens* pasture	162 b	213 b	125 b	159 b	38 b	55 ab
Degraded pasture	156 b	183 bc	121 b	139 c	37 b	46 ab
Mixed forage bank	138 b	161 c	94 c	106 d	47 b	58 b
F-value	14.5***	12.1***	29.6***	28.3***	4.5***	3.0*
CV (%)	22.2	21.2	15.4	14.6	48.5	47.3
Area 2: Dagua (1,350 masl)						
Native forest	186 a	214 a	149 a	172 a	42 ab	50 b
Secondary forest	152 b	177 b	115 b	129 b	37 b	48 b
Nat. regen. of degr. land (fallow land)	147 b	171 b	89 c	93 c	59 a	78 a
Improved *Brachiaria decumbens* pasture	142 b	165 b	118 b	141 b	35 b	42 b
Degraded land	97 c	125 c	62 d	71 d	35 b	54 b
Mixed forage bank	86 c	104 d	60 d	68 d	26 b	36 c
F-value	10.7***	9.7***	56.3***	84.9***	3.8*	4.6**
CV (%)	27.9	27.9	15.6	13.7	56.4	55.4

Source: Amézquita et al., 2004.
[a]Research areas, tropical Andean hillsides, Colombia. Carbon Sequestration Project, the Netherlands Cooperation C0-010402.

cut and carrying (97 and 86 Mg·ha^{-1} per 1 m-equivalent, respectively). The SCS data suggest that although native forest is characterized by the highest soil carbon accumulation capacity in this ecosystem, improved pasture systems, and natural regeneration (fallow land and secondary forest) are promising environmental solutions for the recovery of degraded areas, as SCS-improved systems.

Carbon Sequestration Rates

Carbon sequestration rates (Mg C·ha^{-1} per year) were calculated for each land-use system based on their corresponding SCS and duration, with the assumption that the system was converted from degraded land. For reasons of compatibility with other researchers, SCS estimates used for calculating carbon sequestration rates were based on fixed soil depth estimates, expressed as Mg·ha^{-1} per 1 m (Table 13.7).

TABLE 13.7. Estimated carbon sequestration rates (Mg C·ha^{-1} per year) for different land-use systems if converted from degraded land, tropical Andean Hillsides, Colombia[a].

Land-use system	System age (yrs)	SCS (Mg·ha^{-1} per 1 m)	C-seq rate (Mg·ha^{-1} per year)
Area 1: Dovio (1,900 masl)			
Degraded land (reference)[b]	16	125	0
Forage bank for "cut and carrying"	16	161	−2.3
Degraded pasture	16	183	3.6
Improved *B. decumbens* pasture	16	213	5.5
Native forest	>50	262	2.7 [c]
Area 2: Dagua (1,350 masl)			
Degraded land (reference)	16	125	0
Forage bank for "cut and carrying"	16	104	−1.3
Improved *B. decumbens* pasture	14	165	2.9
Nat. regen. of degr. land (fallow land)	16	171	2.9
Secondary forest	25	177	2.1
Native forest	>50	214	1.8 [a]

Source: Amézquita et al., 2004.
[a]Research areas, tropical Andean Hillsides, Colombia. Carbon Sequestration Project, the Netherlands Cooperation C0-010402.
[b]SCS data of the "Degraded land" system from Dagua (area 2) was used as reference for calculating soil carbon sequestration rates for the various land-use systems in both areas.
[c]Age of nonintervened forest was assumed to be 50 years

SOIL CARBON SEQUESTRATION POTENTIAL IN THE TROPICAL ANDEAN HILLSIDES ECOSYSTEM

Soil carbon sequestration rates (Mg C·ha^{-1} per year) were calculated within the SCS estimates and system's age for the range of land-use systems shown in Table 13.7. These, together with expert knowledge on land-use area and degradation level (Table 13.4), allowed a first approximation of carbon sequestration potential for the Tropical Andean Hillsides (Table 13.8).

The data in Table 13.8 present different scenarios for conversion of degraded areas to C-improved land-use systems. Considering the more optimistic land transformation option for each type of degraded area—identified as that with maximum carbon sequestration rate—and using expert

TABLE 13.8. Soil carbon sequestration potential of the tropical Andean Hillsides over a 20-year period: A first approximation.

Present land use	Optional land use (C-improved)	Potential area to be converted to option		Soil C seq pate ($Mg \cdot ha^{-1}$ per year)	Soil C seq potential under each option ($Mg \cdot ha^{-1}$ per 20 years)	Total soil C seq potential under each option over a 20-year period (10^{15} g C)
		%	Mha			
Degraded land	1. Nat. reg of degr. land	22.4	20.6	2.9	58	1.2
	2. Conventional, degraded pasture			3.6	72	1.5
	3. Improved pasture			2.9-5.5	58-110	1.2-2.3
	4. Secondary Forest			2.1	42	0.9
	5. Forest in equilibrium			1.8-2.7	36-54	0.7-1.1
	6. Improved cropland			-	-	-
Degraded pasture	1. Improved pasture	35.0	32.3	–1.3-1.9	-26-38	–0.8-1.2
	2. Natural regeneration to forest in equilibr.			0.6-1.6	12-32	0.4-1.0
	3. Improved cropland			-	-	-
Fallow land	1. Improved pasture			–0.4-2.6	52	0.05
	2. Secondary forest	1.0	0.9	0.24	4.8	0.004
	3. Forest in equilibrium			0.86	17.2	0.016
	4. Improved cropland			-	-	
Secondary forest	1. Forest in equilibrium	4.0	3.8	0.7-1.7	14-34	0.05-0.13
Degraded, nonproductive crops	1. Improved pasture	15.0	13.8	-		-
	2. Nat. regeneration to secondary forest			-		-
	3. Nat. regeneration to forest in equilibrium			-		-
	4. Improved cropland			-		-
Total potential		77.4	71.4			3.7

Sources: Research data or SCS from the "Carbon Sequestration Project, the Netherlands Cooperation C0-010402" in the tropical Andean Hillsides (Amézquita et al., 2004) and expert knowledge information on land-use areas and their degradation stage.
Note: The authors do not have at present available data on SCS for croplands.

knowledge on potential area (Mha) to be transformed into various C-improved options, a maximum potential estimate of carbon sequestration over a 20-year period upon conversion of the Tropical Andean Hillsides Ecosystem is calculated as follows.

- Converting 20.6 Mha of severely degraded land of Tropical Andean Hillsides into improved pasture systems has a maximum carbon sequestration potential of 2.3*1,015 g C over a 20-year period.
- Converting 32.3 Mha of conventional, moderately degraded pasture systems of Tropical Andean Hillsides into improved pasture systems has a maximum carbon sequestration potential of 1.2*1,015 g C over a 20-year period.
- Converting 0.9 Mha of fallow land into improved pasture systems has a maximum carbon sequestration potential of 0.05*1,015 g C over a 20-year period.
- Converting 3.8 Mha of secondary forest into conservation areas without human intervention, to evolve into native forest in equilibrium, has a maximum carbon sequestration potential of 0.13*1,015 g C over a 20-year period.

Because there is no information about soil carbon stocks in cropland, it is difficult to estimate the changes brought about by conversion of cropland to any pasture, silvo-pastoral system, or maintained as natural regeneration to become a secondary forest or forest in equilibrium.

Therefore, considering the best C-sequestration option for each type of degraded area, an estimate of the total carbon sequestration potential of the Tropical Andean Hillsides over a 20-year period is 3.7*1,015 g C (Table 13.8).

CONCLUSIONS

Related to Carbon Sequestration Research on the Tropical Andean Hillsides

1. The information on land-use distribution and land-use change in the Tropical Andean Hillsides Ecosystem since 1950 is scanty. Available information is given by country but not by ecosystem across countries. A special effort must be made to acquire this important information, which is essential for a more precise estimate of carbon sequestration potential.

2. Specific methodological issues related to the proper estimation of SCS need to be considered for any carbon sequestration research project. In par-

ticular: (1) A soil-sampling design taking into account all factors affecting SCS needs to be used to obtain minimum-variance estimates. Variability of SCS estimates depends on land-use type (i.e., higher on degraded land than on improved grass-alone pasture), site conditions (altitude, climate, topography), soil characteristics, and carbon fraction. (2) For statistical comparisons among land-use systems, SCS estimates corrected by bulk density and adjusted to fixed soil mass per sampling point must be used. Estimates thus obtained are more precise than those based on fixed soil depth. (3) SCS estimates for a given land-use system need to be interpreted on the basis of the historic land use.

3. The present study did not address carbon sequestration research on coffee and other perennial crops grown in the Tropical Andean Hillsides. However, given their economic importance to Andean countries, carbon sequestration research on these crops must be given a high priority.

Related to Carbon Sequestration Research in Tropical America

4. Tropical pasture, agro-pastoral, and agro-silvo-pastoral systems are important socioeconomic components in TA across all ecosystems. These are key land-use systems for carbon sequestration research. Therefore, special attention and support must be given to carbon sequestration research in tropical pasture systems across all ecosystems, particularly to those adapted to the region and representing viable economic alternative to farmers. These include improved grass-alone, improved grass-legume, improved grass-legume-trees, improved mixed pasture-crop systems such as agro-pastoral systems combining forage grass, forage legumes, fruit trees, other perennial crops, and short-cycle crops.

5. In order to succeed in the organization and implementation of a carbon sequestration research network in Tropical America, both institutional and technical aspects must be considered. Institutional aspects involve the identification of a scientific leader, network nodes in Latin American biomes, and leading scientists for specific knowledge areas. Also of vital importance for success is the identification of donor agencies representatives and policymakers to collaborate with the network. Technical aspects involve agreement on research objectives and methodology, standards and protocols, sampling methods, laboratory analysis standards and procedures, and methods for calculation of SCS and carbon sequestration rates. These aspects are crucial for a valid comparison of research results.

REFERENCES

Amézquita, E., J. Ashby, E.K. Knapp, R. Thomas, K. Muller-Samann, H. Ravnborg, J. Beltran, J.I. Sanz, I.M. Rao, and E. Barrios. 1998. CIAT's strategic research for sustainable land management on the steep hillsides of Latin America. In F.W.T Penning de Vries, F. Agus, and J. Kerr (eds.), *Soil erosion at multiple scales.* CAB International, London, pp. 121-131.

Amézquita, M.C. 2002. Project objectives, expected products and research methodology. In M.C. Amézquita, F. Ruiz, and B.van Putten (eds.), *Carbon sequestration and farm income: Concepts and methodology.* Internal Publication No. 5. CIAT, Cali, Colombia.

Amézquita, M.C. 2003. Two-year project achievements. Internal Publication No. 9. Fourth International Coordination Meeting. M.C. Amézquita and F. Ruiz (eds.). December 2003. CIAT, Cali, Colombia.

Amézquita, M.C., M. Ibrahim, and P. Buurman. 2004. Carbon sequestration in pasture, agro-pastoral and silvo-pastoral systems in the American tropical forest ecosystem. Proc. 2nd World Congress in Agroforestry Systems, Merida, Mexico, February 2004.

Browder, J.O. 1988. The social costs of rain forest destruction: A critique and economic analysis of the "hamburger debate." *Interciencia* 13(3):115-120.

Buurman, P., M. Ibrahim, and M.C. Amézquita. 2004. Mitigation of greenhouse gas emissions by silvopastoral systems: Optimism and facts. Proc. 2nd World Congress in Agroforestry Systems, Mérida, Mexico, February 2004.

CIAT (Centro Internacional de Agricultura Tropical). 1976-1996. Tropical Pastures Program annual reports. CIAT, Cali, Colombia.

CIAT (Centro Internacional de Agricultura Tropical). 1994. Tropical Lowlands Program annual report 1994. CIAT, Cali, Colombia.

CIAT (Centro Internacional de Agricultura Tropical). 1997-1999. Tropical Lowlands Program annual report 1997-1999. CIAT, Cali, Colombia.

CIAT (Centro Internacional de Agricultura Tropical). 1999. Tropical Forages Project annual report. CIAT, Cali, Colombia.

Departamento Nacional de Estadistica (DANE). 1996. Encuesta Nacional Agropecuaria. Resultados 1995. Bogotá, Colombia.

de Rubinstein, E. and G.A. Nores. 1980. *Gasto en carne de res y productos lácteos por estrato de ingreso en doce ciudades de América Latina.* CIAT, Cali, Colombia.

Ellert, B.H., H.H. Janzen, and T. Entz. 2002. Assesment of a method to measure temporal change in soil carbon storage. *Soil Sci. Soc. Am. J.* 66:1687-1695.

Encarta. 2003. Biblioteca de Consulta. Microsoft Corporation.

Etter A. and W. van Wyngaarden. 2000. Patterns of landscape transformation in Colombia, with emphasis on the Andean region. *Royal Swedish Academy of Sci.* 29:412-439.

FAO. 2000. Food balance sheets. FAO, Rome, Italy.

Gómez, M.E. 2002. Macrocaracterización sub-ecosistema Laderas Andinas y Áreas de Investigación del Proyecto. In M.C. Amézquita, F. Ruiz, and B. van Putten (eds.), *Carbon sequestration and farm income: Concepts and methodology*. Internal Publication No. 5. CIAT, Cali, Colombia, pp. 105-120.

Instituto de Investigacion de Recursos Biologicos A.V. Humboldt—Iavh Ministerio Medio Ambiente, DNP, PNUMA. 1998. Colombia biodiversidad siglo XXI, Santafé de Bogotá, Colombia.

Jones, P. 1993. *Hillsides definition and classification*. CIAT, Cali, Colombia.

Kaimowitz, D. 1996. *Livestock and deforestation Central America in the 1980s and 1990s: A policy perspective*. Center for International Forestry Research Special Publication. CIFOR, Jakarta.

Kindersley D. 2004. La Tierra. Capítulo Montañas, Periódicos Asociados Ltda., Fascículo 15. Tecimpre S.A. Bogotá, Colombia, pp. 114-115.

Murgueitio E. 2003. Impacto ambiental de la ganadería de leche en Colombia y alternativas de solución. *Livestock Research for Rural Development* 15(10). Fundación Centro para la Investigación en Sistemas Sostenibles de Producción Agropecuaria (CIPAV), Cali, Colombia. Available online at www.cipav.org.co/lrrd.

Nieuwe Grote Wereldatlas. 1977. Elsevier, Amsterdam, the Netherlands.

Oldeman, L.R., R.T.A. Hakkeling, and W.G. Sombroek. 1990. World map of the status of human-induced soil degradation. ISRIC, in cooperation with Winand Staring Centre-ISSS-FAO-ITC, UNEP.

Pachico, D., J. Ashby, and L.R. Sanint. 1994. Natural resource and agricultural prospects for hillsides of Latin America. Paper prepared for IFPRI 2020-Vision. Workshop, Washington, November 7-10. In Hillsides Program annual report 1993-1994. CIAT, Cali, Colombia, pp. 283-321.

Ramírez C. and R. Ortiz. 1997. Causas de pérdida de biodiversidad. In *Informe Nacional sobre el estado de la biodiversidad Colombia*. Instituto de Investigación de recursos biológicos Alexander von Humboldt, Vol. 2, pp. 33-35.

Rivas, L., D. Pachico, C. Seré, and J. García. 1998. Evolución y perspectivas de la ganadería vacuna en América Latina tropical en un contexto mundial. Proyecto de Evaluación de Impacto. CIAT, Cali, Colombia.

Sader, S.A. and A.T. Joyce. 1988. Deforestation and trends in Costa Rica, 1940 to 1983. *Biotropica* 20(1):11-19.

Sanint, L.R., L. Rivas, M.C. Duque, and C. Seré. 1984. Food consumption patterns in Colombia: A cross sectional analysis 1981. Paper presented at the Internal Workshop of Agricultural Centers on Selected Economic Research Issues in Latin America. CIAT, Cali, Colombia.

Toledo, J. M. 1985. Pasture development for cattle production in the major ecosystems of the tropical American lowlands. Proc. of the XV Intl. Grasslands Congress, Kyoto, Japan.

Valdés, A. and G.A. Nores. 1979. Growth potential of the beef sector in Latin America—survey of issues and policies. 4th World Conference of Animal Production, Buenos Aires, Argentina.

Veldkamp, E. 1993. Soil organic carbon dynamics in pastures established after deforestation in the humid tropics of Costa Rica. PhD thesis, Wageningen Agricultural University, Wageningen, the Netherlands.

Veldkamp, E. 1994. Organic carbon turnover in three tropical soils under pasture after deforestation. *Soil Sci. Soc. Am. J.* 58:175-180.

Vera, R., J.I. Sanz, P. Hoyos, D.L. Molina, M. Rivera, and M.C. Moya. 1993. Pasture establishment and recuperation with undersown rice on the acid soil savannas of South America. CIAT, Cali, Colombia.

Chapter 14

Soil Carbon Storage and Sequestration Potential in the Cerrado Region of Brazil

Mercedes M. C. Bustamante
Marc Corbeels
Eric Scopel
Renato Roscoe

INTRODUCTION

Tropical savannas occupy about 15 percent of the total terrestrial area and store about 13 percent of the total C (vegetation and top 1 m soil) (IPCC, 2000). The cerrado, the principal savanna region south of the equator, represents about 9 percent of the total area of tropical savannas in the world. It occurs entirely within Brazil, mostly in the central region of the country, covering approximately 2 million km^2 (23 percent of the country) (Figure 14.1). The annual precipitation varies from 600 to 2,200 mm, but 65 percent of the area receives between 1,200 and 1,600 mm annually. The rainfall distribution is markedly seasonal with a dry season that lasts from four to seven months in 88 percent of the area, and from five to six months

We would like to acknowledge Dr. Rattan Lal (Ohio State University), Dr. Carlos Cerri (CENA—Universidade de São Paulo), Dr. Christian Feller and Dr. Martial Bernoux (IRD—France), and Dr. Andrew Dowdy (US State Department) for the organization and financial support to the "Workshop on the Potential Soil Carbon Sequestration in Latin America." We are also grateful to Dr. Robert Boddey (Embrapa Agrobiologia) and Dr. Cimélio Bayer (Universidade Federal do Rio Grande do Sul) for their suggestions and comments on this manuscript. This research was partially supported by NASA under "The Large Scale Biosphere-Atmosphere Experiment in Amazonia" (LBA) project ND-07 (UnB/EPA-Assistance Agreement 827291-01), by Embrapa, by the French Found for World Environment (FFEM), and by the French Ministry of Foreign Affairs (MAE).

Carbon Sequestration in Soils of Latin America

doi:10.1300/5755_14

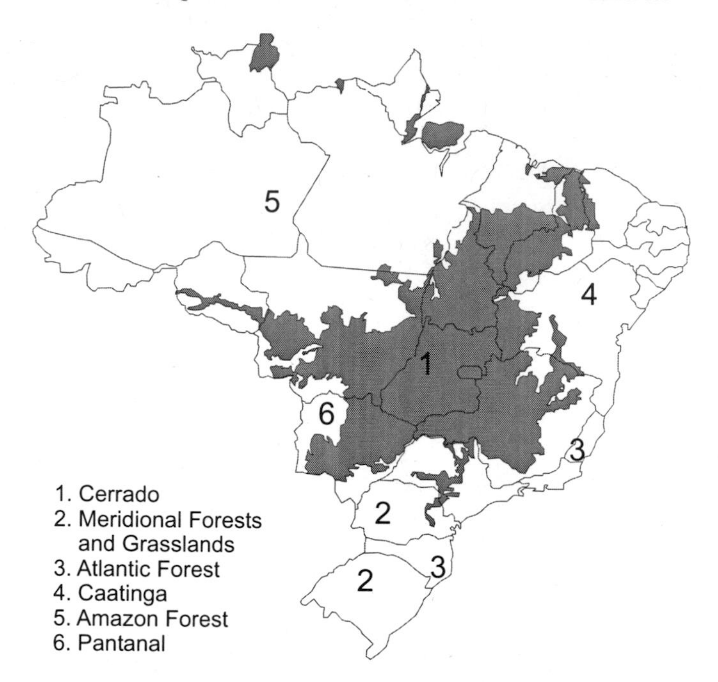

FIGURE 14.1. Geographical distribution of the cerrado biome in Brazil.

in 67 percent of the area. The mean annual temperature varies from 22°C in the south to 27°C in the north of the region (Adámoli et al., 1985).

Defined as wet seasonal savanna, the cerrado consists of a gradient of physiognomies, from the grasslands (locally called "campo limpo") to a sclerophylous forest ("cerradão"). Between these, there are intermediate physiognomies with increasing density of woody species (campo sujo, campo cerrado, and cerrado *stricto sensu*). The type of physiognomy is usually dependent on local climatic conditions (rainfall), soil chemical and physical properties, and human interventions (wood removal, fire, grazing). The region is considered one of the 25 biodiversity hotspots in the world, i.e., areas with high biodiversity but accelerated loss of habitats (Myers et al., 2000), mainly due to the expansion of agricultural activities.

Cerrado soils are dominated by low-activity clays (kaolinite, gibbsite, and iron oxihydroxides) (Adámoli et al., 1985; Reatto et al., 1998). Oxisols and Entisols represent approximately 46 and 15 percent of the area, respectively (Reatto et al., 1998). In general, soils are acidic, with low nutrient concentrations and high aluminum content. Under such conditions, soil organic matter (SOM) is particularly important to physical, chemical, and biological processes related to nutrient cycling, soil aggregation, and plant available water (Resck, 1998; Resende et al., 1996).

LAND-USE HISTORY IN THE CERRADO REGION

Recent estimates indicate that the mean rate of cerrado conversion is approximately 1.1 percent per year, i.e., 2.2 million hectares (Mha) per year (Machado et al., 2004). The authors averaged data for two periods: from 1985 to 1993 with a conversion rate of 1.5 percent per year, i.e., 3 Mha, and from 1993 to 2002 with a conversion rate of 0.67 percent per year, i.e., 1.36 Mha, in both cases considering the total area of the cerrado (2 million km^2).

Since the 1970s the region has been the focus of intense agricultural expansion. Between 1975 and 1996 the area cleared for farming almost doubled, from 34.7 to 64.5 Mha. The main element of these changes was the expansion of planted pastures (mainly *Brachiaria* species), which increased from 16.0 to 49.2 Mha between 1975 and 1996. Presently the cerrado region is one of the most important beef-producing regions in Brazil, with approximately 44 percent of the national herd. However, at least half of the pastures in the cerrado are in advanced stages of degradation due to poor management, especially overgrazing and lack of fertilization (Oliveira et al., 2004).

The area under crop cultivation increased from 6.9 to 8.2 Mha in the same period. The major crops are soybeans (*Glycine max* [L.] Merr.) (for export), maize (*Zea mays* L.), rice (*Oryza sativa* L.), and beans (*Phaseolus vulgaris* L.) (for the internal market). Crop farming in the region is characterized by large farm units, monocultures with high external inputs, and heavy mechanization (Cadavid-Garcia, 1995; Klink et al., 1995). Although annual crops are not the main land-use type in the cerrado, they do induce land clearing, especially in the northern part of the region for the production of soybean. The expansion of soybean in the cerrado was strongly affected by a growing international demand for this crop, but three factors were decisive:

1. Favorable natural conditions of the savannas (e.g., flat landscape, good physical structure of soils);
2. Technological development which made the cultivation of the crop viable (e.g., new soybean varieties); and
3. Investments in transportation infrastructure.

Serious soil degradation (erosion and compaction, loss of organic matter), resulting from inadequate soil management associated with conventional agricultural practices in the cerrado, led to the introduction, in the early 1980s, of no-tillage systems. The area under no-tillage in cerrado was still insignificant in 1989 but increased exponentially since then, and in 1999, there were 3.3 Mha under no-tillage according to the Brazilian Federation on Zero Tillage (FEBRAPDP). Today, approximately 6 Mha of cropland are estimated to be under no-tillage in the cerrado (www.agri.com.br/febrapdp/pd_area_estados.htm).

Land-use changes in the cerrado also include the conversion of native savanna to fast-growing forestry plantations. About 40 years ago planted forests in Brazil occupied a little over 500 hectares. By 1987 Brazil had more than 6 Mha of planted forests, one-third of which were in the state of Minas Gerais (cerrado region) (www.silviminas.com.br). *Eucalyptus* and *Pinus* plantations represent 80 percent of the total planted area. The production is mainly for the pulp and paper industry and for charcoal in the steel industry.

SOIL CARBON STOCKS UNDER NATIVE CERRADO VEGETATION

Considering average values of C stocks for tropical savanna ecosystems as estimated by the International Panel on Climate Change (IPCC, 2000), the C stocks in the cerrado would be about 5.9 Gt C in vegetation and 23.8 Gt C in soil (up to 100 cm). This would represent a stock of about 29 Mg $C \cdot ha^{-1}$ in the vegetation and 117 Mg $C \cdot ha^{-1}$ in soil. However, there is a wide range of vegetation types in the Cerrado, and the use of average values for estimating C stocks may be too overly simplified. Direct estimates based on measurements are not available so far, although some information exists for specific physiognomic forms of the biome.

Abdala (1993) estimated C stocks in different compartments of a cerrado *stricto sensu* from central Brazil. The total C stock in the vegetation plus soil (1 m depth) was 265 $Mg \cdot ha^{-1}$, divided into arboreal (28.5 $Mg \cdot ha^{-1}$) and herbaceous (4 $Mg \cdot ha^{-1}$) strata, litter (5 $Mg \cdot ha^{-1}$), roots and detritus (42.5 $Mg \cdot ha^{-1}$), and SOM (185.0 $Mg \cdot ha^{-1}$). These results emphasize the importance of the soil for C storage in cerrado systems.

Table 14.1 summarizes published values of soil C stocks under different vegetation covers and soil types in the cerrado region. Values for 100 cm soil depth range from ~ 97 to 210 Mg $C \cdot ha^{-1}$. Organic C content tends to increase with clay content, as already shown for several other ecosystems (e.g., Feller and Beare, 1997, for tropical ecosystems; Christensen, 2000, for temperate ecosystems). Grassland vegetations (campo sujo and campo limpo) tend to store more C than vegetation types with higher woody plant

TABLE 14.1. Carbon accumulation ($Mg \cdot ha^{-1}$) in soil layers under different vegetation cover in the cerrado region.

Land use	Area (ha)	Soil layer				Clay content ($g \cdot kg^{-1}$)	Reference
		0 - 20 cm	0 - 40 cm	0 - 100 cm	0 - 200 cm		
Native areas	71,100,000[a] 86,130,000[b]						Machado et al. (2004) Sano et al. (2001)
Cerradão		53.0		148.0	230.0	689 - 808[c]	Lardy et al. (2002)
Cerrado stricto sensu		46.0		143.0	231.0	672 - 697	Lardy et al. (2002)
Cerrado stricto sensu				195.0		600 - 750	Roscoe et al. (2000)
Cerrado stricto sensu				145.0		480	Resck et al. (2000)
Campo sujo		67.0		209.0	297.0	568 - 668	Lardy et al. (2002)
Campo limpo		72.0		198.0	281.0	436 - 589	Lardy et al.(2002)
Gallery forest		62.0		165.0	248.0	626 - 715	Lardy et al. (2002)
Range		53 - 72		143 - 209	230 - 297		
Native pastures	27,144,000[b]						Sano et al. (2001)
Pastured cerrado		45.0		174.0	27.7	716 - 724	Lardy et al. (2002)
Pastured cerrado		31.1	54.7	99.7	n.a	567 - 640	Silva et al. (2004)
Range		31 - 45		100 - 174			
Cultivated pastures	48,000,000[b]						Sano et al. (2000)
12-year-old pasture		54.0		161.0	252.0	488 - 635	Lardy et al. (2002)
Cultivated pastures (range from degraded to fertilized)[d]		30.8 - 39.9	53.9 - 67.9	97.1 - 113	n.a	587 - 610 (0-20 cm) 634 - 654 (80 -100 cm)	Silva et al. (2004)
18-year-old pasture			90.0			480	Resck et al. (2000)
18-year old pasture		42.2	74.8	150.2			Corazza et al. (1999)
Range		31 - 54	54 - 90	97 - 161			

TABLE 14.1 *(continued)*

Land use	Area (ha)	Soil layer			Clay content (g·kg^{-1})	Reference
Croplands						Sano et al. (2001) APDC
Total area	9,500,000					
No-tillage	~ 6,000,000					
NT in the past 6 yrs (4 crop rotations)		55.8 - 59.0			600	Bayer et al. (2004)
CT in the past 6 yrs		54.3			600	Bayer et al. (2004)
NT in the past 15 yrs		47.4	81.2	155.0		Corazza et al. (1999)
CT in the past 15 yrs		37.4	66.6	128.8		Corazza et al. (1999)
			0 -45 cm			
NT (CT and then NT in the past 10 yrs)			99.0[e]		600 - 750	Roscoe and Buurman (2003)
Tillage (30 yrs)			102.0[e]		600 - 750	Roscoe and Buurman (2003)
Range		37 - 59	67 - 115[e]	129 - 182		
Forested areas						
Pinus – 20-year-old (clayey Oxisol)		47.1	61.0		578 - 626	Zinn et al. (2002)
Eucalyptus – 7-year-old (sandy Entisol)		25.1	29.6		162 - 214	Zinn et al. (2002)
Eucalyptus – 7-year-old (loamy Oxisol)		43.3	51.8		337 - 376	Zinn et al. (2002)
Eucalyptus – 12-year-old				148.2		Corazza et al. (1999)

TABLE 14.1 *(continued)*

Land use	Area (ha)		Soil layer			Clay content (g·kg^{-1})	Reference
Eucalyptus – 60-year old (after 20 years of grass-land) – long rotation of 60 years		24.9	41.4	75.6			Maquere (2004)
Eucalyptus – 60-year-old (after 20 years of grassland) – short rota-tion of 6-7 years		32.9	50.4	88.0			Maquere (2004)
Pinus – 20-year-old		26.71[f]	49.1[g]	128.1[h]	172.1	615 - 885	Lilienfein et al. (2001)
Range		25 - 47[f]	30 - 61[g]				

Source: Based on data from Associação de Plantio Direto no Cerrado (www.apdc.com.br).
[a]Estimate for 2002. Considered only the central part of the cerrado region with a total area of 158,000,000 ha based on MODIS images.
[b]Estimates for 1995/1996. Considered the whole cerrado region based on census data.
[c]Includes six different areas (four pure grass and two grass+legumes).
[d]Range of clay content between 0-100 cm depth.
[e]Soil layer is 0-45 cm.
[f]Soil layer is 0-15 cm.
[g]Soil layer is 0-30 cm.
[h]Soil layer is 0-120 cm.

density (gallery forest, cerradão, and cerrado *stricto sensu*), probably due to the high turnover rate of root systems under grasses.

Given its total area and the average C stocks for tropical savannas (IPCC, 2000), the cerrado biome contains about 1 percent of the total C in terrestrial ecosystems (about 30 Gt C), contributing very little to global C stocks (about 2,577 Gt C). On the other hand, eddy covariance data obtained in a native cerrado *stricto sensu* (where vegetation had been burned by fire of low intensity seven years before the experiment) indicated a net uptake of about 2.5 Mg $C \cdot ha^{-1}$ per year (Monteiro, 1995; Miranda et al., 1996). Although these data can not be extrapolated to the entire biome, they illustrate that the potential of C sequestration in the cerrado is comparable to that of most world biomes.

IMPACTS OF LAND USE ON SOIL C STORAGE

Considering the diversity of soil types and climatic conditions in the cerrado region, changes in soil C storage with changes in land use have not been systematically studied. In addition, little information is available about erosion losses. The changes in land use that were more commonly studied are

1. Native cerrado to pasture,
2. Native cerrado to cropland under conventional tillage,
3. Conventional to no-tillage, and
4. Native cerrado to forestry plantations.

In the following section we discuss the impacts of these land-use changes on soil C together with the impact of fire. Fire events are common in the cerrado and occur mainly during the dry season (Kirchoff and Alvalá, 1996). Although it is a natural event in the cerrado, increasing anthropogenic pressure has intensified its occurrence and fire also became part of the management of native pastures (Eiten, 1972; Roscoe et al., 2000).

Native Cerrado to Pasture

Data on C storage with conversion of native savanna into pasture are summarized in Table 14.1. The mean soil C sequestration rate is 1.3 Mg $C \cdot ha^{-1}$ per year. The range of sequestration rates (from –0.87 Mg $C \cdot ha^{-1}$ per year to +3 Mg $C \cdot ha^{-1}$ per year) is illustrated by Silva et al. (2004), who studied soil C storage under different options of pasture management. The data suggest that well-managed, cultivated pastures may provide enough C input to maintain or even increase native C contents (Corazza et al., 1999; Roscoe et al., 2001; Silva et al., 2004). On the other hand, C input from degraded, low-productive pastures may be too low to sustain the high soil C storage under native cerrado. The study by Silva et al. (2004) showed that soil C accumulation under pastures over the previous native stocks occurs only with nutrient inputs through fertilization and legumes. The magnitude of the accumulation is low compared to the initial stocks under native cerrado (100 Mg C·ha^{-1} for 100 cm soil depth). Thus, the available data indicate that the C sequestration potential with the conversion of native cerrado to pasture is rather limited. Moreover, possible benefits on soil C with conversion of native cerrado to pasture are accompanied by large negative impacts of the biodiversity of the cerrado biome. The mean soil C sequestration rate calculated from the published studies (Table 14.1) differs considerably from the rate of about 3 Mg C·ha^{-1} per year recently reported by Fisher and Thomas (2004) for introduced pastures in the eastern plains of Colombia (llanos), thereby extrapolating this figure for the whole of the central lowlands of tropical South America (which represents an area of 8.2 million km^2 including the Amazon basin, the Brazilian shield, and the Orinoco basin) (Fisher and Thomas, 2004) clearly overestimates the C sink for that region. The potential of soil C sequestration under pastures lies more on the recovery of degraded pastures and croplands (Westerhof et al., 1999; Neufeldt et al., 1999; Corazza et al., 1999). Oliveira et al. (2001) showed that grass growth in degraded pastures at three sites in the cerrado region could be recovered through N and P fertilization. Besides, it is generally known that good management of stocking rates is a key component for promoting sustainable pasture growth without risks of degradation.

Native Cerrado to Cropland Under Conventional Tillage

Conventional tillage (plowing and/or harrowing) is considered to be one of the most degradative systems of land use in the cerrado region, which often leads to a reduction of C stocks in soil (Bayer and Mielniczuk, 1999; Resck et al., 2000). Evidence of soil degradation and reduction of water holding capacity and infiltration as a result of conventional tillage have been widely reported for the cerrado (Silva et al., 1994; Lepsch et al., 1994; Resende et al., 1996; Resck, 1998; Resck et al., 2000). Soil degradation processes that affect soil fertility and crop productivity may cause indirect effects on soil C storage through reduced C inputs into the soil. Silva et al. (1994) sampled 220 cerrado topsoils (0 to 15 cm) from soybean fields that were continuously cultivated with a heavy disk harrow. The authors observed severe losses from 41 percent (clayey soils with >30 percent clay) to

80 percent (sandy soils with <15 percent clay) of the initial SOM contents after five years of cultivation. However, C losses are not always observed upon cultivation of native cerrado. For example, Freitas et al. (2000) and Roscoe and Buurman (2003) did not observe changes in the SOM stocks (0 to 40 cm) of a clayey Dark Red Latosol after 25 and 30 years of maize-bean cropping with conventional tillage. Similarly, Lilienfein and Wielcke (2003) reported no significant changes in C content of a clayey Oxisol after 12 years of maize cropping in rotation with soybean under conventional tillage. These studies suggest a high stability of SOM in cerrado soils, which has to be attributed to the high contents of iron and aluminum oxihydroxides in these soils (Resende et al., 1997). In the cerrado Oxisols most of the organic C is concentrated in the clay + silt size fraction (Freitas et al., 2000; Roscoe and Machado, 2002). Therefore, it is anticipated that the large C losses from topsoil under intensive tillage with soybean monocropping (Silva et al., 1994) largely occurred through erosion. Control of soil erosion has been the primary incentive for the adoption of no-tillage practices by farmers in the cerrado region. Based on the few available studies, it also seems that the effect of conventional tillage in reducing soil C stocks is stronger in sandy and loamy soils than in clayey soils.

Conventional Tillage to No-Tillage

The adoption of no-tillage practices is wide spread in the cerrado region. In addition to not tilling the soil, no-tillage systems in the cerrados are often characterized by more intensive cropping. By suppressing tillage practices farmers can sow the main commercial crop earlier in the season, allowing for the planting of a second crop during the same growing season. If climatic conditions are favorable, this second crop is grown as a commercial crop and is harvested, but if conditions are not appropriate it is used as a cover crop to protect the soil from soil erosion. No-tillage systems with increased cropping intensity are referred to as direct seeding mulch-based cropping systems (DMC). Due to a better use of natural resources (radiation, water, and nutrients) as a result of continuous cropping throughout the whole rainy season, plant biomass production in DMC systems can be up to three times higher than in conventional soybean monocropping (Séguy et al., 2003; Scopel et al., 2004). Thus, they represent a significant increase in C input to the soil. DMC systems are accompanied by an increase in the types of crops grown in the region. The choice of the first commercial crop (soybean, maize, rice [*Oryza sativa* L.] or cotton [*Gossypium hirsutum* L.]) and of the second crop (maize, sorghum [*Sorghum bicolor* (L.) Moench.], millet [*Pennisetum glaucum* (L.) R. Brown], beans) opens many opportunities for the diversification of cropping systems. For example, new systems integrating pasture and crops have recently been developed: grasses (often *Brachiaria* sp.) are sown in association with the main commercial crop to achieve a dense soil cover at harvest of the main crop (Kluthcouski et al., 2000). The resulting pasture is an excellent cover crop and can be maintained during a few years. Before pasture degradation, the field is returned to cropland by direct sowing of a commercial crop into the mulch of pasture residues after a total-herbicide application. Despite the new options with DMC systems, cropping systems in the cerrados are not always that diversified. A survey made in the southern region of the federal state of Goiás (Scopel et al., 2003) showed that two types of cropping systems cover about 77 percent of the area with DMC systems. Maize or soybean with fallow represented 32 percent of the area while soybean followed by maize or sorghum or millet as a second crop represented 45 percent of the area.

Although DMC systems have spread quickly in the cerrados, relatively little information is available on their functioning, especially with respect to SOM dynamics (Resck et al., 2000). Higher C stocks in NT than in CT were reported by Corazza et al. (1999), Bayer et al. (2004), and Oliveira et al. (2004), but this was not observed by Roscoe and Buurman (2003) and Freitas et al. (2000). These contrasting results are probably related to the diverse cropping practices associated with no-tillage. It is clear that no-tillage systems with a second crop have more potential to sequester C than no-tillage systems with only one crop per year. In addition, differences in soil texture also play a role. Clayey soils have a larger proportion of stabilized SOM which will be less affected upon tillage, compared to sandy soils.

Bayer et al. (2004) observed that the particulate SOM fraction increased by 37 to 52 percent in a cerrado Oxisol under no-tillage (0 to 20 cm) compared to conventional tillage. This fraction is considered to be particularly sensitive to changes in soil management. Besides, as pointed out by Mielniczuk et al. (2003) and Roscoe and Boddey (2004), organic C accumulation in soil under no-tillage systems seems to occur only when nitrogen is not limiting. These conclusions are also supported by Lovato et al. (2004) and Sisti et al. (2004) working with subtropical soils from Brazil. Data for cerrado soils are not available so far, although results based on simulation modeling, as discussed in the next section, indicate the importance of nitrogen supply to C soil storage.

Native Cerrado to Forestry Plantation

Available data on C stocks under forestry plantations in the cerrado region are not conclusive about the potential of soil C accumulation under this type of land use. Corazza et al. (1999) estimated an accumulation rate of 1.2 Mg C·ha^{-1} per year (0 to 100 cm) after 12 years of *Eucalyptus* cultivation on land cleared from its native vegetation. However, the clay content of the topsoil under eucalypt (72 percent) was considerably higher than that under native cerrado (49 percent). Other studies indicated no changes or even losses in soil C stocks under forestry plantations. Lilienfein et al. (2001) reported no significant differences in soil C storage (0 to 200 cm) between native cerrado (179 Mg C·ha^{-1}) and a 20-year-old *Pinus caribaea* plantation (172 Mg C·ha^{-1}). However, under *Pinus,* the organic layer stored more C (95.0 Mg C·ha^{-1}) than under cerrado (1.2 Mg C·ha^{-1}). Compared to native cerrado, Zinn et al. (2002) observed soil C losses (0 to 60 cm depth) of 0.4 Mg C·ha^{-1} per year under a 20-year-old *Pinus* stand grown on an Oxisol, and 1.3 Mg C·ha^{-1} per year under a 7-year-old *Eucalyptus* stand grown on an Entisol.

Fire

Fire affects the dynamics of the vegetation, particularly the grass/woody biomass ratio. Sato et al. (1998) reported a reduction in the woody cerrado vegetation of 27 percent and 38 percent, after three cycles of prescribed fires (every two years), respectively, in the middle and at the end of the dry season. The high tree mortality rates suggest that this fire regime is changing the physiognomy of the vegetation to a more open form, with the grasses as the major component of the herbaceous layer. This change in physiognomy subsequently favors the occurrence of more intense fires (Miranda et al., 1996). Surface fires, which consume the fine fuel of the herbaceous layer, are the most common in savannas. The load of the fine fuel varies with the degree of woodiness. The fine fuel of the herbaceous layer ranged from 85 to 97 percent from woodland savannas to open savannas in central Brazil (Miranda et al., 2003).

The effect of increasing fire incidence on SOM has been barely assessed so far (Roscoe et al., 2000). Breyer (2001) observed that ecosystem CO_2 fluxes measured by eddy covariance were lower in a cerrado frequently burned (1.4 Mg C·ha^{-1}) than in an area protected from fire (2.6 Mg C·ha^{-1}). The reduced aboveground biomass as a result of fire (Eiten, 1972, 1992) is expected to have an important impact on C inputs to the soil and thus on soil C storage. However, Roscoe et al. (2000), comparing a plot of cerrado *stricto sensu* burned 12 times during 21 years to a plot protected from fire, did not observe significant changes in soil C stocks, although the woody vegetation was significantly reduced.

MODELING CHANGES IN SOIL ORGANIC C AND TOTAL N

Changes in soil C levels can, in the simplest terms, be predicted by calculating the net balance between C inputs to the soil and C mineralization from SOM (Paustian et al., 1997). In this section, we predicted and analyzed changes in potential C storage under different cropping practices following conversion of cerrado savanna land based on the assumptions that (1) existing decomposition models represent SOM decomposition and its controls reasonably well and (2) C inputs to soil can be adequately assessed by simulating crop total biomass and grain production. Data from long-term experiments and model analyses have clearly shown that C inputs derived from plants have a major influence on soil C levels (Paustian et al., 1997).

For this simulation study we used the G'DAY model (Generic Decomposition and Yield, described in detail by Comins and McMurtrie, 1993). G'DAY is a linked plant-soil model that incorporates the well-established CENTURY organic matter decomposition model (Parton et al., 1993). The plant submodel in G'DAY represents the C and N content in foliage, wood (including stems, branches, and coarse roots), and fine roots. The soil submodel contains four litter pools of C and N (structural and metabolic, both above- and belowground) and three SOM pools of C and N (active, slow, and passive). Processes represented include plant C assimilation, plant N uptake, allocation, tissue senescence and N resorption, litter and SOM decomposition, soil N mineralization and immobilization, N input by atmospheric deposition, biological fixation and chemical fertilization, and N loss by leaching or gaseous emission.

Model Simulations and Parameterization

G'DAY was run for the following scenarios of crop management upon clearing of native savanna land: (1) soybean monoculture (with fallow) under conventional tillage; (2) soybean monoculture (with fallow) under no-tillage; (3) soybean with millet as cover crop under no-tillage; and (4) continuous soybean-maize cropping (two crops per year) under no tillage. Mean annual temperature (22°C) and solar radiation (19.5 $MJ \cdot m^{-2}$ per day) were used as the climatic driving variables under the assumption that crop growth was not water limited during the growing season. Based on the mean monthly precipitation and potential evapotranspiration, an average annual moisture reduction factor of 0.7 was introduced for decomposition. Net primary productivity was simulated based on radiation use efficiency (RUE).

For each crop the value for RUE was adjusted to reflect the local crop varieties and growing conditions, and to match site-specific crop productivity. The fraction of the total plant production allocated belowground was based on crop-specific average values for root production obtained from the literature. The model was first run to equilibrium for a native savanna vegetation to determine the initial soil C and N pools. G'DAY was parameterized for natural savanna (cerrado *sensu stricto*) based on plant biomass data from an experimental site described in Castro and Kauffman (1998). General and non-site-specific parameters were those from earlier G'DAY model testing (Comins and McMurtrie, 1993; Halliday et al., 2003). Decomposition rates of soil C pools (active, slow, and passive) were decreased by 25 percent to account for soil depth effects (0 to 40 cm). Tillage effects were simulated by transferring a fraction of aboveground crop residue material (80 percent) into the soil, and by increasing the decomposition rates of the soil C pools. In this study, they were set 1.2 times higher relative to no-tillage in the month after the tillage operation.

Modeling results

Figure 14.2 shows modeled changes in SOC and soil total N for the four crop management practices on a typical clayey soil (20 percent silt, 55 percent clay) during 30 years after conversion of native cerrado vegetation.

As expected, the potential soil C storage was highest under cropping systems with two crops per year. Simulated C and N stocks under soybean with millet as cover crop or maize as a second crop were comparable to those under native vegetation. In contrast, soil C and N contents under soybean monoculture with bare fallow were simulated to decline by approximately 30 percent after 30 years. Simulated effects of tillage on soil C storage were small. It can be deduced that the potential for soil C storage from a change of soybean monoculture to soybean-millet or soybean-maize is about 1 Mg $C \cdot ha^{-1}$ per year.

Model simulations indicated that with conversion to agricultural land a yearly C input of about 8.5 Mg $C \cdot ha^{-1}$ per year was necessary to maintain the initial soil C levels under the native savanna. These C inputs include both crop residues and roots from soybean and millet or maize. The C input under the soybean-fallow system (assessed to be about 4.2 Mg $C \cdot ha^{-1}$ per year) was clearly insufficient to sustain these C levels.

The modeling results also illustrate that gains in soil C were related to gains in soil N (Figure 14.2b): i.e., gains in plant biomass production were sustained by increased N inputs and reduced N losses under cropping systems with two crops per year. This confirms the findings by, e.g., Sisti et al.

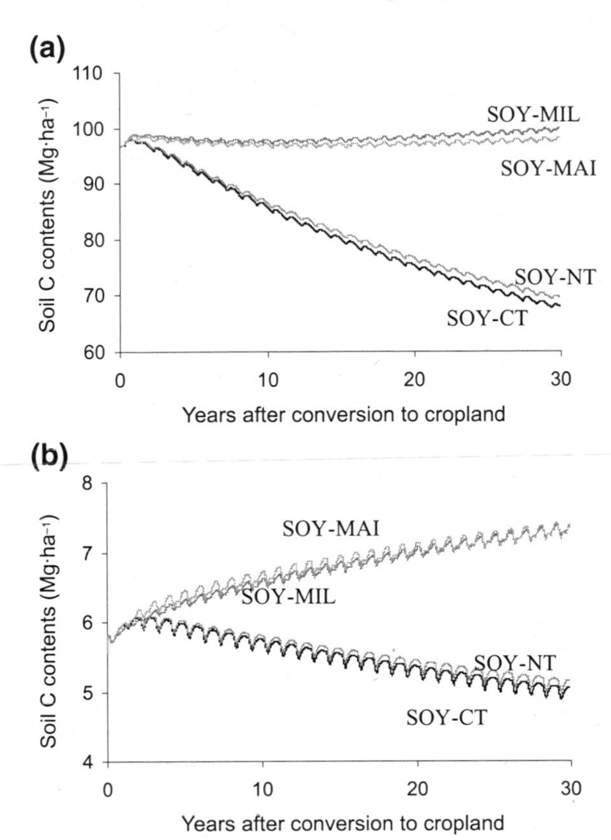

FIGURE 14.2. Modeled changes in soil organic C (a) and soil total N (b) contents in the 0-40 cm soil layer for soybean-bare fallow under conventional tillage (SOY-CT); soybean-bare fallow under no-tillage (SOY-NT); soybean with millet as cover crop under no-tillage (SOY-MIL); and soybean-maize cropping under no-tillage (SOY-MAI). Time 0 represents steady-state conditions under native savanna vegetation (type cerrado *stricto sensu*).

(2004) that long-term accumulation of soil C can be expected only when the net N balance of the cropping systems is positive.

ADOPTION OF BETTER AGRICULTURAL PRACTICES AND POTENTIAL OF SOIL C SEQUESTRATION

Two improved land-use and management practices will substantially contribute to an enhanced potential of soil C sequestration in the cerrado region: (1) adoption of DMC systems and (2) reformation of degraded pasture.

According to the model simulations presented in the previous section (Figure 14.2), the maximum rate of C sequestration with adoption of DMC is about 1 Mg C·ha^{-1} per year. Considering a total area of 6.5 Mha (3.5 Mha under conventional tillage and an estimated 3 Mha under NT, but with only one crop per year) that can be converted into DMC, the potential for soil C sequestration with DMC in the cerrado can be estimated at 6.5 Tg C per year. The productivity of degraded pastures can be increased by improved management and nitrogen cycling (Oliveira et al., 2004). Soil C sequestration with conversion of native savanna into pasture ranged from –0.87 Mg C·ha^{-1} per year to +3 Mg C·ha^{-1} per year (Table 14.1). These data include both productive and degraded pastures. Overall, the data suggest that soil C levels under well-managed pastures are above those under native cerrado vegetation, while soil C levels under degraded pastures are equal or below those under native cerrado. On the basis of this, we can assume that as much as1.5 Mg C·ha^{-1} per year can be stored in soil with the restoration of degraded pastures into productive pastures. On the other hand, it is estimated that degraded pastures represent approximately 24 Mha (50 percent of the total area of cultivated pastures) in the cerrado region. Therefore, the soil C sequestration potential with restoration of degraded pastures is estimated at 36 Tg C per year. Combined, both improved management practices allow the sequestration of about 42 Tg C per year, at least during the first years after their adoption.

REFERENCES

Abdala, G.C. 1993. Análise energética de um cerrado e sua exploração por atividade de carvoejamento rústico. MSc thesis, Universidade de Brasília, Brasília, Brasil.

Adámoli, J., J. Macedo, L.G. Azevedo, and J. Madeira Neto. 1985. Caracterização da região dos cerrados. In W.J. Goedert (Ed.), *Solos dos cerrados: Tecnologias e estratégias de manejo*. EMBRAPA-CPAC, Nobel.

Bayer, C., L. Martin-Neto, J. Mielniczuk, and A. Pavinato. 2004. Armazenamento de carbono em frações lábeis da matéria orgânica de um Latossolo Vermelho sob plantio direto. *Pesq. Agropec. Bras.* 39: 677-683.

Bayer, C. and J. Mielniczuk. 1999. Dinâmica e função da matéria orgânica. In G.A. Santos and F.A.O. Camargo (Eds.), *Fundamentos da matéria orgânica do solo: ecossistemas tropicais e subtropicais*. Genesis, Porto Alegre-RS, Brasil, pp. 9-26.

Breyer, L.M. 2001. Fluxos de energia, carbono e água em áreas de cerrado sensu stricto submetidas a diferentes regimes de queima. PhD thesis, Universidade de Brasília, Brasil.

Cadavid-Garcia, E.A., 1995. Desenvolvimento econômico sustentável do cerrado. *Pesquisa Agropecuária Brasileira* 30(6): 759-774.

Castro, E.A. and J.B. Kauffman. 1998. Ecosystem structure in the Brazilian cerrado: A vegetation gradient of aboveground biomass, root mass and consumption by fire. *J. Trop. Ecol.* 14: 263-283.

Christensen, B.T. 2000. Organic matter in soil—structure, function and turnover. DIAS Report No. 30, Plant Production, Tjele.

Comins, H.N. and R.E. McMurtrie. 1993. Long-term response of nutrient-limited forests to CO2 enrichment: Equilibrium behavior of plant-soil models. *Ecological Applications* 3: 666-681.

Corazza, E.J., J.E. Silva, D.V.S. Resck, and A.C. Gomes. 1999. Comportamento de diferentes sistemas de manejo como fonte ou depósito de carbono em relação a vegetação de cerrado. *Revista Brasileira de Ciências do Solo* 23: 425-432.

Eiten, G. 1972. The cerrado vegetation of Brazil. *The Botanical Review* 38(2): 201-341.

Eiten, G. 1992. Natural Brazilian vegetation types and their causes. *Anais da Academia Brasileira de Ciências* 64: 35-65.

Feller, C. and N.H. Beare. 1997. Physical control of soil organic matter dynamics in the tropics. *Geoderma* 79: 69-116.

Fisher, M.J. and R.J. Thomas. 2004. Implications of land use change to introduced pastures on carbon stocks in the central lowlands of tropical South America. *Environment, Development and Sustainability* 6: 111-131.

Freitas, P.L., P. Blancaneaux, E. Gavinelli, M.C. Larré-Larrouy, and C. Feller. 2000. Nível e natureza do estoque orgânico de latossolos sob diferentes sistemas de uso e manejo. *Pesq. Agropec. Bras.* 35: 157-170.

Halliday, J.C., K.R. Tate, R.E. McMurtrie, and N.A. Scott. 2003. Mechanisms for changes in soil carbon storage with pasture to *Pinus radiata* land-use change. *Global Change Biology* 9: 1294-1308.

IPCC (Intergovernmental Panel on Climate Change). 2000. *Land use, Land-use change, and forestry: Special report of the IPCC.* Cambridge University Press, Cambridge, UK.

Kirchhoff, V.W.J.H. and P.C. Alvalá. 1996. Overview of an aircraft expedition into the Brazilian cerrado for the observation of atmospheric trace gases. *J. Geophys. Res.* 101: 23973-23981.

Klink, C.A., R.F. Macedo, and C.C. Mueller. 1995. De grão em grão o Cerrado perde espaco. Cerrado: impactos do processo de ocupacão. Documento para discussão. WWF-PRODECER, Brasília.

Kluthcouski J., T. Cobucci, H. Aidar, L.P. Yokoyama, I.P. Oliveira, J.L.S. Costa, J.G. Silva, L. Vilela, A.O. Barcellos, and C.U. Magnabosco. 2000. Integração lavoura/pecuária pelo consórcio de culturas anuais com forrageiras, em áreas de lavoura, nos sistemas direto e convencional. Embrapa Arroz e Feijão, Circular Técnica No. 38.

Lardy, L.C., M. Brossard, M.L. Lopes, and J.-Y. Laurent. 2002. Carbon and phosphorus stocks of clayey Ferralsols in cerrado native and agroecosystems, Brazil. *Agriculture, Ecosystems and Environment* 92: 147-158.

Lepsch, I.F., J.R.F. Menk, and J.B. Oliveira. 1994. Carbon storage and other properties of soils under agriculture and natural vegetation in São Paulo State, Brazil. *Soil Use and Management* 10: 34-42.

Lilienfein, J. and W. Wielcke. 2003. Element storage in native, agri-, and silvicultural ecosystems of the Brazilian savanna: I. Biomass, carbon, nitrogen, phosphorus and sulphur. *Plant and Soil* 254: 425-442.

Lilienfein, J., W. Wielcke, R. Thomas, L. Vilela, S.C. Lima, and W. Zech. 2001. Effects of *Pinus caribaea* forests on the C, N, P and S status of Brazilian savanna Oxisols. *Forest Ecology and Management* 147: 171-182.

Lovato, T., J. Mielniczuk, C. Bayer, and F. Vezzani. 2004. Adição de carbono e nitrogênio e sua relação com os estoques no solo e com o rendimento do milho em sistemas de manejo. *Revista Brasileira de Ciência do Solo,* 28: 175-187.

Machado, R.B., M.B. Ramos Neto, P.G.P. Pereira, E.F. Caldas, D.A. Gonçalves, N.S. Santos, K. Tabor, and M. Steininger. 2004. Estimativas de perda da área do Cerrado brasileiro. Technical report. Conservation International, Brasília, DF.

Maquere, V. 2004. Utilisation des terres (cerrado, pâturage, eucalyptus) et stockage de matière organique dans les sols du Brésil. Mémoire D.E.A., IMA-PG, Paris.

Mielniczuk, J., C. Bayer, F.M. Vezzani, T. Lovato, F.F. Fernandes, and L. Debarba. 2003. Manejo de solo e culturas e sua relação com os estoques de carbono e nitrogênio do solo. Tópicos em Ciência do Solo—Vol. 3, Viçosa, MG: Sociedade Brasileira de Ciência do Solo. pp. 209-247.

Miranda, A.C., H.S. Miranda, J. Lloyd, J.A. Grace, J.A. McIntyre, P. Meir, P. Riggan, R. Lockwood, and J. Brass 1996. Carbon dioxide fluxes over a cerrado sensu stricto in central Brazil. In Gash et al. (Eds.), *Amazonian deforestation and climate*. John Wiley & Sons, New York, pp. 353-363.

Miranda H.S., M.M.C. Bustamante, and A.C. Miranda. 2003. The fire factor. In O.S. Oliveira and R.J. Marquis (Eds.), *The cerrados of Brazil: Ecology and natural history of a neotropical savanna.* Columbia University Press, New York: pp. 55-68.

Monteiro, J.M.G. 1995. Fluxos de CO_2 em um cerrado sensu stricto. MSc thesis, Universidade de Brasília, Brasil.

Myers, N., R.A. Mittermeier, C.G. Mittermeier, G.A.B. Fonseca, and J. Kent. 2000. Biodiversity hotspots for conservation priorities. *Nature* 403: 853-858.

Neufeldt, H., M.A. Ayarza, D.V.S. Resck, and W. Zech. 1999. Distribution of water-stable aggregates and aggregating agents in cerrado Oxisols. *Geoderma* 93: 85-99.

Oliveira, O.C., I.P. Oliveira, E. Ferreira, B.J.R. Alves, C.H.B. Miranda, L. Vilela, S. Urquiaga, and R.M. Boddey. 2001. Response of degraded pastures in the Brazilian cerrado to chemical fertilization. *Pasturas Tropicales* 23: 14-18.

Oliveira, O.C., I.P. Oliveira, S. Urquiaga, B.J.R. Alves, and R.M. Boddey. 2004. Chemical and biological indicators of decline/degradation of *Brachiaria* pastures in the Brazilian cerrado. *Agriculture, Ecosystems and Environment* 103(2): 289-300.

Parton, W.J., J.M.O. Scurlock, D.S. Ojima, T.G. Gilmanov, R.J. Scholes, D.S. Schimel, T. Kirchner, J.-C. Menaut, T. Seastedt, E. Garcia Moya, A. Kamnalrut, and J.I. Kinyamario. 1993. Observations and modeling of biomass and soil organic matter dynamics for the grassland biome worldwide. *Global Biogeochemical Cycles* 7: 785-809.

Paustian, K., H.P. Collins, and E.A. Paul. 1997. Management controls on soil carbon. In E.A. Paul, K. Paustian, E.T. Elliott, and C.V. Cole (Eds.), *Soil organic matter in temperate agroecosytems.* CRC Press, Boca Raton, FL, pp. 15-49.

Reatto, A., J.R. Correia, and S.T. Spera. 1998. Solos do Bioma Cerrado: aspectos pedologicos. In S.M. Sano and S.P. Almeida (Eds.), *Cerrado: ambiente e flora.* EMBRAPA-CPAC, Planaltina, pp. 47-87.

Resck, D.V.S. 1998. Agricultural intensification systems and their impact on soil and water quality in the cerrados of Brazil. In: R. Lal (Ed.), *Soil quality and agricultural sustainability.* Ann Arbor Press, Chelsea, pp. 288-300.

Resck, D.V.S., C.A. Vasconcelos, L. Vilela, and M.C.M. Macedo. 2000. Impact of conversion of Brazilian cerrados to cropland and pasture land on soil carbon pool and dynamics. In R. Lal, J.M. Kimble, and B.A. Stewart (Eds.), *Global climate change and tropical ecosystems.* Adv. Soil Sci., CRC Press, Boca Raton, FL, pp. 169-196.

Resende, M., N. Curi, S.B. Rezende, and G.F. Correa. 1997. *Pedologia: base para a descrição de ambientes.* 2nd ed. Viçosa, NEPUT.

Resende, M., J.C. Ker, and A.F.C. Bahia-Filho. 1996. Desenvolvimento sustentado do Cerrado. In V.H. Alvarez, L.E.F. Fontes, and M.P.F. Fontes (Eds.), *O solo nos grandes domínios morfoclimáticos do Brasil e o desenvolvimento sustentado.* SBCS-UFV, Viçosa pp. 169-199.

Roscoe, R. and R. Boddey. 2004. A matéria orgânica do solo no sistema plantio direto. In R. Roscoe, F.M. Mercante, and R.P. Scorza-Júnior (Eds.), *Modelagem matemática e simulação da matéria orgânica do solo no sistema plantio direto.* Dourados, Embrapa Agropecuária Oeste.

Roscoe, R. and P. Buurman. 2003. Effect of tillage and no-tillage on soil organic matter dynamics in density fractions of a cerrado Oxisol. *Soil and Tillage Research* 70: 107-119.

Roscoe, R., P. Buurman, E.J. Velthorst, and J.A.A. Pereira. 2000. Effects of fire on soil organic matter in a "cerrado sensu-stricto" from southeast Brazil as revealed by changes in δ13C. *Geoderma* 95: 141-160.

Roscoe, R., P. Buurman, E.J. Velthorst, and C.A. Vasconcellos. 2001. Soil organic matter dynamics in density and particle size fractions as revealed by the 13C/12C isotopic ratio in a cerrado's Oxisol. *Geoderma* 104: 185-202.

Roscoe, R. and P.L.O.A. Machado. 2002. *Fracionamento físico do solo em estudos de matéria orgânica.* Dourados, Embrapa Agropecuária Oeste.

Sano, E.E, A.O. Barcellos, and H.S. Bezerra. 2000. Assessing the spatial distribution of cultivated pastures in the Brazilian savanna. *Pasturas Tropicales* 22: 2-15.

Sano, E.E, E.T. Jesus, and H.S. Bezerra. 2001. Mapeamento e quantificação de áreas remanescentes do Cerrado através de um sistema de informações geográficas. *Sociedade & Natureza* 13(25): 47-62.

Sato, M.N., A.A. Garda, and H.S. Miranda. 1998. Effects of fire on the mortality of woody vegetation in Central Brazil. In D.X. Viegas (Ed.), *Proceedings of the*

14th Conference of Fire and Forest Meteorology II 16-20th Nov. 1998 University of Coimbra, Portugal, pp. 1785-1792.

Scopel, E., N. Doucene, S. Primot, J.M. Douzet, A. Cardoso, and C. Feller. 2003. Diversity of direct seeding mulch based cropping systems (DMC) in the Rio Verde region (Goias, Brazil) and consequences on soil carbon stocks. Producing in harmony with nature, II World Congress on Sustainable Agriculture proceedings, Iguaçu, Brazil, August 10-15.

Scopel, E., B. Triomphe, M.F. dos Santos Ribeiro, L. Séguy, J.E. Denardin, and R.A. Kochhann. 2004. Direct seeding mulch-based cropping systems (DMC) in Latin America. New directions for a diverse planet, Proceedings of the 4th International Crop Science Congress, Brisbane, Australia, September 26-October 1.

Séguy, L., S. Bouzinac, E. Scopel, and M.F.S. Ribeiro. 2003. New concepts for sustainable management of cultivated soils through direct seeding mulch based cropping systems: The CIRAD experience, partnership and networks. Producing in harmony with nature, II World Congress on Sustainable Agriculture proceedings, Iguaçu, Brazil, August 10-15.

Silva, J.E., J. Lemainski, and D.V.S. Resck. 1994. Perdas de matéria orgânica e suas relações com a capacidade de troca catiônica em solos da região de cerrados do oeste baiano. *Revista Brasileira de Ciência do Solo* 18: 541-547.

Silva, J.E., D.V.S. Resck, E.J. Corazza, and L. Vivaldi. 2004. Carbon storage in clayey Oxisol cultivated pastures in the "cerrado" region, Brazil. *Agriculture, Ecosystems and Environment* 103: 357-363.

Sisti, C.P.J., H.P. Santos, R. Kohhann, B.J.R. Alves, S. Urquiaga, and R.M. Boddey. 2004. Changes in carbon and nitrogen stocks in soil under 13 years of conventional or zero tillage in southern Brazil. *Soil and Tillage Research* 76: 39-58.

Westerhof, R., P. Buurman, C. van Griethuysen, M. Ayarza, L. Vilela, and W. Zech. 1999. Aggregation studies by laser diffraction in relation to plowing and liming in the cerrado region in Brazil. *Geoderma* 90: 277-290.

Zinn, Y.L., D.V.S. Resck, and J.E. Silva. 2002. Soil organic carbon as affected by afforestation with *Eucalyptus* and *Pinus* in the cerrado region of Brazil. *Forest Ecology and Management* 166: 285-294.

Chapter 15A

Potential of Carbon Sequestration in Soils of the Atlantic Forest Region of Brazil

Robert M. Boddey
Claudia P. Jantalia
Michele O. Macedo
Octávio C. de Oliveira
Alexander S. Resende
Bruno J. R. Alves
Segundo Urquiaga

INTRODUCTION

When the Portuguese commander Pedro Alvarez Cabral first sighted Brazil on Easter Sunday in the year 1500, the Atlantic coast of the land he christened "Santa Cruz" was covered with a resplendent tropical forest which occupied approximately 100 million ha (Mha), an area 12 times that of Portugal. Until 1961, when the Brazil's capital was relocated from Rio de Janeiro to Brasília, it was the Atlantic forest region which stretched from the state of Rio Grande do Norte in the northeast to Rio Grande do Sul in the extreme south that hosted virtually all of the country's agricultural and pasture production. The history of the anthropogenic devastation of this forest region, which to a large degree was the result of completely nonsustainable agricultural activities, has been described in detail by Dean (1995).

Few data exist on the area of this region that was used for food production for domestic consumption until the start of the twentieth century, but the first major export crop was sugarcane *(Saccharum officinarum),* first introduced by the European settlers in the coastal region (São Vincente) of what is now the state of São Paulo as early as 1529 (Machado et al., 1987). Sugar production was estimated to be approximately 10,000 Mg in 1600 and did not exceed 20,000 Mg until the start of the nineteenth century (Gal-

Carbon Sequestration in Soils of Latin America

doi:10.1300/5755_15

loway, 1989). This crop was grown mainly on flat or gently sloping land, and usually after two planting cycles (typically one plant crop and two or three ratoons) the land was abandoned and new land cleared. It is estimated that by 1700 the area under sugarcane plantation and subsequently abandoned may have totaled 100,000 ha. Owing to the rather long period of cropping (10 to 15 years), little of this area would have reverted to mature forest and probably would have remained as rough pasture or scrub. However, a further 120,000 ha may have been deforested for fuel for the mills (Dean, 1995).

While the area of the Atlantic forest region cleared for sugarcane is not altogether negligible, the crop that was responsible for a much greater devastation of this forest was that for which Brazil is famous: coffee *(Coffea arabica)*. This crop was introduced to Brazil in the late eighteenth century, and it is estimated that production in 1790 was a little over 1 Mg of beans (Dean, 1995). Coffee is a crop that does not tolerate waterlogging but does require a wet climate (1,300 to 1,800 mm per year), and it was soon found that the hill slopes in the regions near Rio de Janeiro were ideal for its cultivation. To provide nutrients for this crop, the practice adopted was to clear virgin forest with its thick layer of plant litter. In many other parts of the world coffee was grown under the canopy of larger trees, but in Brazil the forest was cleared and burned, and the seedlings were planted in the remaining ash. The result was that after one cycle of coffee (10 to 15 years) the soils were depleted of nutrients and unable to sustain coffee production. Soil erosion on the steep slopes was severe because it was customary to plant coffee in wide rows (4 to 5 m) and to clean all interrow vegetation. Hence, new areas of virgin forest were cleared for renewal of the plantations. As Dean (1995) laments, "Thus coffee marched across the highlands, generation by generation, leaving nothing in its wake but denuded hills." By the mid-nineteenth century Brazil produced 70 percent of the world's coffee. Dean (1995) estimates that in the first century of coffee production (1788-1888), a minimum of 720,000 ha of virgin forest were destroyed along with burning of 3 million Mg of forest biomass. After most of the Paraíba Valley inland from Rio had been devastated, coffee cultivation was moved to the hilly areas of Minas Gerais and São Paulo, and later to Espírito Santo. In the second half of the twentieth century large areas of forest of northern Paraná (the state immediately to the south of São Paulo) were also cleared for coffee planting.

During the twentieth century, deforestation of this region continued at an ever-increasing rate, partly due to the expansion of coffee, as described previously, and partly for timber extraction, but the most devastating cause was the production of charcoal for iron smelting which grew apace after the establishment of the nationalized steel industry in the 1940s. As is today the case in Amazônia, this process was exacerbated by the construction of new roads through heretofore almost impenetrable forest. For example, aerial photographs showed that in 1945 in the extreme south of Bahia state there were approximately 2 Mha of native Atlantic forest. In the three years following 1974, when the main road from Rio de Janeiro to Salvador was constructed through this region, 40 percent of this forest was destroyed, giving way mainly to the establishment of pastures. By 1990, satellite surveys showed that the area of native forest had been reduced to only 165,000 ha, or 8 percent of the original forest area.

Deforestation continues even during the first decade of the twenty-first century in many regions. The NGO SOS Mata Atlântica estimates that forest cover in the state of Rio de Janeiro declined from 27 percent of the total area of the state in 1985 to 17 percent in 2000, and today is estimated at 15 percent (www.sosmatatlantica.org.br). Over the whole area of the Atlantic forest region, this organization estimates that only 6 percent of the forest now remains of the 100 Mha present when Cabral first landed in 1500.

This historical introduction serves to explain why the soil carbon (C) stocks of this region are so depleted today in comparison with those present under the remnants of undisturbed forest. Bernoux et al. (2002) calculated the potential total C stock of Brazilian soils under native vegetation to a depth of 30 cm for the whole country. Although these data were not reported by region or biome, it is clear that for the Atlantic forest region these estimates would be far higher than those actually present today except under the remnants of this native vegetation. This is due not only to the stimulation of the mineralization of soil organic matter (SOM) exacerbated by tillage but also to widespread and serious erosion losses, especially in the mountainous and hilly areas which constitute the greatest part of this region.

This large depletion of the potential soil C stocks suggests that there is, at least theoretically, a huge potential for restocking the soils with C, but this potential is certainly limited by the economic, political, and fiscal environment extant in this region. This review collates the available, although often inadequate, data on land use state by state within the Atlantic forest biome, and assesses the viable potential for increasing soil C stocks under these different land-use systems based on a few point studies performed by various research groups in the region.

STATES OF THE FEDERATION WITHIN THE BIOME "ATLANTIC FOREST"

Only very recently has the Instituto Brasileiro de Geografia e Estatística (IBGE) produced a map and data on biomes based on the earlier vegetation maps. These data, displayed in Table 15A.1, are not entirely compatible for the all–South America classification utilized by the editors of this book. IBGE, obviously, considers only Brazil, so the small area of Argentina classified as Atlantic forest by the editors is not considered here. Also according to IBGE, 32 percent of São Paulo state is from the vegetation classification regarded as belonging to the cerrado biome. This area of Cerrado (almost 8 Mha) is almost completely surrounded by Atlantic forest, and we have considered 100 percent of São Paulo state as Atlantic forest. On the other hand, IBGE lists 14 percent of Mato Grosso do Sul (5 Mha) and 3 percent of Goiás (1 Mha) as Atlantic forest, but these areas were not included, as they have been considered as part of the cerrado and included in that biome by Bustamante et al. (Chapter 14). Likewise, the total area of this biome in the

TABLE 15A.1. Areas of different states of the Brazilian Federation which belong to the Atlantic forest biome.

State (abbreviation)	Total area (ha × 1,000)	Area belonging to the Atlantic forest biome: ha × 1,000	Area belonging to the Atlantic forest biome: % of entire state
Rio Grande do Sul (RS)	28,175	10,425	37
Santa Catarina (SC)	9,535	9,535	100
Paraná (PR)	19,931	19,532	98
São Paulo (SP)	24,821	16,878	68
Rio de Janeiro (RJ)	4,370	4,370	100
Minas Gerais (MG)	58,653	24,047	41
Espírito Santo (ES)	4,608	4,608	100
Bahia (BA)	56,469	10,729	19
Sergipe (SE)	2,191	1,117	51
Alagoas (AL)	2,777	1,443	52
Pernambuco (PE)	9,831	1,671	17
Paraíba (PB)	5,644	452	8
Rio Grande do Norte (RN)	5,280	264	5
Mato Grosso do Sul (MS)	35,712	5,000	4
Goiás (GO)	34,009	1,020	3
Total	302,006	111,091	100

Source: Based on data from IBGE.

northeastern states of Rio Grande do Norte, Paraíba, Pernambuco, Alagoas, and Sergipe amounts to less than 5 Mha, just 4.7 percent of the total, and the potential for increasing soil C stocks in this area was not considered except for the area (approximately 1.2 Mha) planted to sugarcane.

LAND USE

Sources of Data

The principal land-use data for the different states were obtained from the 1995 agricultural census carried out by IBGE. These data on land use are the latest comprehensive data for each municipal district (5,560 districts in total). The next census to be conducted in 2006 is to be based on data for 2005. Several problems of methodology affect the quality of the results. First, the data reported are based solely on questionnaires answered by the land owners as they are interviewed by personnel trained by IBGE. This introduces a certain bias into the collected data; for example, a land owner is not likely to admit that he has illegally destroyed an area of virgin forest.

One of the other main problems concerns the definition of pastures. There are no subdivisions beyond what are termed "natural" pastures and "planted" pastures. At the time of the initial European colonization, in the Atlantic forest region, there were only very small areas of natural grasslands, but the data from the 1995 census show very large areas which landowners call natural pastures (IBGE, 1996). What this actually means is that the landowner did not plant the pasture, and also believes that neither did his predecessor(s). A very large proportion of the deforested land, particularly the hillsides, was colonized spontaneously by grasses distributed by wind, seeds falling from cattle trucks, cattle dung, etc. In the states north from Paraná the grasses of these "unplanted pastures" are mostly "capim gordura" (molasses grass, *Melinis minutiflora*), "jaraguá" *(Hyparrhenia rufa)*, and "colonião" (guinea grass, *Panicum maximum*). None of these three grasses are native to Brazil, all being imported accidentally (often as bedding on slave ships) from Africa, but are considered as indigenous by most landowners and even by many agronomists and pasture/grazing specialists.

The biome map does not give data at the municipal level for the different biomes. IBGE groups municipal districts into microregions, and in the 1995 census give data for land use both for municipal districts and microregions as well as larger units (macroregions and states). For the states of RS, MG, and BA, which are not 100 percent Atlantic forest but have large regions of this biome, the methodology used was to compare the data for land use at the microregion level from the 1995 census, and then classify each microregion on the 2004 biome map as belonging to the Atlantic forest biome or not. Totals of the different land use for all microregions within the Atlantic forest biome were then computed.

For areas of specific crops mentioned (soybean, maize, sugarcane, coffee, etc), the data were taken from IBGE and from issues of the monthly "Systematic Survey of Agricultural Production" (LSPA, 2004). The LSPA does not give total areas and yields of crops by municipal district or any unit of area less than the whole state. To calculate the areas planted to different crops in the states that are not entirely within the Atlantic forest region (BA, MG, and RS), data are available online at the time of writing for 2002 using IBGE-SIDRA, the automated system for data recovery of IBGE data (www.sidra.ibge.gov.br). The SIDRA system does not provide land-use data for anything but agricultural crops, so for pastures, forest, etc., the data from the 1995 land-use census were used (IBGE, 1996).

Data on Land Use by Region

With the exception of the state of São Paulo, the total land area of the Atlantic forest region in each state, as calculated from the 1995 IBGE census (Table 15A.2), are lower than those totals for each state from the IBGE biome data (Table 15A.1). In the case of São Paulo, the area of 7.94 Mha classified as cerrado vegetation has been included, which results in a total area of 24.82 Mha for this state, considerably above the total area computed from the 1995 census data (17.37 Mha). These larger values for the areas using the biome data are because these data cover the whole area of the state, while the 1995 census data include only areas classified as rural properties. For example, urban areas (very significant in São Paulo and Rio de Janeiro), national parks, areas owned by the military, and lakes and rivers are not included in the census data. For the purposes of this study on potential for soil C sequestration, the fact that these areas are not included would appear to be of very minor importance.

A further reason for this discrepancy (111 Mha total biome area versus 82 Mha from the census data) is that the relatively minor areas of the Atlantic forest biome in the five states of the northeast region of Brazil (SE, AL, PE, PB, and RN) and the two states in the central western area (MS and GO) amounting to a total of 10.97 Mha (Table 15A.1) were not included.

TABLE 15A.2. Land use (thousands of ha) in the Atlantic forest region by state of the federation.

Land use	Rio Grande do Sul	Santa Catarina	Paraná	São Paulo	Rio de Janeiro	Minas Gerais	Espírito Santo	Bahia	Total
Permanent crops[a]	152	127	311	1,369	79	837	635	1,061	4,571
Annual crops[b]	3,724	1,444	4,789	3,888	258	1,151	193	347	15,795
Crop areas in fallow	187	154	390	228	38	288	53	164	1,503
Natural pastures	3,176	1,779	1,377	2,006	901	7,455	763	3,038	20,496
Planted pastures	328	560	5,300	7,056	644	3,255	1,058	2,692	20,893
Natural forest	1,088	1,349	2,082	1,352	323	2,119	372	1,427	10,111
Planted forest	265	562	713	597	26	690	173	181	3,207
Unused, but productive land	129	140	259	155	39	382	92	462	1,659
Unusable land[c]	484	499	725	719	107	756	150	320	3,761
Total	9,534	6,613	15,947	17,369	2,416	16,934	3,489	9,692	81,994

Source: Data for 1995 from IBGE, 1996.

[a]Including principally fruit tree crops (citrus, banana, apples), coffee, cacao, and cassava (manioc).

[b]Includes sugarcane which, while perennial, is mostly harvested annually.

[c]Land which is seasonally waterlogged, bare rock, or too steep for any kind of agricultural or forestry production.

SOIL CARBON ACCUMULATION UNDER DIFFERENT LAND USES

Introductory Comments

As will be described in Synthesis, Scenarios, and Conclusions, a few main land uses and crops dominate most of the area of Atlantic forest region, and some studies at specific sites have been made on the consequences of the land uses or, more important, land-use changes, on the accumulation or loss of soil organic C (SOC). It was decided to assess possible changes in SOC stocks based on what changes in land use and management might occur over the next 20 years (2005 to 2025).

The general classifications of land use will follow those used by IBGE (Table 15A.2), and where changes in land use are associated with specific crops or cropping systems, these will be discussed individually.

Pasture

The categories of "natural" and "planted" pastures add up to slightly over 50 percent of the whole area of the Atlantic forest region surveyed by the 1995 census (IBGE, 1996) (Table 15A.2). The 20.5 Mha identified as natural pastures must almost entirely be classified as degraded, and mainly on hillsides. However, much of the planted pastures, which are probably mostly *Brachiaria* spp., are also degraded. There are considerable areas of *Brachiaria* on the tablelands in the south of Bahia, which probably account for most the 2.7 Mha of planted pasture in this region, and to this could be added a large portion of the 1.06 Mha of planted pastures in Espirito Santo, also on these tablelands and deforested only since the 1970s. However, the pastures have in general been very badly managed with overgrazing and no chemical fertilization at all, and most show a fairly advanced state of degradation.

In the states of São Paulo and Paraná there are over 12 Mha of planted pastures. There is a considerable amount of intensive animal production (both meat and dairy) in these states, so some portion of these pastures are fertilized and maintained as productive, but no data are available on the areas that could be considered productive and well managed and those badly managed and hence degraded. This distinction is important in that there are several studies in the Amazon region (Buschbacher et al., 1988; Choné et al., 1991; Feigl et al., 1995; Trumbore et al., 1995; Neill et al., 1997; Koutika et al., 1997; Bernoux et al., 1999) and a few in the cerrado region of Brazil (Corazza et al., 1999; da Silva et al., 2000) which show that SOC stocks under well-managed pastures can increase until they exceed those originally under the native vegetation.

There appear to be only two studies on changes in SOC stocks due to the introduction of *Brachiaria* spp. in the Atlantic forest region. The only study of this nature conducted in any region in Brazil which was based on sampling a long-term experiment, as opposed to a chronosequence, was that conducted by Tarré et al. (2001). They studied the effects of nine years of continuous grazing at three stocking rates of beef cattle (two, three, and four head per ha) of grass-alone pastures of *Brachiaria humidicola* and the same grass in a mixed sward with the forage legume *Desmodium ovalifolium* at a site on the tablelands of the extreme south of Bahia (Itabela). The soil was sampled to a depth of 100 cm at the end of the study and all samples analyzed for ^{13}C abundance. The main findings were (1) there was no apparent significant effect of the animal stocking rates on the SOC stocks, (2) there was no appreciable C derived from the *Brachiaria* (C_4 carbon) below 40 to 50 cm in this soil, and (3) under the grass-alone *B. humidicola,* the soil C accumulation rate over a nine-year period was 0.66 Mg C·ha^{-1} per year to a depth of 100 cm, and in the mixed pasture this rate was approximately doubled to 1.17 Mg C·ha^{-1} per year. These pastures were fertilized annually with P and K at low rates (less than 25 kg P_2O_5 or 15 kg K_2O per ha, respectively) but with no N fertilizer additions.

The other study, which as yet has not been published in full, was also on the tablelands in this region but some 300 km south of Itabela in the north of the state of Espirito Santo (de Campos, 2003). Here the study was based on a chronosequence. These tablelands are, as the name suggests, extremely flat, except where cut by watercourses, and the soil texture was very uniform between the areas of forest and *Brachiaria decumbens*. This gives some confidence to the results which indicated that under a light grazing but with no fertilizer additions at all, the total SOC stock to 100 cm depth increased from 62.0 Mg C·ha^{-1} under the forest, to 70.8 Mg C·ha^{-1} after 22 years of pasture, a mean of 0.4 Mg C·ha^{-1} per year. The stock of forest (C_3)-derived C decreased to 56.4 Mg C·ha^{-1} and the soil accumulated 14.5 Mg of C_4-C·ha^{-1} from the *Brachiaria*.

These data, which may be considered quite reliable, are, however, site specific. Both are on the tablelands north of Espirito Santo and south of Bahia. The soils are of low fertility and of sandy texture (both Ultisols with increasing clay content down the profile). The SOC stocks would be greater on soils of higher clay content, and it is probable that the quantity (not proportion) of C accumulated under productive *Brachiaria* pastures in such soils would be greater. On the other, hand these increases in SOC stocks under the pastures were compared to native Atlantic forest. There is no realistic scenario that more forest in this region will be felled and converted to pasture. If new pastures are formed it will be on lands that have been used for cropping under conventional tillage, or the reform of degraded pastures. The possible scenarios are discussed in the last section.

The areas classified in the 1995 IBGE census as natural or planted pastures amount to 41.4 Mha (Table 15A.2). This is over half of the entire area classified as rural properties in the Atlantic forest region, and by far the largest single "land use" by area in this region followed by annual crops (15.7 Mha), natural forest (10.1 Mha), and permanent crops (4.6 Mha). Much of this "pasture" is on hilly/mountainous land and is severely degraded with soil C stocks far below those under the original forest. There obviously has been much erosion and loss of soil and SOC, but no quantitative estimates seem to be available. From the point of view of SOC sequestration, this vast area obviously holds the greatest potential, but what change of land use would be desirable, economically viable, or find favor with policymakers is discussed in the last section.

Reforestation is obviously one alternative. The species most planted for timber, charcoal, and cellulose production are *Pinus, Araucaria,* and *Eucalyptus*. However, in a search of the literature and Internet, we could find no data for the impact of these plantations on SOC stocks which would be useful to estimate the impact of the installation of such plantations on the areas of degraded hillsides which cover this large area in the Atlantic forest region. One of the research teams at EMBRAPA Agrobiologia (Franco and de Faria, 1997) developed the use of fast-growing leguminous trees to recover degraded areas. On hillsides that have been severely eroded, or cut for engineering (e.g., road) works, these tree seedlings (principally species of *Acacia, Mimosa, Enterolobium, Samanea, Albizia,* etc.) inoculated with both preselected strains of rhizobium and endomycorrhizal fungi can produce essentially full ground cover in 18 months.

Two unpublished studies recently conducted by the authors of this chapter show the long-term effects of the introduction of these species on the recovery of SOC stocks in two different situations. The first study was at a sloping site where in 1989 a large amount of topsoil was removed (to a depth of 40 cm) to construct an irrigation dam at the field station of EMBRAPA Agrobiologia. Originally the slope was covered by guinea grass *(Panicum maximum),* and sampling of this in 2004 showed that from 0 to 60 cm depth the mean SOC stock was 84.3 Mg $C{\cdot}ha^{-1}$, and to 100 cm, 116.7 Mg $C{\cdot}ha^{-1}$. Sampling was made in trenches to a depth of 100 cm and soil bulk density was evaluated down the profile as has been described by Sisti et al. (2004). Three trenches were dug under each of the three tree species and under the grass. If it was assumed that the top 40 cm of this soil were removed, the SOC stocks were drastically reduced to 44.5 Mg $C{\cdot}ha^{-1}$ in the remaining 60 cm. The planting of *Mimosa caesalpiniifolia* 15 years earlier caused an increase in SOC stocks to 65.7 Mg C·ha^{-1}, but *Acacia auriculiformis* and *Pseudosamanea guachapele* were considerably more effective and increased SOC stocks to, respectively, 99.5 and 94.6 Mg C·ha^{-1}. This means that if our simulation of the decapitation of this soil is valid, the three tree species increased SOC stocks by between 21 and 55 Mg C·ha^{-1} in a period of 15 years.

The second study near Angra dos Reis on the coast of Rio de Janeiro was a steep slope (50 percent) which had been deforested and the topsoil removed for use in foundations of a shopping center. The recovery operation by planting trees (principally *Acacia mangium, A. holosericea, Mimosa caesalpiniifolia*) was conducted in 1991. Here part of the original deforested hillside from which soil had been removed was left unrecuperated, and in 2004 was still almost devoid of vegetation, and some 1000 m further east along the hillside was an area of the original Atlantic forest. All three sites were sampled in triplicate as before, but to a depth of only 60 cm. The degraded unrecuperated hillside had a mean SOC stock (0 to 60 cm) of 68.9 Mg C·ha^{-1}. The SOC stock under the mature undisturbed Atlantic forest was 110.5 Mg C·ha^{-1}, and the soil under the area recovered using the legume trees had a SOC stock 96.4 Mg·ha^{-1}. In only 13 years this technology had recovered SOC stocks from 69 to 96 Mg C·ha^{-1}, a mean of 2.1 Mg C·ha^{-1} per year.

These results show clearly the enormous potential for recovering SOC stocks by reforestation of the denuded hillsides of the Atlantic forest region.

Natural Forest

In the 1995 census data, slightly over 10 Mha of land was classified as natural forest, although this excludes national parks and biological reserves. This is over 10 percent of the Atlantic forest region surveyed in this census, but only a small proportion would be remaining primary forest, the rest being, at best, remnants of the forest from which valuable timber has been removed, and at worst secondary regrowth with a low biodiversity and biomass which would not be classified as Atlantic forest by the NGO SOS Mata Atlantica. These latter areas in many cases are just abandoned areas which are no longer grazed or subject to burning, and slowly some of the woody species have recovered.

In the Atlantic forest region, many remnants of remaining primary forest are preserved, but deforestation (illegal in this region) has not entirely ceased (see www.sosmatatlantica.org.br). If these areas are preserved and protected and no significant climate change occurs, it would be expected that the SOC stocks would be stable over time. In the case of the more damaged forest or the secondary regrowth, if these areas are protected there should be a gradual increase in above- and belowground biomass and probably an increase in plant biodiversity along with a gradual increase in SOC stocks. As there are no data on the state of recovery of these areas, no estimate of potential SOC sequestration can be made.

Planted Forest

IBGE does not regularly report on the area of planted forest, and the latest data available for individual municipal districts or microregions comes from the 1995 census (IBGE, 1996). However, the site of the Brazilian Society for Silviculture (Sociedade Brasileira de Silvicultura [SBS], www.sbs.org.br) lists areas of the two most important tree genera (*Pinus* and *Eucalyptus*) state by state (Table 15A.3). In the three states which are not entirely within the Atlantic forest biome, Bahia, Minas Gerais, and Rio Grande do Sul, the SBS and the 1995 census data are as follows: For Bahia the 1995 census data indicate that in the Atlantic forest region of this state there were 181,000 ha of planted forest. This area of Bahia is the coastal plain and planted forests are principally *Eucalyptus,* so that the 213,400 ha of this species registered by the SBS is probably close to the area of planted forest in the region today. There is likely to be some underestimation, as the area of *Eucalyptus* is still expanding rapidly. For Minas Gerias the census registered 690,000 ha of planted forest in the Atlantic forest region, much lower than that for the whole state (1.54 Mha). As there is a great deal of planted forest in the cerrado region of this state, this value from the census is probably quite close to today's reality. The SBS (259,900 ha) and the census data (265,200 ha) for the Atlantic forest region are very close for Rio Grande, which means that either large areas of the northern (Atlantic forest) region of this state have been planted to forest since 1995, or that most silviculture activity is in this region of the state and has not changed much. Assuming this latter hypothesis to be true, the total area of planted forest in the Atlan-

TABLE 15A.3. Area of planted forest (*Pinus* spp. and *Eucalyptus*) in the Atlantic forest region by state (year 2000).

Plant	Brazil	Rio Grande do Sul	Santa Catarina	Paraná	São Paulo	Rio de Janeiro	Minas Gerais	Espírito Santo	Bahia	Total
Pinus spp.	1,840.1	136.8	318.1	605.1	202.0	0.0	143.4	0.0	238.4	1,643.8
Eucalyptus	2,965.9	115.9	41.6	67.0	574.2	0.0	1,535.3	152.3	213.4	2,699.7
Total	4,805.9	260.8	366	673	778	0	1685	155	459	4,343.5

Source: Data for 2000 from www.sbs.org.br/area_platada.htm.

tic forest region becomes approximately 3.12 Mha, of which 1.26 are *Pinus* spp. and the remaining 1.85 Mha *Eucalyptus*.

There is a considerable amount of literature concerning the amount of C accumulated in aboveground biomass of *Pinus* and *Eucalyptus* plantations, but a search of the literature has not revealed any studies in the Atlantic forest region on changes in SOC stocks associated with planted forests. It is clear from studies in other regions that if primary forest were to be replaced by planted forest there would be a short-term loss of SOC associated with soil preparation for planting, which would then be followed by some recovery of SOC stocks. It is unlikely that SOC would eventually reach the stocks under the native forest, as there would be harvesting and replanting at intervals of several years. The results of studies in the Cerrado region are inconclusive. Corraza et al. (1999) registered accumulation of SOC under a plantation of *Eucalyptus* compared to the native cerrado. Lilienfein et al. (2001) and Lilienfein and Wilke (2003) found no significant difference between SOC stocks under cerrado vegetation or a 20-year-old plantation of *Pinus caribaea,* and Zinn et al. (2002) registered losses of SOC compared to the native Cerrado under a 20-year-old *Pinus caribaea* plantation and a 7-year-old *Eucalyptus camaldulensis* plantation. This work is much more fully discussed by Bustamante et al. (Chapter 14).

However, such plantations in the Atlantic forest region are invariably introduced in areas of degraded pastures/hillsides or on land that has been under conventional tillage for many years, and hence with SOC stocks considerably depleted in comparison with the original native forest.

Permanent Crops

Of the 4.8 Mha in the Atlantic forest region classified as permanent crops, the largest portion of this area is covered by coffee (1.96 Mha) followed by citrus (740,000 ha) and cacao (600,000 ha) (Table 15A.4). The only other permanent crop in this region to exceed 100,000 ha is banana (210,000 ha).

The state with largest area of coffee in Brazil is Minas Gerais (1.18 Mha) of which is an estimated 815,000 ha are within the Atlantic forest biome. Owing to the frequent occurrence of frost in winter, no coffee is grown in Rio Grande do Sul or Santa Catarina. The great majority of the area of coffee in Minas Gerais, São Paulo (227,000 ha), and Paraná (126,000 ha) is Arabica *(Coffea arabica)*; as this crop requires free drainage but high rainfall (usually 1,300 to 1,800 mm per year), in the Atlantic forest region it is grown mainly on sloping lands, often at altitudes above 500 m. The traditional practice is to have widely spaced plants and maintain the area be-

TABLE 15A.4. Area (ha × 1,000) occupied by permanent crops in Brazil and the Atlantic forest biome of Brazil.

Crop	Brazil	Bahia	Minas Gerais	Espírito Santo	Rio de Janeiro	São Paulo	Paraná	Santa Catarina	Rio Grande do Sul	Total area Atlantic forest	% in region
Apples	31.7	0.0	0.0	0.0	0.0	0.2	1.8	16.4	13.4	31.7	100.0
Banana	529.3	34.2	25.0	22.3	25.9	61.0	9.8	29.7	0.9	208.7	39.4
Black pepper	26.1	1.1	0.0	2.5	0.0	0.0	0.0	0.0	0.0	3.6	13.7
Cacao	677.6	578.0	0.0	21.4	0.0	0.0	0.0	0.0	0.0	599.4	88.5
Cashew nuts	673.8	19.5	0.0	0.0	0.0	0.0	0.0	0.0	0.0	19.5	2.9
Cassava	1,752.1	150.8	70.0	16.8	10.4	36.7	110.9	28.4	88.9	513.0	29.3
Coconut	283.3	74.1	2.8	15.3	4.1	0.0	0.0	0.0	0.0	96.3	34.0
Coffee	2,602.3	85.7	815.4	623.5	12.7	227.0	126.0	0.0	0.0	1,890.3	72.6
Grapes	69.0	0.0	0.91	0.0	0.0	12.4	6.5	3.67	38.5	57.4	83.3
Guarana	14.0	5.8	0.0	0.0	0.0	0.0	0.0	0.0	0.0	5.8	41.5
Oranges	824.0	43.0	41.0	2.9	7.3	586.0	14.3	9.7	27.2	731.3	88.7
Pineapple	61.9	4.7	12.8	3.4	2.4	3.5	0.0	0.0	0.3	27.1	43.8
All permanent crops	7,887.8	996.9	967.0	708.2	62.7	926.8	269.3	84.1	169.1	4,184.1	

Source: For the states entirely within the Atlantic forest biome, the data for crop area were taken from LSPA (2004), but for the states Bahia, Minas Gerais, and Rio Grande do Sul, which have large areas not included in this biome, the data were calculated from the microregions within the biome using values obtained from the select regions in the Atlantic forest biome map.

tween rows clear of vegetation (weeds, grass, or an intercrop). This means that these areas are frequently prone to erosion. Practices are slowly changing with the introduction of intercrops and grasses to reduce such soil losses, and also the system of high-density planting where after two or three years the plant canopy completely covers the soil. However, no data are available on how widespread these changes in management are or on their possible impact on SOC stocks.

The state of Espirito Santo also has a large area of coffee (625,000 ha), but much of this is the robusta coffee *(C. canephora)* of inferior quality and lower price. However, the crop is much more tolerant to drought, temporary waterlogging, and high temperature and is widely grown in the tableland areas of the north of the state and in the valleys in the south. Almost all coffee in the Atlantic forest region of Bahia is *C. canephora* as well, again much of it grown on the tablelands which dominate the extreme south of this state, contiguous with the north of ES.

High chemical fertilizer inputs are generally utilized, and the mean addition has been estimated to be 407 kg·ha^{-1} (total fertilizer; Anonymous, 1998), and heavy use of insecticides, fungicides, and herbicides is also standard practice. There is a growing movement, albeit as yet on a small scale, for the conversion of some areas of coffee for organic production. The export market has a high demand for this product, and prices are very favorable. Eliminating the use of industrial fertilizers and pesticides in coffee growing is a major challenge. The solutions usually involve intercropping of the coffee, often with green-manure legumes, which not only help to control weeds but also provide some input of biologically fixed nitrogen, and the diversification aids controls of plant disease and insect attack. It seems evident that this change in practice would favor SOC accumulation, and the elimination of the large quantities of industrial fertilizers would also reduce overall greenhouse gas (GHG) emissions. As yet no studies in this respect seem to have been published.

Approximately 80 percent of citrus (oranges) production is concentrated in São Paulo (586,000 ha). Again this crop is generally with high agrochemical inputs. Most oranges are used for juice production, the greater part of which is exported. There is no clear trend in change of the area of this crop, but from 1994 to 2000 yields increased from 97,000 to 124,000 fruits per ha. Banana production is common throughout the region, but again there are no indications of any major management changes envisaged for the next 20 years which would make any significant impact on SOC stocks.

Cacao *(Theobroma cacao)* in the Atlantic forest region is largely grown in Bahia (580,000 ha) with a far smaller area (21,000 ha) in the north of ES. Traditional practice is still the most widespread with the cacao bushes planted under the canopy of the Atlantic forest. Most valuable timber trees are usually removed along with others that have very dense shade so that the plant biological diversity is considerably reduced in comparison with the climax forest. However, available data suggest that SOC stocks under cacao are similar to those under the native forest. There have been serious threats to the cacao industry owing to intense competition, especially from West African countries, and over the past 10 to 15 years the spread of the fungal disease *(Crinipellis perniciosa)* known as witch's-broom. Clones of fungus-tolerant trees and better management have considerably reduced the problem caused by witch's-broom, but over the past ten years a little more than 100,000 ha of cacao has been abandoned, and generally the forest was cleared to make way for pastures and crops. Although well-managed pastures in this region of Bahia can increase SOC stocks (see below and Tarré et al., 2001), good management is not generally practiced, and cropping in this region is invariably under conventional tillage. Since 1994, the fall in price and the problem with witch's-broom has led to a fall in production from 330,000 to 170,000 Mg. At present the area under this crop seems to be stable, but further reductions in international prices or other disease problems could mean that the area dedicated to this crop could be reduced with serious consequences for a considerable loss of biological diversity and serious detriment for the region's SOC stocks.

Temporary Crops

Annual Grain Crops

Temporary, or annual, crops occupy the second largest proportion (15.8 Mha, 19.3 percent) of the area of the Atlantic forest region after "pastures" according to the 1995 IBGE census (Table 15A.5). Unlike forest and pasture species, the monthly LSPA reports list state by state the areas and yields of annual crops, and totals for each municipal district are available at the time of writing for 2002 on the IBGE database (www.sidra.ibge.gov.br). The data show that the two most important crops by area in the Atlantic forest region are soybean (*Glycine max,* 7.8 Mha) and maize (*Zea mays,* 6.0 Mha), and most of the soybean is grown in the southern region (mainly Rio Grande do Sul and Paraná), with some 700,000 ha in São Paulo. The third most important crop by area in the Atlantic forest region is sugar cane (>3.6 Mha), but as this is a semiperennial and is managed in a completely different manner to the annual grain crops, this is discussed in the next section.

In this region much of the soybean is grown in rotation with wheat *(Triticum aestivum)* and oats *(Avena sativa)* or sometimes barley *(Hordeum vulgare),* so that the total area under these winter cereals (2.6 Mha) cannot

TABLE 15A.5. Area (ha × 1,000) occupied by the most important[a] temporary crops in Brazil and the Atlantic forest biome of Brazil.

Crop	Brazil	Bahia	Minas Gerais	Espírito Santo	Rio de Janeiro	São Paulo	Paraná	Santa Catarina	Rio Grande do Sul	Total area	% in Atlantic forest region
Barley	154.3	0.0	0.0	0.0	0.0	0.0	54.4	4.0	96.0	154.3	100.0
Common bean	4,337.6	64.6	227.4	29.8	6.3	199.2	507.9	138.5	227.4	1,401.1	32.3
Cotton	1,130.6	0.2	1.8	0.0	0.0	83.6	47.2	0.0	0.0	132.9	11.8
Groundnuts	94.9	0	0.5	0.0	0.0	72.3	3.9	4.7	4.0	85.4	90.0
Maize	12,781.5	48.8	501.3	49.6	11.4	1,064.9	2,427.0	813.0	1,133.7	6,049.7	47.3
Oats	299.4	0.0	0.0	0.0	0.0	0.0	229.0	20.0	49.0	298.0	99.5
Potatoes	135.6	0.0	0.0	0.6	0.0	30.0	28.9	8.6	0.0	68.0	50.2
Rice	3,732.6	0.0	44.9	3.5	3.2	35.0	68.2	151.8	96.5	403.0	10.8
Sorghum	853.0	0.0	0.4	0.0	0.0	85.7	2.5	0.0	5.4	93.9	11.0
Soybean	21,351.1	0.0	45.2	0.0	0.0	723.2	3,926.7	315.5	2,771.7	7,782.3	36.4
Sugar cane	5,672.7	44.4	74.8	66.4	170.7	2,799.2	393.2	16.4	74.8	3,639.9	64.2
Tobacco	452.7	12.5	1.4	0.0	0.0	0.2	58.9	147.0	220.5	440.5	97.3
Wheat	2,658.6	0.0	0.0	0.0	0.0	53.4	1,302.7	85.0	709.0	2,150.1	80.9
Total	53,201.9	158.0	896.3	149.9	191.6	5,146.5	8,991.4	1,557.6	5,167.4	22,699.1	

Source: For the states entirely within the Atlantic forest biome, the data for crop area were taken from LSPA (2004), but for the states Bahia, Minas Gerais, and Rio Grande do Sul, which have large areas not included in this biome, the data were calculated from the microregions within the biome using values obtained from the select regions into Atlantic Forest Biome map.
[a]Crops which occupy less than 80,000 ha in the entire country not included (garlic, onion, rye, sunflower, tomato, triticale).

be added to the 13.8 Mha of maize and soybean when calculating the area under annual crops. When soybean cropping started to expand in the southern region in Brazil in the 1970s, it was almost always in rotation with wheat. This continuous sequence of soybean in summer followed by wheat in winter led to the buildup of crop diseases, especially root diseases of wheat, which led some farmers to diversify and plant maize one year in two or three as the summer crop, and occasionally oats in winter. At the same time, the plowing and disking of the soil twice a year led to a rapid decline in SOM reserves and, on the more sandy soils of the region, led to severe erosion. At the end of the 1970s a few innovative farmers started using no-till (NT) agriculture which slowly spread through the region such that by 1992 it occupied approximately 2 Mha in the whole country (~35 Mha of mechanized annual grain crop production). However, owing to greatly reduced expenditure on diesel fuel and other advantages, the system has spread rapidly since that date such that it is estimated for the 2003/2004 season approximately 21 Mha were managed under NT (Figure 15A.1), and in the southern region over 80 percent of soybean-based crop rotations are managed with the NT (personal communication from local IBGE agents). The system is truly NT, and if farmers move out of conventional tillage (CT), they generally go to complete NT; "intermediate" tillage systems, such as "conservation tillage" widely practiced in the United States and other parts of the world are not often adopted. Much more detail about the

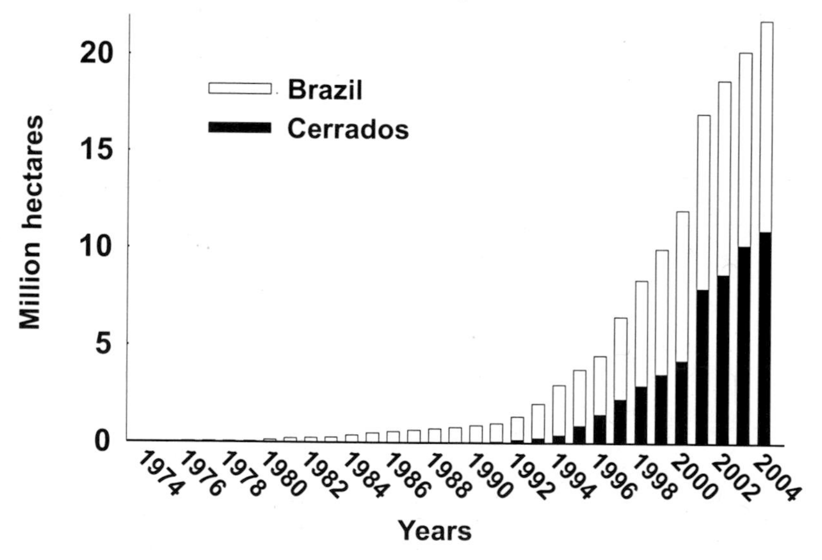

FIGURE 15A.1. Growth in the area of annual crop production under zero-tillage in Brazil.

adoption of the NT system in Brazil is given by Derpsch (2002), Landers et al. (2003), and Boddey et al. (2003).

We can thus assume that soybean-based crop rotations, which may occasionally include maize as a summer crop, frequently include wheat or oats (*Avena sativa* L. or *A. strigosa* L.) as a winter crop, and less frequently barley or very occasionally green manure crops such as oil radish (*Raphanus sativus* L. var. *oleifera*), vetch (*Vicia sativa* L.), or lupins (*Lupinus albus* L.), occupy an area of approximately 13 Mha. Almost all of this is concentrated in the three states of the southern region and the remainder in São Paulo.

There seems to be a consensus in the United States that the conversion of grain cropping from conventional to no-till or conservation tillage will lead to a recovery in the SOC stocks which have been depleted by up to a century of CT. For example, Kern and Johnson (1993) estimated that if no-till and other conservation tillage techniques were adopted in 76 percent of the area of grain production in the United States until the year 2020, the SOC stocks in this area could increase by 277 to 452 million tons (Tg) in the top 15 cm of soil. However, not all studies in Brazil have shown that soil C stocks increase when no-till replaces CT. In one of the earliest studies in the area, Muzilli (1983) compared the concentrations of SOM in the 0 to 30 cm depth interval under CT and NT for three different crop sequences: (1) continuous soybean/wheat, (2) continuous maize-wheat, and (3) a two-year rotation of soybean-wheat followed by maize-wheat. This study was conducted at two sites in the north of Paraná state on an Alfisol and in the south on an Oxisol over periods of five and four years, respectively. At the site in the north of Paraná, organic C concentrations were higher only in the top 5 cm of the soil under NT than CT, and at the other site higher only under NT for the continuous maize-wheat. In a further study in Paraná, Sidiras and Pavan (1985) examined SOC concentrations to a depth of 60 cm in an Oxisol and an Alfisol after five years under CT and NT. In these experiments the crop rotation consisted of two years of soybean-wheat followed by one year of a mixed sward of rye grass *(Lolium multiflorum)* and perennial soybean *(Glycine wightii)*. In both experiments SOC concentrations were higher under NT than CT to depths of 40 and 20 cm for the Oxisol and the Alfisol, respectively. In neither of these two studies was soil bulk density measured, so that SOC stocks were not compared between tillage systems. More recently, studies conducted on longer-term experiments by Machado and Silva (2001), Freixo et al. (2002), and Sisti et al. (2004), working at sites at the soybean research center of EMBRAPA in Londrina (Paraná) and the wheat center in Passo Fundo (Rio Grande do Sul), showed that even after 13 years of continuous wheat-soybean, there were no significant differences in SOC stocks to depth of 40 or 100 cm between NT and CT management. However, many other studies conducted by the research team at the Federal

University of Rio Grande do Sul (UFRGS), working near Porto Alegre, found higher C concentrations and/or SOC stocks in many studies on different crop rotations managed under NT than under CT (Bayer and Mielniczuk, 1997; Bayer and Bertol, 1999; Amado et al., 2001; Bayer et al., 2000, 2002; Lovato et al., 2004; Diekow et al., 2005). For a summary of the results of all the studies based on long-term experiments in the southern region see Table 15A.6.

The study of Sisti et al. (2004), however, appears to provide an explanation of this disparity between the results of different studies. In all of the studies by the Porto Alegre group (UFRGS), a winter green-manure legume or a summer forage legume was included in the rotations (Table 15A.6). In most of the other studies no legume was included or the only legume was soybean. Results obtained using the ^{15}N natural abundance (Shearer and Kohl, 1986; Boddey et al., 2000) and the ureide abundance technique (Herridge, 1982) have shown that while soybean is able to obtain 70 to 80 percent of its N requirement from biological N_2 fixation (BNF), the proportion of N exported from the field as grain is usually slightly greater than this, such that soybean makes no overall contribution to the soil N reserves (Alves et al., 2002, 2003). The SOM generally has a C:N ratio of between 11 and 13, so that to "sequester" 1 Mg of C it is necessary to have an input of approximately 80 kg N. In the studies where green-manure legumes were incorporated into the rotation, the crop and its residues were not removed from the field, so that there was a overall N gain from BNF. Sisti et al. (2004) compared three crop rotations managed under either NT or CT, and found no significant difference between SOC stocks under the different tillage systems in a continuous sequence of wheat/soybean, but in the other two rotations where vetch as a winter crop was included before maize, either one year in two or one year in three, SOC stocks were approximately 17 $Mg{\cdot}ha^{-1}$ higher to a depth of 100 cm after 13 years of cropping under NT than under CT. The results of the study by Diekow et al. (2005) led to similar conclusions. They showed that after 14 years of CT the soil C stocks (0 to 17.5 cm) originally under a native pasture declined from 39.0 to 30.4 Mg $C{\cdot}ha^{-1}$. Subsequently, when the soil was left fallow (weeds controlled with herbicide) for an additional 17 years (1983 to 2000) SOC stocks further decreased to 24.9 Mg $C{\cdot}ha^{-1}$. The effects of three different crop sequences managed under NT on the recovery of SOC stocks were examined: (1) a continuous oats-maize sequence, (2) an intercrop of lablab *(Lablab purpureum)* with maize, or (3) an intercrop of pigeon pea *(Cajanus cajan)* with maize. All were managed with or without N fertilizer—120 kg $N{\cdot}ha^{-1}$ from 1983 until 1994, and 180 kg $N{\cdot}ha^{-1}$ thereafter. The continuous oats-maize under NT increased SOC stocks even with N fertilizer to only 29.2 Mg $C{\cdot}ha^{-1}$, which indicated no recovery of SOC stocks since the previous 14

TABLE 15A.6. Summary of studies on long-term field experiments which compare crop rotations/sequence managed under zero and conventional tillage with respect to changes in soil carbon stocks.

Site	Soil type	Crop rotation/ sequence	Sampling depth (cm)	Age of experiment (yr)	Total difference in soil C stock[a] ($Mg \cdot ha^{-1}$)	Source
Lages (SC)		Maize, soybean, common bean, wheat, oats [b]	0 – 20	8	8.5	Bayer and Bertol (1999)
Eldorado do Sul (RS)	Paleudult (US tax)	Oats/maize	0 – 30	9	4.6	Bayer et al. (2000)
Eldorado do Sul (RS)	Paleudult (US tax)	Oats+vetch/maize+cowpea	0 – 30	9	6.4	Bayer et al. (2000)
Santa Maria (RS)	Paleudalf	Black oat+vetch/maize	0 – 20	8	0.7[c]	Amado et al. (2001)
Santa Maria (RS)	Paleudalf	Azevém+vetch/maize	0 – 20	8	1.24[c]	Amado et al. (2001)
Santa Maria (RS)	Paleudalf	Velvet bean + maize	0 – 20	8	5.42[c]	Amado et al. (2001)
Santa Maria (RS)	Paleudalf	Jack bean + maize	0 – 20	8	0.65[c]	Amado et al. (2001)
Passo Fundo (RS)	Oxisol	Wheat/soybean	0 – 100	13	0.5	Sisti et al. (2004)
Passo Fundo (RS)	Oxisol	Wheat/soybean – vetch/maize	0 – 100	13	16.9	Sisti et al. (2004)
Passo Fundo (RS)	Oxisol	Wheat/soybean – oats/soybean – vetch/maize	0 – 100	13	16.7	Sisti et al. (2004)
Passo Fundo (RS)	Oxisol	Wheat-soybean	0 – 30	13	–1.3	Sisti et al. (2004)
Passo Fundo (RS)	Oxisol	Wheat/soybean – vetch/maize	0 – 30	13	5.4	Sisti et al. (2004)
Passo Fundo (RS)	Oxisol	Wheat/soybean – oats/soybean – vetch/maize	0 – 30	13	9.1	Sisti et al. (2004)
Eldorado do Sul (RS)	Paleudult	Oats/maize no N	0 – 17.5	17	1.7[d]	Diekow et al. (2005)
Eldorado do Sul (RS)	Paleudult	Oats/maize with N	0 – 17.5	17	4.3[d]	Diekow et al. (2005)
Eldorado do Sul (RS)	Paleudult	Lablab+maize no N	0 – 17.5	17	13.9[d]	Diekow et al. (2005)
Eldorado do Sul (RS)	Paleudult	Lablab+maize with N	0 – 17.5	17	18.6[d]	Diekow et al. (2005)
Eldorado do Sul (RS)	Paleudult	Pigeon pea+maize no N	0 – 17.5	17	13.4[d]	Diekow et al. (2005)
Eldorado do Sul (RS)	Paleudult	Pigeon pea +maize with N	0 – 17.5	17	20.5[d]	Diekow et al. (2005)
Eldorado do Sul (RS)	Paleudult	Oats/maize no N	0 – 107.5	17	2.9[d]	Diekow et al. (2005)

TABLE 15A.6 *(continued)*

Site	Soil type	Crop rotation/ sequence	Sampling depth (cm)	Age of experiment (yr)	Total difference in soil C stock[a] ($Mg \cdot ha^{-1}$)	Source
Eldorado do Sul (RS)	Paleudult	Oats/maize with N	0 – 107.5	17	7.2[d]	Diekow et al. (2005)
Eldorado do Sul (RS)	Paleudult	Lablab+maize no N	0 – 107.5	17	20.3[d]	Diekow et al. (2005)
Eldorado do Sul (RS)	Paleudult	Lablab+maize with N	0 – 107.5	17	29.4[d]	Diekow et al. (2005)
Eldorado do Sul (RS)	Paleudult	Pigeon pea+maize no N	0 – 107.5	17	28.4[d]	Diekow et al. (2005)
Eldorado do Sul (RS)	Paleudult	Pigeon pea+maize with N	0 – 107.5	17	33.2[d]	Diekow et al. (2005)
Eldorado do Sul (RS)	Paleudult	Oats/maize no N	0 – 30	13	6.8	Lovato et al. (2004)
Eldorado do Sul (RS)	Paleudult	Oats/maize with N	0 – 30	13	3.2	Lovato et al. (2004)
Eldorado do Sul (RS)	Paleudult	Vetch/maize no N	0 – 30	13	6.3	Lovato et al. (2004)
Eldorado do Sul (RS)	Paleudult	Vetch/maize with N	0 – 30	13	5.8	Lovato et al. (2004)
Eldorado do Sul (RS)	Paleudult	Oat + vetch/maize + cow-pea no N	0 – 30	13	7.2	Lovato et al. (2004)
Eldorado do Sul (RS)	Paleudult	Oat + vetch/maize + cow-pea with N	0 – 30	13	7.7	Lovato et al. (2004)
Cruz Alta (RS)	Oxisol	Wheat-soybean	0-30	17	–4.1	Jantalia (2005)
Cruz Alta (RS)	Oxisol	Wheat-soybean	0-100	17	–3.8	Jantalia (2005)
Cruz Alta (RS)	Oxisol	Wheat/soybean – vetch/maize	0-30	17	3.6	Jantalia (2005)
Cruz Alta (RS)	Oxisol	Wheat/soybean – vetch/maize	0-100	17	8.8	Jantalia (2005)
Londrina (PR)	Oxisol	Lupin/Maize-oat/soybean-wheat/soyben	0-20	6	0.58	Zotarelli (2005)
Londrina (PR)	Oxisol	Lupin/Maize-oat/soybean-wheat/soyben	0-80	6	1.05	Zotarelli (2005)

[a]Difference between ZT and CT management unless otherwise indicated.
[b]Order of crop sequence not given.
[c]The reference is fallow/maize in ZT.
[d]The reference is bare soil without cropping.

years of CT. However, where the legumes were present in the intercrops, even without N fertilizer SOC stocks to 17.5 cm depth reached 38.3 to 38.8 Mg C·ha^{-1}, almost equal (difference not significant at $P < 0.05$) to those under the native grassland 31 years earlier (39 Mg C·ha^{-1}). With the addition of fertilizer N the SOC stocks under the two treatments of maize intercropped with legumes significantly exceeded those present under the grassland in 1969.

Both of these studies by Diekow et al. (2005) and Sisti et al. (2004) demonstrate the very significant impact of N_2-fixing legumes on the accumulation of SOC under NT, and the data of Sisti et al. (2004) showed that this did not occur under CT, nor when soybean was the only legume present, presumably owing to the zero or negative N balance of this crop caused by the export of very large quantities of N from the field in the grain (high N harvest index). Another important discovery common to these two studies was that under NT, where legumes were present and there were large accumulations of soil C under NT, much of this SOC was found at considerable depth in the soil. Diekow et al. (2005) sampled the soil to 107.5 cm and found that up to 24 percent of the overall SOC losses from the fallow and oat-maize sequence, and up to 63 percent of gains in the maize/legume intercrops fertilized with N, occurred at depths below 17.5 cm. Similarly, Sisti et al. (2004) found that between 46 and 68 percent of the difference in SOC stocks between NT and CT in the rotations which included vetch occurred at 30 to 85 cm depth. A similar result was observed under winter wheat in Nebraska by Doran et al. (1998). These results suggest that agricultural activity can have major impacts on SOC stocks at depths well below the "plow layer" (which under NT does not exist), and that the general, and much more convenient, sampling of soils for change in SOC to depths of 20, 30, or 40 cm may lead to considerable errors, either positive or negative, of the actual quantities of CO_2 removed from, or liberated to, the atmosphere when tillage practices are changed.

We consulted local IBGE personnel and others to ascertain certain details concerning the adoption of NT and the most common crop rotations used in the southern region of Brazil where almost all of the mechanized grain production of the Atlantic forest region is localized (personal communications from Sr. Lamar Sakis, Cooplantio, Passo Fundo; Sr. Jorge Myryczka, IBGE agent, Londrina, Paraná). In summary, in Rio Grande do Sul, where approximately 90 percent of annual crops are planted under the NT system, it is estimated that 60 percent of farmers practice a rotation of the summer crops of soybean and maize with wheat, oats, or barley in winter, and 40 percent plant soybean every summer with the same winter crops. Only approximately 2 percent of farmers use legumes or oil radish in their rotations. In Paraná it is estimated that 92 percent of soybean, 66 percent of maize, and between 83 and 89 percent of wheat, oats, and barley are planted with NT. Much of the oat crop in the southern region does not appear in the IBGE statistics (Table 15A.5), as much is knocked down with a knife roller or desiccated with herbicides at the milk stage solely for use as straw to improve the NT system. As no grain is produced the values for area and yield do not feature in the IBGE statistics.

These data suggest that the impact on SOC stocks of the change from CT to NT, which has happened on an increasing scale since 1992 (Figure 15A.1), has as yet been minimal, as the results from several long-term experiments (Table 15A.6) show that these systems do not accumulate SOC owing to their low yield and the lack of a net N input. At present there is a considerable reluctance to substitute maize for soybean as a summer crop, as maize grain prices are too low. Wheat is marginally profitable, varying from year to year, but oats can be planted broadcast with no fertilizer inputs, and no significant quantities of herbicides or other pesticides are necessary.

The further diversification of the soybean-based crop rotations is recommended by farmers' NT organizations (e.g., Clube da Minhoca, Amigos da Terra, and the national federation of NT organizations, FEBRAPDP) as well as the federal and state research organizations such as EMBRAPA (federal) and IAPAR (Paraná state) and extension service personnel. The recommendation is based mainly on the observation that the more diverse the rotations the lower is the incidence of diseases and pests, and if legumes other than soybean are introduced this will mean that SOC stocks will increase by the kinds of values registered in long-term studies (Table 15A.6). If international prices of crops other than soybean increase, and farmers wish to lower their expenses on pesticides and N fertilizer, then the adoption of winter legumes is likely to increase.

Sugarcane

The area listed in Table 15A.5 for sugarcane in the Atlantic forest region is slightly over 3.6 Mha. Because of the relatively small areas of Atlantic forest in the northeastern states of Sergipe, Alagoas, Pernambuco, Paraiba, and Rio Grande do Norte (total of less than 5 Mha), land use in these states was not calculated. However, the Atlantic forest region in these states is all within 100 km or so of the Atlantic coast, and the most important single crop in this region is sugarcane. Using the data from the LSPA (2004), the total sugarcane area in these states is 1.2 Mha, which added to the 3.6 Mha of this crop in the more southerly areas of the Atlantic forest region, total almost 5 Mha of the 5.6 Mha for the whole country.

Sugarcane in Brazil is a crop which is very friendly to the global atmosphere in that approximately half of the area is dedicated to the production of bioethanol (national production approximately 13 billion liters) for use in light vehicles (cars and vans) either mixed 24 percent with gasoline or used in the hydrous form (95 percent ethanol) to fuel Brazil's still-considerable fleet of ethanol-powered vehicles. At the end of the 1980s approximately 4 million light vehicles were powered by hydrated ethanol, so that in 1988 the proportion of all fuel (gasohol or bioethanol) used by Otto cycle engines was 57 percent on a gasoline energy equivalent. After 1988, owing to low gasoline prices, ethanol consumption declined, but even today approximately 40 percent of all fuel used by light vehicles is from ethanol. Recent studies have shown that to produce 9 energy units of bioethanol requires just one unit of fossil fuel (Macedo, 1998; energy balance 9:1), and this compares extremely favorably with bioethanol made from maize in the United States (energy balance 1.34:1, Shapouri et al., 2002) or biodiesel made from canola (rape methyl-ester [RME], energy balance 1.6:1, Armstrong et al., 2002) in Europe.

With the recent large increases in the price of crude oil on the international market, bioethanol from sugarcane is looking increasingly attractive, and this combined with the recently developed "Flexfuel" Otto-cycle engines, which can run on hydrated ethanol or gasohol, means that bioethanol consumption is likely to increase very significantly and with it the area planted to sugarcane in the Atlantic forest region. Volkswagen do Brasil (Brazil's largest car manufacturer) estimates that by 2007 all their light vehicles will be equipped with such Flexfuel engines and the other major car manufacturers (GM, Fiat, and Ford) already produce cars equipped with Flexfuel engines and are also planning expansion. A recent article in the magazine *Globo Rural* (www.globorural.globo.com) predicted that the area under sugarcane would expand to at least 7 Mha by 2007.

The impact of an expansion of the sugarcane area on SOC stocks is difficult to estimate. The areas into which this crop will expand are likely to be flat areas of pastures which are of low productivity or degraded. These areas are probably not especially low in SOC stocks, but there seem to be few studies which have investigated the impact of this land-use change on SOC. One recent unpublished study by de Campos (2003) showed that when an area of native Atlantic forest on the tablelands of northern Espirito Santo (Ultisol, 80 percent sand) was deforested in 1980 and a *Brachiaria decumbens* pasture was established, over a period of 22 years of mainly light grazing and no fertilizer inputs, SOC stocks increased from 62 $Mg{\cdot}ha^{-1}$ to 70.8 $Mg{\cdot}ha^{-1}$ (on an equal mass of soil basis, Neill et al., 1997) in the first 100 cm depth. In 1990 this pasture was plowed up and planted to sugarcane, and in 2002 SOC stocks had decreased to 50.9 $Mg\ C{\cdot}ha^{-1}$ to the same depth. Although the sugarcane was fertilized with N (the ratoon crops received 80 kg $N \cdot ha^{-1}$), the lowering of the SOC stocks was almost certainly due to the intensive tillage used to establish the crop, and again when the crop was twice renovated at five-year intervals. Although this soil was more sandy and prone to SOM loss due to tillage operations than would be the case in many areas in São Paulo, where over 60 percent of Brazil's sugarcane is grown, the standard practice of intensive tillage every five to six years to renovate the crop suggests that SOC stocks would probably be lower, or unchanged, under this crop than under the pasture it replaced.

One major change in management of sugarcane that has been introduced in recent years is green cane harvesting. From the 1940s until recently, sugarcane was burned immediately before harvest to facilitate manual harvesting. In the past decade there has been increasing pressure from environmental agencies, especially in São Paulo state, to abandon the preharvest burning to reduce atmospheric pollution. As green (unburned) cane is so much more difficult to hand cut (estimated at three times the labor input, Boddey, 1993), machine harvesters have been imported and more recently national machines have been developed, and in these areas it is estimated that each mechanical harvester has replaced 80 to 100 workers (http://www.petruscommodities.com.br/infopetrus/08-04/mecanizacao.htm). There are no long-term studies on the impact of the change to green cane harvesting in the state of São Paulo, but the team at EMBRAPA Agrobiologia established an experiment in 1983 in Pernambuco, which is still being evaluated every year, on the effects of preharvest burning versus trash conservation, N fertilizer additions (0 and 80 kg $N \cdot ha^{-1}$ for the plant and ratoon crops, respectively), and the annual addition of 80 m^3 of vinasse (distillery waste) on cane yields, crop N accumulation, and SOM (de Resende et al., 2005). The data show that in the first cycle (1983-1992), for the last five ratoons cane yields were 24 percent higher (12.5 Mg $cane \cdot ha^{-1}$ per year) where trash was preserved than in plots where it was burned. In the second cycle (1992-1999), the plots where residues were conserved produced on average 45 percent more cane (16 $Mg \cdot ha^{-1}$ per year). It was found that the conservation of cane residues had two main benefits: conservation of soil moisture and the preservation of SOM. The data from this trial showed that after 16 years the SOC content of the top 10 cm of soil under the burned plots was 1.16 percent compared to 1.33 percent where trash was conserved. In the 10 to 20 cm layer, these values were 1.10 and 1.28 percent, respectively, together representing a difference of 4.28 Mg $C \cdot ha^{-1}$, or a mean increase of 270 kg $C \cdot ha^{-1}$ per year.

Relatively short-term studies at two sites in São Paulo have been reported, both close to the city of Ribeirão Preto, one on an Oxisol (Hapludox) and the other on an Entisol (Quartzipsamment) (de Campos, 2004).

The accumulation of trash in the unburned cane fields reached, respectively, 4.5 and 3.6 Mg dry matter·ha^{-1} after four years without burning. The author concluded that an annual rate could be calculated for C accumulation from these data, but the decomposable fraction was probably achieving steady-state by this time. He also observed an increase of approximately 1 Mg C·ha^{-1} per year in the soil during this period, but there was no replanting of the cane during this period. When cane is replanted heavy tillage and deep plowing are used which lead to large mineralization losses of SOM. For this reason, the difference between the SOC stocks under burned and green cane is not likely to be as high at 1 Mg C·ha^{-1} per year, but the values from the long-term EMBRAPA Agrobiologia experiment in Pernambuco may be lower than a mean for the Atlantic forest region, in that this area often has years with low yields due to lack of rainfall, and mean yields for the region are only about 60 percent of those for the state of São Paulo.

Cassava

Cassava *(Manihot esculenta)* is classified as a "long-term temporary crop" as it is not harvested annually, but usually from 18 months to 2 years after planting. It occupies an area second only to coffee in Brazil (1.75 Mha) among crops classified as long term or permanent, and approximately 30 percent of this area is in the Atlantic forest region. It is widely grown by resource-poor (RP) farmers, and tillage is generally by hand (hoe, etc.) or conventional plowing. This crop does not easily lend itself to NT, as the planting material is setts (stem pieces). Furthermore, use of NT is much less rarely employed by RP farmers and in regions other than the south (PR, SC, and RS). The prospects for much change in the area or management of this crop over the next two decades are minimal, so that significant changes in SOC stocks under this crop are not to be expected.

SYNTHESIS, SCENARIOS, AND CONCLUSIONS

Introduction

In this section we examine two possible scenarios for the potential increase in SOC stocks in the Atlantic forest region for the period 2005 to 2025. The first is based on the opinion of the authors of what is the likely outcome for changes in SOC stocks in the present and possible future economic and political climate in Brazil. The second is a estimate of what potential changes in soil C stocks could occur based on purely technical criteria, regardless of economic viability and political will. The discussion refers only to SOC stocks, and not C sequestered in aboveground biomass of forests, etc., or the very considerable economies in greenhouse gas emissions that will be, or could be, due to the growing of biofuel crops. Apart from the very large bioethanol program based on sugarcane that is certain to expand very considerably over the next two decades, the Brazilian government is also initiating biodiesel production from castor oil *(Ricinus communis)* grown in the semiarid northeastern states, from African oil palm *(Elaeis guineensis)* principally in the Amazon region, and perhaps soybean or sunflower oil may also be exploited for this purpose in the near future (www.biodieselecooleo.com.br/noticias/, accessed January 27, 2005).

Most of the studies cited which report changes in SOC stocks due to changes in land use or modifications of agricultural practice were based on long-term experiments and a lesser number were based on chronosequences. The former are almost always more reliable, as they are usually based on randomized plots such that if statistically significant differences between treatments are detected, the effect of any spatial variations in SOC stocks across the studied areas before the experiment was installed are included in the residual variance, such that these differences are really due to the different management of the treatments. Results from chronosequences, such as those reported here for the recovery of degraded areas with fast-growing leguminous trees or the study of de Campos (2003) on pasture and sugarcane, are inherently less reliable, as they depend on nonreplicated adjacent areas where there is no certainty that SOC stocks were identical across all areas at the time of installation of the different crops/treatments.

When land use is changed, the SOC stocks begin to change until a new steady state is achieved. If the new land use is constant in productivity (i.e., not, for example, a pasture that is installed and then heavily grazed with no further nutrient inputs), then the new steady state of SOC is attained when the rate of decomposition (C release as CO_2) of the belowground C attains the same rate as that of the belowground deposition of C in crop residues (Boddey et al., 2002). These two fluxes are exceedingly difficult to measure in the field, as belowground CO_2 emission from decomposing residues is almost impossible to distinguish from root respiration, and belowground C deposition involves the evaluation of belowground primary productivity, which is exceedingly labor-intensive and fraught with methodological problems (Steen, 1984; Santantonio and Grace, 1987).

The studies reported on the response of SOC stocks to land-use change normally have generated a value for the change in SOC stock to a defined depth over a certain number of years. Similar to almost all of the studies, at least on the impact of the introduction of NT in southern Brazil, these studies have been for periods of less than 20 years; these data cannot be extrapolated from mean annual rates of SOC accumulation, as in the first years the rate of accumulation will be higher than in later years as the SOC stocks approach asymptotically the new (and unknown) steady state. This was admirably illustrated for soil N stocks in the very long-term (since 1845) studies in the Rothamsted Broadbalk plots in the United Kingdom, reported by Jenkinson (1991). In the tropical or subtropical Atlantic forest region the attainment of the new steady state will be much more rapid than in the temperate climate of Britain, as deposition of residues will be higher in most of the cropping systems owing to the yearround cropping of high-yielding crops, and decomposition rates will be much higher because of the higher temperatures and yearround rainfall in this region. For this reason, the estimates of the accumulation of soil C between 2005 and 2025 that we used in this study are lower than extrapolations of mean annual rates. However, the estimates made are very subjective, as we have no data at all on possible eventual steady state SOC stocks or the rate at which this limit is being approached.

The impact of the different land uses on SOC stocks is discussed in order of the area under each land-use change, as reported in Table 15A.2.

Scenario A

"Natural" Pastures

We have assumed that the area of 20.5 Mha classified as "natural" pastures consists principally of the area of abandoned hillsides left behind after deforestation for coffee or charcoal, etc. No data are available on land use versus slope, so this affirmation cannot be tested. We assume that some of these abandoned areas may be reforested, but as tree plantations, as there is little or no financial return from attempts to recover the original vegetation of the Atlantic forest. We have optimistically assumed that 10 percent of the area might be recovered over the 20-year period with reforestation using fast-growing legume and nonlegume tree species, either native or exotic, similar to those used in the recovery of degraded areas by the team at EMBRAPA Agrobiologia. Such reforestation would be a priority where gallery forest has been totally removed, to protect water resources which already is law in the whole of Brazil, but only now being slowly implemented. The forest code dictates that areas within 30 m of small streams and up to 600 m at either side of large lakes and rivers must be reforested or protected, and while this is happening on only a very small scale at present, it is almost certain to increase over the next two decades.

Our two studies in Seropédica and Angra dos Reis, both in the state of Rio de Janeiro, examined the increase in SOC stocks after, respectively, 15 and 13 years. The problem with extrapolation from these studies is that in the areas planted to the fast-growing legume trees the soil was decapitated, and hence much lower in SOC than areas of degraded grassland on the hillsides, but no data are available for this at either site or from elsewhere. At Seropédica the increase in SOC stocks was between 21 and 55 Mg C to a depth of 60 cm over 15 years, and at Angra the increase was 27 Mg C·ha^{-1} for the same depth interval. The SOC stock under the neighboring guinea grass (*Panicum maximum*) at Seropédica was 84 Mg C·ha^{-1}, and under the area of native forest near the Angra site 110.5 Mg C·ha^{-1}. If we assume that the fast-growing leguminous trees at Angra would eventually accumulate SOC to the same stocks as under the original forest, degraded pastures typically have soil SOC of approximately 80 Mg C·ha^{-1} (to 60 cm), then in the 20-year interval we might expect an increase in SOC stocks of approximately 25 Mg C·ha^{-1} using this technology, if only approximately 85 percent of the final new steady state is reached.

Of course, this estimate is highly subjective. Existing SOC stocks under the degraded hillsides/pastures are not known. Soil texture has a major influence on SOM stocks, and considerable evidence has been accumulated to indicate that in general there is a good general linear relationship between silt + clay and SOM contents (Feller and Beare, 1997). An enormous amount of sampling, analysis, and mapping would be required to come up with a reliable estimate of what increase in SOC stocks would be achieved over a 20-year period, as the area of hillsides in the Atlantic forest region is so heterogeneous, but using our suggested estimate of 25 Mg C·ha^{-1} results in an overall gain, if only 10 percent of the area were restored with this technology, of 50 Tg C in this period.

Planted Pastures

Planted pastures in this region are almost entirely composed of species of *Brachiaria*. If it is assumed that these pastures represent areas that are flat or on gently sloping land and generally actually used for grazing, then their recovery for this purpose can be achieved by lime and fertilizer addition, the main nutrients necessary being probably N and P (de Oliveira et al., 2001; Macedo, 2002; Boddey et al., 2004). In the cerrado region, deep plowing and undersowing the new pasture with new seeds of *Brachiaria* has become a common practice (de Oliveira et al., 1996), and more recently the "Santa Fé" system which relies on pasture restoration using NT has as well (Kluthcouski et al., 2000). Although tillage will temporarily lower SOC stocks, ample evidence shows that after a few years of pasture SOC stocks recover and then exceed those originally under the native Amazon forest (Feigl et al., 1995; Trumbore et al., 1995; Koutika et al., 1997; Neill et al., 1997; Bernoux et al., 1999) or of the cerrado (Fisher et al., 2004). The data reported by Tarré et al. (2001) and reported previously showed that in the south of Bahia SOC stocks (0 to 100 cm) increased by 6 Mg C·ha^{-1} over a nine-year period when a grass-alone *Brachiaria humidicola* pasture was carefully managed. Soil under a lightly grazed but unfertilized *B. decumbens* pasture in the north of Espirito Santo gained 9 Mg C·ha^{-1} in 22 years in the same depth interval. Both these soils of the tablelands of this region (northern ES and the extreme south of BA) are of very sandy texture (80 to 90 percent sand, 0 to 20 cm), so that in many regions farther south on soils with lower sand contents are likely to be able to accumulate greater soil SOC stocks. These comparisons were between native climax forest and pastures established immediately following deforestation. The 20.9 Mha classified by IBGE as planted pasture is mainly degraded, and results from the cerrado region confirm, what seems to be self-evident, that degraded pastures have lower SOC stocks than the original native vegetation (Fisher et al., 2004). This is another reason to believe that the values of SOC accumulation published by Tarré et al. (2001) and de Campos (2003) would be lower than those to be expected if degraded *Brachiaria* pastures throughout the Atlantic forest region were to be restored with or without tillage.

We have assumed, owing to the continuing growth in both the national and international demand for Brazilian beef, that there will be considerable incentives in the next decade or so for large areas (40 percent) of the degraded *Brachiaria* pasture of the regions to be restored. We suggest that the impact of this might be terms of a growth in SOC stocks of 15 Mg C·ha^{-1} over the 20-year period, amounting to a total C accumulation of 126 Tg.

Soybean-Based Crop Rotations

In the Atlantic forest region the three southern states (RS, SC, and PR) are responsible for 90 percent of the soybean production, most of the remainder being in São Paulo. In these states our contacts with agronomists working of the IBGE gave us information to suggest that in this region approximately 90 percent of the soybean-based crop rotations are already under NT. They also reported that owing to buildup of plant disease (especially root diseases) continuous wheat/soybean is rare, but most diversification of crops in the rotations is achieved with the addition of oats in winter and maize in summer. The data accumulated from most of the recent studies from long-term experiments in the southern region (Table 15A.6) indicate that if no green-manure legume is introduced into the rotation or high N fertilizer levels are not used, then accumulation of SOC under NT will be very limited. Hence for the period 2005 to 2025 the big gain in soil C stocks under crop rotations under NT will come from the introduction of winter legumes (vetch, lupin, etc.) into the rotations, and not because NT is replacing CT, as this transition is already almost complete in this area.

It is estimated that subsidies by the rich developed countries (principally the European Union, United States, and Japan) on agricultural commodities amount to approximately US$350 billion per year (*The Independent,* UK daily newspaper, September 10, 2003). At the Doha round of talks of the World Trade Organization in 2003, these countries promised to reduce these subsidies, and there is some hope that progress may be made in this direction in the next decade or so. We have assumed that if the developed countries are forced to diminish their subsidies this will lead to an increase in the demand for soybean, and perhaps maize and wheat. Although a great deal of agricultural expansion and intensification will probably occur in the cerrado region, it is likely that some unprofitable degraded pasture land in the southern region will be converted to soybean-based crop rotations under NT (SBCRNT), and we have assumed that this might amount to 2 Mha. Owing to complete lack of physical disturbance of the soil, degraded pastures are not especially low in soil C stocks. Thus, we do not expect any significant positive or negative impact of this increase in SBCRNT on soil C stocks.

However, with this addition the total area under SBCRNT in the Atlantic forest region would total 15.8 Mha. At present, all the state and federal research and extension organizations and the farmers' NT associations and their national federation (FEBRAPDP) all recommend diversification of crops used in SBCRNT, which mostly involves the introduction of winter green-manure legumes. If we assume that on 50 percent of the area a winter legume is introduced before maize and wheat or oats is the winter crop before the soybean (the latter is often already the case), then we estimate from the various long-term experiments using these kind of rotations that SOC stocks would increase by approximately 20 Mg $C{\cdot}ha^{-1}$. The very recent studies of Sisti et al. (2004) and Diekow et al. (2005) showed that a large portion of the C accumulated in the soil beneath NT-managed rotations which contained a leguminous green-manure component was at depths between 30 and 80 cm, and in this estimate we assume that a significant portion of the accumulated C was below a depth of 30 to 40 cm, which was the greatest depth sampled in other studies (see Table 15A.6).

If it is assumed that until 2025 leguminous green-manure crops were introduced in 50 percent of the area under NT-managed rotations, we estimate that this would amount to an accumulation of 158 Tg of C over the 20-year period (Table 15A.7).

TABLE 15A.7. Possible land-use change and its consequences for soil carbon stocks in the Atlantic forest region (AFR) based on the authors' predictions of the future political and economic climate in Brazil.

Land-use system	Total present area of system[a]	Management change	Area affected by change (Mha)	Change in soil C stock 2005-2025 (Mg C·ha^{-1})	Total change in soil C stock for AFR (Tg)
"Natural" pastures	20.50	Recovery of 10% of the area using fast-growing legume trees	2.0	+25.0	+50
Planted pastures	20.89	Recovery of 40% of the pastures using fertilizers with or without tillage	8.4	+15.0	+126
Rotations of annual crops	13.80	a. Increase in area at the expense of degraded pastures	2.0	0.0	0
		b. Introduction of green-manure legumes in rotations under zero-tillage (50% of area)	7.9	+20.0	+158
Sugarcane	4.80	a. Increase in area at expense of degraded pastures (only on flat land)	4.0	0.0	0
		b. Increase from 20% of area under trash conservation to 60%	5.3	+5.0	+26.4
Planted forest	3.20	Increase in area at the expense of degraded pastures	3.0	0.0	0
Coffee	1.96	20% of the area dedicated to organic production	0.4	+10.0	+4.0
Cacao	0.58	Decrease in area by 20%	0.12	−10.0	−1.2
Total					+363

[a]Present area occupied by this land-use system in the AFR.

Sugarcane

As mentioned before, owing to a projected increase in the demand for bioethanol and possible lowering of international tariff barriers on sugar, it is estimated that just in the state of São Paulo an extra 1.5 Mha of sugarcane will be planted until 2007. It seems inevitable that until 2025 much larger areas will be planted to this crop and most will be areas dedicated at present to pasture, much of it of low productivity or degraded. We have assumed that the area of sugarcane in the Atlantic forest region until this date will increase by 4.0 Mha, while there are likely also to be large increases in the cerrado region. Although sugarcane is a perennial crop and renewed on average only every five to six years, the tillage used at present for this renewal is extremely intensive and our own long-term study in the NE region of Brazil (de Resende et al., 2005) would suggest that SOC stocks are not likely to increase, unless the area used for expansion is one where annual crops have been grown under CT for many years. As we expect most of the expansion to be at the expense of degraded or low-productivity pastures, we do not expect the increase of area per se to increase SOC stocks.

On the other hand, approximately 20 percent (30 percent in São Paulo) of the area of sugarcane is at present under trash conservation, that is, the practice of preharvest burning has been abandoned. We have assumed that until 2025 continued pressure from environmental organizations to abandon burning will lead to 60 percent of the expanded area of 8.9 Mha to be under trash conservation. As much of the gain of SOC due to this practice is lost when the plantations are reformed with intensive tillage, we assume that this will lead to an mean increase in SOC stocks of only 5 Mg $C \cdot ha^{-1}$, giving a total soil C gain of 26.4 Tg for the whole region until 2025.

Planted Forest

It seems inevitable that the planting of *Eucalyptus* in the north of Espirito Santo, the south of Bahia, and perhaps in significant areas of the Atlantic forest region of Minas Gerais will increase considerably over the next 20 years. The present area of planted forest in this region is approximately 3.2 Mha, and we have assumed that this area may almost double until 2025, reaching 6.2 Mha. However, to facilitate easy harvesting most of these plantations will probably be installed on flat areas such as the tablelands of north ES and the extreme south of Bahia, and will replace degraded pastures. As discussed earlier, as degraded pastures, unlike areas used for many years of annual cropping with conventional tillage, do not have exceptionally low SOC stocks, there is no evidence that belowground stocks of C will be affected. We conclude, therefore, that while aboveground C stocks in timber are certain to increase with the increased area under planted forest, there may be no impact at all on SOC stocks.

Coffee

Brazil is already the world's largest coffee producer, and in recent years there has been some increase in area (2.0 Mha in 1994 to 2.4 Mha in 2003) owing to stiff international competition; there is no clear indication whether this area would change in the next 20 years. Also, no data are available to assess the impact of any increase or decrease in area on SOC stocks, which would depend on what present land use was subject to change. We could conclude, however, that the expansion of organic coffee almost certainly would increase SOC stocks as the system employed is to grow other crops such as leguminous green-manures between the rows instead of the usual practice of maintaining bare soil between the rows. As international prices for organically produced coffee are very attractive, we have assumed that 20 percent of the present area (0.4 Mha) may be converted to this system, and that in a 20-year period this would cause a mean increase in SOC stocks of 10 Mg $C \cdot ha^{-1}$, resulting a total C accumulation of 4 Tg.

Cacao

Cacao is predominantly planted under existing Atlantic forest in the south of Bahia. This forest is considerably depleted in biodiversity, but authorities consulted from CEPEC/CEPLAC, the cacao research organization, consider that this depletion of biodiversity has little impact on SOC stocks. However, as described in an earlier section, the trend in the area dedicated to this crop is downward, and we have assumed that until 2025 this area will be reduced by 20 percent. Although some of this area will be converted to pastures, other areas may be used for cropping, and at present NT is almost unknown north of São Paulo in the Atlantic forest region. We have assumed that there will a mean loss of 10 Mg $C \cdot ha^{-1}$, but as the present area under cacao is only 580,000 ha, the impact is estimated to be only a loss of 1.2 Tg of SOC.

Scenario B

In this scenario we have made optimistic estimates of land-use change that would favor accumulation of SOC if it became an important priority for government policy and financing for this became available as a principal objective. The estimates of the impact of land-use change on SOC stocks on a mean per ha basis are arrived at from the (often meager) available data, and we have not changed them except when considering tillage reduction under sugarcane. However, we have tried to make optimistic but realistic estimates of how much land is available for the land-use changes available, these being in many cases considerably above our Scenario A estimates, which we felt were those likely to occur given the present and predicted future national and international political and economic climate.

With regard to the "natural" pastures, we have assumed in this scenario that 50 percent of this area (instead of only 10 percent under Scenario A) of what is probably mainly degraded hillsides would be recovered using fast-growing legume trees. This would make a huge impact on SOC accumulation for this land use, with an estimate of 256 Tg over the 20-year period (Table 15A.8).

As we have suggested that a considerable area of the planted (mainly *Brachiaria* spp.) pastures would be transferred to other uses (soybean-based crop rotations, planted forest, and sugarcane), we have not upgraded this estimate.

For soybean-based crop rotations under NT (SBCRNT) we have assumed that the area would undergo a greater expansion in this scenario (+4 Mha versus +2 Mha under Scenario A) and that green-manure legumes would be introduced into 90 percent of the rotations (as opposed to 50 percent under Scenario A) before 2025. This would also make a large impact on soil C accumulation, leading to an increase from 158 to 320 Tg.

We have assumed that the area under sugarcane will increase by 5 Mha in the Atlantic forest region during the next 20 years. It is also assumed that the practices of NT and trash conservation will be adopted throughout most of the area (90 percent) in this scenario. The NT for sugarcane physically disturbs the soil far more than NT for annual seeded crops, as a furrow of approximately 30 cm wide and deep is excavated to accommodate the setts, but this still disturbs the soil far less than the several heavy tillage operations used traditionally. For this reason we assume that the gain in SOC stocks per ha under this combined system of NT and trash conservation would be 10 Mg C·ha^{-1} as opposed to only 5 Mg C·ha^{-1} in Scenario A, where only trash conservation was considered. With the increased area and the increased portion under improved management, it is estimated that soil C stocks under this crop would increase by 78 Tg as opposed to only 26.7 Tg under Scenario A.

We have assumed that 50 percent of coffee in the Atlantic forest region will be produced organically, which increases the potential C accumulation in soil under this crop to 10 Tg, as opposed to only 4 Tg under Scenario A.

TABLE 15A.8. Potential land-use change and its consequences for soil carbon stocks based on technically viable solutions but ignoring the possible future political and economic climate in Brazil.

Land-use system	Total present area of system[a] (Mha)	Management change	Area affected by change (Mha)	Change in soil C stock 2005-2025 (Mg C·ha^{-1})	Total change in soil C stock for AFR (Tg)
"Natural" pastures	20.50	Recovery of 50% of the area using fast-growing legume trees	10.25	+25.0	+256
Planted pastures	20.89	Recovery of 40% of the pastures using fertilizers with or without tillage	8.4	+15.0	+126
Rotations of annual crops	13.80	a. Increase in area at the expense of degraded pastures	4.0	0.0	0
		b. Introduction of green-manure legumes in rotations under zero-tillage (90% of area)	16.0	+20.0	+320
Sugarcane	4.80	a. Increase in area at expense of degraded pastures (only on flat land)	5.0	0.0	0
		b. Increase from 20% of area under trash conservation to 90%	7.8	+10.0	+78
Planted forest	3.20	Increase in area at the expense of degraded pastures	3.0	0.0	0
Coffee	1.96	50% of the area dedicated to organic production	1.0	+10.0	+10.0
Cacao	0.58	No change in area	0	−10.0	0
Total					790

[a]Present area occupied by this land-use system in the AFR.

We have assumed that the area under planted forest is not different from that in Scenario A, which in any case we suggested would have no significant impact on SOC stocks. We also have assumed that, unlike in Scenario A where we assumed that the area under cacao would decrease by 20 percent, in this scenario the area will not change, meaning that there will be no negative impact on soil C stocks under this crop.

Conclusions

It is apparent that even with what we suggest is a realistic estimate of possible changes in SOC stocks (Scenario A) that this area of Brazil, which has been so heavily exploited and holds very large areas of degraded soils and pastures and only small remnants (~6 percent) of the original Atlantic forest, has a large potential for increase in SOC stocks. It may be opportune here to offer a word of caution concerning the use of the phrase "carbon sequestration." *The New Shorter Oxford Dictionary* (Brown, 1993) defines the general use of the word "sequester" as "separate and reject" or "eliminate" or "exclude." When used in chemistry, which might be the root of its use in soil science, the word is defined as "to form a stable complex." In either case the use of the phrase "carbon sequestration" implies that the removal of the CO_2 from the atmosphere is not easily reversible. This is not compatible with the concept of arriving at a new steady state of SOC stocks due to a management change, which will just as easily be reversed if the management change is reversed or altered. However, if restored pastures become degraded, reforested hillsides are again deforested, or NT operations are abandoned in favor of CT, this "sequestered" C will again be released. Policymakers should be made aware of this, so that higher productivity systems with minimal soil disturbance, once installed, must be maintained if higher SOC stocks are to be sustained. Furthermore, once a new steady state of SOC has been attained, no further mitigation of CO_2 emissions is possible on that land.

From the point of view of mitigating CO_2 emissions by agricultural activities, a surer method is the use of biofuels. In this case each Mg of biofuel C produced will leave in the geological strata a certain proportion of that Mg of C (depending on the energy balance). This substitution is a real and permanent gain.

REFERENCES

Alves, B.J.R., R.M. Boddey, and S. Urquiaga. (2003). The success of BNF in soybean in Brazil. *Plant and Soil* 252:1-9.

Alves, B. J. R., L. Zotarelli, R. M. Boddey, and S. Urquiaga. 2002. Soybean benefit to a subsequent wheat cropping system under zero tillage. In *Nuclear techniques in integrated plant nutrient, water and soil management*. IAEA, Vienna, pp. 87-93.

Amado, T.J.C., C. Bayer, F.L.F. Eltz, and A.C. Brum. (2001). Potencial de culturas de cobertura em acumular carbono e nitrogênio no solo no plantio direto e a melhoria da qualidade ambiental. *Revista Brasileira da Ciência do Solo* 25:189-197 (in Portuguese).

Anonymous (1998). Lavoura Orgânica. *Manchete Rural* (Rio de Janeiro) 11:37-38 (in Portuguese).

Armstrong, A.P., J. Baro, J. Dartoy, A.P. Groves, J. Nikkonen, and D.J. Rickeard. (2002). *Energy and greenhouse gas balance of biofuels for Europe—an update*. Prepared by the CONCAWE Ad Hoc Group on Alternative Fuels, European Community, Brussels.

Bayer, C. and I. Bertol. (1999). Características químicas de um Cambissolo húmico afetadas por sistemas de preparo, com ênfase à matéria orgânica. *Revista Brasileira da Ciência do Solo* 23:687-694 (in Portuguese).

Bayer, C. and J. Mielniczuck. (1997). Nitrogênio total de um solo submetido a diferentes métodos de preparo e sistemas de culturas. *Revista Brasileira da Ciência do Solo* 21:235-239 (in Portuguese).

Bayer, C., J. Mielniczuck, T.J.C. Amado, L. Martin-Neto, and S.V. Fernandes. (2000). Organic matter storage in a sandy clay loam Acrisol affected by tillage and cropping systems in southern Brazil. *Soil & Tillage Research* 54:101-109.

Bayer, C., J. Mielniczuck, L. Martin-Neto, and P.R. Ernani. (2002). Stocks and humification degree of organic matter fractions as affected by no-tillage on a subtropical soil. *Plant and Soil* 238:133-140.

Bernoux, M., M. de C.S. Carvalho, B. Volkoff, and C.C. Cerri. (2002). Brazil's soil carbon stocks. *Soil Science Society of America Journal* 66:888-896.

Bernoux, M., B.J. Feigl, C. Cerri, A.P.A. Geraldes, and S.V. Fernanandes. (1999). Carbono e nitrogênio em solo de uma cronossequência de floresta tropical-pastagem de Paragominas. *Scientia Agricola* 56:1-12 (in Portuguese).

Boddey, R.M. (1993). "Green" energy from sugar cane. *Chemistry and Industry* (London) 10(May 17):355-358.

Boddey, R.M., R. Macedo, R.M. Tarré, E. Ferreira, O.C. Oliveira, C. de P. Rezende, R.B. Cantarutti, J.M. Pereira, B.J.R. Alves, and S. Urquiaga. (2004). Nitrogen cycling in *Brachiaria* pastures: The key to understanding the process of pasture decline. *Agriculture, Ecosystems and Environment* 103:389-403.

Boddey, R.M., M.B. Peoples, B. Palmer, and P.J. Dart. (2000). Use of the ^{15}N natural abundance technique to quantify biological nitrogen fixation by woody perennials. *Nutrient Cycling in Agroecosystem* 57:235-270.

Boddey, R.M., S. Urquiaga, B.J.R. Alves, and M. Fisher. (2002) Potential for carbon accumulation under *Brachiaria* pastures in Brazil. In *Agricultural practices and policies for carbon sequestration in soil* (J. Kimble, R. Lal and R.F. Follet, eds.). Lewis Publishers, Boca Raton, FL, pp. 395-408.

Boddey, R.M., D.F. Xavier, B.J.R. Alves, and S. Urquiaga. (2003). Brazilian agriculture: The transition to sustainability. *Journal of Crop Production* 9:593-621.

Brown, L., ed. 1993. *The New Shorter Oxford Dictionary*. Oxford University Press, Oxford, UK.

Buschbacher R., C. Uhl, and E.A.S. Serrão. (1988). Abandoned pastures in eastern Amazonia: II. Nutrient stocks in the soil and vegetation. *Journal of Ecology* 76: 682-699.

Choné, T., F. Andreux, J. C. Correa, B. Volkoff, and C. Cerri. (1991). Changes in organic matter in an oxisol from the central Amazonian forest during eight years as pasture, determined by ^{13}C isotopic composition. In *Diversity of environmental biogeochemistry* (J. Berthelin, ed.). Elsevier, Amsterdam, pp. 397-405.

Corazza, E.J., J.E. da Silva, D.V.S. Resck, and A.C. Gomes. (1999). Comportamento de diferentes sistemas de manejo como fonte ou depósito de carbono em relação à vegetação de Cerrado. *Revista Brasileira da Ciência do Solo* 23:425-432 (in Portuguese).

da Silva, J.E., D.V.S. Resck, E.J. Corazza, and L. Vivaldi. (2004). Carbon storage in clayey Oxisol cultivated pastures in the "cerrado" region, Brazil. *Agriculture, Ecosystems and Environment* 103:357-363.

Dean, W. (1995) *With broadax and firebrand: The destruction of the Brazilian Atlantic forest.* University of California Press, Berkeley, CA.

de Campos, D.C. (2004). Potencialidade do sistema de colheita sem queima da cana-de-açúcar para o sequëstro de carbono. PhD thesis, Escola Superior de Agricultura, Luiz Queiroz (ESALQ), Universidade de São Paulo, Piracicaba, SP, Brazil. (in Portuguese).

de Campos, D.V-B. (2003). Uso da técnica de ^{13}C e fracionamento físico da matéria orgânica em solos sob cobertura de pastagens e cana de açúcar na região da mata atlântica. PhD thesis, Universidade Federal Rural do Rio de Janeiro, Seropédica, RJ, Brazil (in Portuguese).

de Oliveira, I.P., J. Kluthcouski, L.D. Yokoyama, L.G. Dutra, T. de A. Portes, A.E. da Silva, B. da S. Pinheiro, E. Ferreira, E. da M. de Castro, C.M. Guimarães, J. de C. Gomide, and L.C. Balbino. (1996). Sistema Barreirão: Recuperação/ renovação de pastagens degradadas em consórcio com culturas anuais. Documento 64, Embrapa Arroz e Feijão, Goiânia, GO, Brazil (in Portuguese).

de Oliveira, O.C., I.P. de Oliveira, E. Ferreira, B.J.R. Alves, C.H.B. Miranda, L. Vilela, S. Urquiaga, and R.M. Boddey. (2001). Response of degraded pastures in the Brazilian cerrado to chemical fertilization. *Pasturas Tropicales* 23:14-18.

Derpsch, R. (2002). Sustainable agriculture. In *The environment and zero-tillage* (H.M. Saturnino and J.N. Landers, eds.). APDC, Brasília, Brazil, pp. 31-51.

de Resende, A.S., S. Urquiaga, B.J.R. Alves. and R.M. Boddey. (2005) Long-term effects of pre-harvest burning and nitrogen and vinasse applications on yield of sugar cane and soil carbon and nitrogen stocks on a plantation in Pernambuco, N.E. Brazil. (submitted).

Diekow, J., J. Mielniczuck, H. Knicker, C. Bayer, D.P. Dick, and I. Kogel-Knabner. (2005). Soil C and N stocks as affected by cropping systems and nitrogen fertilisation in a southern Brazil Acrisol managed under no-tillage for 17 years. *Soil & Tillage Research* 81:87-95.

Doran, J.W., E.T. Elliott, and K. Paustian. (1998). Soil microbial activity, nitrogen cycling, and long-term changes in organic carbon pools as related to fallow tillage management. *Soil & Tillage Research* 49:3-18.

Feigl, B.J., J. Melillo, and C. Cerri. (1995). Changes in the origin and quality of soil organic matter after pasture introduction in Rondônia (Brazil). *Plant and Soil* 175:21-29.

Feller, C. and M.H. Beare. (1997). Physical control of soil organic matter dynamics in the tropics. *Geoderma* 79:69-116.

Fisher, M.J., S.P. Braz, R.S.M. dos Santos, S. Urquiaga, B.J.R. Alves, and R.M. Boddey. (2004). Another dimension to grazing systems: Soil carbon. Paper presented at the 2nd International Symposium on Grassland Ecophysiology and Grazing Ecology, October 11-14, 2004, Universidade Federal de Paraná, Curitiba, PR, Brazil (CD-ROM).

Franco, A.A. and S.M. de Faria. (1997). The contribution of N_2-fixing tree legumes to land reclamation and sustainability in the tropics. *Soil Biology and Biochemistry* 29:897-903.

Freixo, A.A., P.L.O. de A. Machado, H.P. dos Santos, C.A. Silva, and F. de S. Fadigas. (2002). Soil organic carbon and fractions of a Rhodic Ferrasol under the influence of tillage and crop rotation systems in southern Brazil. *Soil &. Tillage Research* 64:221-230.

Galloway, J. H. (1989) *The sugar cane industry: An historical geography from its origins to 1914.* Cambridge University Press, Cambridge, UK.

Herridge, D.F. (1982). Relative abundance of ureides and nitrate in plant tissues of soybean as a quantitative assay of nitrogen fixation. *Plant Physiology* 70:1-6.

IBGE—Instituto Brasileiro de Geografia e Estatística. (1996). Censo Agropecuário 1995-1996. Available online at www.ibge.gov.br (in Portuguese).

Jantalia, C.P. (2005) Estudo de sistemas de uso do solo e rotações de culturas em sistemas agrícolas brasileiros: dinâmica de nitrogênio e carbono no sistema solo—planta—atmosfera. PhD thesis, Universidade Federal Rural do Rio de Janeiro, Seropédica, RJ, Brazil.

Jenkinson, D.S. (1991). The Rothamsted long-term experiments: Are they still of use? *Agronomy Journal* 83:2-10.

Kern, J.S. and M.G. Johnson. (1993). Conservation tillage impacts on national soil and atmospheric carbon levels. *Soil Science Society of America Journal* 57:200-210.

Kluthcouski, J., T. Cobucci, H. Aidar, L.P. Yokoyama, I.P. de Oliveira, J.L. da Silva Costa, J.G. da Silva, L. Vilela, A. de O. Barcellos, and C de U. Magnabosco. (2000). Sistema Santa Fé—Tecnologia Embrapa: Integração lavoura-pecuária pelo consórcio de culturas anuais com forrageiras, em áreas de lavoura, nos sistemas direto e conventional. Circular Técnica 38. Embrapa Arroz e Feijão, Goiânia, GO, Brazil (in Portuguese).

Koutika, L.-S., F. Bartoli, F. Andreux, C. Cerri, G. Burtin, T. Choné, and R. Philippy. (1997). Organic matter dynamics and aggregation in soils under rain forest and pastures of increasing age in the eastern Amazon basin. *Geoderma* 76:87-112.

Landers, J.N., M.C. de Oliveira, P.L. de Freitas, and F. Paladini. (2003). Land-use intensification through zero tillage as a viable solution for mitigating de-forestation. *II World Congress on Conservation Agriculture.* Proceedings Federação Brasileira de Plantio Direto na Palha, Ponta Gross, PR, Brazil, pp. 83-86.

Lilienfein, J. and W. Wilke. (2003). Element storage in native, agri-, and silvicultural ecosystems of the Brazilian savanna: I. Biomass, carbon, nitrogen, phosphorus, and sulfur. *Plant and Soil* 254:425-442.

Lilienfein, J., W. Wilke, L. Vilela, S. de C. Lima, R.J. Thomas, and W. Zech. (2001). Effects of *Pinus caribaea* plantations on the C, N, P and S status of Brazilian savanna Oxisols. *Forest Ecology and Management* 147:171-182.

Lovato, T., J. Mielniczuk, C. Bayer, and F. Vezzani. (2004) Adição de carbono e nitrogênio e sua relação com os estoques no solo e com o rendimento do milho em sistemas de manejo. *Revista Brasileira de Ciência do Solo* 38:175-187.

LSPA. (2004). Levantamento Sistemático da Produção Agrícola: Pesquisa mensal de previsão e acompanhamento das safras agrícolas no ano civil. Available online at www.ibg.gov.br (in Portuguese).

Macedo, I. de C. (1998). Greenhouse gas emissions and energy balances in bio-ethanol production and utilization in Brazil (1996). *Biomass and Bioenergy* 14:77-81.

Macedo, M.C.M. (2002). Degradação, renovação e recuperação de pastagens cultivadas: Ênfase sobre a região dos cerrados. In *Proceedings Simpósio sobre Manejo Estratégico da Pastagem* (J.A. Obeid, O.G. Pereira, D.M. da Fonseca, and D. do Nascimento Júnior, eds.). November 14-16, 2002, Universidade Federal de Viçosa, Viçosa, MG, Brazil. (in Portuguese).

Machado, G.R., W.M. da Silva, and J.E. Irvine. (1987). Sugar cane breeding in Brazil: The Copersucar Program. In *Copersucar International Sugarcane Breeding Workshop*. Cooperativa de Produtores de Cana, Açúcar e Álcool do Estado de São Paulo Ltda, São Paulo, SP, Brazil, pp. 215-232.

Machado, P.L.O. de A. and C.A. Silva. (2001). Soil management under no-tillage systems in the tropics with special reference to Brazil. *Nutrient Cycling in Agroecosystems* 61:119-130.

Muzilli, O. (1983). Influência do sistema de plantio direto, comparado ao convencional, sobre a fertilidade da camada arável do solo. *Revista Brasileira da Ciência do Solo* 7:95-102 (in Portuguese).

Neill, C., J. Melillo, P.A. Steudler, C.C. Cerri, J.F.L. Moraes, M.C. Piccolo, and M. Brito. (1997). Soil carbon and nitrogen stocks following forest clearing for pasture in the southwestern Brazilian amazon. *Ecological Applications* 7:1216-1225.

Santantonio, D. and J.C. Grace. (1987). Estimating fine root production and turnover from biomass and decomposition data: A compartment flow model. *Canadian Journal of Forest Research* 17:900-908.

Shapouri, H., J.A. Duffield, and M. Wang. (2002). The energy balance of corn ethanol: An update. U.S. Department of Agriculture, Office of the Chief Economist, Office of Energy Policy and New Uses. Agricultural Economic Report No. 814.

Shearer, G.B. and D.H. Kohl. (1986). N_2-fixation in field settings: Estimations based on natural ^{15}N abundance. *Australian Journal of Plant Physiology* 13:699-756.

Sidiras, N. and M.A. Pavan. (1985). Influência do sistema de manejo do solo no seu nível de fertilidade. *Revista Brasileira da Ciência do Solo* 9:249-254 (in Portuguese).

Sisti, C.P.J., H.P. dos Santos, R.A. Kochhann, B.J.R. Alves, S. Urquiaga, and R.M. Boddey. (2004). Change in carbon and nitrogen stocks in soil under 13 years of conventional or zero tillage in southern Brazil. *Soil & Tillage Research* 76:39-58.

Steen, E. (1984). Variation of root growth in a grass ley studied with a mesh bag technique. *Swedish Journal of Agricultural Research* 14:93-97.

Tarré, R., R. Macedo, R.B. Cantarutti, C. de P. Rezende, J.M. Pereira, E. Ferreira, B.J.R. Alves, S. Urquiaga, and R.M. Boddey. (2001). The effect of the presence of a forage legume on nitrogen and carbon levels in soils under *Brachiaria* pastures in the Atlantic forest region of the South of Bahia, Brazil. *Plant and Soil* 234:15-26.

Trumbore, S.E., E.A. Davidson, P.B. de Camargo, D. Nepstad, and L.A. Martinelli. (1995). Belowground cycling of carbon in forests and pastures of eastern Amazonia. *Global Biogeochemical Cycles* 9:515-528.

Zinn, Y.L., D.V.S. Resck, and J.E. da Silva. (2002). Soil organic carbon as affected by afforestation with *Eucalyptus* and *Pinus* in the Cerrado region of Brazil. *Forest Ecology and Management* 166:285-294.

Zotarelli, L. (2005) Influência do sistema de plantio direto e convencional com rotação de culturas na agregação, acumulação de carbono e emissão de oxido nitroso, num latossolo vermelho distroférrico. PhD thesis, Universidade Federal Rural do Rio de Janeiro, Seropédica, RJ, Brazil.

Chapter 15B

Potential of Soil Carbon Sequestration for the Brazilian Atlantic Region

F. F. C. Mello
C. E. P. Cerri
M. Bernoux
B. Volkoff
C. C. Cerri

INTRODUCTION

The Brazilian Atlantic region (Figure 15B.1) includes the south and southeastern Brazilian states (Rio Grande do Sul, Santa Catarina, Parana, Sao Paulo, Rio de Janeiro), part of Minas Gerais state, and small portions of the northeastern states (Espirito Santo, Bahia, Sergipe, Pernambuco). It encompasses the oldest and most important Brazilian agricultural ecosystems. Covering an area of 1.1 Mkm^2, the region comprises 13 percent of Brazil's territory and 50 percent of its population.

The area is recognized worldwide because of its high level of diversity. Besides the immense animal diversity, a record number of woody species varieties, with as many as 454 species per hectare, are reported in southern Bahia. The forest, called "Mata Atlantica" or Atlantic forest, is the typical vegetation of the Brazilian Atlantic region. Compared to other Brazilian areas, the Atlantic forest represents the most endangered biome: In terms of remaining forest areas, it covers only 16 percent (20 percent in the study area) of the forested area that was documented in the year 1500. This decline can be attributed to the European immigration to South America, including the exploitation of Brazil-wood (*Caesalpinia echinata* Lam.) followed by mining of gold and cultivation of sugar cane (*Saccharum* ssp.).

The importance of the region was also recognized by the United Nations (UN). Parts of the area are included in the Mata Atlantica Biosphere Re-

Carbon Sequestration in Soils of Latin America

doi:10.1300/5755_16

FIGURE 15B.1. Location of the Brazilian Atlantic region.

serve, which was declared an International Biosphere Reserve by UNESCO in 1991 and designated as a World Heritage site in 1999. As early as 1988, the Atlantic forest was declared a national heritage by the federal constitution of Brazil (Fundaçâo SOS Mata Atlântica and INPE, 2002). Yet information on the soil organic carbon (SOC) sink capacity is not known for the Atlantic region. Therefore, the main objective of this study was to estimate the potential of soil carbon sequestration for the Brazilian Atlantic region, and to discuss land-use and management practices that will enhance and stabilize SOC sequestration.

DESCRIPTION OF THE STUDY AREA

The region is characterized by a wet climate, either tropical or subtropical, and forest is the major vegetation of this biome.

Physiographic

The major feature of the region is the mountainous landscape, which extends from 30°S near Porto Alegre to 13°S near Salvador. It corresponds to the eastern slopes of the Brazilian highland facing the Atlantic Ocean. The highest summits, over 1,600 m high, are located in the Serra do Mar, whereas the elevation ranges from approximately 500 to 1,000 m in most other areas. Pre-Cambrian rocks predominate the mountain chain. Sedimentary low-plateaus ("tabuleiros") and marine deposits occur along the coastal zone, between Rio de Janeiro and Recife.

In the west, the Brazilian Atlantic region is directly connected to the central Brazilian plateau. In the north, prolongation of the central Brazilian ancient plateaus constitutes the coastal mountain range. In the center, at approximately the latitude of Rio de Janeiro, is a distinct transition with the central Brazilian plateau. This transition occurs over a wide area characterized by a hilly topography of about 1,000 masl and pre-Cambrian rocks. Some high slopes of Serra da Mantiqueira are >1,800 masl. This extension of the Brazilian Atlantic mountain is also connected to the southern part of these mountains which form the Serra do Mar.

The western plateau, which is an extension of the central Brazilian plateau, occupies the inerior regions of the Sao Paulo and Parana states on the eastern side of the Parana River. The altitude ranges from 600 masl in the east to 300 masl in the west. The plateau corresponds to the sedimentary Parana basin, of which the upper strata is composed of various sandstone types and extensive intrusions of eruptives rocks, mainly basalts. In the southern part of the region, i.e., in Santa Catarina and Rio Grande do Sul, the basaltic plateau reaches the coastal area. Its elevation is about 1,000 m or more in the east, and about 500 m in the west. Deep, large valleys running from east to west (Iguaçu, Uruguay Rivers) dissect it.

Climate

Due to its latitude and its mean altitude over 600 masl, the dominant climate in the region is subtropical/temperate (Figure 15B.2). The tropical climate occurs only in latitudes north of Belo Horizonte. The tropical portion of the Atlantic Brazilian region is characterized by a wet climate (Af) in most of the coastal zone of Bahia (1,200 to 1,800 mm of rainfall distributed throughout the year), and a semihumid climate (Am) inland as well as toward the south, with an average annual rainfall from 1,000 to 1,500 mm and a dry period occurring from May through September.

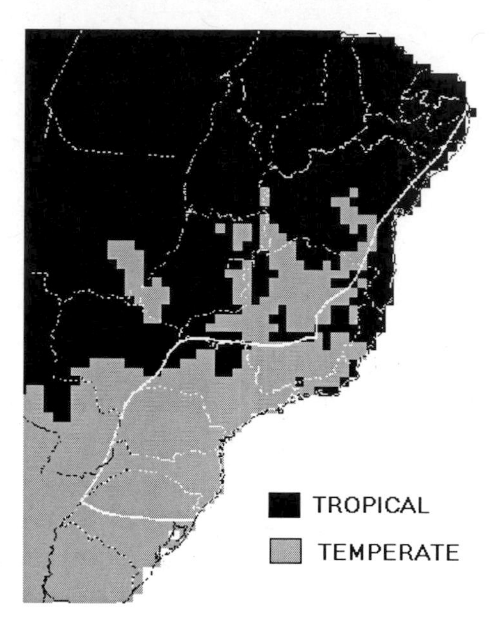

FIGURE 15B.2. General climate in the Atlantic region. (*Source:* Adapted from FAO SDRN, 1997.)

The subtropical (or temperate) part of the Brazilian Atlantic is subdivided in three regions: (1) the northern region, which has an annual rainfall of 1,500 to 2,000 mm and as high as 4,000 mm in the Serra do Mar, and a dry winter (Cw climate according to Koeppen's classification); (2) the central and southeastern zone, which also has annual rainfall of 1,500 to 2,000 mm, but has hot summers and a dry season from April through September, and variations between the wettest and driest months are less than those for the first zone (Cfa climate according to Koeppen's classification); and (3) the southeastern zone, which corresponds to the southern part of the basaltic plateau, and has annual rainfall of 1,300 mm to 2,000 mm with cool summers and no defined dry season (Cfb climate according to Koeppen's classification).

Vegetation

Forest is the major vegetation of the entire region. It represents the Brazilian Atlantic forests and is composed of several types of tropical and subtropical moist broadleaf forests (Figure 15B.3).

(a) Evergreen forests (or Atlantic Evergereen forests) extend from Bahai to the Serra do Mar, in the coastal Bahia region, and in the central and southern coastal slopes of Brazilian Atlantic mountains.
(b) Semideciduous forests (or semievergreen rain forests) cover most of the remaining areas of the region.
(c) Brazilian Araucaria moist forests (or mixed ombrophyllous forests) are in the eastern part of the southern basaltic plateau.
(d) Savannas are represented by the cerrado vegetation in the central western border of the Brazilian Atlantic region.
(e) Campos vegetation occurs as inclusions in the middle of the forest, mainly in the southern basaltic plateau.

The data in Table 15B.1 show that the 1.1 million km^2 of the Brazilian Atlantic forest region was originally dominated by seasonal semideciduous forest (V5) and Atlantic evergreen forest (V3), together covering more than

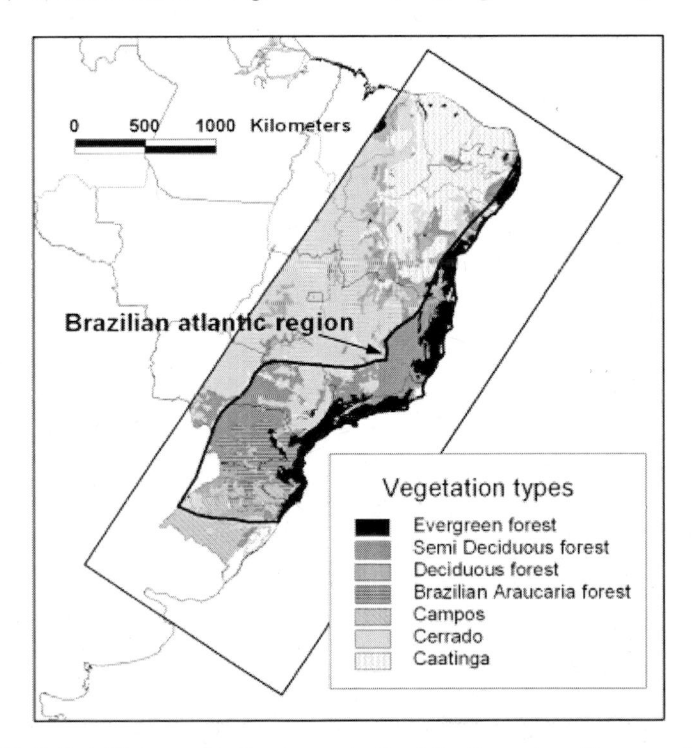

FIGURE 15B.3. Main vegetation types in the Atlantic region. (*Source:* Adapted from IBGE, 1988.)

TABLE 15B.1. Areas of selected native vegetation types in states of the Atlantic region.

		Area originally covered by selected native vegetation (in 1000 km^2)						
State	**Percent of state area**	**V5**	**V3**	**V6**	**V9**	**V7**	**V4**	**Other**
Espirito Santo	100	10	33					1
Parana	100	78	7	91	13	5		1
Rio de Janeiro	100	17	21		<1			3
Santa Catarina	100		27	45		14	7	<1
Sao Paulo	100	90	46	4	99			<1
Alagoas	51	3	9					2
Rio Grande do Sul	51	3	1	30		57	37	9
Sergipe	44	4			4			2
Minas Gerais	38	167	19	1	30		6	
Bahia	16	14	60		4		7	5
Pernambuco	12	4	7					1
Paraiba	5	1			1			
Total	43	392	231	170	152	76	58	26

Source: Adapted from Bernoux et al., 2002.
Note: V5: Semideciduous forest; V3: Evergreen Atlantic forest; V6: Brazilian araucaria forest; V9: Cerrado; V7: Campos; V4: Deciduous forest.

half of the total area and extending into most of the states. The mixed ombrophyllous forest (V6) represents the third largest vegetation type, and it primarily occurs in the southern region (Parana, Santa Catarina, and Rio Grande do Sul). Savanna vegetation or cerrado (V9) is the dominant vegetation in Sao Paulo state and also occurs in parts of Minas Gerais state. South savanna or campos vegetation (V7) and seasonal deciduous forest (V4) occur mainly in the southern part of Brazil.

In the description of the vegetation map of IBGE (1988), seasonal semidecidual forest is related to the climate of two seasons in the area, a rainy season and a dry season (with mean temperature of 21°C), or with a short dry period accompanied by pronounced low temperature in the subtropical area (with mean temperatures of 15°C). Dominant arboreal elements occur, which are adapted to the cold or dry season whereby the proportion of deciduous trees ranges from 20 to 50 percent. This vegetation system occurs predominantly at dissected plateau areas that divide the drainage of the Amazon river and cover the lower hillsides of Serra do Mar and Mantiqueira, as well as the river basins Paraguai and Paraná.

This vegetation system consists of alluvial, lowland, submontane, and montane formations. The mixed ombrophyllous forest (or conifer forest) exclusively covers the south Brazilian plateau regions, with distinct disjunctions in high areas of Serra do Mar and Mantiqueira. It prevails in an ombrophyllic climate without a dry season, with an annual average temperature of 18°C. The temperature is <15°C for three to six months.

The arboreal formations of the southern plateau areas enclose two Brazilian floras; the Afro-Brazilian and the temperate Austro-Brazilian flora with a predominance of conifers *(Araucaria angustifólia).* These areas were intensively exploited by the timber industry and subsequently used for agriculture and pastures. Rizzini (1997) defined the Atlantic forest as a vegetation system covering the coastal contour from Rio Grande do Sul to the northeast in an ocean strip, including some eastern ranges such as Serra do Mar and Mantiqueira, as well as the states of São Paulo, Minas Gerais, Rio de Janeiro, and Espírito Santo.

Soils

Most soils in the region are highly weathered, due to the persistence of wet, tropical, and subtropical climates and the soil age. Intense weathering is indicated by the presence of very deep saprolites in the crystalline basement rocks. Also, geomorphic evidence shows that many features of the Atlantic Brazilian region landscape are directly connected to equivalent features in the central Brazilian region and the same ancient erosional and depositional surfaces, and comparable old soils are found in both regions.

Soils are mainly Ferralsols, derived from either sedimentary, volcanic, or crystalline rocks. Ferralsols are always associated with Acrisols. The dominant clay mineral of these soils is kaolinite. These are normally unsaturated, with high exchangeable Al^{+3}. In the wet subtropical climatic environment, vermiculite-Al is normally associated with kaolinite. Consequently, exchangeable Al^{+3} contents are very high and many soils have allic properties. Another consequence of the relatively cold and humid subtropical environment is the increase in soil organic matter (SOM) concentration in the soil profile and the frequent occurrence of umbric horizons.

In the Brazilian Atlantic region, the nature of the soils changes according to the parent material, the altitude, and the geomorphic or landscape position. The general distribution of the soil types is depicted in the Brazilian Soil Map (EMBRAPA, 1981).

In the north, the dominant soil is Latossolo Amarelo (Xanthic Ferralsol) on the low coastal plateaus ("tabuleiros") and Latossolo Vermelho Amarelo (Haplic Ferralsol) on the high plateau where it is associated with Podzolico

Vermelho Amarelo distrófico (Ferric Acrisols) on the hilly regions of mountain slopes. Podzolico vermelho amarelo eutrofico (Ferric Luvisols) occurs in the dryer regions (Jequitinhonha Valley).

In the central-eastern mountain and hilly zone, which is underlain by crystalline basement, Latossolo vermelho amarelo distrofico (Haplic Ferralsol) is the dominant soil type. It is associated with Podzolico vermelho amarelo distrofico (Ferric Acrisol) and Cambissolo distrofico (Dystric Leptosol).

In the central-western part, (i.e., the sedimentary and basaltic plateau of Sao Paulo and Parana), the dominant soils are Latossolo vermelho escuro and Latossolo roxo (Rhodic Ferralsol) on the predominantly basaltic substrate, and Podzolico vermelho amarelo distrofico (Ferric Acrisol) and Latossolo vermelho Escuro (Rhodic Ferralsol) on sandstones. Podzolico Vermelho Amarelo eutrofico (Haplic Lixisol) occurs in the Marilia region on calcic sandstone parent material.

In the southern basaltic plateau, Latossolo Roxo and Latossolo Vermelho Escuro (Rhodic Ferralsol), and Terra Roxa Estruturada distrofica (Rhodic Nitosol) occur on the plateau, while Solo Litolico eutrofico (Eutric Leptosol) occurs on the basaltic slopes. Latossolo bruno distrofico (Humic Ferralsol), and its associated Solo Litolico húmico distrofico (Humic Cambisol) in the slopes, occur on the higher eastern parts of the basaltic plateau.

Carbon Pools Under Native Conditions

For the Atlantic Forest, Bernoux and Volkoff (Chapter 4) estimated a total C pool of about 5,657 Tg for 0 to 30 cm depth (Table 15B.2), which corresponds to a potential sink capacity under native vegetation. The mean C pool for the studied area was estimated at 5.5 kg $C \cdot m^{-2}$, and represents a 10 percent higher value compared to the overall mean for the Latin American region. Detailed information about C pools under selected vegetation and main soil types, and the respective states are given in Tables 15B.2 and 15B.3.

Table 15B.4 reports the total and mean C pools by state according to the simplified soil classification proposed by Bernoux et al. (2002) based on criteria recommended by IPCC/UNEP/OECD/IEA (1997) such as soil texture, base saturation, and soil water status. The IPCC/UNEP/OECD/IEA (1997) proposed six categories that are characterized as high-activity clay (HAC) mineral soils, low-activity clay (LAC) mineral soil, sandy soils, volcanic soils, wet soils, and organic soils. Definitions of the Brazilian soil classification system were taken into account to separate HAC (soils with cation-exchange capacity $\geq$ 24 $cmol_c \cdot kg^{-1}$ clay) from LAC (soils with cation-exchange capacity < 24 $cmol_c \cdot kg^{-1}$ clay). The original Brazilian soil

TABLE 15B.2. Carbon pools under selected native vegetation types in different states of the Brazilian Atlantic region.

	Total C pool under native vegetation (Tg)							
State	**V5**	**V3**	**V6**	**V9**	**V7**	**V4**	**Other**	**Total**
Espirito Santo	43	164					8	215
Paraná	316	44	707	49	28		4	1148
Rio de Janeiro	72	137		1			15	225
Santa Catarina		129	344		96	28	1	598
Sao Paulo	365	212	31	385			2	995
Alagoas	11	42					6	60
Rio Grande do Sul	11	4	272		433	161	50	931
Sergipe	14			12			8	34
Minas Gerais	688	93	10	113		20		925
Bahia	54	340		17		31	15	457
Pernambuco	16	34					4	54
Paraiba	5	1		4				9
Total	1599	1200	1363	582	557	240	117	5657

Source: Adapted from Bernoux et al., 2002.

TABLE 15B.3. Mean C pool under selected native vegetation types in different states of the Brazilian Atlantic region.

	Percent of state	Mean pool under native vegetation ($kg{\cdot}m^{-2}$)							
State	**area**	**V5**	**V3**	**V6**	**V9**	**V7**	**V4**	**Other**	**Mean**
Espirito Santo	100	4.1	4.9					5.1	4.1
Parana	100	4.0	6.1	7.8	3.7	6.1		4.9	5.8
Rio de Janeiro	100	4.1	6.6		3.9			4.7	5.4
Santa Catarina	100		4.7	7.6		6.8	4.0	4.5	6.4
Sao Paulo	100	4.1	4.6	8.4	3.9			4.9	4.0
Alagoas	51	4.0	4.6		3.6			3.3	4.2
Rio Grande do Sul	51	3.9	4.1	9.2		7.5	4.3	5.5	6.7
Sergipe	44	3.7			3.5			3.8	3.6
Minas Gerais	38	4.1	4.9	8.8	3.8		3.5	4.9	4.1
Bahia	16	4.0	5.7		3.9		4.1	3.3	5.1
Pernambuco	12	3.8	4.9					3.8	4.4
Paraiba	5	3.7	4.5		3.2			3.3	3.5
Total area	43	4.1	5.2	8.0	3.8	7.3	4.2	4.5	5.5

Source: Adapted from Bernoux et al., 2002.

TABLE 15B.4. Soil organic carbon pool by state and soil type.

State	Variable	S1	S2	S3	S4	S5	S6	Total
Espirito Santo	Pool (Tg)	4	134	61	9	7	1	215
	Area (1000 km)	1	27	15	2	2	<0	45
	Mean (kg C·m^{-2})	5.1	5.1	4.1	5.6	4.2	41.8	4.8
Paraná	Pool (Tg)	195	445	484	5	6	13	1148
	Area (1000 km)	27	63	103	1	1	<0	196
	Mean (kg C·m^{-2})	7.3	7.0	4.7	6.3	5.3	41.8	5.9
Rio de Janeiro	Pool (Tg)	2	68	86	5	11	52	225
	Area (1000 km)	1	15	21	1	2	2	41
	Mean (kg C·m^{-2})	4.5	4.7	4.1	5.3	5.0	26.7	5.4
Santa Catarina	Pool (Tg)	135	186	262	12	4		598
	Area (1000 km)	17	22	52	2	1		93
	Mean (kg C·m^{-2})	8.2	8.3	5.1	6.1	3.6		6.4
Sao Paulo	Pool (Tg)	1	567	396	26	5		995
	Area (1000 km)	<0	129	103	12	1		245
	Mean (kg C·m^{-2})	2.8	4.4	3.8	2.3	5.0		4.1
Alagoas	Pool (Tg)	1	23	34	2	<0		60
	Area (1000 km)	<0	5	9	1	<0		14
	Mean (kg C·m^{-2})	2.8	5.0	4.0	4.0	3.7		4.3
Rio Grande do Sul	Pool (Tg)	292	447	185	1		6	931
	Area (1000 km)	46	53	37	<0		2	138
	Mean (kg C·m^{-2})	6.4	8.4	5.0	6.3		2.7	6.7
Sergipe	Pool (Tg)	3	4	19	2	1	5	34
	Area (1000 km)	1	1	5	1	0	1	10
	Mean (kg C·m^{-2})	3.2	4.4	3.6	4.4	4.6	3.1	3.6
Minas Gerais	Pool (Tg)	11	581	333	<0			925
	Area (1000 km)	2	131	89	<0			222
	Mean (kg C·m^{-2})	5.0	4.4	3.8	1.9			4.2
Bahia	Pool (Tg)	36	198	151	10	4	58	457
	Area (1000 km)	8	39	37	2	1	3	90
	Mean (kg C·m^{-2})	4.6	5.1	4.1	4.5	4.9	21.3	5.1
Pernambuco	Pool (Tg)	<0	28	25	<0		1	54
	Area (1000 km)	<0	6	6	<0		<0	12
	Mean (kg C·m^{-2})	4.1	5.0	3.9	5.0		3.3	4.4

TABLE 15B.4 *(continued)*

State	Variable	S1	S2	S3	S4	S5	S6	Total
Paraíba	Pool (Tg)			8	1		<0	9
	Area (1000 km)			2	<0		<0	3
	Mean (kg C·m^{-2})			3.7	2.2		3.3	3.5
Total area	Pool (Tg)	680	2681	2044	73	44	135	5657
	Area (1000 km)	102	491	479	21	10	9	1112
	Mean (kg C·m^{-2})	6.7	5.5	4.3	3.5	4.4	15.2	5.1

map lists no soil type corresponding to volcanic and organic soils. Based on these observations, soil types can be theoretically split into HAC soils, LAC soils, sandy soils, and wet soils. But, the Brazilian order named Latossolos, which pertains to LAC soils, alone covers 38.8 percent of the country. The Brazilian Latossolos correspond to well-drained Oxisols in the U.S. Soil Taxonomy, and to Ferralsols in the FAO-UNESCO (FAO-UNESCO) soil map legend. Then, the LAC soils were divided into LAC-Latossolos (S2) and LAC-non-Latossolos (S3). The other soil categories were named HAC soils (S1), sandy soils (S4), and wet soils (S5). An additional category contains all the soils that did not match with one of the other groups, and were marked as other soils (S6). Two soil types (S2 and S3) cover 87 percent of the total area and contribute 84 percent of the total C pools. States where HAC soils (S1) are prevalent contain higher mean soil C pools than those where these soils are absent.

ACTUAL LAND USE

The deforested area was calculated by subtraction of the remaining areas from original forest areas (Table 15B.5) taken from Conservation International of Brazil (2000) and Fundaçâo SOS Mata Atlântica and INPE (2002).

On the basis of the description given previously; the information on land use for the entire Atlantic region was taken from the databases of IBGE and SIDRA (Table 15B.6). Information on pastures was obtained from ANUALPEC (FNP, 2003). The sum of the production areas of all states is defined as the production area of the Atlantic region.

In order to obtain a more accurate estimate, production data of the used agricultural crops as well as the data of the remaining areas of the year 2000 were compared. More recent estimates of the pasture areas and the reforested areas were taken from 2002 and 1996, respectively.

TABLE 15B.5. Original areas and remaining and deforested areas of the Atlantic forest.

State	Original area (Mha)	Area of the state (Mha)	Atlantic forest in the state (%)	Remaining areas (Mha)	Deforested area (Mha)
Alagoas	1.4	2.75	51	0.07	1.33
Bahia	9.0	56.25	16	2.63	6.37
Espírito Santo	4.4	4.40	100	1.40	3.00
Minas Gerais	22.3	58.68	38	4.19	18.11
Paraíba	0.2	4.00	5	0.05	0.15
Pernambuco	1.2	10.00	12	0.09	1.11
Paraná	19.5	19.50	100	3.92	15.58
Rio de Janeiro	4.1	4.10	100	0.84	3.26
Rio Grande do Sul	13.7	26.86	51	2.13	11.57
Santa Catarina	9.3	9.30	100	3.00	6.30
Sergipe	1	2.27	44	0.09	0.91
São Paulo	23.9	23.90	100	3.00	20.90
Total	110	222		21.41	88.59

Source: Adapted from Conservation International do Brasil, 2000; and from Fundaçâo SOS Mata Atlantica and INPE, 2002.

TABLE 15B.6. Areas under different land uses in the Atlantic region in 2000.

Land use	Area	
	Mha	**% of total**
Forest	21,396,553	19.45
Well-managed pasture	27,993,940	25.45
Degraded pasture	27,993,940	25.45
Annual crops under conventional tillage	6,855,078	6.23
Annual crops under no-tillage	8,022,030	7.29
Sugarcane with burning	2,415,863	2.20
Sugarcane without burning	1,035,370	0.94
Reforestation	3,093,012	2.81
Fruitculture	2,690,374	2.45
Horticulture/gardening	999,350	0.91
Others	7,504,490	6.82
Total	110,000,000	100

Source: Adapted from FNP, 2003; IBGE-SIDRA, 2004.

The actual land use was grouped in eight categories to give a general characterization of the productive area of the Atlantic region: (1) forest, (2) annual crops, (3) sugarcane, (4) pasture, (5) reforestation, (6) fruit plantation, (7) horticulture/gardening, and (8) others or nonagricultural areas.

SCENARIOS AND POTENTIAL OF SOIL CARBON SEQUESTRATION

The carbon sequestration potential was estimated by multiplication of carbon accretion rate for the soil under each land use with the corresponding area.

The Atlantic region presents four main groups of agricultural uses/practices which can be changed to a recommended land use: (1) convert sugarcane areas harvested by burning (manual cut) into areas without burning and mechanical harvest, (2) convert conventional tillage into no-till systems for production of grain crops, (3) increase reforested areas, and (4) recover degraded pasture areas into productive pastures.

It is possible to increase the amount of carbon sequestered by increasing the area under recommended land use. Hence, a scenario about changes of the actual land use was established (Figure 15B.4), and considers the following new land-use practices: (1) adoption of conservative agricultural practice, such as no-till in areas where annual crops are cultivated conventionally or where soil loss and organic matter degradation occur; (2) mechanized harvesting of sugarcane in areas where burning is still practiced; (3) restoration of degraded pastures or converting these to other crops, such as sugarcane without burning, and (4) reforestation.

Another important factor in support of the proposed scenario is an estimation carried out by FNP (2003). Their calculation shows a decrease in area under pasture in Brazil and an increase in area under meat production. This trend is attributed to restoration of degraded pastures for other land uses (e.g., no-tillage practice for soybean and corn cultivation).

Thus, alternative scenarios are (1) conversion of degraded pasture areas in flat topographical regions to sugarcane cultivation, (2) reforestation of areas unsuitable for mechanization of sugarcane cultivation, and (3) increase in cultivated areas under no-till soybean and corn cultivation through conversion of degraded pastures, as estimated by FNP (2003). The following estimates were based on the assumption that pastures will be converted to other land uses in the same proportion as it is occurring presently, and (4) degraded pastures are converted to productive pastures.

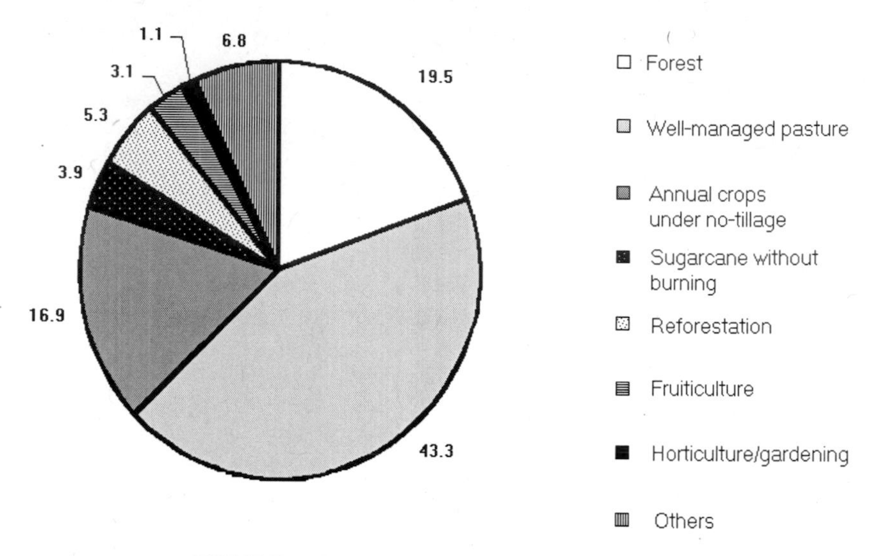

FIGURE 15B.4. Possible land-use scenarios.

Conversion of Sugarcane with Burning to Sugarcane Without Burning

Brazil is the largest producer of sugarcane worldwide, with about 26 percent of the world cultivated area and 29 percent of world production (FAO, 2004). Considering the Atlantic region, area under sugarcane in Sao Paulo is 2.5 Mha (FNP, 2004), which represents 72 percent of the total Atlantic area.

Sugarcane is one of the most important crops in Brazil. In contrast to fossil fuels, it produces an efficient alternative as a renewable fuel. Further, 100 percent of the production technology of sugarcane is indigenous to Brazil.

Most of the sugarcane harvest is still done manually. It is costly and is a significant source of CO_2 emissions to the atmosphere, because it involves burning of sugarcane leaves. Furthermore, most of the cultivated areas are located in regions unsuitable for mechanical practices due to steep slope gradient >15 percent. In Sao Paulo state, only 44 percent of the area under sugarcane is cultivated on sites suited to mechanical harvesting (Toledo et al., 1991). Of this, 30 percent of the area is mechanically harvested. Thus, the mechanized harvesting can be extended to the additional 14 percent of the cultivated area.

According to new environmental laws, sugarcane must be totally harvested without burning and, consequently, by mechanized operations. Hence, land-use changes are necessary in areas unsuitable for mechanized sugarcane harvest. It implies alterations of an area of about 1.9 Mha. One alternative would be to convert the unsuitable areas to forest plantations, mainly by *Pinus* species, because the production areas of such forest plantations have been static.

In consideration of the projected national demand for the year 2020, wood production must be increased by 37 Mm^3 of *Pinus* wood *(Pinnus elliottii)* (Paim, 2003). Thus, sugarcane production would have to be concentrated on level degraded pasture areas, where mechanized harvesting is feasible. In this context, land-use changes required would involve conversion of degraded pastures to sugarcane cultivation for mechanized harvesting and conversion of sugarcane on steep lands to forest plantations.

Estimates of the carbon sequestration potential in consideration of these scenarios/assumptions are given in Table 15B.7. The carbon sequestration

TABLE 15B.7. Potential of soil carbon sequestration in the 0 to 20 cm layer for the Atlantic region.

Land use	Total area (Mha)	Soil carbon sequestration rate ($Mg \cdot ha^{-1}$ per year)	Source	Potential of soil C sequestration (Tg per year)
Sugarcane burning to without burning	3.30	1.62	1	5.35
Degraded pasture to sugarcane without burning	1.93	0.1 - 0.8	1	0.19 - 1.54
Existent sugarcane without burning	1.03	1.62	1	1.67
Conventional to no-tillage	6.85	0.2 - 0.8	2, 5	1.37 - 5.58
Degraded pasture to no-tillage	3.80	0 - 0.71	6, 2	0 - 2.70
Existent no-tillage	8.02	0.2 - 0.8	2, 5	1.60 - 6.41
Degraded to well-managed pasture	19.65	2.71	3, 4	53.25
Existent well-managed pasture	28.00	2.71	3, 4	73.88
Sugarcane with burning to reforestation	1.93	0.66	4	1.27
Degraded pasture to reforestation	0.78	0 - 1.63	6, 4	0.078 - 1.27
Existent reforestation	3.10	2.42	4	7.50

Source: (1) Cerri, Bernoux, Cerri, and Feller, 2004; (2) Cerri, Bernoux, Feller, et al., 2003; (3) Manfrinato, 2002; (4) Campos, 1998; (5) IPCC, 2000; (6) Szakács, 2003.

rates were taken from Cerri, Bernoux, Cerri, and Feller (2004) and extrapolated and upscaled to the whole study area.

Conversion from Conventional Grain Crop Production to a No-Till System

Land area under no-till farming is rapidly increasing throughout Brazil. According to Federação Brasileira de Plantio Direto na Palha (FEBRAPDP, 2004), about 22 Mha of conventional tillage were converted to no-till system in Brazil by 2003. About 50 percent of the total area under no-till farming in Brazil was located in the Atlantic region in 2001 (Figure 15B.5).

In a no-till system, chemical and physical soil properties are improved, which enhances carbon sequestration capacity and offsets CO_2 emissions (Lal, 1997). Séguy et al. (2001) concluded that cultivation practices such as plowing for monocultures of annual crops cause severe losses of organic matter. The soil organic matter content is controlled by edapho-climatic factors.

The Atlantic region has a high potential for expansion of the no-till system. Presently, only some states (e.g., Sao Paulo, Santa Catarina, Rio Grande do Sul, and Paraná) have adopted no-till farming. These states have additional cropland areas in which no-till can be adopted (Figure 15B.6).

The potential of soil C sequestration under a no-till system is shown in Table 15B.7. Estimates of C sequestration rates were taken from those pub-

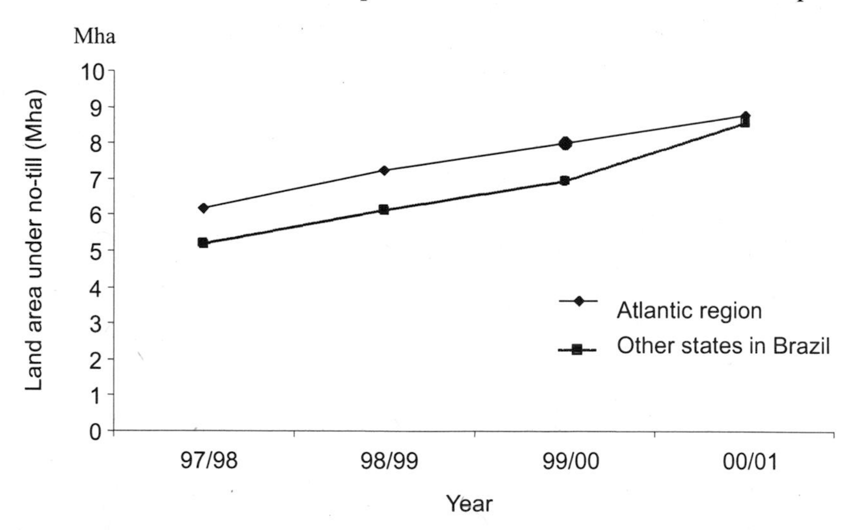

FIGURE 15B.5. Temporal changes in no-till areas in the Atlantic region and other Brazilian states. (*Source:* Adapted from FEBRAPDP, 2004.)

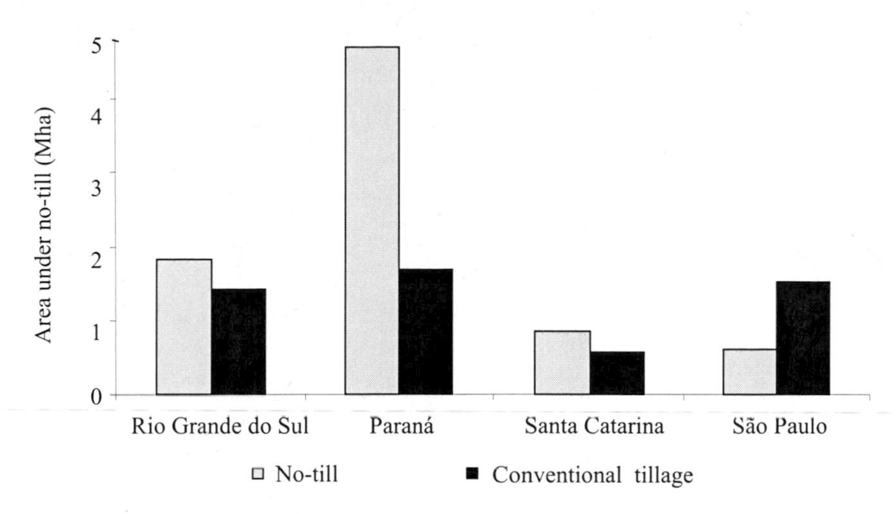

FIGURE 15B.6. Cropland area under conventional and no-till farming in principal grain-producing states of the Atlantic region. (*Source:* Adapted from FEBRAPDP, 2004.)

lished by Cerri, Bernoux, Feller, et al. (2004) and IPCC (2000), and were extrapolated to the entire Atlantic region. A strong possibility exists of achieving the C sequestration potential, because the conversion from conventional tillage to no-till needs only routine changes in crop management compared to the complex process of afforestation or reforestation.

Restoration of Degraded Pastures

Total area under pasture represents the largest area among different land uses in the Atlantic region, covering almost 56 Mha. About 50 percent of this area is prone to some degree of soil degradation (Zimmer et al., 1994) (Figure 15B.4). Soil C sequestration rates in pastures were taken from those published by Manfrinato (2002) and Campos (1998), and were extrapolated to the entire Atlantic region.

Using the estimates reported by FNP (2003), decline in the area under pasture provides an opportunity to expand cropland under no-till farming and convert to other restorative land uses. Assuming that the entire area under pasture land would be degraded, approximately 6.5 Mha are available for conversion to improved land use with high C sequestration/sink capacity.

Therefore, C sequestration potential will be achieved through restoration of an estimated degraded area of about 19.5 Mha, over and above the 28 Mha already under improved pastures (Table 15B.7).

Reforestation

Reforestation can be done in areas where mechanized harvest of sugarcane is not feasible, and it could increase the forest area by 2 Mha. Moreover, if pastures were also reforested, total reforested area would increase to 2.7 Mha. Reforestation would increase the wood production area in the Atlantic region by about 85 percent, and it would also enhance soil C sequestration by 18 to 34 percent (Table 15B.7). Carbon sequestration rates for forested areas were taken from those published by Campos (1998), and extrapolated to the entire Atlantic region.

The data in Table 15B.7 show a large range in rates of SOC sequestration because of difference in pasture species, severity of soil degradation, forest species, and soil type. For example, conversion of a pasture, at the initial stage of degradation by weed infestation, into no-till farming may lead to little or no SOC sequestration. However, conversion of a pasture at an advanced stage of degradation, with a large proportion of bare soil in between, can have a high rate of SOC sequestration. In general, the SOC sequestration rate is low when slightly degraded pasture land is converted to no-till crop production, and high when severely degraded pasture is restored to no-till farming.

CONCLUSIONS

The potential of SOC sequestration in the Atlantic region, in the 0 to 20 cm soil layer, ranges from 144 to 154 Tg C per year (150 Tu Tg per year). Total carbon sequestration over 20 years is 3 Pg C. The estimation shows the importance of improving the use of the already deforested areas in the Atlantic region, which corresponds to 88 Mha. This region is home to 50 percent of Brazil's population and includes some of the largest cities of Brazil.

REFERENCES

Bernoux, M., Carvalho, M. C. S., Volkoff, B., and Cerri, C. C. 2002. Brasil's soil carbon stocks. *Soil Science Society of America Journal,* 66, 888-896.

Campos, D. C. 1998. Influência da mudança do uso da terra sobre a matéria orgânica do solo no município de São Pedro. Dissertação (Mestrado), Escola Superior de Agricultura Luiz de Queiroz, Universidade de São Paulo, Brazil.

Cerri, C. C., Bernoux, M., Cerri, C. E. P., and Feller, C. 2004. Carbon cycling and sequestration opportunities in South América: The case of Brasil. *Soil Use and Management,* 20, 248-254.

Cerri, C. C., Bernoux, M., Feller, C., Campos, D. C., de Luca, E. F., and Eschenbrenner, V. 2004. Canne à sucre: l'exemple du Brésil. Canne à sucre et sequestration du carbone. Academie d'Agriculture meeting, March 17.

Conservation International do Brasil, Fundação SOS Mata Atlântica, Fundação Biodiversitas, Instituto de Pesquisas Ecológicas, Secretaria do Meio Ambiente do Estado de São Paulo, SEMAD/Instituto Estadual de Florestas–MG. 2000. Avaliação e ações comunitárias para a conservação da biodiversidade da mata atlântica e campos sulinos. Brasília, Ministério do Meio Ambiente, Secretaria de Biodiversidade e Florestas.

EMBRAPA. 1981. Mapa de Solos do Brasil. Escala 1:5.000.000. EMBRAPA, Rio de Janeiro.

FAO SDRN. 1997. Global climate maps. FAO, Rome. Available online at www.fao.org.

FAO. 2004. FAOSTAT data. Available online at www.fao.org.

FEBRAPDP. 2004. Federação Brasileira de Plantio Direto na Palha. Available online at www.febrapdp.org.

FNP Consultoria & Agroinformativos. 2003. *ANUALPEC 2003, anuário da pecuária brasileira.*

FNP Consultoria & Agroinformativos. 2004. *AGRIANUAL 2004, anuário da agricultura brasileira.*

Fundação SOS Mata Atlântica & INPE. 2002. Atlas dos remanescentes florestais da mata atlântica, período 1995-2000, relatório final. São Paulo.

IBGE. 1988. Mapa de Vegetação do Brasil. IBGE, Rio de Janeiro.

IBGE. 2004. Sistema IBGE de Recuperação Automática–SIDRA. Available online at www.sidra.ibge.gov.br.

IPCC. 2003. *Land use, land use change and forestry: A special report of the IPCC.* Cambridge University Press, Cambridge, UK.

IPCC/UNEP/OECD/IEA. 1997. *Revised 1996 IPCC Guidelines for national greenhouse gas inventories* Vol. 1, *Reporting instructions,* Vol. 2, *Workbook,* Vol. 3, *Reference manual.* Intergovernamental Pannel on Climate Change, United Nations Environment Programme, Organization for Economic Co-Operation and Development, International Energy Agency, Paris.

Lal, R. 1997. Residue management, conservation tillage, and soil restoration for mitigating greenhouse effect by CO_2 enrichment. *Soil Till. Res.,* 43, 81-107.

Manfrinato, W. A. 2002. Estoques de carbono no solo em uma cronosequência de floresta-pastagem em Guaraqueçaba (PR). Dissertação (Mestrado), Escola Superior de Agricultura Luiz de Queiroz, Universidade de São Paulo, Brazil.

Paim, A. 2003. A potencialidade inexplorada do setor florestal brasileiro. Associação dos engenheiros florestais do Espírito Santo (AEFES). Available online at www.aefes.org.

Rizzini, C.T. 1997. Tratado de Fitogeografia do Brasil: aspectos ecológicos, sociológicos e florísticos. Âmbito Cultural Edições, Brazil.

Séguy, L., Bouzinac, S., and Maronezzi, A. C. 2001.Um dossiê do Plantio Direto: sistemas de cultivo e dinâmica da matéria orgânica. CIRAD, Agronorte Pesquisas, Grupo Maeda, ONG TAFA/ FOFIFA/ANAE, Goiânia, Goiás.

Szakács, G. G. J. 2003. Seqüestro de carbono nos solos: Avaliação do potencial dos solos arenosos sob pastagens, Anhembi-Piracicaba/SP. Dissertação (Mestrado), Centro de Energia Nuclear na Agricultura (CENA), Universidade de São Paulo, Brazil.

Toledo, P. E. N., Yoshii, R. J., and Otani, M. N. 1991. Avaliação do potencial de uso de colheitadeiras de cana-de-açúcar no Estado de São Paulo. *Informações Econômicas, São Paulo,* 21(6), 13-20.

Zimmer, A. H., Macedo, M. C. M., Barcellos, A. O., and Kichel, A. N. 1994. Estabelecimento e recuperação de pastagens de *Brachiaria.* In Simpósio sobre manejo de pastagem, 11, Piracicaba, São Paulo.

Chapter 16

The Potential for Soil Carbon Sequestration in the Pampas

Martín Díaz-Zorita
Daniel E. Buschiazzo

INTRODUCTION

The Pampas region is a vast plain of approximately 130 Mha located in the southern part of Brazil, Uruguay, and Argentina (Figure 16.1). In Argentina, it covers almost 60 Mha located along the central part of the country, with 35 Mha suitable for cropping. According to rainfall and soil quality patterns (Viglizzo et al., 2001), the Argentine Pampas can be divided in five subregions: rolling Pampas, inland or central Pampas, southern Pampas, flooding Pampas, and Mesopotamian Pampas (Figure 16.2).

The climate is warm temperate with adequate to less than adequate rainfall for normal crop production. The mean temperature ranges between 18°C and 14°C in the north and in the south of the Pampas, respectively (Hall et al., 1992). In part of the region, the temperature and the frost-free period are adequate for growing double crops (i.e., soybean [*Glycine max* (L.) Merrill] or maize [*Zea mays* L.] planted after wheat [*Triticum aestivum* L.] crops). Rainfall amounts are highly variable between years. Most rainfall occurs between October and April (spring to fall seasons), and the long-term average ranges between 500 and 1000 mm in the southwest and in the northeast of the region, respectively (Hall et al., 1992).

LAND USE

Most of the region was originally covered by grasslands interrupted only by gallery forests along the margins of the Paraná, Uruguay, and la Plata Rivers. The area under agriculture increased exponentially between 1870

Carbon Sequestration in Soils of Latin America

doi:10.1300/5755_17

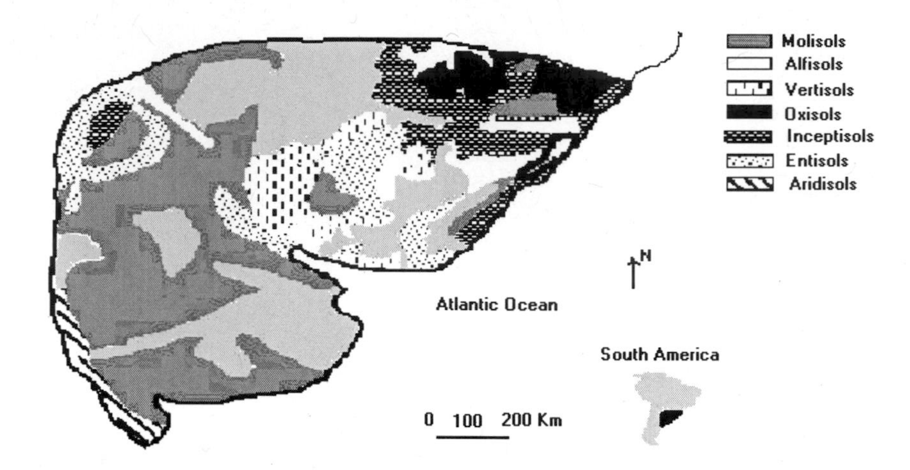

FIGURE 16.1. Location of the Pampas region of South America showing the distribution of soil orders.

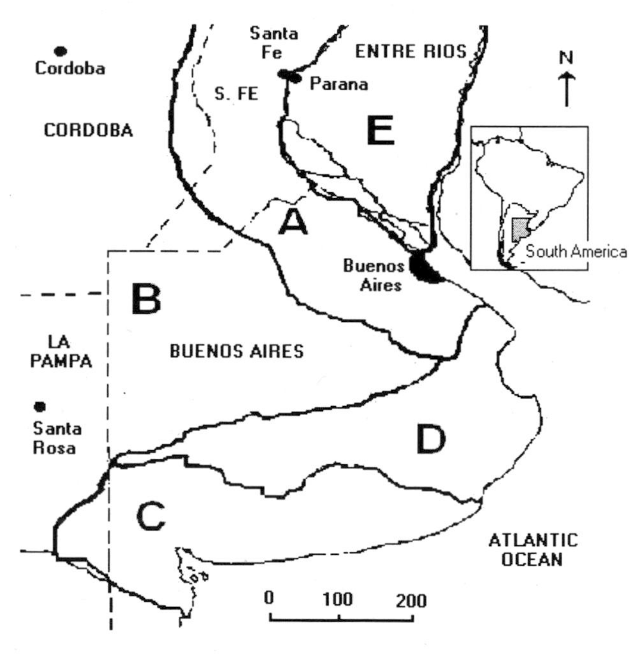

FIGURE 16.2. The Pampas region in Argentina showing boundaries for the subregions (solid lines) and provinces (broken lines): A = rolling Pampas, B = inland or central Pampas, C = southern Pampas, D = flooding Pampas, and E = Mesopotamian Pampas.

and 1940, and, at present, almost 50 percent of the area is under agricultural practices (Table 16.1).

In Argentina, perennial pastures comprising alfalfa (*Medicago sativa* L.), fescue (*Festuca arundinacea* L.), and other grasses under direct-grazing livestock systems cover approximately 8 Mha, and the rest of the area is under annual crop production (approximately 12 Mha of soybean, 2.4 Mha of maize, 6.3 Mha of wheat, 1.8 Mha of sunflower [*Helianthus annus* L.], and other crops) (Argentine Secretary of Agriculture, Livestock, Fisheries and Food, 2004). In the rolling Pampas and in the eastern part of the inland Pampas, almost 100 percent of the area is under annual crop sequences, while the portion of area covered by pastures increases toward the boundaries of the region. The flooded Pampas are mostly under native grasslands for cattle production systems.

In southern Brazil, a continuous soybean/winter wheat rotation is widely practiced and, in the absence of no-till practices, causes significant water erosion problems. Barley (*Hordeum vulgare* L.), rice (*Oryza sativa* L.), sorghum (*Sorghum bicolor* [L.] Moench), maize and sugarcane (*Saccharum* sp. L.) are also cultivated, but to a lesser extent. Since 1972, the use of no-till operation has increased, accounting for almost 9.5 Mha of the cropped land in the states of Paraná, Santa Catarina, and Rio Grande do Sul.

SOILS AND SOIL DEGRADATION

The most frequently cultivated soils are Mollisols with prevailing udic and thermic water and temperature regimes (mostly Chernozems, Kastano-

TABLE 16.1. Land use and land-use change for the Pampas during the twentieth century.

Land use	1880 (Mha)	1939 (Mha)	1996 (Mha)
Natural ecosystems	52.3	29.1	13.7
Plantations (tree crops)	0	0	0
Forest fallow	0	0	0
Shifting cultivation	0	0	0
Pastures	27	4.8	7.6
Cropland	147	12.6	20.4
Mine land	0	0	0
Urban land	6.1	12.0	16.8
Others	0	0	0
Total area	58,525	58,535	58,525

Source: Adapted from Tim, 2003.

zems, and Phaeozems after the FAO classification system), and are susceptible to erosion by water and wind. Erosion by water predominates in the rolling and the Mesopotamian Pampas, and in the southern states of Brazil where intensive rainfalls, silty topsoil textures, and long slopes predominate (Hall et al., 1992). Soil inorganic carbon contents are not relevant in the A horizons of the soils of this region. Wind erosion due to the combination of tillage operations and strong winds during the dry season (winter) is frequently observed in the western part of the inland and the southern Pampas (Buschiazzo et al., 1999).

Nutrient depletion, under both agricultural and livestock production systems and soil structural decline, are extended soil degradation problems in the Argentine Pampas (Table 16.2). The amount of nutrients harvested in grains and other agricultural products is lower than that applied by fertilizers. On average, only the equivalent of 40 percent of the P in the grains of the major crops of the Pampas is applied yearly with the fertilizers. Thus, the area with extractable P deficiency increased from less than 50 percent to almost 80 percent of the Pampas between the late 1970s and the end of the 1990s (Montoya et al., 1999; García, 2001). However, due to the use of better crop production practices (i.e., genotypes, pesticides, herbicides, etc.) grain yields and livestock production per unit area are increasing annually (Argentine Secretary of Agriculture, Livestock, Fisheries and Food, 2004).

In the central and the eastern parts of the region, flooding events are very common, especially since the mid-1970s, reducing the area under production. Soil acidification and heavy metal toxicities are not widespread in the Pampas and affect less than 3 Mha. High Ca uptake rates by alfalfa pastures under intensive livestock and dairy production systems in light-textured soils justify most of the sites with liming requirements in the Pampas. Soils

TABLE 16.2. Extent of soil degradation in the Pampas.

Process	Land affected area (Mha)
Erosion by water	11.7[a]
Erosion by wind	8.8[a]
Elemental toxicity	1
Acidification	3
Nutrient depletion	46.0
Soil structural decline	34.0
Others[b]	20.0

[a]Data from Panigatti, 1988.
[b]Includes soil compaction and water flooding.

with toxicities are commonly described within several urban lands and surrounding industrialized areas (Torri and Lavado, 2002).

Soil organic C content depends on soil texture, increasing with the increase in proportion of fine-sized fractions (clay and silt) (Buschiazzo et al., 1991). The SOM of the top layer decreases from almost 40 $g \cdot kg^{-1}$ to less than 15 $g \cdot kg^{-1}$ from north-east to south-west following a textural trend from fine to coarse textures (Alvarez and Lavado, 1998). Nitrogen and, to a lesser extent, phosphorus reserves are low for normal crop production mostly all along the Pampas region. In localized areas, sulfur and several micronutrients (i.e., boron, chloride, zinc) are deficient for obtaining high-yielding crops.

The Pampas have experienced a significant loss of C since the early 1900s, mainly in areas where grain cropping is the main activity (i.e., rolling Pampas). The depletion in soil organic C contents in the Pampas is mostly attributed to a combination of soil erosion processes and negative C budgets. On average, the introduction of agricultural practices based on intensive tillage operations reduced the SOC contents in the soils from the Pampas by 30 to 50 percent (Alvarez, 2001; Casas, 2003; Sá et al., 2001b). The potential of soil management practices to increase the sequestration of CO_2 is associated mainly with the buildup and maintenance of organic matter in agricultural soils based on the use of no-till and improved fertilization practices. Cropping sequences, including pastures in rotation with annual crops, also contribute to restore soil organic C contents and are widely adopted production systems toward the marginal areas of the Pampas (Hall et al., 1992; Buschiazzo et al., 1998). However, neither high productivity agriculture under no-till practices nor crop sequences with pastures allow the restoration of soil organic C contents to the level of the native ecosystem (Buschiazzo et al., 1991; Hevia et al., 2003).

SOIL/CROP MANAGEMENT

No-till and crop sequences integrating cereals (i.e., maize, wheat) and oilcrops (i.e., soybean, sunflower) are recommended management practices for sustainable land use in most of the Pampas (Plate 16.1). But toward the marginal boundaries of the region (i.e., western part of the inland Pampas, southwest of the southern Pampas in Argentina), the rotation of annual crops and perennial pastures under livestock production are recommended (Plate 16.2). Abundant research and extension reports show that the adoption of continuous no-till cropping practices, also in rotation with pastures, reduce soil erosion, facilitate soil organic C conservation, consolidate soil structure, and reduce the risk of soil compaction, improving soil productiv-

PLATE 16.1. No-tillage in the Pampas region, Argentina: (a) summer crops planter and (b) soybean crop established on maize residues.

PLATE 16.2. Alfalfa direct grazing operation for livestock production in the Pampas region, Argentina.

ity (Díaz-Zorita et al., 2002). In the Mesopotamic Pampas, terracing and no-till cropping systems are required soil management practices for reducing water erosion processes.

The efficient management of soil organic matter, and related properties, depends not only on the lack of soil disturbance but also on the C input to the soil based on above- and belowground productivity. In most of the Pampas, the use of N and P fertilizers are recommended for agronomical optimum crop production. In the case of soybean and other legumes (i.e., peanuts [*Arachys hipogea* L.], alfalfa), efficient and yearly use of inoculants with selected strains of rizobia is also recommended.

High C sequestration rates contribute not only to mitigating the greenhouse effect but also to maintaining high soil quality. Increasing soil organic C contents in the semiarid and subhumid inland Pampas increases wheat grain yields (Díaz-Zorita et al., 1999; Alvarez et al., 2002).

POTENTIAL OF SOIL CARBON SEQUESTRATION

The effects of tillage, crop sequences, and recommended crop management practices under no-till systems have been studied since the 1970s at different locations within the Pampas region. In general, it is concluded that soil C contents can be increased by limiting the number of cultural operations and increasing the dry matter production of the crops, basically cereals (Sá et al., 2001b). But crop nutrient requirements must be met to achieve optimum crop yields (Buschiazzo et al., 1998).

Several researchers concluded that changes in soil organic C under no-till are closely related to the production of C (Díaz-Zorita and Grove, 2001; Alvarez, 2004; Fontanetto and Keller, 2001; Sá et al., 2001b). However, the total amount of C required for increasing the soil C pool and the rates of soil C sequestration with recommended management practices are not well established.

The use of no-till practices is progressively increasing. However, only 20 percent of crops are planted without tillage (Table 16.3). Thus, we estimated that approximately 22 Mhas of the Argentine Pampas are degraded and have a high potential to sequester C if appropriate crop and soil management practices are adopted. Using the data of 24 reports covering most of the agroecological conditions of the Pampas (Table 16.4), the use of high-productivity no-till practices would increase the soil C pool in the 0 to 20 cm soil layer by approximately 9.7 (± 4.1) $Mg \cdot ha^{-1}$. Thus, the potential of the soils from the Pampas for soil C sequestration based in the adequate

TABLE 16.3. Average of land under no-tillage agriculture in the Pampas.

	Province (%)				
Crop	Santa Fe	Entre Rios	Buenos Aires	La Pampa	Córdoba
Maize	61	62	32	30	63
Wheat	50	55	21	25	63
Soybean	66	65	49	40	78
Sunflower	25	39	15	20	43

Source: Adapted from Alvarez and Mulin, 2004.

TABLE 16.4. Soil organic carbon (SOC) pools, bulk density (BD) values, and SOC sequestration capacity of no-till in relation to tilled soils in long-term field experiments in the Pampas.

	Location (province)[a]	SOC ($g \cdot kg^{-1}$)		BD ($Mg \cdot m^{-3}$)		SOC ($Mg \cdot h^{-1}$)[b]			Cseq rate ($Mg \cdot ha^{-1}$ per year)[c]	Source
		Tilled	No-tilled	Tilled	No-tilled	Tilled	No-tilled	Difference		
1	Marcos Juárez (C)	18.1	18.6	1.31	1.33	47.42	49.48	2.05	0.4	Andriulo and Cordone (1998)
2	Dorila (LP)	8.5	10.8	1.26	1.31	21.42	28.30	6.88	1.4	Quiroga et al. (1998)
3	Manfredi (C)	10.5	16.5	1.27	1.16	26.67	38.28	11.61	2.3	Dardanelli (1998)
4	Tres Arroyos (BA)	51.5	52.0	1.15	1.22	118.45	126.88	8.43	1.7	Bergh (1998)
5	Rafaela (SF)	16.5	19.1	1.16	1.25	38.28	47.75	9.47	1.9	Fontanetto and Vivas (1998)
6	Rafaela (SF)	15.4	18.4	1.16	1.27	35.73	46.74	11.01	2.2	Fontanetto and Vivas (1998)
7	Rafaela (SF)	21.8	27.6	1.31	1.24	57.12	68.45	11.33	2.3	Fontanetto and Keller (1998)
8	Anguil (LP)	11.3	15.4	1.2	1.25	27.12	38.50	11.38	2.3	Sagardoy et al. (2001)
9	Daireaux (BA)	12.3	16.8	1.33	1.32	32.72	44.35	11.63	2.3	Díaz-Zorita and Grove (2001)
10	Rosario (SF)	14.2	14.9	1.4	1.45	39.76	43.21	3.45	0.7	Morras et al. (2001)
11	Rivadavia (BA)	8.2	10.5	1.20	1.25	19.68	26.25	6.57	1.3	Díaz-Zorita et al. (2002)
12	Pergamino (BA)	19.1	22.2	1.24	1.26	47.18	55.72	8.55	1.7	Alvarez and Alvarez (2000)
13	Parana (ER)	18.4	24.4	1.24	1.26	45.45	61.24	15.80	3.2	Benintende et al. (1995)
14	Parana (ER)	17.5	21.3	1.24	1.26	43.23	53.46	10.24	2.0	Benintende et al. (1995)
15	Rafaela (SF)	21.8	27.0	1.24	1.26	53.85	67.77	13.92	2.8	Fontanetto and Keller (2001)
16	Gral.Villegas (BA)	8.8	10.2	1.31	1.38	23.06	28.15	5.10	1.0	Díaz-Zorita (1999)
17	Balcarce (BA)	37.0	45.8	1.30	1.17	96.20	107.17	10.97	2.2	Cabria and Culot (2001)
18	Balcarce (BA)	29.0	31.4	1.44	1.46	83.52	91.69	8.17	1.6	Crespo et al. (2001)
19	Marcos Juárez (C)	14.0	19.6	1.31	1.33	36.68	52.14	15.46	3.1	Moron (2004)
20	Balcarce (BA)	23.5	28.9	1.44	1.46	67.68	84.39	16.71	3.3	Fabrizzi et al. (2003)
21	San Jorge (SF)	13.8	15.2	1.40	1.50	38.64	45.60	6.96	1.4	Pidello et al. (1995)
22	Chabas (SF)	13.6	14.8	1.40	1.50	38.08	44.40	6.32	1.3	Pidello et al. (1995)
23	Bordenave (BA)	5.7	7.5	1.38	1.34	15.73	20.10	4.37	0.9	Kruger (1996)
24	Manfredi (C)	10.4	16.2	1.25	1.3	26.00	42.12	16.12	3.2	Buschiazzo et al. (1998)
Mean								9.69	1.9	

[a]C = Córdoba, LP = La Pampa, BA = Buenos Aires, SF = Santa Fe, and ER = Entre Ríos provinces.
[b]Estimated from the 0 to 20 cm soil layer.
[c]Five years was the duration considered in all experiments in order to calculate SOC sequestration rate.

use of high-production agricultural and livestock systems on degraded soils is approximately 213 × 10^6 Mg. This represents an average 25 percent increase in the mean C contents of degraded soils. Differences among sites are mainly attributed to differences in crop productivity.

The time required for the soil organic C to reach equilibrium depends on the annual C input determined by the mean temperature and soil water regimes (Alvarez et al., 1998). Consequently, the levels of annual C additions needed to maintain the equilibrium increase with increases in the average annual temperature (Andriulo et al., 1999). In the northern part of the Pampas, about 4.0 Mg·ha^{-1} of C are required for maintaining the soil organic C contents, while approximately 2.8 Mg·ha^{-1} are required in the southern part of the region. Andriulo and Cordone (1998) concluded that in the rolling Pampas an annual mean production of 4.0 Mg C·ha^{-1} produced a 10 Mg C·ha^{-1} increment, while 3.0 Mg C·ha^{-1} per year resulted in no increase in soil organic carbon content. In the southwest part of Buenos Aires province, Rosell et al. (1994) estimated that the net yearly mineralization rate was 1.14 Mg C·ha^{-1} and mean annual production of 5.0 Mg of crop residues per ha is required for maintaining soil organic carbon content (Andriulo et al., 1991).

The mean accumulation rate of C for Argiudolls under no-till practices in the southern part of the Pampas has been estimated at 0.9 Mg·ha^{-1} per year (Crespo et al., 2001). In the southern part of Santa Fe province toward the central part of the Pampas, Casas (2003) calculated that the C sequestration rate varies depending on the duration of the no-till production cycle following cultivation and the soil type. In Hapludolls, the C sequestration rate after six years of no-till practices may be higher than that in Argiudolls, 0.5 Mg·ha^{-1} per year and 0.4 Mg C·ha^{-1} per year, respectively. During the first seasons of no-till practices after tillage cultivation, the rate of C sequestration was estimated to vary between 0.8 to 1.6 and 0.7 to 1.2 Mg C·ha^{-1} per year for Hapludolls and Argiudolls, respectively (Casas, 2003). In the rolling Pampas, the equilibrium soil organic C contents is reached after a period of about five years (Andriulo et al., 1999). On average, for the first six years of no-till practices in soils from the Pampas, Steinbach and Alvarez (unpublished) estimated a C sequestration rate of approximately 0.5 Mg·ha^{-1} per year, with a greater relative increment in regions with lower C contents, basically within the semiarid region.

The adoption of no-till practices in Southern Brazil can increase soil C storage at a rate of 0.8 Mg C per year for the 0 to 20 cm soil layer with a potential of C sequestration of 9.37 to 12.54 Tg per year for the region (Sá et al., 2001a; Bayer et al., 2000). A new equilibrium or steady-state level is predicted to occur approximately 40 years after the adoption of no-till with high inputs of crop residues (Sá et al., 2001b).

POLICY CONSIDERATIONS

Increasing the adoption of restorative land use and recommended crop and soil management practices requires linking C sequestration with reducing the risk of soil losses due to water and wind erosion. Several provinces promoted the use of soil conservation practices through tax incentives, but such programs have been discontinued.

Widespread adoption of high-productivity cereals (maize, sorghum, wheat, etc.) in rotation with other crops is required for ascertaining adequate soil residue cover for diminishing erosive process and as a C source. In this case, policies are needed that promote the production and use of such crops to compensate for low returns compared with other crops.

No-till production practices not only reduce soil C losses and improve soil organic C storage and enhance productivity, but also reduce the use of fuel. Many policy strategies are required for promoting the use of this production system in all crops as well as the use of highly productive perennial pastures in soils of low agricultural productivity.

Finally, the amount of C that can be produced by crops is in general limited by availability of nutrients. Presently, the overall P balance in the Pampas is negative, and the high cost of fertilizers limits its use in the region. Policies that enhance adequate availability of nutrients must be implemented.

CONCLUSIONS

The Pampas region has the potential of being one of the major soil C sinks of the world because most of the area is degraded due to intensive tillage cultivation practices with significant C losses during the twentieth century. Predominant climate and agricultural practices with most of the area turning into no-till agricultural systems provide adequate conditions for reversing this trend while increasing the amount of C stored in the soil.

Observations made on long-term tillage and crop sequences show that the use of high-productivity crops under no-till practices can increase the soil organic C about 10 $Mg \cdot ha^{-1}$ at a rate of 1.9 $Mg \cdot ha^{-1}$ per year during the first five years after the adoption of no-till systems. In southern Brazil, the adoption of no-till practices can increase the C storage at a mean rate of 0.8 Mg C per year with a potential C sequestration of 9.37 to 12.54 Tg per year for the region. These goals can be realized by adoption of recommended management practices, including cropping sequences based on cereals and the intensive use of fertilizers.

REFERENCES

Alvarez, C.R. 2004. Impacto del manejo del suelo sobre el carbono y las propiedades físicas en el norte de Buenos Aires: Ensayos de larga duración y lotes de producción (pp. 39-44). In *Simposio: Fertilidad 2004, Fertilidad de suelos para una agricultura sustentable.* Impofos Cono Sur, Rosario, SF, Argentina.

Alvarez, C. and E. Mulin. 2004. *El Gran Libro de la Siembra Directa.* Clarin, Buenos Aires, Argentina.

Alvarez, R. 2001. Estimation of carbon losses by cultivation from soils of Argentine Pampa using the century model. *Soil Use Manage.* 17: 62-66.

Alvarez, R. and C. R. Alvarez. 2000. Soil organic matter pools and their associations with carbon mineralization kinetics. *Soil Sci. Soc. Am. J.* 64: 184-189.

Alvarez, R., C.R. Alvarez, and H.S. Steinbach. 2002. Association between soil organic matter and wheat yield in humid pampa of Argentina. *Commun. Soil Sci. Plant Anal.* 33:749-757.

Alvarez, R. and R.S. Lavado. 1998. Climate, organic matter and clay content relationships in the Pampa and Chaco soils, Argentina. *Geoderma* 83: 127-141.

Alvarez, R., M.E. Russo, P. Prystupa, J.D. Scheiner, and L. Blotta. 1998. Soil carbon pools under conventional and no-tillage systems in the Argentine rolling Pampa. *Agron. J.* 90:138-143.

Andriulo, A. and G. Cordone. 1998. Impacto de labranzas y rotaciones sobre la materia orgánica de suelos de la región pampeana húmeda (pp. 65-96). In J. L. Panigatti, H. Marelli, D. Buschiazzo, and R. Gil (eds.), *Siembra Directa.* Hemisferio Sur, Buenos Aires, Argentina.

Andriulo, A.E., J.A. Galantini, J.O. Iglesias, R.A. Rosell, and A.E. Glave. 1991. Sistemas de producción con trigo en el sudoeste bonaerense: Materia orgánica y nitrógeno total edáficos (pp. 78-80). In *Actas XIII Congreso Argentino de la Ciencia del Suelo.* Asociación Argentina de la Ciencia del Suelo Bariloche, Río Negro, Argentina.

Andriulo, A., J. Guérif, and B. Mary. 1999. Evolution of soil carbon with various cropping sequences on the rolling Pampas: Determination of carbon origin using variations in natural 13C abundance. *Agronomie* 19:349-364.

Argentine Secretary of Agriculture, Livestock, Fisheries and Food. 2004. Available online at www.sagpya.gov.ar.

Bayer, C., J. Mileniczuk, T.J.C. Amado, L. Martin-Neto, and S.V. Fernández. 2000. Organic matter storage in a sandy clay loam Acrisol affected by tillage and cropping systems in southern Brazil. *Soil and Tillage Research* 54: 101-109.

Benintende, M., O. Borgetto, and S. Benintende. 1995. Mineralización de nitrógeno y contenido de biomasa microbiana en diferentes sistemas de laboreo. *Ciencia del Suelo* 13: 98-100.

Bergh, R. 1998. Evaluación de sistemas de labranzas en el centro-sur bonaerense (pp. 223-236). In J. L. Panigatti, H. Marelli, D. Buschiazzo, and R. Gil (eds.), *Siembra Directa.* Hemisferio Sur, Buenos Aires, Argentina.

Buschiazzo, D.E., S. Aimar, and T. Zobeck. 1999. Wind erosion in soils of the semiarid Argentinian Pampas. *Soil Science* 164: 133-138.

Buschiazzo, D.E., J.L. Panigatti, and P.W. Unger. 1998. Tillage effects on soil properties and crop production in the subhumid Argentinean Pampas. *Soil Till. Res.* 49: 105-116.

Buschiazzo, D.E., A.R. Quiroga, and K. Stahr. 1991. Patterns of organic matter distribution in soils of the semiarid argentinean Pampas. *Z. Pflanzenernähr. Bodenk.* 154: 437-441.

Cabria, F.N. and J.P. Culot. 2001. Efecto de la agricultura continua bajo labranza convencional sobre características físicas y químicas en Udoles del sudeste bonaerense. *Ciencia del Suelo* 19: 1-10.

Casas, R. 2003. El aumento de la materia orgánica en suelos argentinos: el aporte de la siembra directa y la rotación de cultivos (pp. 9-16). In *Rotaciones en SD Dic.:* AAPRESID.

Crespo, I., L.I. Picone, Y.E. Andreoli, and F.O. García. 2001. Poblaciones microbianas y contenido de carbono y nitrógeno del suelo en sistemas de siembra directa y labranza convencional. *Ciencia del Suelo* 19: 30-38.

Dardanelli, J. 1998. Eficiencia en el uso del agua según sistemas de labranzas (pp. 107-115). In J. L. Panigatti, H. Marelli, D. Buschiazzo, and R. Gil (eds.), *Siembra Directa.* Hemisferio Sur, Buenos Aires, Argentina.

Díaz-Zorita, M. 1999. Efectos de seis años de labranzas en un Hapludol del noroeste de Buenos Aires, Argentina. *Ciencia del Suelo* 17: 31-36.

Díaz-Zorita, M., D.E. Buschiazzo, and N. Peinemann. 1999. Soil organic matter and wheat productivity in the semiarid Argentine Pampas. *Agron. J.* 91: 276-279.

Díaz-Zorita, M., G.A. Duarte, and J.H. Grove. 2002. A review of no-till systems and soil management for sustainable crop production in the subhumid and semiarid Pampas of Argentina. *Soil Till. Res.* 65: 1-18.

Díaz-Zorita, M. and J.H. Grove. 2001. Rotaciones de cultivos en siembra directa y las propiedades de suelos de la pampa arenosa (pp. 235-238). In J. L. Panigatti, D. Buschiazzo, and H. Marelli (eds.), *Siembra Directa II.* Ediciones INTA, Buenos Aires, Argentina.

Fabrizzi, K.P., A. Morón, and F.O. García. 2003. Soil carbon and nitrogen organic fractions in degraded vs. non-degraded Mollisols in Argentina. *Soil Sci. Soc. Am. J.* 67: 1831-1841.

Fontanetto, H. and O. Keller 1998. Las siembra directa de forrajeras en el centro de Santa Fe (pp. 287-300). In J. L. Panigatti, H. Marelli, D. Buschiazzo and R. Gil (eds.), *Siembra Directa.* Hemisferio Sur, Buenos Aires, Argentina.

Fontanetto, H. and O. Keller. 2001. Variación en las propiedades químicas de un suelo argiudol bajo siembra directa continua. *Revista Argentina de Producción Animal* 21(Supl. I): 103-104.

Fontantetto, H. and H. Vivas. 1998. Labranzas en el centro de Santa Fe (pp. 275-286). In J. L. Panigatti, H. Marelli, D. Buschiazzo, and R. Gil (eds.), *Siembra Directa.* Hemisferio Sur, Buenos Aires, Argentina.

García, F.O. 2001. Phosphorus balance in the Argentine Pampas. *Better Crops Int.* 15: 22-24.

Hall, A.J., C.M. Rebella, C.M. Ghersa, and J.P. Culot. 1992. Field-crop systems of the Pampas (pp. 413-450). In C.J. Pearson (ed.), *Field crop ecosystems,* Elsevier, Amsterdam.

Hevia, G.G., D.E. Buschiazzo, E.N. Hepper, A.M. Urioste, and E.L. Antón. 2003. Organic matter accumulation in size fractions of soils of the semiarid Argentina: Effects of climate, soil texture and management. *Geoderma* 116: 265-277.

Krüger, H. 1996. Sistemas de labranza y variaciones de propiedades químicas en un Haplustol Entico. *Ciencia del Suelo* 14: 53-55.

Montoya, J., A.A. Bono, A. Suárez, N. Darwich, and F. Babinec. 1999. Cambios en el contenido de fósforo asimilable en suelos del este de la Provincia de La Pampa, Argentina. *Ciencia del Suelo* 17: 45-48.

Morón, A. 2004. Efecto de las rotaciones y el laboreo en la calidad del suelo (pp. 29-36). In *Simposio: Fertilidad 2004, Fertilidad de suelos para una agricultura sustentable.* Impofos Cono Sur. Rosario, SF, Argentina.

Morrás, H., C. Irurtia, C. Ibarlucea, M. Lantin, and R. Michelena. 2001. Recuperación de suelos pampeanos degradados mediante siembra directa y subsolado (pp. 263-278). In J. L. Panigatti, D. Buschiazzo, and H. Marelli (eds.), *Siembra Directa II.* Ediciones INTA, Buenos Aires, Argentina.

Panigatti, J. 1988. Erosión (pp. 47-54). In *El deterioro del ambiente en Argentina.* FECIC, Buenos Aires, Argentina.

Pidello, A., E.B.R. Perotti, G.F. Chapo, and L.T. Menendez. 1995. Materia orgánica, actividad microbiana y potencial redox en dos Argiudoles tipicos bajo labranza convencional y siembra directa. *Ciencia del Suelo* 13: 6-10.

Quiroga, A., O. Ormeño, and H. Otamendi. 1998. La siembra directa y el rendimiento de los cultivos en la region semiárida pampeana central (pp. 237-244). In J. L. Panigatti, H. Marelli, D. Buschiazzo, and R. Gil (eds.), *Siembra Directa.* Hemisferio Sur, Buenos Aires, Argentina.

Rosell, R.A., J.A. Galantini, J. Iglesias, and A.E. Glave. 1994. Comparison of measured and modeled organic matter turnover in a Pampean Haplustoll under two cropping systems (pp. 701-706). In N. Senesi and T.M. Miano (eds.), *Humic substances in the global environment and implications on human health.* Elsevier, Dordrecht, the Netherlands.

Sá, J. C. de M., C. C. Cerri, W. A. Dick, R. Lal, S. P. V. Filho, M. C. Piccolo, and B. E. Feigl. 2001a. Carbon sequestration in a plowed and no-tillage chronosequence in a Brazilian Oxisol (pp. 466-471). In D.E. Stott, R.H. Mohtar, and G.C. Steinhardt (eds.), *Sustaining the global farm.* Selected Papers from the 10th International Soil Conservation Organization Meeting, May 24-29, West Lafayette, IN.

Sá, J. C. de M., C. C. Cerri, W. A. Dick, R. Lal, S. P. V. Filho, M. C. Piccolo, and B. E. Feigl. 2001b. Organic matter dynamics and carbon sequestration rates for a tillage chronosequence in a Brazilian Oxisol. *Soil Science Society of America Journal* 65: 1486-1499.

Sagardoy, M., H.E. Gómez, F.A. Montero, C. Zoratti, and A.R. Quiroga. 2001. Influencia del sistema de siembra directa sobre los microorganismos del suelo (pp. 69-81). In J. L. Panigatti, D. Buschiazzo, and H. Marelli (eds.), *Siembra Directa II.* Ediciones INTA, Buenos Aires, Argentina.

Tim, J. 2003. Variabilidad climática y cambios en el uso de la tierra en la región pampeana argentina. Graduation thesis, Facultad de Ciencias Exactas y Naturales, UNLPam, Santa Rosa La Pampa, Argentina.

Torri, S.I. and R.S. Lavado. 2002. Distribución y disponibilidad de elementos potencialmente tóxicos en suelos representativos de la provincia de Buenos Aires enmendados con biosólidos. *Ciencia del Suelo* 20: 98-109.

Viglizzo, E.F., F. Lértora, A.J. Pordomingo, J.N. Bernardos, Z.E. Roberto, and H. Del Valle. 2001. Ecological lessons and applications from one century of low external-input farming in the pampas of Argentina. 2001. *Agric., Ecosyst. Environ.* 83: 66-81.

Chapter 17

Effects of Environmental and Management Practices on the Potential for Climatic Change Mitigation in the Pampas of Argentina

Juan A. Galantini
Ramón A. Rosell

INTRODUCTION

A continuous uptake and emission of greenhouse gasses (GHGs), mainly carbon dioxide (CO_2), occurs on the soil surface. Change in the atmospheric CO_2 concentration because of the dynamics (sequestration or fixation, stability, and decomposition) of the massive soil organic carbon (SOC) pool and other sources are a cause of the global climatic changes.

The worldwide expansion of agriculture and industry in recent decades has been accompanied by an increase in the oxidation of organic carbon from soils and fossil fuels with a parallel increase of CO_2 concentration in the atmosphere from approximately 315 µmols in 1960 to 370 µmols in 2003 (Mullen et al., 1999; IPCC, 2001). The 1996 report of the International Panel of Climate Change (IPCC) states that there is "clear evidence that human activities have affected concentrations, distributions and life cycles of the GHGs."

The SOC level reflects the long-term balance between additions and losses of organic carbon. Its equilibrium content is a function of the amount and composition of crop residues, plant roots, and other organic materials returned to the soil, and of the rate of SOC decomposition. Gregorich et al. (1996) and Hassink and Whitmore (1997) observed that net rate of SOC ac-

The authors thank all colleagues and institutions that contributed to the preparation of this chapter: Bordenave INTA Experimental Station, Comisión de Investigaciones Científicas (CIC, La Plata), CONICET (PICT 0305/00), Department of Agronomy (UNS), Asociación de Cooperativas Argentinas (ACA–Cabildo), and CONICET.

Carbon Sequestration in Soils of Latin America

doi:10.1300/5755_18

cumulation also depends on the size and capacity of the reservoir. The long-term balance is disrupted by soil cultivation, whereby the carbon in soil organic matter (SOM) was exposed to oxidative processes through tillage practices. Thus, cultivation is in fact a form of "mining" soil nutrients to make them easily available for plant uptake, which also makes them vulnerable to losses into the environment. When the amounts of carbon entering the soil exceeds the loss to the atmosphere, oxidative processes are slowed by management changes which retain more crop residues on or near the soil surface, the SOC pool increases, and vice versa.

Adoption of improved agricultural practices has the potential to increase the amount of carbon sequestration in cropland soils. Biological, physical, and/or chemical protection is also provided by formation and stabilization of soil aggregates. Complex organo-mineral compounds slow microbial degradation and stabilize an SOC pool that may remain in soil for hundreds of years.

Conventional agriculture normally reduces the SOC pool of the soil plow or surface layer. Conservation tillage systems have been studied as an alternative to conventional moldboard plowing to decrease water and wind erosion and to maintain and/or to increase the SOC pool. Conservation tillage systems (e.g., no-till or minimum till) retain crop residue as mulch on the soil surface. Within a cropping system, the equilibrium level of SOC is related linearly to the amount of crop residue returned to the soil (Rasmussen and Collins, 1991; Paustian et al., 1997).

The South American Meadow region (Biome G) is located in the Pampas covering several provinces, including Buenos Aires, Santa Fe, and Entre Ríos, in Argentina, Uruguay, and southern parts of Brazil. Cropping is the predominant land use, with cultivation of soybeans *(Glycine max),* corn (*Zea mays*), wheat *(Triticum aestivum),* barley *(Hordeum vulgare),* and other grain and cereals.

Figure 17.1 presents an estimation of the area, SOC content, and mean SOC density (kg SOC·m^{-2} to 1 m depth) in different soil orders found in the Buenos Aires, Santa Fe, and Entre Ríos provinces of Argentina. Mollisols represent the most abundant soil order, with over 350,000 km^2, while Aridisols, Alfisols, Vertisols, and Entisols represent between 6 and 11 percent of each of the Argentinean Biome G soils. Estimated SOC content in these soils is 5,418 Tg, of which 85 percent is in the Mollisol order.

Agriculture in these regions began in the second part of the nineteenth century and has gradually increased during recent decades. The expansion of agriculture is a consequence of such various actions as introduction of pasture-crops rotation, use of improved tillage systems, increase in crop yield potential, and use of specific pesticides.

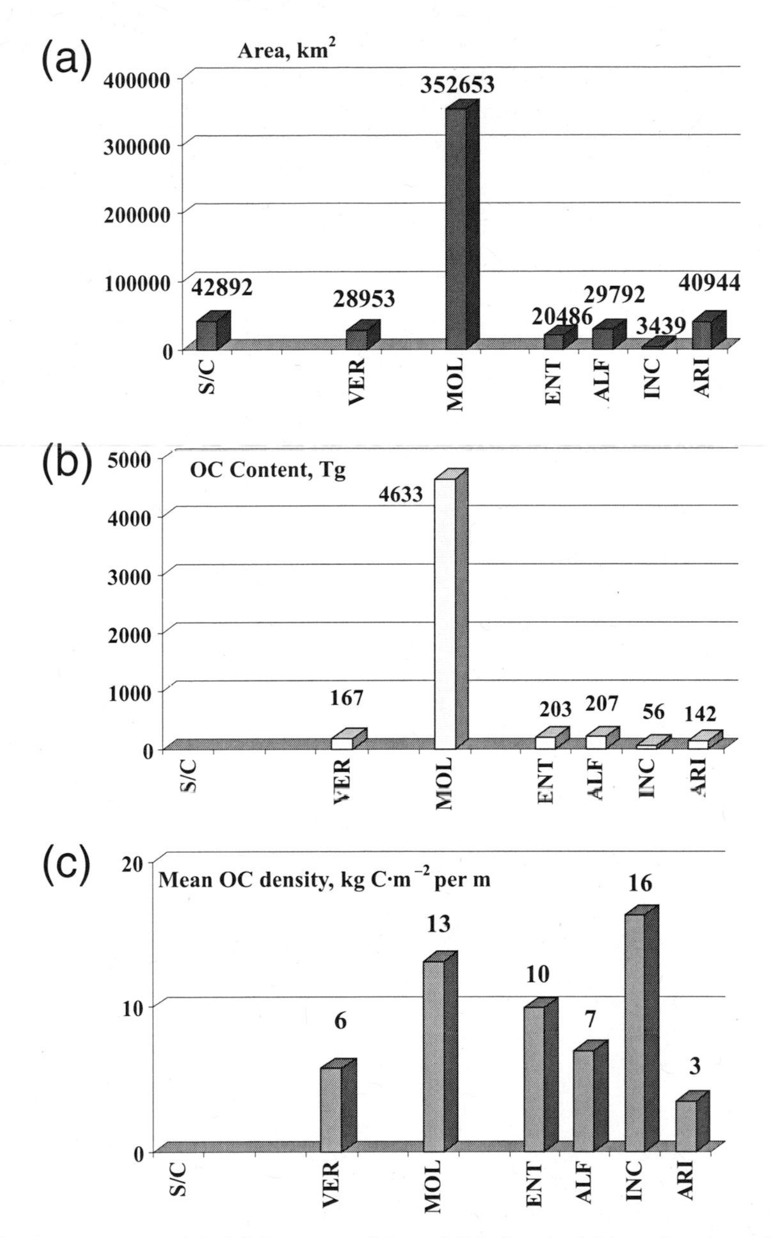

FIGURE 17.1 Area (a), SOC content (b), and OC density (c) in soil orders of Argentine Biome G. (*Source:* Data from SIG Atlas de Suelos de la República Argentina [INTA] and Rosell and Galantini, 1998.)

Expansion of agriculture was accompanied by a loss of SOC and plant nutrients, physical problems (compaction, low stability, etc.), and erosion processes (by water in wet regions and by wind in dry regions).

The negative effects of loss of SOC necessitated the adoption of soil erosion control measures such as conservational or no-till (NT) systems, increase in fertilizer applications, and crop residue inputs to achieve high yields, etc.

Data from the Encuesta Nacional Agropecuaria (ENA) reported that 50 to 60 percent of the area under wheat and corn was seeded under NT in 2000, and the highest percent of NT was observed for soybean.

Figure 17.2 shows temporal changes in area and production during the past two decades. There is a gradual decrease of the area under cereals. In contrast, there was an increase in the area under oil-producing crops (soybean). The reasons for such changes can be explained as follows: animal production was shifted to marginal soil areas; NT was introduced intensively; double-cropping systems (mainly wheat-soybean) were adopted; and fertilizer application was increased.

Knowledge of the changes in land use and cropping systems are important to assessing the SOC dynamics and CO_2 emissions from agricultural practices.

Soils CO_2 sink capacity is related to SOC pool and its dynamics, and depends on several factors, including soil texture, soil depth, temperature:precipitation ratio, etc.

The equilibrium carbon level, based on the input (plant residue) and output (soil organic matter oxidation), is the most important factor to determine strategies for achieving soil sink capacity. Management practices are important to determine soil function as a sink or source of atmospheric CO_2.

ENVIRONMENTAL FACTORS RELATED TO SINK CAPACITY

Soil Texture

The fine components (silt + clay) of soils favor SOC accumulation because of their favorable effects on the protection mechanisms. The SOC adsorption on mineral particles (Oades, 1988), the pore isolation from the soil air (Tisdall and Oades, 1982), and protection from microorganism are very important mechanisms which protect SOC against decomposition (Van Veen and Kuikman, 1990).

Hassink, Bouwman, Zwart, and Brussard (1993) identified two mechanisms related to SOC protection in soils of different textures. In clay soils, the organic matter is protected within the small pores. High clay content fa-

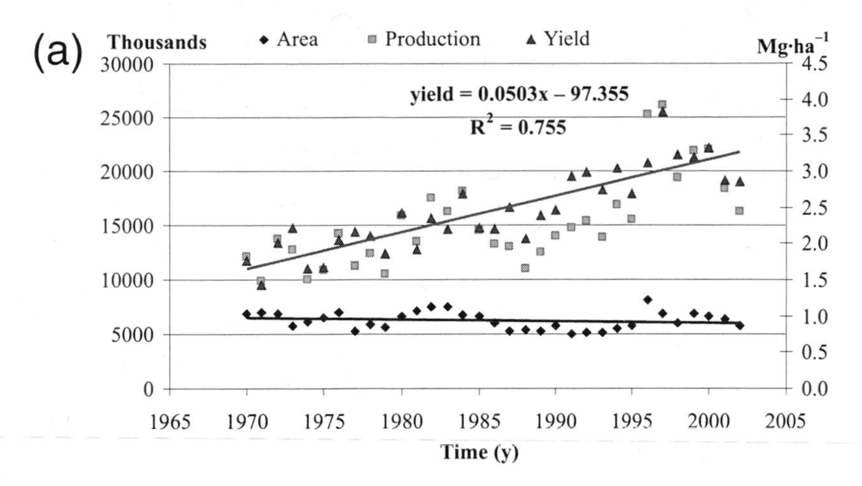

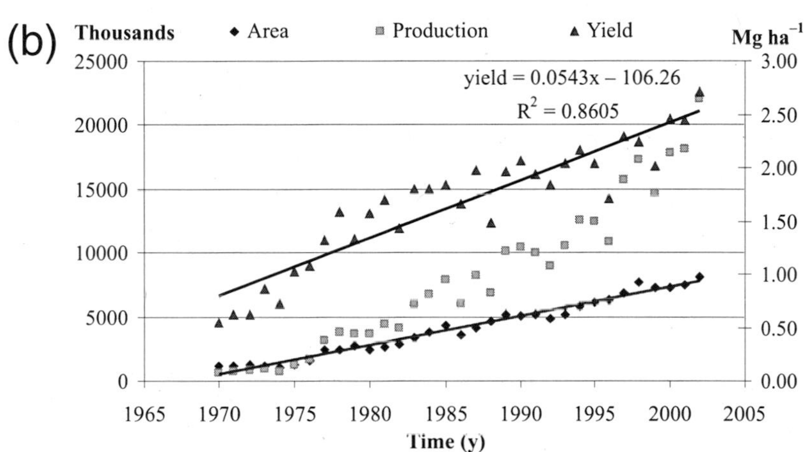

FIGURE 17.2. Cropped area, total production, and mean yield of (a) cereals (corn and wheat) and (b) oilseeds (soybean and sunflower) in Argentine Biome G during 1970-2002. (*Source:* Adapted from data from ENA [Encuesta Nacional Agropecuaria].)

cilitates formation of small pores and SOC protection due to the fact that bacteria movement requires pores larger than three times the diameter of microorganisms (Van Veen and Kuikman, 1990; Hassink, Bouwman, Zwart, and Brussard, 1993). In contrast, the organic matter protection in sandy soils is due to the formation of organo-mineral complex. Soil texture affects water retention, gas diffusion, and microorganisms' habitats (Stott et al., 1986).

There is a high correlation between the quantity of fine fraction and SOC content (Buschiazzo et al., 1991; Galantini, 1994). This trend also depends upon the degree of SOM decomposition (Hassink, Bouwman, Zwart, and Brussard, 1993). For that reason, the SOM fractions associated with the minerals are amorphous with a high degree of transformation. The less transformed SOM in the coarser (sandy) soil is formed with semitransformed residues of high C:N ratios. Accordingly, the humified or transformed SOC is generally associated with the fine mineral fractions (MOC, Figure 17.3a).

Mineral OC is often similar in soils under cultivation or pasture, suggesting that both systems have similar formation of organo-mineral complexes. In contrast, particulate organic carbon (POC) decreases sharply under cultivation (Galantini, 1994). Similarly, the less transformed or POC shows high variability but a similar trend (Figure 17.3b).

The SOC content of profiles (down to 0.75 to 0.90 m) of noncultivated soils for more than 25 years was significantly correlated to the fine fraction (0 to 100 μm) content, with a range of 33 to 117 $Mg \cdot ha^{-1}$ (Figure 17.4). These values showed the plot variability and the importance of the soil texture on the SOC stabilization when the other factors were constant.

Other factors impact the equilibrium level of SOC in arid regions. Water availability is the most important limitation to plant productivity (carbon input) and SOM decomposition (carbon mineralization).

Soil texture plays an important role in the protection of organic material against microbial and biochemical decomposition. The formation of an organo-mineral complex is lower in sandy than clay soils, and is the reason for high correlation between texture and SOM.

Another important effect of the texture is on soil water balance, affecting rainfall dependence of plant growth and SOM decomposition. Galantini and Rosell (2002) found different accumulation patterns of SOM fractions depending on soil texture and depth (Figure 17.5). Sandy soils favor the accumulation of undecomposed or particulate organic matter (POM) mainly in the upper soil layers. This differential accumulation modified the SOM fraction relationships (Figure 17.6). In arid and semiarid environments, the accumulation of undecomposed organic materials are more evident on the soil surface due to the presence of debris that represent an important reserve of carbon. The crop residue inputs and humification conditions are better under humid than dry conditions.

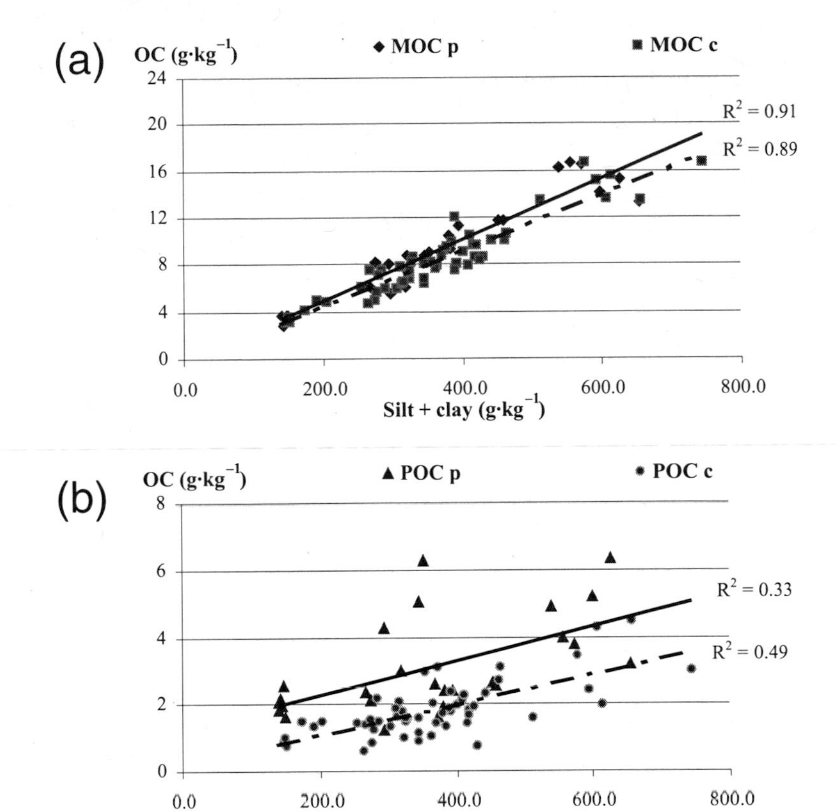

FIGURE 17.3. Organic carbon (OC) in (a) mineral-associated (MOC) and (b) particulate (POC) fractions in soils with different texture under pasture (p) and crops (c). (*Source:* Adapted from Galantini, 1994.)

Soil Depth

The subsoil in the south and southwestern parts of Buenos Aires province has a calcareous layer (caliche) at variable depths between 0.2 and 1.5 m. The epigenic material was deposited during the Quaternary period, having high calcium carbonate content which limits the soil depth. Consequently, soil fertility may be low. Ten soils with variable depth of the caliche layer between 0.3 to 1.0 m were studied during 2001. The soils were distributed in the "Criadero de Semillas" (seed breeding area) of the Asociación

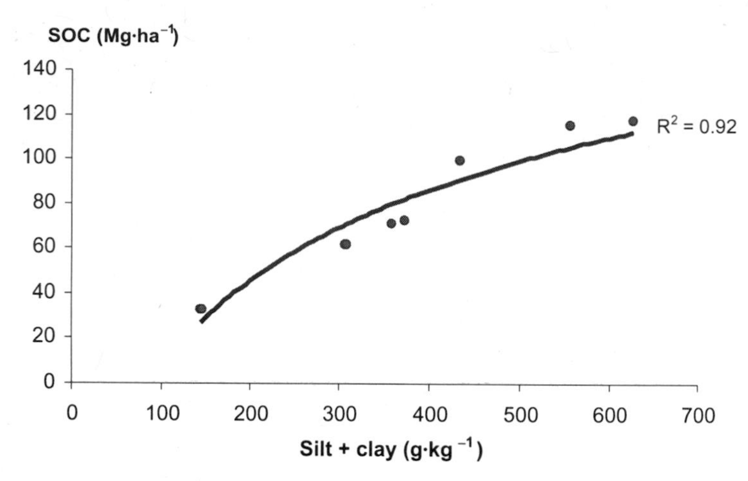

FIGURE 17.4. OC content in Pampean soils (0-1.0 m) with different silt + clay contents.

de Cooperativas Argentinas (ACA) near Cabildo, 30 km northeast of Bahía Blanca in Buenos Aires (BA) province. Predominant soils of this region belong to a complex of Entic and Typical Haplustolls, sandy loamy family, mixed, thermic, with a variable horizon sequence (A-AC-C-Cca).

Figure 17.7 shows the variability found in the distribution of SOC and inorganic carbon (SIC) in these soils. The capacity to sequester carbon was low in shallow soils due to the small amount of carbonaceous (humic substances, root, plant residues, stubble, etc.) materials in the upper horizons because of the low available nutrients and water capacity. The SIC was relatively high when compared with SOC and is an important carbon reserve. For that reason, a 0.3 m deep soil may contain more total carbon than a 1.0 m deep soil. It will be important to validate this hypothesis because it may be related to carbon sequestration and the carbon cycling process.

Temperature and Moisture

The evolution of a soil carbon pool is a function of the climate conditions. High soil temperature and moisture favor plant growth and SOC formation. Knowledge of the equilibrium level of these two parameters is useful to understanding the SOC content of the Pampas and Chaco semiarid regions (Alvarez and Lavado, 1998). Crespo and Rosell (1990) observed a good SOC–precipitation correlation in a moisture sequence study from

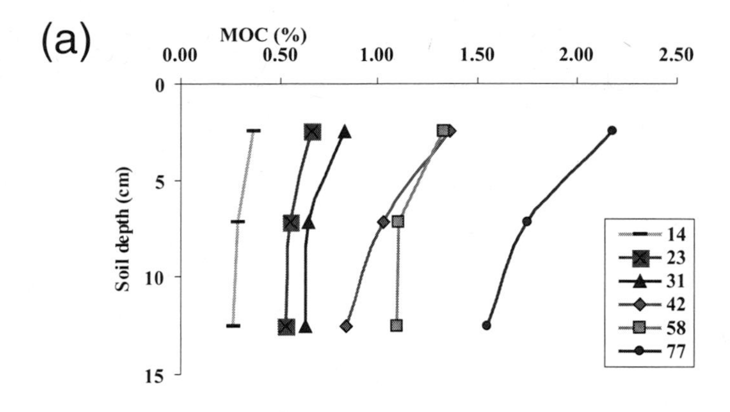

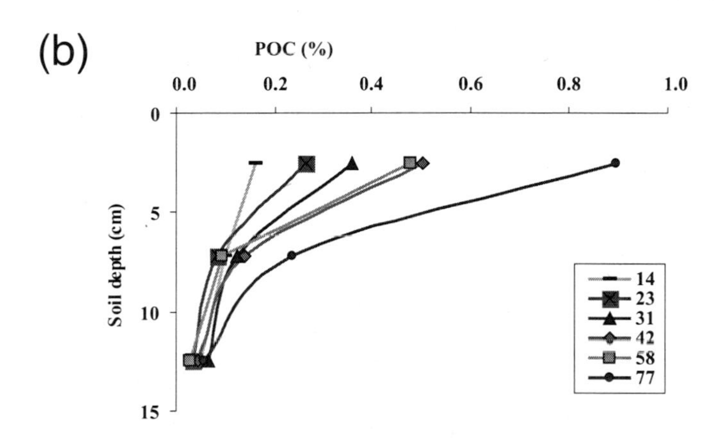

FIGURE 17.5. Mineral-associated (MOC) (a) and particulate organic carbon (POC) (b) distribution in soils with different fine fraction (<0.1 mm) contents.

eastern (Necochea, BA, more than 900 mm yearly precipitation) to western (Cochillo Có, La Pampa, less than 400 mm yearly precipitation) semihumid and semiarid Pampas. Jenkinson and Ayanaba (1977) reported that the decomposition rate of organic materials was higher in Nigeria than in the United Kingdom. The decomposition rate was found to be two times higher in Australia (Oades, 1988) and three to four times higher in tropical soils than in soils of the temperate region.

The relationship between climatic factors (e.g., temperature and moisture regimes) and SOC and clay (particles <2μm) contents has been reported by Jenny et al. (1949).

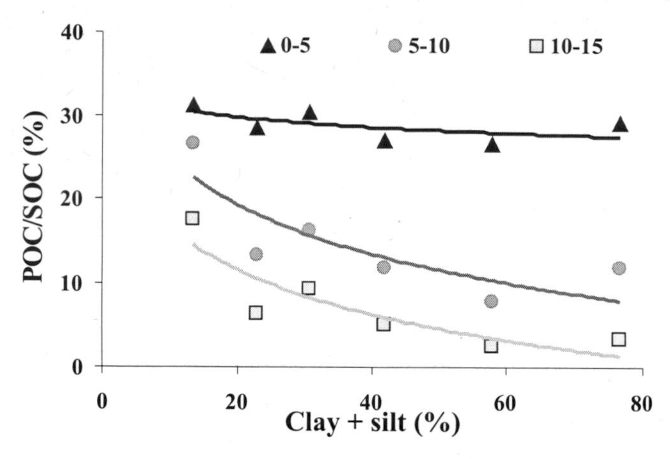

FIGURE 17.6. Particulate/total soil organic carbon ratio at different soil depths in soils with different textures.

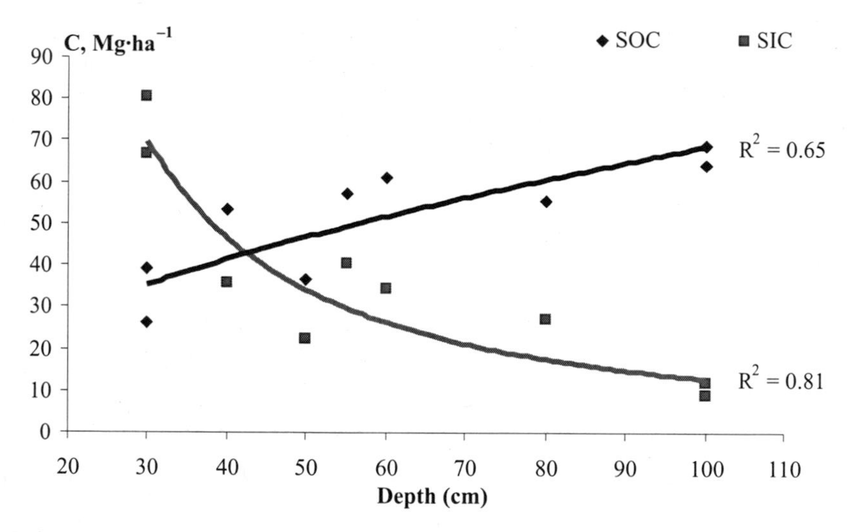

FIGURE 17.7. Soil organic (SOC) and inorganic (SIC) carbon in soils with a calcareous layer at different depths.

In soils of the Pampa and Chaco Plains, Alvarez and Lavado (1998) determined the importance of climatic factors on soil genesis and SOC levels. They reported the highest correlation between SOC (at 0 to 0.50 m depth) with precipitation and temperature within 800 to 1,000 mm and 14 to 16°C ranges, respectively. They observed that a better index was obtained by combining precipitation (P) and temperature (T) into a P/T ratio with regard to the climatic effects on SOC content. The P/T ratio explained 69 percent of the variation in carbon content in the studied soils. The resulting equations were as follows:

$$SOC\ (g{\cdot}kg^{-1}) = 153\ T^{-092} \qquad R^2 = 0.58 \qquad (17.1)$$

$$SOC\ (g{\cdot}kg^{-1}) = 0.00012\ P^{1.7} \qquad R^2 = 0.75 \qquad (17.2)$$

$$SOC\ (g{\cdot}kg^{-1}) = 0.024\ P/T^{1.6} \qquad R^2 = 0.69 \qquad (17.3)$$

Similar trends of SOC with the P/T ratio have been reported for other regions of the world (Theng et al., 1989; Tate, 1992). The SOC pool is a direct consequence of plant debris inputs (Cole et al., 1993) and the SOM decomposition.

MANAGEMENT EFFECTS ON SOC SEQUESTRATION

Increasing problems of environmental quality in arable lands and the long-term productivity of agroecosystems are indicative of the need to develop and improve management strategies that maintain and protect quality of the soil and other natural resources. Depending on site characteristics, changes in management may lead to changes in quantity and quality of SOM (Janssen, 1984; Campbell et al., 1999).

The SOC content in cropland is strongly correlated with crop and soil management practices. These practices include crop species and rotation, tillage methods, fertilizer rate, manure application, pesticide use, irrigation and drainage, and soil and water conservation (Paustian et al., 1997). These practices affect the input of crop residues and addition of organic amendments, and output through decomposition and transportation into aquatic ecosystems by leaching, runoff, and soil erosion.

Direct measurement of short-term losses or gains of SOM resulting from differences in external factors, such as land use, is not possible due to the generally high antecedent value of soil carbon. Therefore, approaches based on the characterization of active SOM components with comparatively rapid turnover rates have been suggested as a useful and more sensitive measure of SOM change (Sparling, 1992).

Improved agricultural practices have great potential to increase the amount of carbon sequestration in cropland soils. There are several options for maintaining and/or increasing the soil carbon level during the cropping phase with conservation or no tillage, crop rotations, and fertilizer application. The effect of these management practices is discussed in the following sections.

Tillage System

One of the most pronounced effects of continuous no-tillage (NT) is the redistribution and stratification of SOC within the soil profile. Several studies have documented that conservation tillage maintains SOC at a high level and improves soil structure. Also, with time, crop yields may increase substantially from what they used to be with more intensive tillage systems (Dick, 1983; Follett and Peterson, 1988; Papendick and Parr, 1997). This accumulation of SOM under NT compared to conventional tillage (CT) enhances soil quality, improves soil fertility, and sequesters carbon.

Soil tillage changes the composition or distribution of compounds along with the change in size distribution of aggregates (Balesdent et al., 2000). No tillage improves several soil parameters (higher SOM and nutrient concentration, lower bulk density, higher water holding capacity, etc.) compared to CT. Deleterious effects of NT can be the soil compaction and increased N_2O emissions.

Stratification of the labile fraction of the SOC is another soil characteristic related to the NT effect (Krüger, 1996). The POC fraction increases sharply in the surface horizons and follows a different depth distribution than CT. Consequently, less mobile nutrients (e.g., phosphorus) increase in the surface layer and availability may be affected by soil moisture conditions. In contrast, CT does not exhibit any stratification because of the mixing of the surface horizon during plowing.

Crop Rotation

Crop rotation is an important technique to improve soil properties such as nutrient availability, pest control, porosity increase, etc. However, socio-economical and technical reasons (cost of inorganic fertilizers and pesticides) have favored adoption of monoculture. Crop rotations can maintain or increase SOC levels, especially when legumes and a longer cropping period are included in the cycle (Galantini et al., 2002; Miglierina et al., 2000).

Casas (2003) calculated the SOM losses as percentage of the initial values in continuously tilled and crop-pasture rotations for 17 soil series of the Pampean region. He observed that continuous tillage produced higher SOM losses than crop-pasture rotation in all soils (Figure 17.8). Crop rotation could reduce the SOM losses in these soils to about 10 percent.

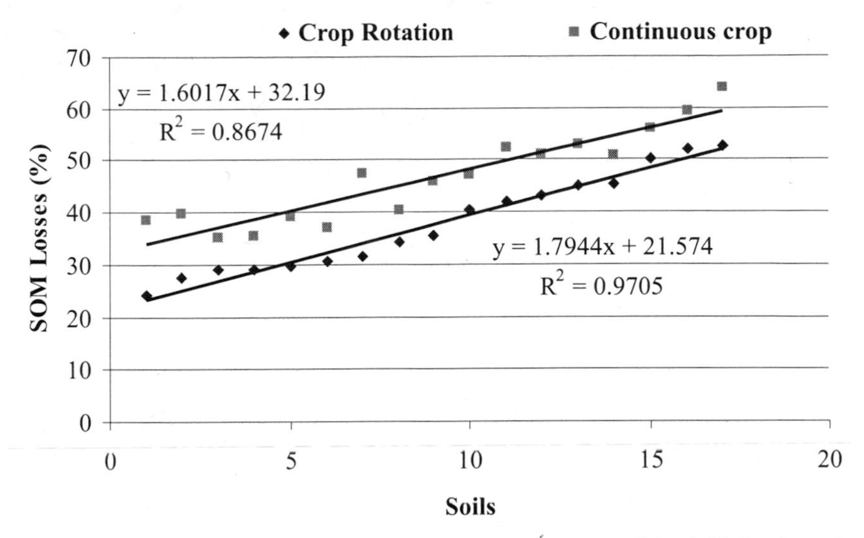

FIGURE 17.8. Soil organic matter losses, in percentage of the initial values, in different Pampean soils with crop rotation and continuous crop. (*Source:* Adapted from Casas, 2003.)

Miglierina et al. (2000) studied the sequences WW (continuous wheat), WG (wheat–grazed natural pastures), and WL (wheat–legumes) for 15 years in an Entic Haplustoll (0 to 0.21 m) from INTA Experimental Station, Bordenave, Argentina. There were differences in the SOC content (Table 17.1). The WW treatment had the lowest and the WL the highest SOC level, and the SOC level in the WG system was intermediate between the other two. Grant et al. (2001) indicated that SOC accumulation produced during crop rotations acted synergistically, increasing the SOC sequestration of all rotations.

In this case, fertilizer application in traditional and continuous wheat systems increased the SOC content between 5 to 9 percent, while incorporation of a legume in the rotation systems increased it to about 14 percent.

Organic fractions in the SOC pool have different sensitivity to production systems. Table 17.2 compares 18 production systems for their respective initial or noncultivated soil with those after nine years in the Bordenave INTA, Argentina (Galantini, 1994). The mean loss of total SOC was about 10 percent under pasture rotations and 30 percent under annual crop rotations. Assessment of different SOC fractions indicated that the humified, mineral-associated SOC was the most resistant to decomposition with a variation of ±15 percent. In contrast, the POC incurred the largest losses in all rotations. These results confirm the importance of POC as a sensitive in-

TABLE 17.1. Organic carbon content in Haplustolls under different production systems.

Depth (m)	Production system					
	WW		WG		WL	
	nf	f	nf	f	nf	f
0 - 0.21	**40.2**	**43.7**	**43.3**	**45.5**	**44.0**	**51.3**
Effect of:						
Fertilizer	3.5 Mg·ha^{-1} (8.7%)		2.2 Mg·ha^{-1} (5.1%)		5.96 Mg·ha^{-1} (13.5%)	
Rotation	–3.1	–1.8	Traditional system		+0.7	+5.8

Source: Adapted from Miglierina et al., 2000.
Note: WW, continuous wheat; WG, wheat–natural grasses; WL, wheat–legume; fertilized (f) and nonfertilized (nf).

TABLE 17.2. Mean OC fraction content (Mg·ha^{-1}) and change in different crop rotations.

Crop rotation	SOC	Δ (%)	MOC	Δ (%)	POC	Δ (%)
Pasture-crops	23.4	–9.5	19.8	14.5	3.7	–51.3
Oat+Vicia-wheat	24.8	–20.5	20.7	0.0	3.9	–60.8
Wheat-crops	19.5	–30.4	16.3	–10.8	3.2	–66.1
Corn	32.5	–13.0	26.5	0.5	6.0	–44.5
Nonfertilized	23.6	–22.4	20.0	–2.9	3.5	–63.0
Fertilized	22.4	–20.8	18.4	–0.1	4.0	–55.4

Source: Adapted from Galantini, 1994.
Note: SOC, soil organic carbon; MOC, mineral-associated OC; POC, particulate organic carbon; 18 production systems were analyzed and grouped. The comparisons were made between the production system and its reference soil, and calculated as percentage of change (Δ%).

dicator of the susceptibility to degradation and the mineral nutrient potential availability. All rotations produced a significant loss in the POC fraction. The mean loss of POC due to cultivation was 59 percent, representing 12 percent of the total SOC pool.

Galantini and Rosell (1997) studied the effect of wheat-sunflower (WS) and pasture-crops (PaC) rotations on SOM quantity and quality (Table 17.3). They reported that humic acid content did not change with cropping, indicating the stability of these humic substances. In contrast, the humified organic carbon (HOC), mainly fulvic acid and humine, decreased drastically in the WS crop rotation compared with the reference control. The decline was mainly due to the low residue production in the WS treatment.

TABLE 17.3. Distribution and change of the soil organic carbon fractions ($Mg{\cdot}ha^{-1}$) in different soils.

Crop rotation	Humified (HOC) C-HA	C-FA	Humine	Total	Particulate (POC)	Total (SOC)
Ref	5.42a	3.45a	8.03a	16.89a	7.11a	24.00a
PaC	6.07a	3.96a	7.30a	17.33a	3.50b	20.83b
WS	5.23a	1.99b	3.93b	11.16b	2.18c	13.34c

Source: Adapted from Galantini and Rosell, 1997.
Note: Ref, natural grass or reference soil; PaC, 5 years pasture–5 years cereals crops; WS, wheat–sunflower rotation; C-HA and C-FA, carbon in humic and fulvic acid fractions, respectively. In each column different letters indicate significant differences.

Both POC and SOC contents followed a similar decreasing trend in reference control, PaC, and WS treatments. Losses in WS treatments exceeded 40 percent compared with the reference control. It is likely that the coarse fraction organic carbon constitutes one of the most labile fractions in soils. In all cases the most drastic SOC losses in different components were observed in the WS continuous cropping treatment. Pastures must be introduced in the rainfed cropping cycle under semiarid conditions to minimize these losses.

Krüger et al. (2004) studied the SOC distribution in different soil horizons of undisturbed and cultivated Hapludols. They reported the highest level of SOC in undisturbed soil and the lowest in the continuous cropping and during the first pasture-crop rotation cycle (Rot 1) (Figure 17.9). Each rotation cycle (Rot 1, Rot 2, and Rot 3) consisted of five-year grasses; five-year crops (wheat, sunflower, etc.) increased SOC in all horizons studied. In this soil, the effect of mixed pastures on SOC content can be observed up to 35 to 40 cm (B3 horizon) depth, indicative of an important strategy of CO_2 sequestration in the subsoil horizons.

The grass-crop rotation system sequesters SOM, mainly the labile fraction, during pasture cycles, and it is oxidized during cropping cycles. The dynamics of SOC fractions in a Typic Argiudol of Balcarce (BA) are shown in Figure 17.10. Casanovas et al. (1995) concluded that labile SOC fractions were early indicators of crop rotation effects on soil biochemical properties.

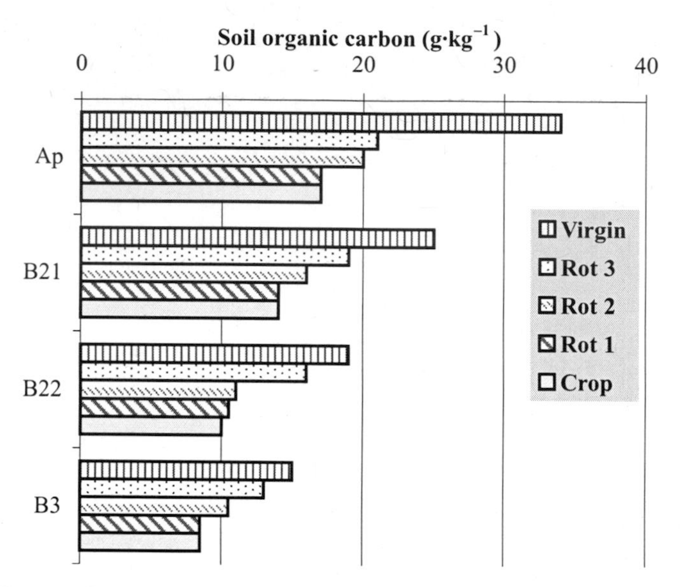

FIGURE 17.9. Organic carbon at different horizons in 0, 1, 2, and 3 ten-year grass-crops rotation cycles and virgin soils. (*Source:* Adapted from Krüger et al., 2004.)

Fertilizers and Organic Manure

Fertilizer applications affect CO_2 emissions during the production, formulation, transport, and use of products. In general, each mole of N applied releases 1.4 moles of CO_2-C (Schlesinger, 2000). For that reason, the application of high amounts of N may result in negative sequestration or CO_2-C emissions.

In order to formulate a fertilizer strategy, 25 experiments with winter cereals and N applications were conducted in the semiarid Pampean region (unpublished data). Fertilizer application increased dry matter production and enhanced SOC sequestration. The net SOC sequestration was zero (CO_2-C production equal to CO_2-C sequestration) at about 70 kg N·ha^{-1} fertilizer use, leading to negative sequestration at higher fertilization rates (Figure 17.11). Use of organic manures also increased CO_2-C sequestration, but at a lower level than use of chemical fertilizers. Large amounts of manure are needed to maintain a productive nutritional balance (Schlesinger, 2000). The risk of a negative CO_2-C sequestration balance must be taken into account while formulating recommendations for fertilizer and manure application.

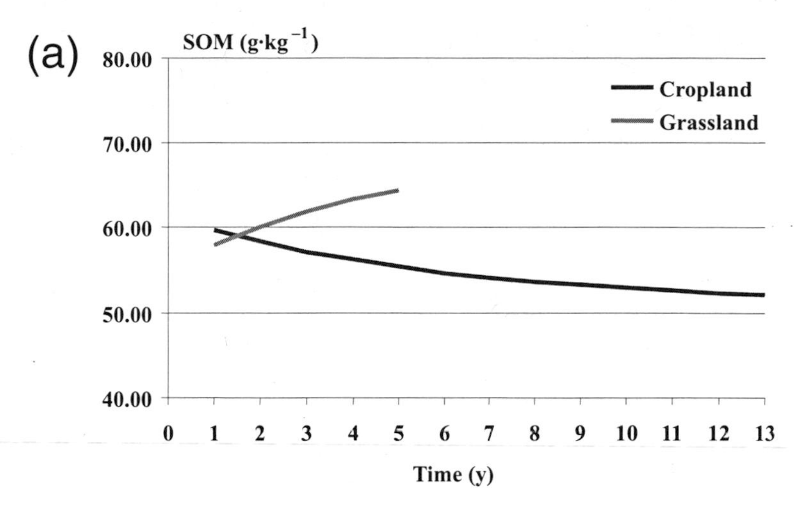

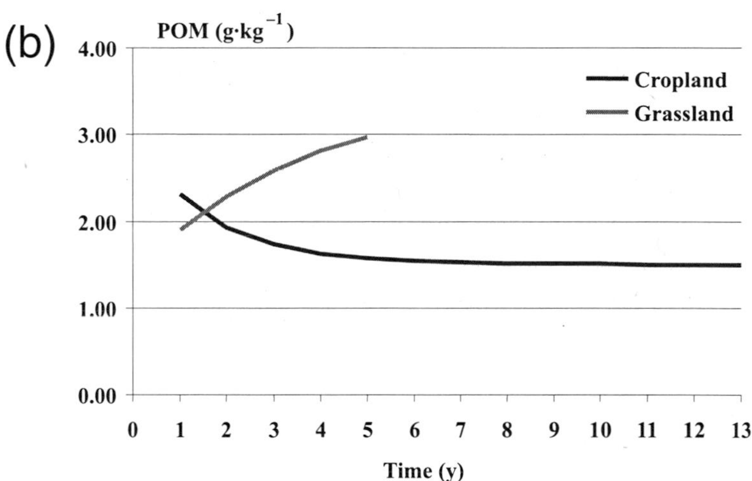

FIGURE 17.10. Total and particulate SOM dynamics during grassland and cropland periods. (*Source:* Adapted from Casanovas et al., 1995.)

These results showed that SOC can be increased by 10 to 20 percent with crop rotation inclusions, by 5 to 10 percent with adequate fertilizer application, and by 10 to 20 percent using no-till systems. Combining these practices, SOC can be increased about 20 to 30 percent. The increase occurs in the upper layer, which contain more than 50 percent of the SOC present in the 0 to 1 m depth. Improving 10 percent SOC content in the upper layer by management practices will produced an increase of around 5 percent of

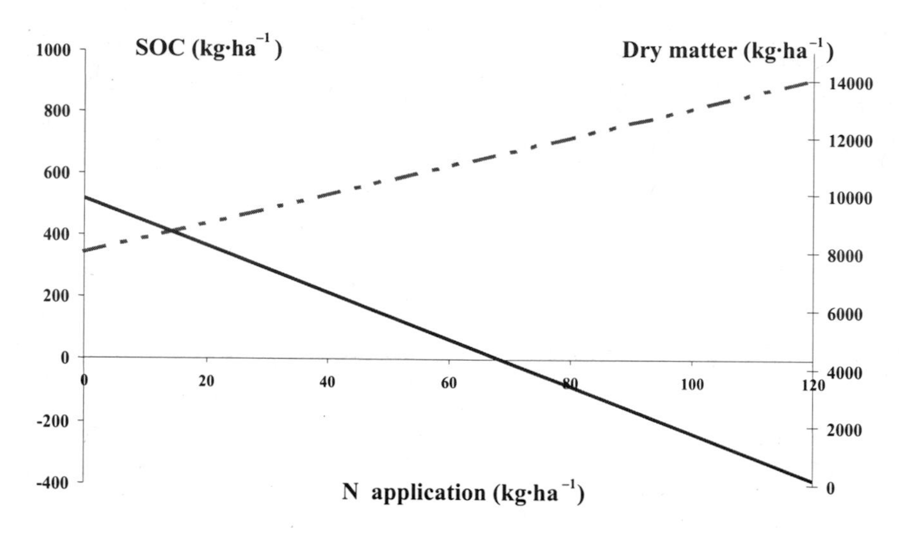

FIGURE 17.11. Straw production (broken line) and the net fixation of OC to the soil (solid line) in 25 experiments with N application in the Pampean region.

the SOC present in the 0 to 1 m layer. Taking into consideration the distribution of soil orders and their SOC density (near 52 Mha with 5,400 Tg of SOC), the increase of 10 percent in the upper layer will sequester about 270 Tg of SOC in the whole of Argentine Biome G.

CONCLUSIONS

The Biome G of Argentina covers 529,000 km^2 and contains about 5400 Tg of SOC, mostly in the Mollisol soil order.

The main land use is croplands, especially annual crops, whose area has increased steadily due to economic reasons.

The most feasible strategy to enhance SOC balance is through adoption of more efficient production systems, increasing inputs of organic materials, and decreasing decomposition rate of SOM. Both environment and management practices have a strong influence on the SOC pool.

Adoption of no-till can increase SOC by 12 percent in the Pampas soils through increase in the POC fraction. Incorporation of legumes in the crop rotation and moderate use of fertilizer can also improve the SOC levels in these soils. The combination of these recommended management practices have synergistic effects on the SOC increase.

The SOC sequestration capacity is also related to the inherent properties of these soils and the interaction of the organic and inorganic carbon.

Carbon sequestration and the net carbon balance are influenced by soil properties (e.g., clay content and nutrient release by organic matter), climate, management, crop rotation, nondestructive soil tillage, etc. All these properties must be considered in evaluating soil carbon sequestration potential.

BIBLIOGRAPHY

Alvarez, R., and R.S. Lavado. 1998. Climate, organic matter and clay content relationships in the Pampa and Chaco soils, Argentina. *Geoderma* 83:127-141.

Andreux, F. 1996. Humus in world soils. In *Humic substances in terrestrial ecosys tems* (Ed. A. Piccolo). Elsevier, Amsterdam, pp. 45-100.

Balesdent, J., C. Chenu, and M. Balabane. 2000. Relationship of soil organic matter dynamics to physical protection and tillage. *Soil Till. Res.* 53:215-230.

Buschiazzo, D.E., A.R. Quiroga, and K. Stahr. 1991. Patterns of organic matter accumulation in soils of the semiarid Argentinian Pampas. *Z. Pflanzenem. Bodenk.* 154:347-441.

Campbell, C.A., V.O. Biederbeck, B.G. McConkey, D. Curtin, and R.P. Zenter. 1999. Soil quality-effect of tillage and fallow frequency: Soil organic matter quality as influenced by tillage and fallow frequency in a silt loam in southwestern Saskatchewan. *Soil Biol. Biochem.* 31:1-7.

Casanovas, E.M., G.A. Studdert, and H.E. Echeverría. 1995. Materia orgánica del suelo bajo rotaciones de cultivos: II. Efecto de los ciclos de agricultura y pastura. *Ciencia del Suelo* 13:21-27.

Casas, R. 2003. El aumento de la material orgánica en suelos argentinos: El aporte de la siembra directa. *XI National Congress of AAPRESID (Argentina)* I:155-168.

Cole, C.V., K. Paustian, E.T. Elliott, A.K. Metherell, D.S. Ojima, and W.J. Parton. 1993. Analysis of agroecosystem carbon pools. *Water Air Soil Poll.* 70:357-371.

Crespo, M.B., and R.A. Rosell. 1990. Change of properties of humic substances in an edaphic climosequence. *Agrochimica* 34:193-200.

Dick, W.A. 1983. Organic carbon, nitrogen, and phosphorus concentrations and pH in soil profiles as affected by tillage intensity. *Soil Sci. Soc. Am. J.* 47:102-107.

Follett, R.F., and G.A. Peterson. 1988. Surface soil nutrient distribution as affected by wheat–fallow tillage systems. *Soil Sci. Soc. Am. J.* 52:141-147.

Galantini, J.A. 1994. Modelos de simulación de la dinámica de la materia orgánica en suelos de la región semiárida bonaerense. Master's thesis, Universidad Nacional del Sur, Bahía Blanca, Argentina.

Galantini, J.A., and R.A. Rosell. 1997. Organic fractions, N, P, and S changes in a semiarid Haplustoll of Argentine under different crop sequences. *Soil Till. Res.* 42:221-228.

Galantini, J.A., and R.A. Rosell. 2002. Secuestración de carbono en suelos de la región semiárida bonaerense. XVII Congreso Argentino de la Ciencia del Suelo, Puerto Madryn.

Galantini, J.A., R.A. Rosell, A.E. Andriulo, A.M. Miglierina, and J.O. Iglesias. 1992. Humification and N mineralization of crop residues in semi-arid Argentina. *Science Total Environ.* 117/118:263-270.

Galantini, J.A., R.A. Rosell, G. Brunetti, and N. Senesi. 2002. Dinámica y calidad de las fracciones orgánicas de un Haplustol durante la rotación trigo-leguminosas. *Ciencia del Suelo* 20:17-26.

Gee, G.W., and J.W. Bauder. 1986. Particle-size analysis. In *Methods of soil analysis,* Part I (Ed. A. Klute). Madison, WI.

Grant, R.F., N.G. Juma, J.A. Robertson, R.C. Izaurralde, and W.B. McGill. 2001. Long-term changes in soil carbon under different fertilizer, manure, and rotation. *Soil Sci. Soc. Am. J.* 65:205-214.

Gregorich, E.G., B.H. Ellert, C.F. Drury, and B.C. Liang. 1996. Fertilization effects on soil organic matter turnover and corn residue C storage. *Soil Sci. Soc. Am. J.* 60:472-476.

Hassink, J. 1995. Prediction of the non-fertilizer N supply of mineral grassland soils. *Plant Soil* 176:71-79.

Hassink, J., L.A. Bouwman, K.B. Zwart, J. Bloem, and L. Brussaard. 1993. Relationships between soil texture, physical protection of organic matter, soil biota, and C and N mineralization in grassland soils. *Geoderma* 57:105-128.

Hassink, J., L.A. Bouwman, K.B. Zwart, and L. Brussaard. 1993. Relationships between habitable pore space, soil biota and mineralization rates in grassland soils. *Soil Biol. Biochem.* 25:47-55.

Hassink, J., and A.P. Whitmore. 1997. A model of the physical protection of organic matter in soils. *Soil Sci. Soc. Am. J.* 61:131-139.

IPCC. 2001. *Climate change: The scientific basis.* Intergovernment Panel on Climate Change. Cambridge University Press, Cambridge, UK.

Janssen, B.H. 1984. A simple method for calculating decomposition and accumulation of "young" soil organic matter. *Plant Soil* 76:297-304.

Jenkinson, D.S., and A. Ayanaba. 1977. Decomposition of 14C labeled plant material under tropical conditions. *Soil Sci. Soc. Am. J.* 41:912-915.

Jenny, H. 1941. *Factors of soil formation: A system of quantitative pedology.* McGraw-Hill, New York.

Jenny, H., S.P. Gessel, and F.T. Bingham. 1949. Comparative study of decomposition rates of organic matter in temperate and tropical regions. *Soil Sci.* 68:419-432.

Krüger, H.R. 1996. Sistemas de labranzas y variación de propiedades químicas en un Haplustol entico. *Ciencia del Suelo* 14:53-55.

Krüger, H., S. Venanzi, and J.A. Galantini. 2004. Rotación y cambios en propiedades químicas de un Hapludol Tipico del sudoeste bonaerense bajo labranza. XIX Congreso Argentino de la Ciencia del Suelo, Paraná.

Mann, L. 1986. Changes in soil organic carbon storage after. *Soil Sci.* 142:279-288.

Miglierina, A., J. Galantini, J. Iglesias, R. Rosell, and A. Glave. 1995. Rotación y fertilización en sistemas de producción de la región semiárida Argentina: II. Cambios en algunas propiedades químicas del suelo. *Revista Facultad Agronomía* (U.B.A.) 15:9-14.

Miglierina, A., J. Iglesias, M. Landriscini, J. Galantini, and R. Rosell. 2000. The effects of crop rotations and fertilization on wheat productivity in the pampean semiarid region of Argentina: 1. Soil physical and chemical properties. *Soil Till. Res.* 53:129-135.

Mullen, R.W., W.E. Thomason, and W.R. Raun. 1999. Estimated increase in atmospheric carbon dioxide due to worldwide decrease in soil organic matter. *Commun. Soil Sci. Plant Anal.* 30:1713-1719.

Oades, J.M. 1988. The retention of organic matter in soils. *Biogeochemistry* 5:35-70.

Papendick, R.I., and J.F. Parr. 1997. No-till farming: The way of the future for a sustainable dryland agriculture. *Ann. Arid Zone* 36:193-208.

Parton, W., D. Schimel, C. Cole, and D. Ojima. 1987. Analysis of factors controlling soil organic matter levels in Great Plains grasslands. *Soil Sci. Soc. Am. J.* 51:1173-1179.

Paustian, K., H.P. Collins, and E.A. Paul. 1997. Management controls on soil carbon. In *Soil organic matter in temperate agroecosystems: Long-term experiments in North America* (Eds. E. Paul, K. Paustian, E. Elliott, and C. Cole). CRC Press, Boca Raton, FL, pp. 343-351.

Rasmussen, P.E., and H.P. Collins. 1991. Long-term impacts of tillage, fertilizer, and crop residue on soil organic matter in temperate semiarid regions. *Adv. Agronomy* 45:93-134.

Rosell, R., and J. Galantini. 1998. Soil organic carbon dynamics in native and cultivated ecosystems of South America. In *Advances in soil science: Management of carbon sequestration in soil* (Eds. R. Lal, J. Kimble, R. Follett, and B. Stewart). CRC Press, Boca Raton, FL, pp. 11-33.

Schlesinger, W.H. 2000. Carbon sequestration in soils: Some cautions amidst optimism. *Agriculture Ecosys. Environ.* 82:121-127.

Sparling, G.P. 1992. Ratio of microbial biomass carbon to soil organic carbon as a sensitive indicator of changes in soil organic matter. *Aust. J. Soil Res.* 30:195-207.

Stott, D.E., L.F. Elliott, R.I. Papendick, and G.S. Campbell. 1986. Low temperature or low water potential effects on the microbial decomposition of wheat residue. *Soil Biol. Biochem.* 18:577-582.

Strong, D.T., P.W.G. Sale, and K.R. Helyar. 1999. The influence of the soil matrix on nitrogen mineralization and nitrification: IV. Texture. *Aust. J. Soil Res.* 37:329-344.

Tate, K.R. 1992. Assessment, based on a climosequence of soils in tussock grassland, of soil carbon storage and release in response to global warming. *J. Soil Sci.* 43:697-707.

Theng, B.K.G., K.R. Tate, and P. Sollins. 1989. Constituents of organic matter in temperate and tropical soils. In *Dynamics of soil organic matter in tropical ecosystems* (Eds. D.C. Coleman, J.M. Oades, and G. Uehara). Niftal Proyect Publ., pp. 5-32.

Tisdall, J.M., and J.M. Oades. 1982. Organic matter and water-stable aggregates in soils. *J. Soil Sc.* 33:141-163.

Van Veen, J.A., and P.J. Kuikman. 1990. Soil structural aspects of decomposition of organic matter by micro-organisms. *Biogeochemistry* 11:213-233.

PART III: SOIL CARBON ASSESSMENT METHODS

Chapter 18

Laser-Induced Breakdown Spectroscopy and Applications for Soil Carbon Measurement

Michael H. Ebinger
David A. Cremers
Clifton M. Meyer
Ronny D. Harris

INTRODUCTION

Loss of soil carbon during the past 150 years has depleted many of the most productive soils and resulted in degradation of marginal soils upon which many depend. The potential to restore carbon stocks in different landscapes through careful land management practices could result in significant improvements to crop production in agricultural lands and overall soil quality improvements in other lands (Lal, 2004; IPCC, 2000). Increases in soil organic carbon (SOC) depend on climate and management practices, and potential increases are estimated from 0 to 150 kg $C \cdot ha^{-1}$ per year in semiarid environments up to 1,000 kg $C \cdot ha^{-1}$ per year in more humid environments (Lal, 2004; Armstrong et al., 2003; West and Post, 2002). Current methods of carbon analysis (e.g., Rossel et al., 2001; Scharpenseel et al., 2001) provide the analytical tools needed to estimate these increases in the SOC pool with some precision. However, advanced analytical methods (e.g., Wielpolski, Chapter 20; Ebinger et al., 2003; McCarty et al., 2002; McCarty and Reeves, 2001; Cremers et al., 2001) offer improved detection, ease of operation, and potential use in the field that could improve precision and accuracy of SOC measurements. In addition, the need for improved accuracy and precision to support national and international policies on carbon emissions and carbon trading may require orders of magnitude more measurements to provide valid support for various positions and land man-

Carbon Sequestration in Soils of Latin America
Published by The Haworth Press, Inc., 2006.
doi:10.1300/5755_19

agement practices. These measurements must be delivered at the lowest cost possible and with well-characterized uncertainties. Current methods fall short on cost-effectiveness as well as accuracy and precision; advanced methods, once fully developed and tested, should optimize the amount of information about SOC pools per dollar spent, and must be designed to keep the cost of assessing carbon to less than 10 percent of the total costs of sequestration practices (DOE, 2004). With carbon trading in the United States and Europe creating a commodity market for sequestered carbon, the need to measure and certify increases (or decreases) in SOC within three to five years of implementing carbon sequestration and management practices, and to do so at the lowest levels of detection, is a pressing issue.

Developing more cost-effective and rapid methods of measuring soil carbon is thus an important need to address aspects of global climate change and terrestrial carbon management issues. Over the past two decades, several advanced analytical methods have been applied to the study of soil carbon. Estimating the retention time and rates of carbon turnover in soils (e.g., Paul et al., 2001), the formation rates of components of soil organic carbon (e.g., Six et al., 2002; Horwath et al., 2001), and the source or history of carbon that comprises soil organic carbon (e.g., Scharpenseel et al., 2001) have begun to be addressed. The composition of soil organic matter has also been evaluated by new applications of instrumental laboratory analyses (e.g., McCarty et al., 2002; Rossell et al., 2001), and total soil carbon has been quantified using infrared spectroscopy (e.g., Ben-Dor and Banin, 1995; Ludwig and Khanna, 2001) or other spectroscopic methods adapted to carbon analysis (e.g., Wielpolinski, Chapter 20; Cremers et al., 2001).

In this chapter we report preliminary calibrations of a new spectroscopic method for measuring total soil carbon that is based on atomic emission spectroscopy and using laser-induced breakdown spectroscopy (LIBS) (Rusak et al., 1997; Moenke-Blankenburg, 1989; Radziemski and Cremers, 1989). In this method, a laser is focused on a solid sample and forms a microplasma that emits light characteristic of the elemental composition of the sample. The emitted light is collected, spectrally resolved, and detected to monitor concentrations of elements via their unique spectral signatures. When calibrated, the method provides quantitative measurements of soil carbon in a sample. The method is readily adaptable to field-portable instrumentation which would provide investigators a means to measure soil carbon in near-real time (Cremers et al., 1996; Yamamoto et al., 1996) or in a laboratory setting with high-throughput analysis. We evaluated the LIBS method for its potential to measure total soil carbon and specifically tested the hypothesis that the LIBS carbon signal is correlated with total soil carbon, which could thereby provide a useful new approach for measuring total soil carbon.

MATERIALS AND METHODS

LIBS Instrumentation and Method

The LIBS method is based on atomic emission spectroscopy (Rusak et al., 1997; Moenke-Blankenburg, 1989; Radziemski and Cremers, 1989). In this method, a laser is focused on a solid sample and forms a microplasma as the sample is ionized. The light from the microplasma is characteristic of the elemental composition of the sample. This emitted light is collected, spectrally resolved, and detected to monitor concentrations of elements via their unique spectral signatures. When calibrated, the method provides quantitative measurements; when applied to soils, carbon and most other elements present in soil samples can be quantified in seconds.

We selected the strong carbon C(I) emission lines at 247.8 nm or 193 nm for this calibration study of LIBS (Ebinger et al., 2003; Cremers et al., 2001; Alkemade et al., 1978; Boutilier et al., 1978). A Nd:YAG laser (Spectra-Physics Lasers, Mountain View, CA) at a wavelength of 1064 nm (50 mJ pulses of 10 ns) was focused with a lens of 50 mm focal length on each soil sample (Figure 18.1). The emitted light was collected by a fused silica fiber optic cable pointed at the plasma from a distance of about 50 mm. A spectrograph of 0.5 m focal length resolved the light that was then detected using a gated-intensified photodiode array detector. For each LIBS analysis a sample was pressed into a disc of about 3 cm then positioned on a track that moved after each set of laser pulses (shots). One hundred shots were directed onto the sample to complete one measurement. Typical measurement areas for the LIBS analysis are 1 to 5 mm^3 per pulse.

Spectra from the 100 shots were collected and averaged into a single spectrum for each measurement. Peak heights or peak areas were recorded for the carbon emission line chosen, and background signals from the spectra were subtracted; this procedure was repeated for each sample. Silicon emission lines at at 251 nm were also collected in order to ratio to the carbon emission lines. In some cases, the C/Si intensity ratio was the preferred carbon measure. However, if collection of spectra could be completed before instrument parameters change (i.e., within about an hour), such as for analyses of intact soil cores, the carbon signal was used without ratioing. Because of shot-to-shot variations in the laser plasma parameters and sampling geometry, measurement precision is often increased by ratioing the analyte signal to the signal from another species, in this case silicon. The concentration of silicon is much greater than carbon in soil samples, thus variation in silicon concentration within and between samples did not perturb the C/Si ratio significantly. The C/Si ratio or the C intensity of the LIBS

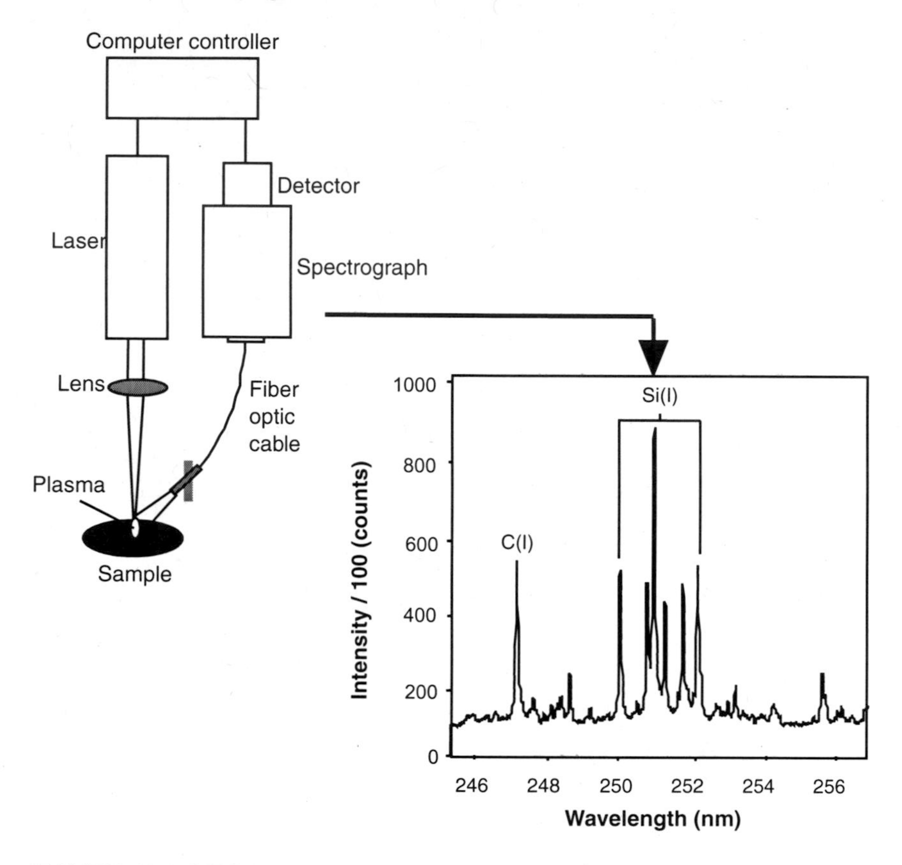

FIGURE 18.1. LIBS instrument in schematic, showing microplasma collection, detection, and spectral resolution of a sample.

signal were the measures used to calibrate with dry combustion data from the same samples.

LIBS Calibration—Discrete Samples

We tested LIBS for carbon analysis by using soils collected during soil surveys in the United States (M. Lee Norfleet, personal communication, 2003). We evaluated the LIBS method for its potential to measure total soil carbon and specifically tested the hypothesis that the LIBS carbon signal is correlated with total soil carbon. If this hypothesis is supported using a wide variety of soils, a new approach for measuring total soil carbon would also be supported.

We measured total soil carbon with the conventional dry combustion method using a Dohrmann DC-180 analyzer (Tekmar–Dohrmann, Mason, OH) or an Elementar VarioMax (Robertson and Paul, 2000). Dry combustion data were used with the C/Si ratio or C intensity data from LIBS measurements to calibrate, verify, and assess performance of LIBS.

The test soils were selected for several characteristics: We required the samples to span coarse, medium, and fine textures; to exhibit various mineralogy; to be subjected to conventional tillage or no-till agriculture; and to be of different morphologies (Table 18.1). In addition, properties such as pH, CEC, and bulk chemistry were incidentally included but not selected for (data not shown). Total soil carbon concentrations from the dry combustion analyses were regressed with the LIBS C/Si ratios or C intensity for each sample, and the results of these calibration curves were the measure of the efficacy of LIBS analysis for soil carbon.

LIBS Analysis of Intact Core Samples

Analysis of discrete samples was the initial aim for the calibration studies. Development of LIBS, however, was initiated to find a new analysis method that could be used in the field or with samples directly obtained from the field with no additional preparation. We modified the LIBS instrument to analyze intact cores of soil extracted with a soil probe (up to 1.3 m in length and 4 cm diameter) in a plastic tube. Soils sampled were the Mesita del Buey soils near Los Alamos National Laboratory (Davenport et al., 1996). After extraction of several 25 to 30 cm of soil cores for this test, a window was cut in the plastic tube to allow access by the laser. The core was then placed on the track of the LIBS instrument and 10 to 100 shots were collected each 1 mm along the length of the core. In this configuration, the cores required less than an hour to scan. Spectra from each step were re-

TABLE 18.1. NRCS soils used in calibration study.

Soil series	Clay (%)	Silt (%)	Sand (%)	Texture class	Till/no-till	Carbon (%, dry combustion)
Otley	37.2 (10.0)	63.8 (2.1)	2.1 (0.3)	silty clay loam	Till	0.8 (0.3)
Vebar	15.7 (1.3)	24.5 (1.5)	59.8 (1.5)	sandy loam	No-till	2.1 (0.8)
MICC17	21.0 (7.2)	32.1 (3.1)	46.0 (9.9)	loam	Till	2.2 (0.3)
ILNL5	29.2 (1.1)	68.9 (0.9)	1.9 (0.2)	silty clay loam	No-till	2.5 (0.5)
MICN16	53.1 (5.7)	30.8 (3.7)	16.1 (3.6)	clay	Till	2.6 (0.6)

Note: Values in parentheses are standard deviations.

tained for quantitative analysis, and after all scans were completed a carbon profile of the core was constructed. Core analysis by LIBS provides a high-resolution distribution of carbon in the extracted core, a feature that cannot be provided using conventional analysis without great difficulty and expense. Data from the LIBS core analysis also contain significant information on the spatial variation of carbon in the soil, another feature that is extremely difficult and costly to provide with conventional analysis.

Data analysis from core measurements was somewhat tedious because of the immense number of shots and spectra collected from a single intact core. We chose to group LIBS data in 2.5 cm depth intervals, resulting in a point estimate from about 25 scans for each interval along the length of the core. We computed the arithmetic and geometric means as well as standard deviations using the LIBS data from each interval. The geometric mean is the better estimator and tends to be less influenced by extreme values because it represents the expected value of a log-normal distribution of the sample data (Isaaks and Srivastava, 1989). The geometric mean from each 2.5 cm interval was plotted with soil depth along the core. In order to calibrate the LIBS measurements with dry combustion data, we removed each 2.5 cm interval of the core after the LIBS analyses were complete, and analyzed these by dry combustion.

RESULTS AND DISCUSSION

Calibration Results—Discrete Samples

Results of previous tests of LIBS for soil carbon analysis were quite satisfactory using a limited number of soil samples (Ebinger et al., 2003; Cremers et al., 2001). Results shown in this volume are the results of preliminary analysis of a more diverse NRCS soil collection as mentioned previously. The intensity of the carbon line alone (i.e., not the C/Si ratio) at 193 nm was plotted with carbon concentration from dry combustion (Figure 18.2). To accomplish this, the samples of each soil series were analyzed sequentially before the instrument parameters drifted. Overall, there was good agreement between LIBS C intensities and carbon concentrations measured from dry combustion, but only when the results were grouped by soil series (Figure 18.2; Table 18.2).

LIBS measurements also showed different relationships between soils from conventionally tilled and no-till sites (Figure 18.3). Carbon/silicon ratios using the 248 nm carbon line and the 251 nm silicon line showed that there was less scatter and slightly more sensitivity from the tilled soils than from the no-till soils.

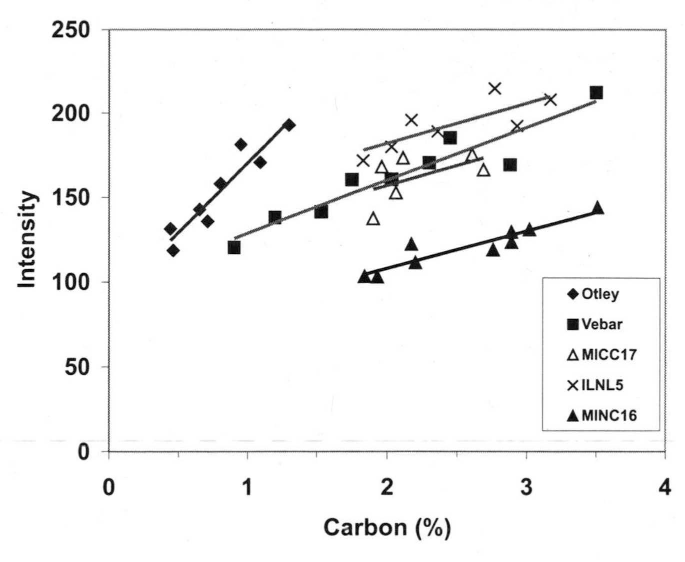

FIGURE 18.2. LIBS versus DC analysis and calibration for NRCS soils.

TABLE 18.2. Regression statistics for NRCS calibration tests using discrete samples.

Soil series	Slope	Intercept	R^2
Otley	81.9	88.4	0.89
Vebar	31.3	97.4	0.90
MICC17	23.8	109.3	0.31
ILNL5	23.9	134.1	0.64
MICN16	22.1	63.8	0.87

Note: LIBS measurements were the dependent variable, and carbon measurement from dry combustions was the independent variable

Core Analysis Results

Analysis of LIBS signal each 1 mm along intact soil cores was surprisingly successful. Data collected and analyzed as described for one core of about 25 cm length show a general decrease in carbon concentration with depth as well as significant variation in carbon distribution throughout the core (Figure 18.4), and plots for other cores from the same site were similar. The raw data were smoothed using the geometric mean computed from 25 LIBS C/Si ratios per 2.5 cm interval. The smoothed data are generally more

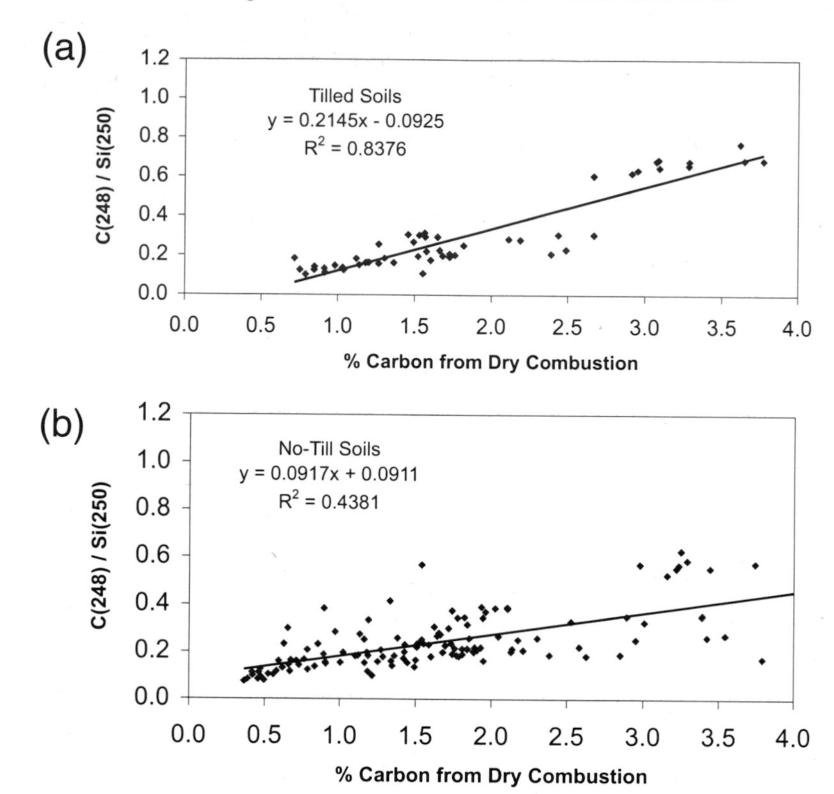

FIGURE 18.3. Calibration curves for soils from (a) till and (b) no-till sites.

comparable to data obtained from dry combustion analysis (Figure 18.5). Standard deviations of the reduced LIBS data also indicate the confidence with which the carbon concentration is estimated. One outlying point (Figure 18.5) and the corresponding depth interval from approximately 15 to 17 cm in the raw data (Figure 18.4) were the result of inclusion of a large root (approximately 1.5 cm diameter) within the extracted core. This outlier clearly was preserved in the reduced LIBS data and proved an obvious anomaly in the calibration plot (Figure 18.6). When the source of the outlier was accounted for, the calibration curve was excellent. The possibility of including large carbon concentrations from roots or other discrete organics illustrates one feature of the potential applicability of LIBS to carbon inventory estimation.

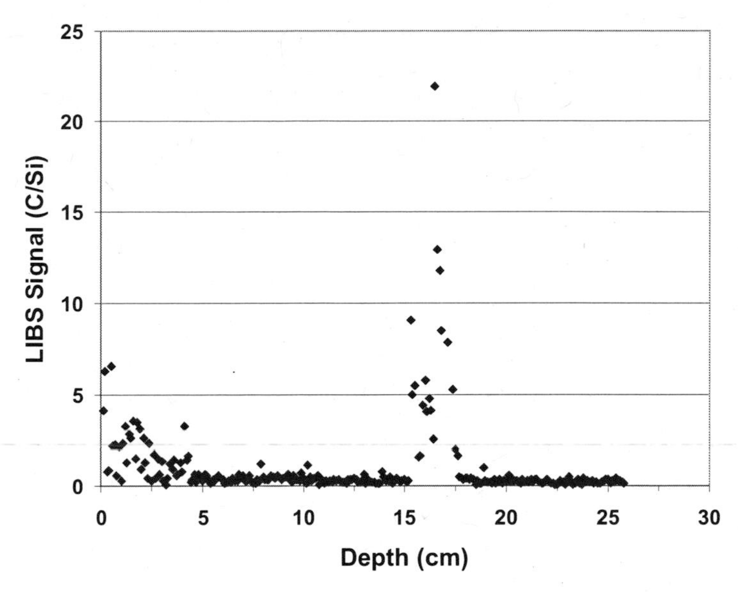

FIGURE 18.4. Raw data from intact soil core showing LIBS C/Si measurements with depth; each point represents 100 shots from a 1 mm interval of the core.

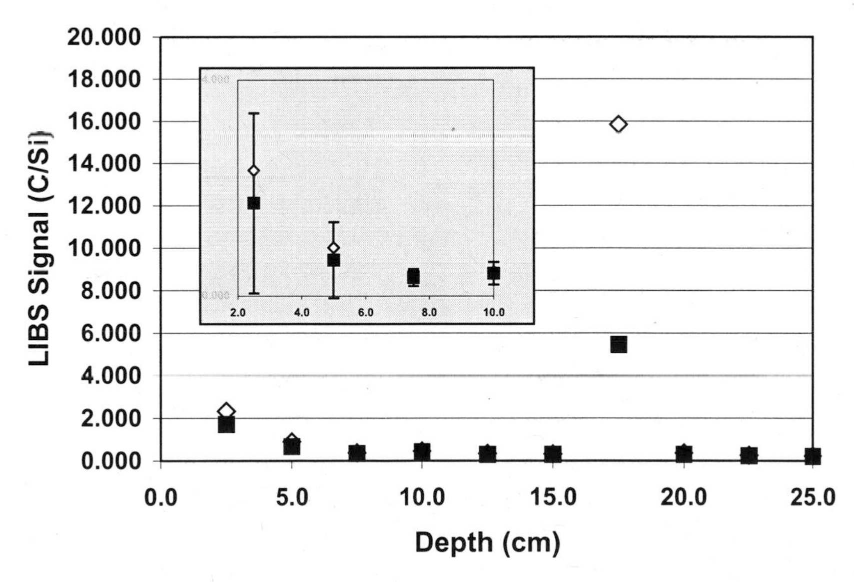

FIGURE 18.5. Reduced data from analysis of intact soil core. Inset shows error bars for 2.5 cm to 10.0 cm depths; open symbols are arithmetic mean; closed symbols are geometric mean.

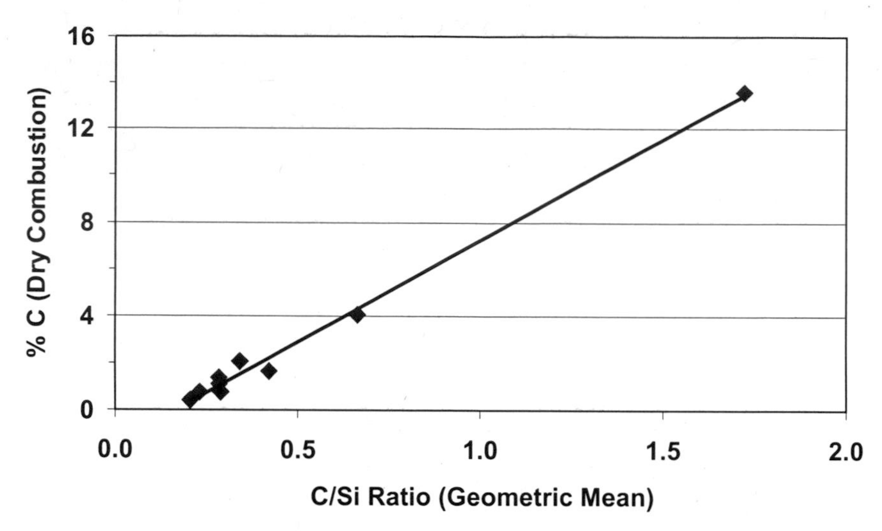

FIGURE 18.6. Calibration of LIBS core analysis data with dry combustion measurements from the same sample; slope = 8.68, intercept = −1.42, and R^2 = 0.997.

Discussion of LIBS Method for Soil Carbon

Discrete Samples

Pressing soil samples into disks is time-consuming but increases the LIBS precision considerably. Cremers et al. (2001) and Ebinger et al. (2003) attempted to use unpressed samples rotated in a quartz vial. Reproducibility of signals was problematic, as was operation of the rotating sample chamber in a field setting. Pressed samples, though better samples for LIBS analysis, require sieving to at least <2 mm, then additional preparation in the press. These steps add time and decrease cost-effectiveness of LIBS analysis of soils as well as remove the analysis from the field. Even with these drawbacks, LIBS analysis of pressed samples could be a significant increase in sample throughput when compared to conventional methods.

Analysis of Intact Soil Cores

The ability to measure the concentration of soil carbon along each 1 mm of an intact soil core is unique to LIBS analysis. The small size of the laser spot and the efficient collection of emitted light from the microplasma allow investigators to incorporate the spatial variability of carbon in soils into estimates of carbon inventories. Conventional methods average the carbon concentration throughout the volume of soil that is analyzed but cannot explicitly incorporate the distribution of carbon into those measurements. LIBS analyses, on the other hand, collect data from much smaller volumes than conventional samples then integrate these discrete measurements in a way that preserves the carbon distribution information. In the upper horizons of a typical soil profile, for example, the carbon concentration and distribution is expected to vary much more than in lower, possibly less reactive horizons. Conventional analysis reduces the information on carbon dynamics to the 2.5 or 5 cm interval commonly used for sampling, whereas the variation of carbon concentration is collected at a much finer scale with LIBS. Figure 18.4 illustrates this point. The upper 2.5 cm of the profile shows considerable variation from point to point, but this variation dampens below about 2.5 cm. Including this spatial variation in soil carbon inventories will improve the estimates and also help to quantify the uncertainties in measuring soil carbon.

The speed with which high-resolution analyses can be conducted and the portability of the instrument are also features that make the method attractive. Currently, data reduction is the most time-consuming part of the core analysis method. But data reduction can be automated to reduce the amount of time required to compute values of carbon concentration from the hundreds of LIBS measurements of a single core. With the advent of smaller LIBS instruments and automated data-reduction routines, core analysis should prove an important tool for in-field analysis of soil cores.

Reproducibility of carbon inventory estimates from intact soil cores will prove challenging, but as the core analysis method evolves, statistical techniques can be applied that cannot be used currently because they require large and/or spatially explicit data sets. Geostatistical analysis of carbon data from intact cores and use of these methods for research design should prove valuable for understanding the distribution of carbon in soils as well as the dynamic nature of soil carbon over time. Quantification of the uncertainties in the measurements via new data analysis methods should also be a significant contribution to a better understanding of carbon dynamics in diverse landscapes.

Nitrogen Analysis by LIBS

Carbon analysis by LIBS has some promising possibilities, but analysis of nitrogen in soils would be an invaluable tool alone or in combination with carbon measurements. Martin et al. (2003, 2002) attempted analysis of nitrogen in soils using LIBS in the ambient atmosphere. They show interesting results that have been reproduced elsewhere (Harris et al., 2004), but incorrect attribution of spectral lines by Martin et al. (2003) to nitrogen discounted the validity of the proposed method. Nitrogen that could be detected by the LIBS instrument discussed would be mostly due to atmospheric nitrogen with a small component from the soil. Without removing the ambient nitrogen from the environment within which the microplasma forms, nitrogen from soils is a very small part of the total signal from the atmosphere. Harris et al. (2004) showed that in an environment free of ambient nitrogen, such as argon gas or a partial vacuum, the nitrogen from soil becomes the dominant signal. Coupled with accurate nitrogen line attribution, the LIBS method using a partial vacuum or a gas that contains no nitrogen holds promise for soil nitrogen measurements.

CONCLUSIONS

Our preliminary analysis of soil carbon by LIBS has shown interesting results. First, it appears that soil carbon can be analyzed quickly and with minimal preparation via LIBS. Discrete samples require some treatment after collection in the field, but to date, soils pressed into disks have provided acceptable calibrations against known and more conventional methods.

Second, analysis of soil carbon from intact soil cores is a technique that needs to be expanded upon quickly. The potential to use the spatial distribution of soil carbon to estimate carbon inventories is one tool that could significantly improve the accuracy and precision of carbon inventory estimates. Data collection is rapid, requiring less than an hour to obtain data from 25 cm of intact core at 1 mm resolution. Data analysis, while cumbersome at present, will benefit greatly from automation. Preliminary data on LIBS analysis of intact soil cores suggests that the LIBS measurements can be reduced and used for carbon inventories in a way similar to conventional data, albeit with much more information on measurement and soil-dependent uncertainties.

Third, the potential to measure nitrogen from soils via LIBS is within current field-portable LIBS technology. With due care in analysis and calibrations, field modification of the LIBS instruments, and proper data reduction, soil nitrogen and soil carbon may be simultaneously analyzed while researchers are in the field taking samples.

Together, these features of LIBS analysis are exciting and promise to significantly improve the way soil scientists, ecologists, and climate researchers measure carbon concentration data. Field testing remains to be completed to show the efficiency of LIBS measurements when compared to conventional methods, the applicability of the type of data generated by LIBS, and comparability to investigations of soil carbon conducted with conventional means. However, the potential for LIBS to deliver accurate, precise, and rapid in-field measurement of soil carbon support the method as one of an emerging suite of valuable new analytical capabilities.

REFERENCES

Alkemade, C.T.J., W. Snelleman, G.D. Boutilier, B.D. Pollard, J.D. Winefordner, T.L. Chester, and N. Omenetto. 1978. Review and tutorial discussion of noise and signal-to-noise ratios in analytical spectrometry: 1. Fundamental principles of signal-to-noise ratios. *Spectrochim. Acta Part B* 33:383-399.

Armstrong, R.D., G. Millar, N.V. Halpin, D.J. Reid, and J. Standley. 2003. Using zero tillage, fertilisers and legume rotations to maintain productivity and soil fertility in opportunity cropping systems in a shallow Vertosol. *Aust. J. Exp. Agriculture* 43:141-153.

Ben-Dor, E., and A. Banin. 1995. Near-infrared analysis as a rapid method to simultaneously evaluate several soil properties. *Soil Sci. Soc. Am. J.* 59:364-373.

Boutilier, G.D., B.D. Polard, J.D. Winefordner, T.L. Chester, and N. Omenetto. 1978. Review and tutorial discussion of noise and signal-to-noise ratios in analytical spectrometry: 2. Fundamental principles of signal-to-noise ratios. *Spectrochim. Acta Part B* 33:401-416.

Cremers, D. A., M. H. Ebinger, D.D. Breshears, P. J. Unkefer, S. A. Kammerdiener, M. J. Ferris, K. M. Catlett, and J. R. Brown. 2001. Measuring total soil carbon with laser-induced breakdown spectroscopy (LIBS). *J. Environmental Qual.* 30:2202-2206.

Cremers, D.A., M.J. Ferris, and M. Davies. 1996. Transportable laser-induced breakdown spectroscopy (LIBS) instrument for field-based soil analysis. *Proc. Soc. Photo Opt. Instrum. Eng.* 2835:190-200.

Davenport, D.W., B. P. Wilcox, and D. D. Breshears. 1996. Soil morphology of canopy and intercanopy sites in a piñon-juniper woodland. *Soil Sci. Soc. Am. J.* 60:1881-1887.

Department of Energy (DOE). 2004. Carbon sequestration technology roadmap and program plan—2004. National Energy Technology Laboratory, DOE Office of Fossil Energy. Available online at www.netl.doe.gov, accessed September 1, 2004.

Ebinger, M. H., M. Lee Norfleet, D. D. Breshears, D. A. Cremers, M. J. Ferris, P. J. Unkefer, M. S. Lamb, K. L. Goddard, and C. W. Meyer. 2003. Extending the applicability of laser-induced breakdown spectroscopy for total soil carbon measurement. *Soil Sci. Soc. Am. J.* 67:1616-1619.

Harris, R. D., D. A. Cremers, M. H. Ebinger, and B. K. Bluhm. 2004. Determination of nitrogen in sand using laser-induced breakdown spectroscopy. *Applied Spectroscopy* 58:770-775.

Horwath, W.R., C. van Kessel, U. Hartwig, and D. Harris. 2001. Use of 13C isotopes to determine net carbon sequestration in soil under ambient and elevated CO2. In R. Lal et al. (eds.), *Assessment methods for soil carbon.* Lewis Publ., Boca Raton, FL, pp. 221-232.

Intergovernmental Panel on Climate Change (IPCC), 2000. *Land use, land use change, and forestry.* Cambridge University Press, Cambridge, UK.

Isaaks, E. H. and R. M. Srivastava. 1989. *An introduction to applied geostatistics.* Oxford University Press, New York.

Lal, R. 2004. Soil carbon sequestration impacts on global climate change and food security. *Science* 304:1623-1627.

Ludwig, B., and P.K. Khanna. 2001. Use of near infrared spectroscopy to determine inorganic and organic carbon fractions in soil and litter. In R. Lal et al. (eds.), *Assessment methods for soil carbon.* Lewis Publ., Boca Raton, FL, pp. 361-370.

Martin, M. Z., S.D. Wullschleger, and C.T. Garten Jr. 2002. Laser-induced breakdown spectrsocopy for total carbon and nitrogen in soils. *Proc. SPIE-International Soc. Opt. Eng.* 4576:188.

Martin, M. Z., S.D. Wullschleger, C.T. Garten Jr., and A.V. Palumbo. 2003. Laser-induced breakdown spectrsocopy for the environmental determination of total carbon and nitrogen in soils. *Appl. Opt.* 42:2072.

McCarty, G.W., and J.B. Reeves III. 2001. Development of rapid instrumental methods for measuring soil organic carbon. In R. Lal et al. (eds.), *Assessment methods for soil carbon.* Lewis Publ., Boca Raton, FL, pp. 371-380.

McCarty, G.W., J.B. Reeves III, V. B. Reeves, R. F. Follett, and J. M. Kimble. 2002. Mid-infrared and near-infrared diffuse reflectance spectroscopy for soil carbon measurement. *Soil Sci. Soc. Am. J.* 66:640-646.

Moenke-Blankenburg, L. 1989. *Laser microanalysis.* John Wiley, New York.

Paul, E.A., S.J. Morris, and S. Böhm. 2001. The determination of soil C pool sizes and turnover rates: Biophysical fractionation and tracers. In R. Lal et al. (eds.), *Assessment methods for soil carbon.* Lewis Publishers, Boca Raton, FL, pp. 193-206.

Radziemski, L.J. and D.A. Cremers. 1989. Spectrochemical analysis using laser plasma excitation. In L.J. Radziemski and D.A. Cremers (eds.), *Applications of laser-induced plasmas.* Marcel Dekker, New York, pp. 295-325.

Robertson, G.P., and E.A. Paul. 2000. Decomposition and soil organic matter dynamics. In O.E. Sala et al. (eds.), *Methods in ecosystem science.* Springer, New York, pp. 104-116.

Rossell, R.A., J.C. Gasparoni, and J.A. Galantini. 2001. Soil organic matter evaluation. In R. Lal et al. (eds.), *Assessment methods for soil carbon.* Lewis Publ., Boca Raton, FL, pp. 311-322.

Rusak, D.A., B. C. Castle, B. W. Smith, and J. D. Winefordner. 1997. Fundamentals and applications of laser-induced breakdown spectroscopy. *Crit. Rev. Anal. Chem.* 27:257-290.

Scharpenseel, H.W., E.M. Pfeiffer, and P. Becker-Heidmann. 2001. Ecozone and soil profile screening for C-residence time, rejuvenation, bomb $_{14}$C photosynthetic $_{13}$C changes. In R. Lal et al. (eds.), *Assessment methods for soil carbon.* Lewis Publ., Boca Raton, FL, pp. 207-219.

Six, J., C. Feller, K. Denef, S.M. Ogle, J.C. de Moraes Sa, A. Albrecht. 2002. Soil organic matter, biota, and aggregation in temperate and tropical soils—Effects of no-tillage. *Agronomie* 22:755-775.

West, T.O., and W.M. Post. 2002. Soil organic carbon sequestration rates by tillage and crop rotation. *Soil Sci. Soc. Am. J.* 66:1930.

Yamamoto, K.Y., D.A. Cremers, M.J. Ferris, and L.E. Foster. 1996. Detection of metals in the environment using a portable laser-induced breakdown spectroscopy instrument. *Appl. Spectr.* 50:222-233.

Chapter 19

The Potential of Spectroscopic Methods for the Rapid Analysis of Soil Samples

J. B. Reeves III
G. W. McCarty
R. F. Follett
J. M. Kimble

INTRODUCTION

Due to the possibility of using C sequestration in soil as a means to remove CO_2 from the atmosphere, and with the potential for payments for such endeavors, there is an increased interest in determining the changes in amounts of C present in soil. Due to differences in the lability of various forms of soil C, there is also interest in determining differences in the composition of the C present. At present, the "gold standard" for determination of total, organic, and inorganic C in soils is combustion, which can also be used to simultaneously determine total N and S depending on the particular instrument and instrument configuration used (Bremner, 1996). However, combustion analysis requires two runs to separate the organic and inorganic C (in the form of carbonates) and cannot determine nonorganic, noncarbonate C, noninorganic C (charcoal, etc.) as different from organic C. It also cannot determine labile as opposed to nonlabile organic C unless extractions are performed and the C content of the extracts determined. Other methods available for determination of soil C include laser-induced breakdown spectroscopy (LIBS) (Ebinger et al., 2003), ground penetrating radar (GPR) for determining root mass (NRCS, 2004), neutron activation spectroscopy (NAS) (Glascock, 2004), and near- (NIRS) and mid-infrared- (MIDIR) reflectance spectroscopy, also called DRIFTS for diffuse reflectance MIDIR Fourier transform spectroscopy (Nguyen et al., 1991; Reeves,

Carbon Sequestration in Soils of Latin America
Published by The Haworth Press, Inc., 2006.
doi:10.1300/5755_20

1996; Janik et al., 1998; Reeves and McCarty, 2001; Reeves et al., 1999, 2001).

Each of these techniques has advantages and disadvantages, and in the final analysis more than one may have to be used to obtain the needed information. For example, LIBS as presently used is very fast (perhaps as many as hundreds of samples per hour) but cannot differentiate organic from inorganic C, and even with projected improvements no claims have been made to ever be able to differentiate forms of organic C. Total C determination is also a feature of NAS. However, GPR may be able to used to determine the physical volume of large roots present in a mass of soil, whereas other methods cannot without extensive sampling. Of all the techniques discussed, only NIRS and DRIFTS offer the possibility of determining the forms of organic C present while simultaneously determining the organic versus inorganic C in soil samples.

One final consideration is the number of samples that may need to be analyzed for payments for C sequestration in soil to become a reality. Within discussion at this workshop, it was estimated that some tens to hundreds of millions of samples may need to be analyzed yearly for Brazil alone. Even ignoring the question of how many combustion units would be required, the cost in gases and disposables alone would be prohibitive, with general cost estimates of several dollars (U.S.) per sample and even more if both organic and inorganic C need to be determined. One advantage of methods such as LIBS, NIRS, and DRIFTS is that there is little or no cost associated with disposables. Also, due to the potential samples involved and other costs associated with sample collection and analysis, there is a need for rapid and on-site analysis of soil samples, as the costs associated with sample collection, storage, and shipping can alone be prohibitive when dealing with the very high potential number of analyses needed for soil carbon trading. In addition, the advent of data-intensive technologies, such as site-specific agriculture and soil management in agriculture, has generated the need for large databases and the analysis of large numbers of samples so that spatial structure in agricultural fields can be characterized.

While combustion is a direct measurement of soil C, methods such as LIBS, NIRS, and DRIFTS are indirect, in that relationships between soil C and spectra, whether they are elemental spectra as in LIBS or molecular spectra as in NIRS and DRIFTS, must be established and maintained. For methods such as NIRS and DRIFTS this is called calibration development, and maintenance is based upon such mathematical techniques as stepwise regression, principal component analysis, and partial least squares regression (PLS). One advantage of NIRS- and DRIFTS-based calibrations is that once developed, a single scan of a sample can be used to determine multiple analytes simultaneously without the need for individual assays for each analyte of interest, thus multiplying the impact of the method on time and cost reduction. A discussion of all the factors involved in developing, using, and maintaining such calibrations is beyond the scope of this chapter, so readers are directed to several books which cover the subject (Burns and Ciurczak, 2001; Williams and Norris, 2001; Roberts et al., 2004).

The specific objective of this chapter is to discuss the advantages, problems, and present state of NIRS and DRIFTS techniques for the determination of soil C.

METHODS

Spectroscopy

Mid-infrared spectra were obtained using a Digilab FTS-60 FTIR equipped with a custom-made (Reeves, 1996) sample transport cell (Figure 19.1) or a Digilab FTS-7000 FTIR equipped with a Pike AutoDiff Autosamples (Plate 19.1). In either case, spectra were collected from 4,000 to 400 cm^{-1} at 4 cm^{-1} resolution using DTGS detectors and a KBr beam splitter. All near-infrared data reported are based on spectra collected using a NIRSystems model 6500 scanning monochromator (Plate 19.2) with spectra scanned

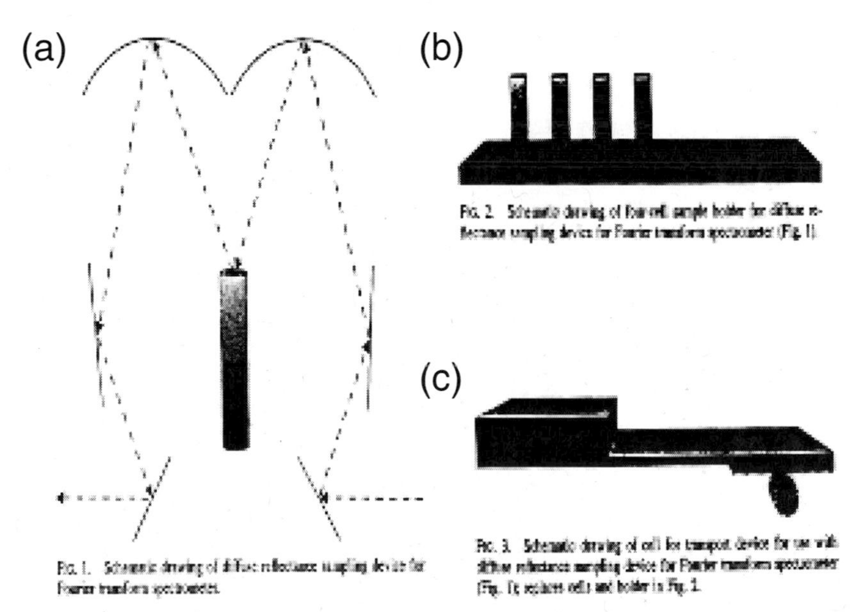

FIGURE 19.1. Diffuse reflectance methods and cell used in the mid-infrared.

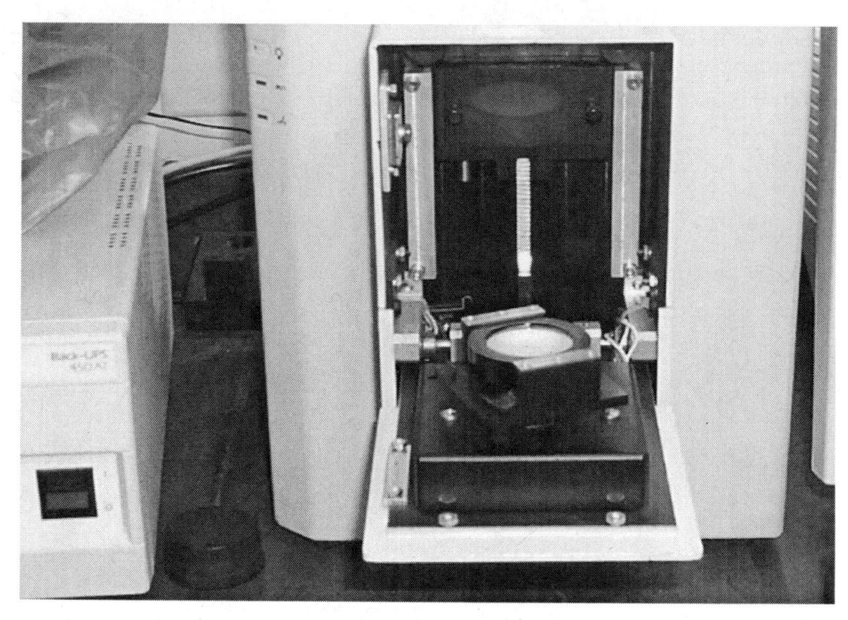

PLATE 19.1. NIRSystems model 6500 scanning monochromator with rotating sample cup.

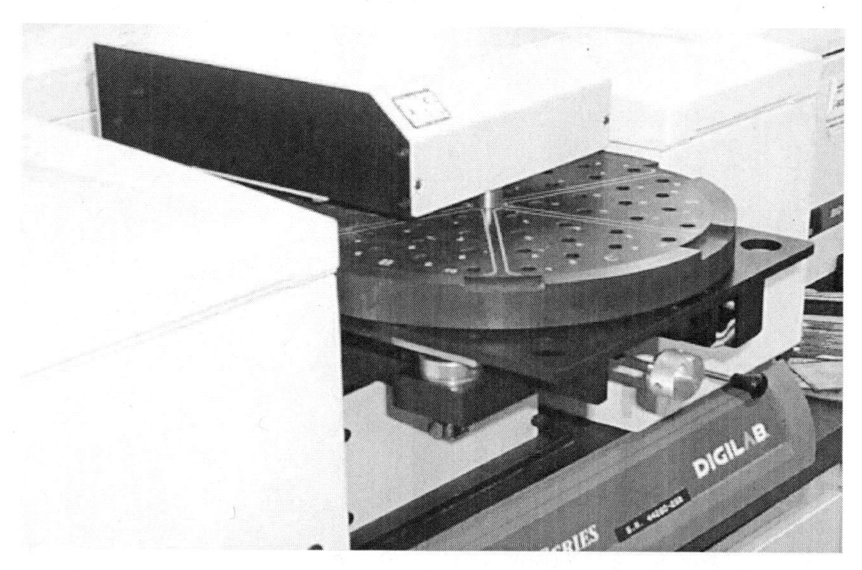

PLATE 19.2. Digilab model 7000 FTIR with Pike Autodiff diffuse reflectance autosampler.

from 400 to 2,500 nm (25,000 to 4,000 cm^{-1}) at a nominal resolution of 10 nm, with only data from 1,100 to 2,500 nm (Pb S detector) used in calibrations due to the lack of improvement in calibration accuracy when including spectral data from 400 to 1,098 nn (Si detector). For all spectra, ground samples were scanned "as is" (64 CO^- added scans per spectra, no dilution with KBr, etc.) unless otherwise noted (Reeves, 2003).

Statistics and Chemometrics

Calibration development was by PLS using Grams/386 PLSPlus V2.1G using a variety of data subsets, spectral point averaging, derivatives, and other data treatments (mean centering, variance scaling, multiplicative scatter correction, and baseline correction) examined. The number of PLS factors to be used in the calibration was determined by the PRESS F-statistic from the one-out cross validation procedure, and a final calibration was developed (PLSplus V2.1G, 1994). Although beyond the scope of this chapter, the basic principles are to accentuate spectral information by the use of spectral pretreatments and then to find relationships between spectral information and sample composition that are sufficiently robust for determination of the composition of future samples from their spectra alone (Naes et al., 2002; PLSplus V2.1G, 1994; other texts previously recommended).

Compositional Determinations

All total, organic, and inorganic C and total N determinations were performed by combustion (Bremner, 1996). Numerous other determinations were also carried out, the methods for which can be found in the References. These determinations include pH, various enzyme measurements, mineralizable N, biomass C and N, active N, and others.

RESULTS AND DISCUSSION

NIR and DRIFTS Spectra

Figures 19.2 to 19.4 show the NIR and DRIFTS spectra of typical soil samples under various conditions. As shown in both Figures 19.2 and 19.3, the spectrum of carbonate minerals can be easily seen in both spectral ranges. The large peaks in the NIR at ~1,950 and 1,400 nm are due to water, as also demonstrated in Figure 19.4. Other than the water and carbonate peaks, soil spectra in the NIR are visually rather featureless, although information is present which can be used to develop calibrations for soil compo-

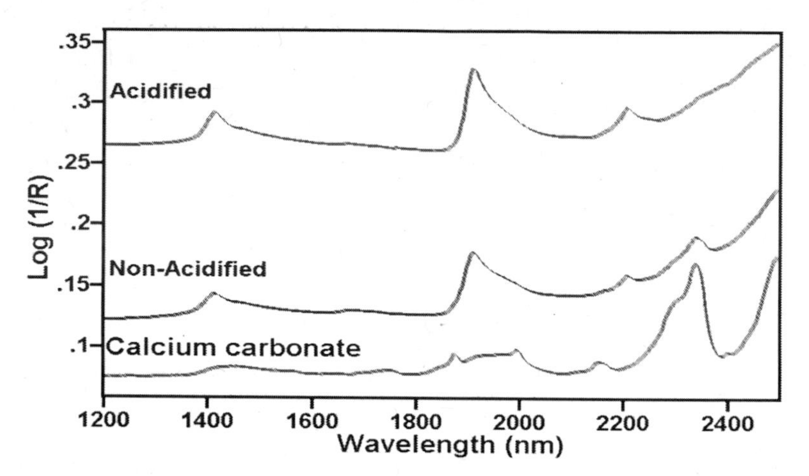

FIGURE 19.2. Near-infrared spectra of acidified and nonacidifed soil and calcium carbonate. (See also color gallery.)

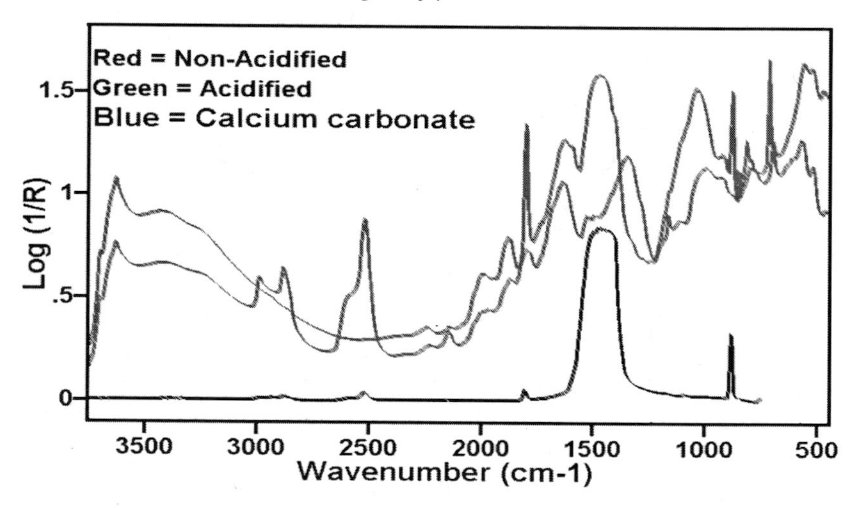

FIGURE 19.3. Mid-infrared spectra of acidified and nonacidified soil and calcium carbonate. (See also color gallery.)

sition, as will be shown with the aid of chemometrics. The DRIFTS spectra of the same samples (Figure 19.3) contain much more visually identifiable information without aid of mathematical transformation. This can be seen even for the carbonate spectra which is relatively prominent in the NIR region. As can be seen, several peaks disappear (~3,000, 2,500, 1,750 cm^{-1}, etc.) on acidification which cannot be seen in the spectrum of pure calcium

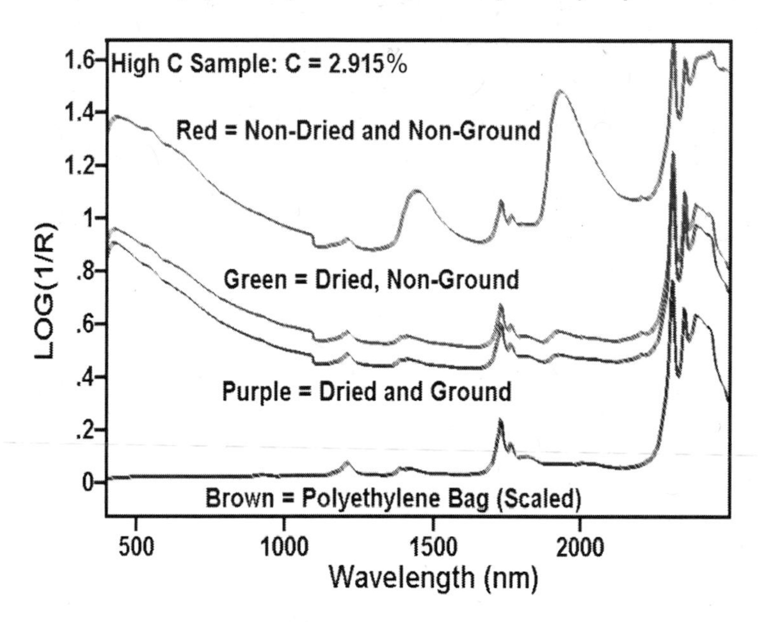

FIGURE 19.4. Near infrared spectra of soils taken under different soil conditions. (See also color gallery.)

carbonate. This divergence in spectral features for a single component results from the variations in MIDIR spectra caused by different concentrations of the component and by the means by which the spectrum is obtained (Reeves, 2003; Reeves et al., 2005). In the NIR, many instrument accessories can be used to obtain a spectrum, including a large sample cell (~50 × 150 mm) which allows moist samples to be scanned using polyethylene bags. The spectra in Figure 19.4 demonstrate both the effects of scanning a moist sample, with absorbances being much higher compared to those for a dried sample, and the results of using the bags (the bag spectrum is added to the soil spectrum). Since each bag spectrum varies a little, calibrations are developed with the bag spectrum included rather than trying to subtract a bag spectrum from each sample spectrum. As a result, chemometric analysis using PLS was able to successfully account for the bag spectrum. Notice also that the spectrum for the ground and dried sample shows overall lower absorbances than the corresponding unground, dried sample. These differences are due to particle size and are commonly seen in a series of samples where particle size, even in ground samples, varies. Finally, similar spectra are not shown for the DRIFTS because MIDIR radiation would be much more strongly absorbed by the polyethylene bag or water. Also, accessories such as the large sample cell are not available for MIDIR spectrometers at this time, although use of a custom-made cell (shown in Figure 19.1c,) has shown that scanning a larger sample area can increase calibration accuracy in the MIDIR, mostly likely due to sample nonhomogeneity even in ground samples (Reeves, 1996). As shown in both Figure 19.1 and Plate 19.1, currently available MIDIR spectrometers are designed only to scan a static sample with the area of lumination being only ~2 mm in diameter.

Carbon Determinations in Western U.S. Soils

In Tables 19.1 and 19.2, results are presented from a set of 237 soils collected over a wide range of the Great Plains region of United States (McCarty et al., 2002). These samples were available as acidified (carbonates removed) and nonacidified samples. As demonstrated, even with the removal of samples identified as outliers by the algorithms within GRAMS software, calibrations based on near-infrared spectra were consistently poorer than those found using DRIFTS spectra (no outliers found). Since

TABLE 19.1. Near- versus mid-infrared results for diverse set of 237 acidified and nonacidified samples from the western United States.

	1-Out X validation		Calibration results		
Assay	R^2	RMSD[a]	R^2	RMSD	RD[b]
Mid-Infrared Spectra[c]					
Organic C[d]	0.965	2.49	0.984	1.68	13.9
Organic C[e]	0.909	4.01	0.957	2.76	22.8
Inorganic C[e]	0.978	1.60	0.990	1.06	17.1
Total C[e]	0.936	3.95	0.971	2.64	14.4
Near-Infrared Spectra (Outliers Removed)[f]					
Organic C[d,g]	0.891	3.57	0.923	2.97	24.6
Organic C[e,h]	0.866	3.89	0.908	3.21	26.6
Inorganic C[e,i]	0.915	2.77	0.969	1.66	26.8
Total C[e,j]	0.884	4.51	0.914	3.90	21.3

[a]RMSD = Root mean squared deviation (%).
[b]RMSD/MEAN as percent.
[c]Spectral data from 4000 to 400 cm^{-1}.
[d]Acidified samples to remove carbonates.
[e]Nonacidified samples.
[f]Spectral data from 1100 to 2498 nm or 9091 to 4003 cm^{-1}.
[g]Nine outliers removed from set of 237.
[h]Six outliers removed from set of 237.
[i]Eleven outliers removed from set of 237.
[j]Five outliers removed from set of 237.

TABLE 19.2. PLS results using three-fourths of samples to develop calibration and remaining one-fourth as an independent test set.

	Calibration results		Test results			
Assay	R^2	RMSD[a]	R^2	RMSD	Bias[b]	RD[c]
Mid-Infrared spectra[d]						
Organic C[e]	0.985	1.62	0.967	2.42	0.552	18.0
Organic C[f]	0.949	3.03	0.942	3.15	0.016	23.4
Inorganic C[f]	0.991	1.11	0.983	1.20	0.328	22.6
Total C[f]	0.967	2.89	0.946	3.42	0.381	18.2
Near-Infrared spectra[d]						
Organic C[e]	0.829	5.54	0.800	5.77	–1.11	42.8
Organic C[f]	0.844	5.28	0.822	5.47	0.084	40.6
Inorganic C[f]	0.968	2.03	0.872	3.14	–0.141	59.2
Total C[f]	0.873	5.71	0.861	5.41	–0.321	28.8

[a]RMSD = root mean squared deviation (%).
[b]Actual – Predicted mean value.
[c]RMSD/MEAN as percent.
[d]See Table 19.1 for spectral ranges.
[e]Acidified samples to remove carbonates.
[f]Nonacidified samples.

the same analyte values and samples were used for all calibration development and the NIR spectrometer sees a much larger sample (rotating cup with a 2 cm spot size, Plate 19.2) compared to the MIDIR (2 mm spot size, nonrotating cup, Plate 19.1), these results support the conclusion that better results can be achieved using DRIFTS. Note, however, that calibrations for organic C appear to be degraded by the presence of carbonates much more in the MIDIR than in the NIR, as shown by the much higher R2 and lower error (0.965 and 2.49 versus 0.909 and 4.01 for the acidified and nonacidified samples, respectively).

A major concern with methods based upon the chemometric approach is whether they can be used to determine values in new samples. In Table 19.2, calibrations were developed using three-fourths of the samples and then used to determine the remaining one-fourth. As demonstrated, results using DRIFTS were again much better than corresponding results in NIR and overall were very similar to those found for the calibration set. These findings indicate that robust DRIFTS calibrations for soil C can be developed for analysis of samples collected within the same geographical location, thus greatly reducing the need for combustion analysis.

C Determinations on a Variety of Data Sets and Instrument Configurations

The data in Table 19.3 are a compilation of results obtained from a variety of studies on soil collected in Maryland (Reeves and McCarty, 2001; Reeves et al., 1999, 2001, 2002). Details on the samples and analytes can be found in the indicated references. One advantage of using NIR spectra is that a greater variety of sample accessories is available compared to the MIDIR. Near-infrared results with various potentially useful accessories are shown in Table 19.3. Results with the use of the fiber-optic probes required the removal of some outliers to achieve results equal to the rotating cup, but overall there was little difference, thus indicating that a fiber-optic probe could be used for scanning samples in situ with good results. The

TABLE 19.3. Calibration results for various spectral configurations.[a]

Data set[b]	n	Cell	R^2	RMSD[c]	RD[d,e]
Near-Infrared spectra[a]					
Tillage study samples	179	Rotating cup	0.964	0.089	6.6
	180	Fiber optics[e]	0.943	0.110	8.3
	174[f]	Fiber optics[e]	0.955	0.086	6.6
Watershed set 1	544	Rotating cup	0.921	0.138	12.5
	524[f]	Rotating cup	0.948	0.110	10.0
Watershed set 2					
Field moist	136	Polybag	0.847	0.138	9.2
Dried only	136	Polybag	0.821	0.149	9.9
Dried and ground	136	Polybag	0.799	0.158	10.5
Dried and ground	136	Rotating cup	0.869	0.128	8.5
Mid-Infrared spectra					
Tillage study samples	180	Custom[g]	0.976	0.071	5.3
Watershed set 1	544	Custom	0.969	0.087	7.9
	529[f]	Custom	0.980	0.069	6.3
Watershed set 2	136	Custom	0.950	0.079	5.3

[a]All samples dried and ground unless otherwise noted, spectral regions as in Table 19.1 unless otherwise noted.
[b]Data sets: Tillage samples = from two sites in Maryland under plow-till and no-till cultivation; watershed = two sets of samples from small watershed at Beltsville, Maryland; western = 237 samples from western United States.
[c]RMSD = Root mean squared deviation.
[d]RD = RMSD/mean.
[e]Spectral region = 1100–1900 (all) + 1900–2300 with every 20 averaged.
[f]Outliers removed according to software recommendations.
[g]Custom transport cell shown in Figure 19.1.

samples under "Watershed 1 and 2" were more diverse in composition, having been taken from the more extensive landscape of a complete watershed as opposed to that limited for cultivated fields, and this shows in the slightly poorer results achieved (Reeves et al., 2002).

The second set consisted of a subset of surface soils from the larger set of 544 and thus should be compared only within that set for different sampling methods. Interestingly, the poorest results were found for the dried and ground samples using the polybags in the large sample transport cell which allows an area 2 × 15 cm to be scanned. It was also noted that the overall absorbance levels for these samples were lower than the others. We believe that two factors affect the results obtained using polybags with dried soils that make this combination undesirable. First, there is potential interference caused by the presence of the bag. Second, it appears that with dried, ground samples, the sample surface becomes more mirrorlike and penetration depth is reduced, resulting in a weaker sample signal. The combined result is that the bag spectrum becomes a more significant interference. Thus, although convenient, the use of polybags to hold soil samples is not recommended unless field moist samples are to be scanned, as results for those samples were about the same in the polybags as for dried, ground samples scanned in the rotating cup. The DRIFTS results were all obtained by use of the dried, ground samples, and these results showed similar variations for the different soil sets as observed for NIR. But, again, DRIFTS was always better in direct set-to-set comparisons.

Noncarbon Calibrations

While the determination of C content is a primary interest for C sequestration, spectroscopy may also be very useful for the determination of other soil composition parameters of value for precision farming, etc. (McCarty et al., 2002). The results in Tables 19.4 through 19.6 are for DRIFTS determinations of a variety of soil parameters for a set of 180 soils from two sites in Maryland (Reeves et al., 1999, 2001). Only results for the DRIFTS are shown, but similar results were obtained for the NIR. As demonstrated in Table 19.4, calibrations for a variety of measures beyond total C were possible, including several measures of biologically active N. These results would indicate that similar determinations for forms of labile versus nonlabile C should be possible. Such determinations would be very informative in indicating how much of the C found in a soil sample would be likely to be sequestered for any length of time. Although the measures of biological activity, as reflected by the enzyme activity measured, are interesting, the question remains as to what is actually being measured. The actual enzymes

TABLE 19.4. Mid-infrared calibrations for various analyses using set of 180 soils from two locations in Maryland.

Assay[a]	R^2	RMSD[b]	RD[c]
Calibration results using all 180 samples			
Loc	0.966	0.061	
Till	0.872	0.358	
pH	0.940	0.159	2.5
Tot-C	0.976	71.3	5.3
Tot-N	0.971	63.9	5.4
Act-N	0.901	25.7	15.5
Bio-M	0.829	17.5	19.3
Min-N	0.174	5.31	50.1
Calibration results using 174 samples			
Arylsulf.	0.720	13.2	23.4
Dehydro.	0.722	3.46	34.3
Nitrif. Pot.	0.771	2.26	24.5
Phos.	0.828	52.2	16.7
Urease	0.831	0.343	18.7

Source: Adapted from Reeves et al., 1999, 2001.
[a]LOC = location, TILL = tillage, TOT = total, ACT = active, BIO = biomass, MIN = mineralizable, Arylsulf = arylsulfatase, Dehydro = dehydrogenase, Nitrif. Pot = nitrification potential, Phos = phosphatase (see References for more information on methods used, etc.).
[b]RMSD = Root mean squared deviation.
[c]RD = RMSD/mean.

are unlikely to be seen due to the very low amounts present, thus these calibrations probably reflect changes or differences in the soil which are correlated with the enzyme levels. Such calibrations are called surrogate calibrations and can be very misleading when the correlations with the surrogate do not hold (Shenk et al., 2001). The fact that the calibration for pH performed so badly when one-third of the samples were used as a test set (Table 19.5) or when samples from one site were used to predict the samples from the second (Table 19.6) also indicated that calibrations for pH (also tillage, Table 19.6) are very sample dependent and probably do not directly determine pH either. Such effects were not seen in either case for total C, total N, active N, or biological N, but were to a smaller extent for the enzyme determinations.

TABLE 19.5. Mid-infrared test set results predicted using calibration based on two-thirds of samples (semirandom).

Assay[a]	R^2	RMSD[b]	RD[c]
Loc	0.988	0.115	
pH	0.730	0.333	5.1
Tot-C	0.964	93.9	6.9
Tot-N	0.946	88.7	7.9
Act-N	0.918	23.8	14
Bio-N	0.794	18.8	19.8
Arylsulf.	0.659	13.8	23.5
Dehydro.	0.668	3.97	38.0
Nitrif. Pot.	0.637	3.56	37.2
Phos.	0.698	60.6	20.3
Urease	0.789	0.458	23.8

Source: Adapted from Reeves et al., 1999, 2001.
[a]LOC = location, TILL = tillage, TOT = total, ACT = active, BIO = biomass, MIN = mineralizable, Arylsulf = arylsulfatase, Dehydro = dehydrogenase, Nitrif. Pot = nitrification potential, Phos = phosphatase (see References for more information on methods used, etc.).
[b]RMSD = root mean squared deviation.
[c]RD = RMSD / mean.

TABLE 19.6. Determination of Loc 2 samples using Loc 1–based calibration using mid-infrared spectra.

Assay[a]	R^2	RMSD[b]	Bias[c]
Tillage	0.134	4.44	4.28
pH	0.338	0.935	–0.849
Tot-C	0.949	1254	–390.3
Tot-N	0.946	179	–49.3
Act-N	0.893	121	–117
Bio-N	0.763	30.6	–21.9

Source: Adapted from Reeves et al., 1999, 2001.
[a]TOT = total, ACT = active, BIO = biomass (see References for more information on methods used, etc.).
[b]RMSD = root mean squared deviation.
[c]RD = RMSD / mean.

Mid-Infrared Calibration Plots

In Figures 19.5 through 19.10, calibration plots derived from various sample sets using DRIFTS are presented. In Figure 19.5, results are shown for a combined set of soils from Maryland and the two sets of acidified and nonacidified Great Plains soils. As can be seen, although a usable calibration using only 10 percent of the total samples could be developed, there

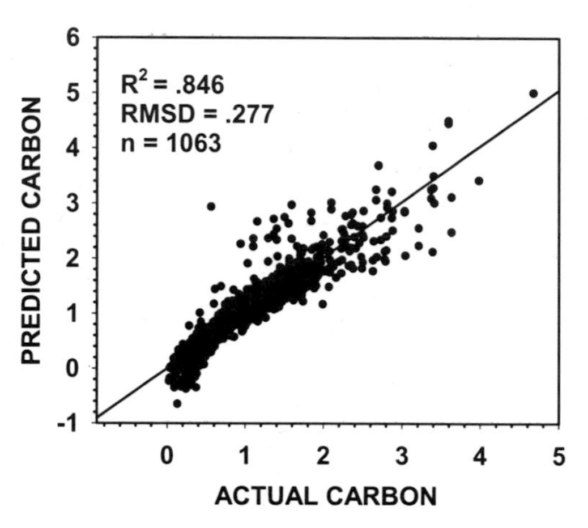

FIGURE 19.5. Mid-infrared calibration for C based on 10 percent of the total sample population selected on the basis of spectral diversity.

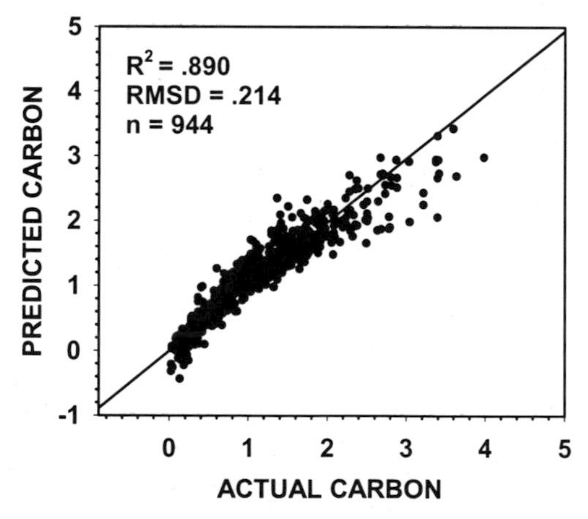

FIGURE 19.6. Mid-infrared calibration for C based on 20 percent of the total sample population selected on the basis of spectral diversity after removal of outliers and high C samples.

were several problems. Samples with C values above 2 percent were determined with less accuracy than those with lower C values, except for those with values near 0 which were generally underpredicted. Using 20 percent of the samples for calibration development and again removing samples with >5 percent C improved the results, but again samples with C >2 percent were poorly predicted. These results may indicate that it is not possible to develop accurate calibrations to cover a wide range of C contents, that the sample set represented by 20 percent of the total set was still not diverse enough for the remaining samples, or some other problem.

In Figures 19.7 through 19.9, the results obtained using a diverse set of samples obtained from across the entire United States are shown. As can be seen, while the calibrations for total C and organic C appear to be able to cover a range of C from 0 to 50 percent, the calibrations were all nonlinear and appeared to have the same problem as seen previously with very low C samples. Also, the calibration of inorganic C (Figure 19.9) was poor with a lot of scatter. Preliminary conclusions from these and the previous results were that calibrations could not be developed to cover a wide range of C content. However, unpublished results with over 1,100 samples from the Brazilian national collection covering the same range of C contents showed that linear calibrations could be developed for such samples, thus indicating another problem. One difference in the U.S. and Brazilian samples is that only the U.S. samples contain appreciable amounts of carbonates. Unpublished work has also demonstrated that the spectra of carbonate varies tremendously in the mid-infrared with concentration, even more than the distortions seen in organic components (Reeves, 2003; Reeves et al., 2005).

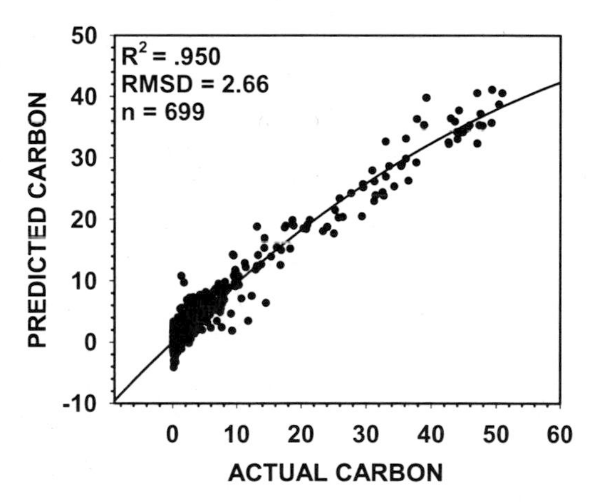

FIGURE 19.7. Mid-infrared calibration for total C for diverse set of soils from across the United States.

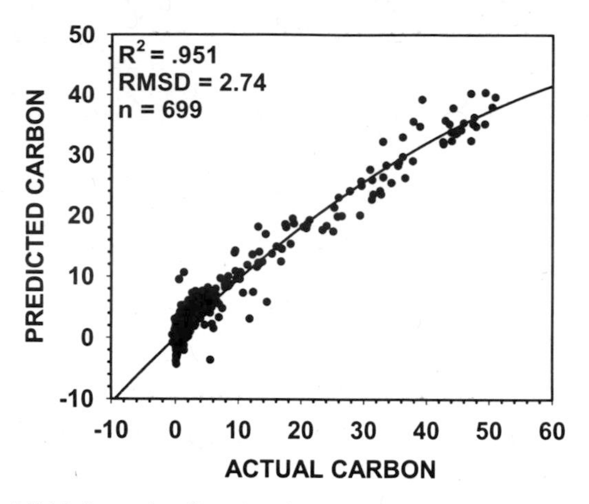

FIGURE 19.8. Mid-infrared calibration for organic C for diverse set of soils from across the United States.

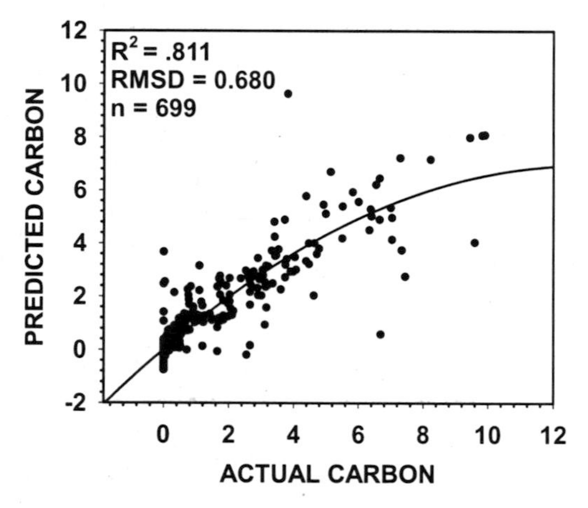

FIGURE 19.9. Mid-infrared calibration for inorganic C for diverse set of soils from across the United States.

Thus, the source of the problem may be that the carbonate spectra is so different in the various samples that the calibrations cannot handle the spectral variation which is caused by specular distortions related to concentration. Finally, even if a linear calibration can be designed to cover a wide range of C contents, calibrations built using a limited range of C contents may well be better for accurately determining C in specific samples. A comprehensive calibration might be first used to classify the sample for more accurate analysis by another more specific calibration. Only further work with more samples can answer this question.

Potential for Spatial Mapping

The data in Figure 19.10 show a comparison of C determinations by conventional means and DRIFTS. A calibration developed using two-thirds of the available samples was used to determine the remaining one-third whose results were used to generate the map shown. As seen, only slight differences exist between the two maps. While two-thirds of the samples were analyzed conventionally for the year shown, in future years only perhaps 10 percent would need to be analyzed for checking the calibration accuracy and/or be added to the existing data set to improve accuracy. Typically, an existing calibration can be expanded to cover similar samples in nearby fields by adding a relatively small number of samples to the original calibration set. This adds any uniqueness found in the new samples to the older set. Reeves et al. (1991) demonstrated this concept in work developing calibrations for forage grown under different environmental conditions.

CONCLUSIONS

Results using a variety of data sets have demonstrated that near- and mid-infrared spectroscopy (DRIFTS) show great potential for the determination of soil composition, especially soil C. While mid-infrared calibrations for soil C appear to be more accurate and robust, the use of DRIFTS for quantitative analysis is not nearly as developed as is NIR. It also may be more difficult to make field-portable MIDIR instrumentation. With reasonable input of resources, spectroscopy has the potential to generate the extensive databases on soil properties needed for generating spatial structure maps for soil properties that can then be used for site-specific management of agricultural lands. Results have also shown that accurate calibrations can be developed using near- or mid-infrared spectra for the determination of a number of compositional parameters, including total C, total N, pH, and many measures of biological activity as reflected by enzyme activities and measures of biologically active N. While efforts at determining the robustness of mid-infrared calibrations indicated that mid-infrared soil calibrations generally perform in a manner similar to NIR calibrations, differences found indicate that the basis for mid-infrared calibrations may at times be different.

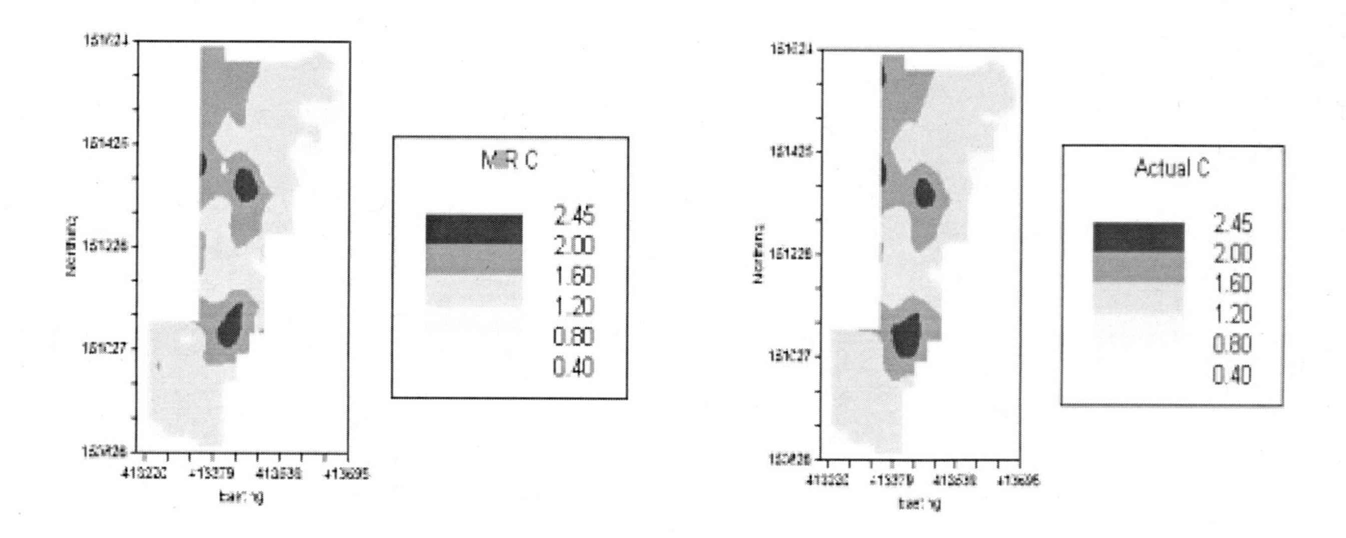

FIGURE 19.10. Results using a calibration developed using two-thirds of samples from a small watershed to develop a calibration to determine the remaining one-third. (See also color gallery.)

REFERENCES

Bremner, J.M. 1996. Nitrogen-total. In *Methods of soil analysis,* Part 3, *Chemical methods,* pp. 1085-1089. J.M. Bigham (Ed.). ASA-CSSA-SSSA, Madision, WI.

Burns, D.A. and E.W. Ciurczak (Eds.). 2001. *Handbook of near-infrared analysis,* 2nd ed. Marcel Dekker, Inc., New York.

Ebinger, M.H., M.L. Norfleet, D.D. Breshears, D.A. Cremers, M.J. Ferris, P.J. Unkefer, M.S. Lamb, K.L. Goddard, and C.W. Meyer, 2003. Extending the applicability of laser induced breakdown spectroscopy for total soil carbon measurement. *Soil Sci. Soc. Am. J.* 67:1616-1619.

Glascock, M.D. 2004. An overview of neutron activation analysis. Available online at www.missouri.edu/~glascock/naa_over.htm.

Janik, L.J., R.H. Merry, and J.O. Skjemstad. 1998. Can mid-infrared diffuse reflectance analysis replace soil extractions? *Aust. J. of Exper. Agric.* 38:681-696.

McCarty, G.W., J.B. Reeves III, V.B. Reeves, R.F. Follett, and J.M. Kimble. 2002. Mid-infrared and near-infrared diffuse reflectance spectroscopy for soil carbon measurement. *Soil Sci. Soc. Am. J.* 66:640-646.

Naes, T., T. Isaksson, T. Fearn, and T. Davies. 2002. *A user friendly guide to multivariate calibration and classification.* NIR Publications, Chichester, West Sussex, UK.

Nguyen, T.T., L.J. Janik, and M. Raupach. 1991. Diffuse reflectance infrared fourier transform (DRIFT) spectroscopy in soil studies. *Aust. J. Soil Res.* 29:49-67.

NRCS. 2004. USDA-NRCS ground penetrating radar program. Available online at www.nesoil.com/gpr.

PLSplus V2.1G. 1994. Galactic Industries Corporation, Salem, NH.

Reeves, J.B., III. 1996. Improvement in Fourier near- and mid-infrared diffuse reflectance spectroscopic calibrations through the use of a sample transport device. *Appl. Spectrosc.* 50:965-969.

Reeves, J.B., III. 2003. Mid-infrared diffuse reflectance spectroscopy: Is sample dilution with KBr neccesary, and if so, when. *American Laboratory* 35(8):24-28.

Reeves, J.B., III, T.H. Blosser, A.T. Balde, B.P. Glenn, and J. Vandersall. 1991. Near infrared spectroscopic analysis of forage samples digested in situ (nylon bag). *J. Dairy Sci.* 74:2664-2673.

Reeves, J.B., III, B.F. Francis, and S.K. Hamilton. 2005. Specular reflection and diffuse reflectance spectroscopy of soils. *Applied Spectroscopy* 59:39-46.

Reeves, J.B., III, and G.W. McCarty. 2001 Quantitative analysis of agricultural soils using near infrared reflectance spectroscopy and a fiber-optic probe. *J. Near Infrared Spectrosc.* 9:25-43.

Reeves, J.B., III, G.W. McCarty, and J.J. Meisenger. 1999. Near infrared reflectance spectrocopy for the analysis of agricultural soils. *J. Near Infrared Spectrosc.* 7:179-193.

Reeves, J.B., III, G.W. McCarty, T.V. Mimmo, V.B. Reeves, R.F. Follett, J.M. Kimble, and G.C. Galletti. 2002. Spectroscopic calibrations for the determination of carbon in soils. 17th WCSS, Thialand, Paper no. 398, 398-1 to 9.

Reeves, J.B., III, G.W. McCarty, and V.B. Reeves. 2001. Mid-infrared diffuse reflectance spectroscopy for the quantitative analysis of agricultural soils. *J. Agric. Food Chem.* 49:766-772.

Roberts, C., J.J. Workman, and J.B. Reeves, III (Eds.). 2004. *NIR in agriculture.* ASA-CSSA-SSSA, Madision, WI.

Shenk, J.S., J.J. Workman, and M.O. Westerhaus. 2001. Application of NIR spectroscopy to agricultural products. In *Handbook of near-infrared analysis,* 2nd ed., pp. 385-431. D.A. Burns and E.W. Ciurczak (Eds.). Marcel Dekker, Inc., New York.

Williams, P. and K. Norris, 2001. *Near-infrared technology in the agricultural and food industries.* Amer. Ass. of Cereal Chemists, Inc., St. Paul, MN.

Chapter 20

In Situ Noninvasive Soil Carbon Analysis: Sample Size and Geostatistical Considerations

Lucian Wielopolski

INTRODUCTION

The consequences of global warming, promoted by anthropogenic CO_2 emissions into the atmosphere, are partially mitigated by the photosynthesis of terrestrial ecosystems that act as an atmospheric CO_2 scrubber, thereby sequestering carbon belowground, in the topsoil, and in the aboveground litter and vegetation. Soil carbon sequestration is one of several programs that encompass the study of point-source capture from industrial activities, CO_2 conversion, and storage in geological formations and the oceans. Sequestration in terrestrial ecosystems, albeit being recognized as a short-term solution, provides an immediate, low-cost solution. Soil carbon sequestration includes planting trees, employing no-till farming, preserving forests, reclaiming land, and other agricultural practices. Carbon sequestration also is promoted by putting forward a system for national and international "carbon credits" that can be negotiated and traded between CO_2 producers and sequesters.

A central theme of research on terrestrial carbon sequestration over the past decade evolved around the recognition that there is a large terrestrial sink for the atmospheric CO_2 (Sarmiento and Wofsy, 1996). The magnitude of this sink was estimated to be on the order of 1.5 ± 1.0 Gt C per year for 1980-1989 (Schimel et al., 1996), and as having increased to >2.5 Gt C per year in the past decade, which is about 25 percent of the annual CO_2 emission into the atmosphere. Thus, the potential consequences of the steady increase in atmospheric emissions are partially mitigated by photosynthesis

This work was supported by the U.S. Department of Energy under contract No. DE-AC02-98CH10886.

Carbon Sequestration in Soils of Latin America
Published by The Haworth Press, Inc., 2006.
doi:10.1300/5755_21

in plants that removes CO_2 from the atmosphere and sequesters it in soil. The corollary benefits of carbon sequestration are the increased quality and fertility of the soil, and a reduction in erosion. In addition, under private emission-trading strategy, farmers would be able to sell carbon "credits" to a CO_2-emitting industry. Estimates of this market in the United States are about $1 to 5 billion per year for the next 30 to 40 years (Rice, 2004). These two issues, namely, carbon sequestration and trade with carbon credits, require the development of novel nondestructive instrumentation to facilitate a better understanding of the belowground carbon processes, while at the same time providing a direct quantitative measure or an index of the belowground carbon stores for the carbon credit trade.

The current standard method for quantifying carbon in soil involves extracting a core and analyzing samples from it (Taylor et al. 1991). This traditional approach is labor-intensive, slow, destructive, and, consequently, very limited in its utility and scope. Three newly emerging methods to measure carbon in soil in situ are laser-induced breakdown spectroscopy (LIBS) (Cramers et al., 2001), near- and mid-infrared spectroscopy (McCarty et al., 2002), and inelastic neutron scattering (INS) (Wielopolski et al., 2004). The first two methods present improvements over traditional core sampling, but both are destructive. In the first case, small volumes of about 50 microliters from an extracted core are vaporized, and the spectral emission is measured. In the second case, a sensor mounted on a tip of a shank ploughs through the soil at a set depth and senses the carbon to a depth of few millimeters. The third method, INS, is based on fast neutron scattering and concurrent measurement of characteristic gamma rays induced in carbon. This process is denoted as $^{12}C(n, n',\gamma)\,^{12}C$, in which n and n' are the incident and scattered neutrons, respectively (Wielopolski et al., 2000, 2003). The INS method is nondestructive and can be used repetitively for measurements at exactly the same spot over extended periods or, alternatively, employed in a continuous scanning mode over large areas. The basic methodology and its use were described by Csikai (1987) and Nargolwalla and Przybylowicz (1973), respectively. An INS system to measure carbon in soil was reported by Wielopolski et al. (2004). The INS system, with its calibration and attributes, is briefly described in the subsequent sections with special attention to the distinctly large volumes it can measure.

ATTRIBUTES OF THE INS SYSTEM

The INS system consists of a neutron generator (NG), an electrical device that can generate neutrons based on the (d,t) reaction, a shadow shielding placed between the NG and the detectors, and a NaI scintillation detection system. The original INS system was calibrated by placing the NG and the detection assembly directly on the soil in a sandpit filled with a homogeneous mixture of sand and carbon at 0, 2.5, 5, and 10 percent by weight. This setup, shown in Plate 20.1, generated the calibration line presented in Figure 20.1 (Wielopolski et al., 2004). The current field-deployable INS system was modified to include three detectors instead of one, and mounted on a stand about 30 cm above the ground. The height of the stand was optimized to maximize the carbon signal's yield and was designed to accommodate the various residues left on the ground in no-till fields, thus making it suitable as a field-scanning system when mounted on a mobile cart; the new setup is shown in Plate 20.2. The new system's configuration improved the sensitivity to carbon by a factor of 7.3, with a concurrent increase in the background by 6.2, thus improving its overall performance by about a factor of 3 $[7.3/\sqrt{(6.2)}]$.

Based on the calibration shown in Figure 20.2, the INS results were compared with those from chemical analysis of the top 5 cm of core samples taken from different sites; the findings are summarized in Table 20.1. It is important to point out that all the results are expressed in terms of carbon

PLATE 20.1. The original INS system placed on the ground of sandpit filled with sand and 10 percent carbon by weight.

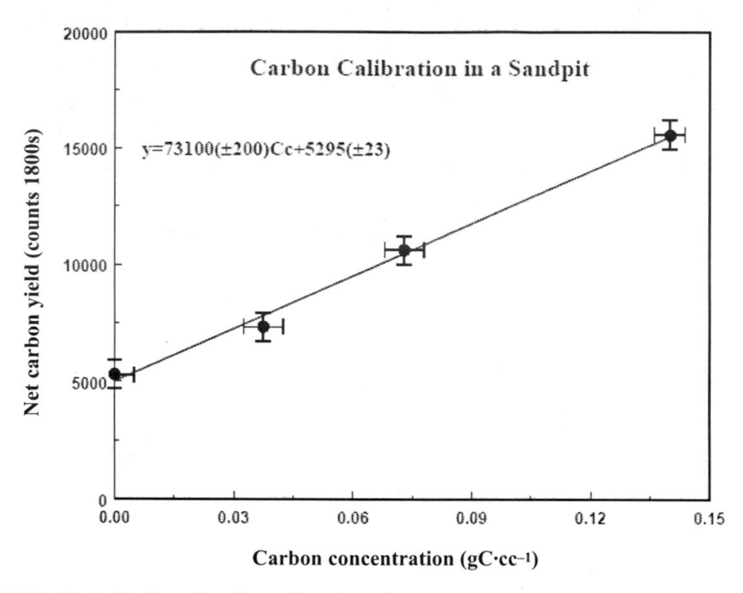

FIGURE 20.1. Calibration line of carbon in a sandpit at 0, 2.5, 5, and 10 percent carbon by weight and soil bulk density 1.4 g·cc^{-1}.

PLATE 20.2. An INS system mounted on a stand for in-field stationary measurements.

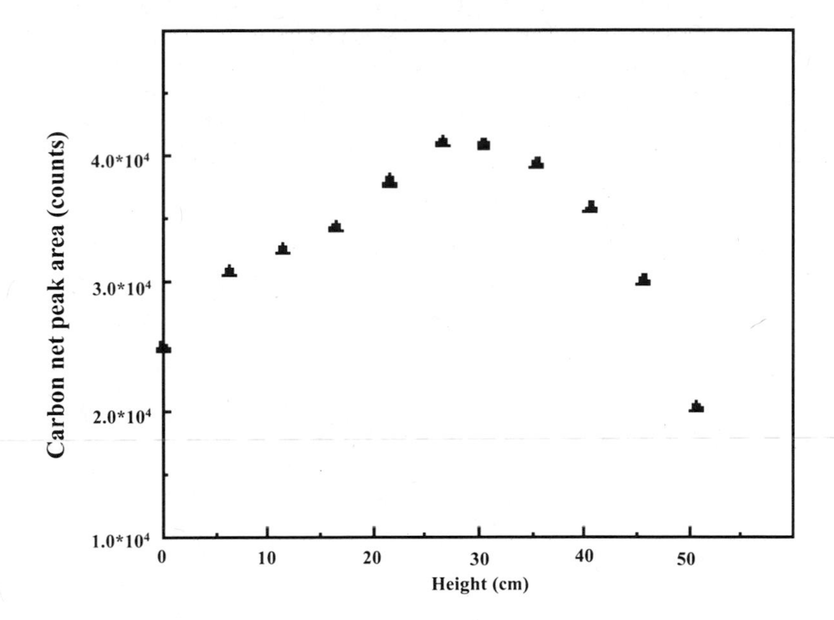

FIGURE 20.2. Net carbon yield as a function of the source and height of the detection system above the ground level.

TABLE 20.1. Comparison of INS results versus chemical analysis of the top 5 cm of soil.

Soil site	INS (g C per cc)	Chem. analysis (mean g C per cc)	Ratio
Pine stand (w,l)	0.099 ± 0.005	—	
Pine stand (w/o,l)	0.079 ± 0.005	0.073 ± 0.021	1.082
Oak forest (w/o,l)	0.072 ± 0.004	0.085 ± 0.017	0.847
Sand patch	0.026 ± 0.003	0.025 ± 0.002	1.040
Sandy soil	0.091 ± 0.007	0.104 ± 0.019	0.875
Sand pit (Cal.)	0.0	0.0004 ±	

density (g C·cc^{-1}) because the INS instrument basically "sees" a constant volume, making the carbon signal proportional to the number of carbon atoms present in that volume.

The INS results in Table 20.1 can be expressed mathematically as an integral of carbon concentration, $C_c(x,y,z)$, with a weight function, w(x,y,z), that depends on the distribution of neutron flux, the reaction rates, the gamma-ray yield, and the efficiency of gamma-ray counting. Thus, in general, the carbon yield can be expressed as

$$C_Y = k\, C_c(x,y,z)w(x,y,z)dxdydz \qquad (20.1)$$

where k is a proportionality constant and the integration is carried out over the volume seen by the detectors. When the profile of carbon depth is constant, i.e., uniformly distributed with depth, Equation 20.1 simplifies to represent the mean value for carbon multiplied by the integrated transfer function of the system. Remember that, by definition, the mean value of carbon concentration in soil is an imaginary value, such that if it had been uniformly distributed in a given volume it would yield the same amount of total carbon as the true distribution in that volume. Thus, in general, the weighted mean of soil carbon is given in Equation 20.2:

$$M_{Cc} = \frac{C_c(x, y, z)w(x, y, z)dxdydz}{dxdydz} \qquad (20.2)$$

where M_{Cc} is the weighted mean carbon concentration, C_c is its spatial distribution, and the integration is performed over the entire volume of interest. Since the system sees a constant volume combining Equations 20.1 and 20.2, its response can be calibrated in terms of the mean carbon concentration. Clearly, this calibration depends on the carbon profile in the soil, so that further investigation is needed about the sensitivity of the calibration to the variations in the profile's parameters. The significance of this calibration increases with an increase in the sampled volume, which is relatively large, hundreds of liters, for the INS system.

The volume sampled by the INS system is defined by the intersection of the space irradiated by neutrons with that subtended by the solid angle of the detectors. Because of the propagation of the nuclear radiation, this space is semi-infinite; however, for any practical purposes 90 or 95 percent of the signal is derived from a finite sample size. To assess the size of that volume, a Monte Carlo Neutron Photon (MCNP) probabilistic transport code (Breismeister, 1993; Wielopolski et al., 2005) first was applied to estimate the volume in soil in which the neutron flux is reduced to the 50 percent and 10 percent levels; these volumes were 11.5 and 277 liters, and assuming a soil bulk density of 1.4 $g{\cdot}cc^{-1}$, they correspond to 17.5 and 416 kg, respectively. Figure 20.3 shows the neutron isoflux distribution in soil, with a projection of specific isoflux contours in Figure 20.4.

An additional consequence of sampling large volumes of soil and, in particular, with a large footprint, such as that of the INS system, is the averaging effect on the lateral variability of the carbon distribution. Although spatial and temporal heterogeneities of in-field measurements are well recognized, these are minimized or ignored on the account of normal statistics. It is assumed that (1) the observations are spatially or temporarily independent of each other, and (2) the mean values are based on normally dis-

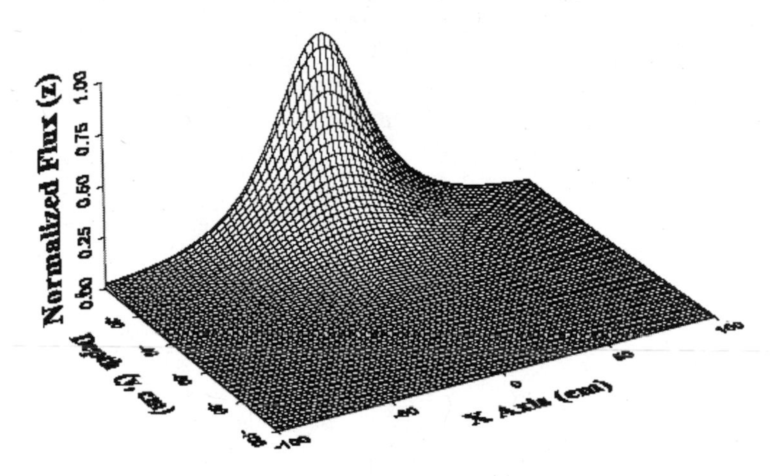

FIGURE 20.3. Three-dimensional display of the relative neutron flux in soil from a point-source position 30 cm above the ground. The neutron flux is normalized at (0,0).

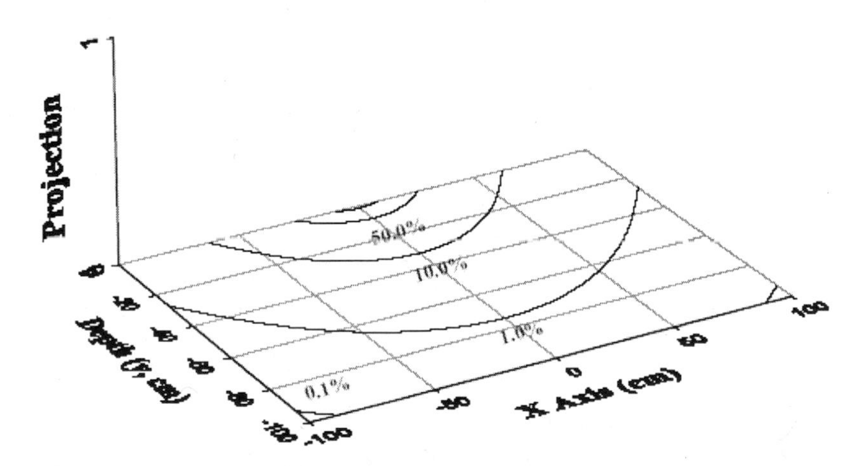

FIGURE 20.4. Isoflux contour lines at 0.1, 1.0, 10.0, 50, and 90 percent levels for a source 30 cm above the ground.

tributed sets of observations. However, the validity of these two assumptions is seldom investigated. The largeness of the sampled volume also affects error propagation in estimates of the field parameters, thus affecting the number of samples required for a desired level of confidence in the estimates. For example, 10,000 points, representing carbon atoms in a unit volume, uniformly distributed in space and normally distributed with depth are shown in Figures 20.5 and 20.6, respectively. The sampling volume is denoted in these figures by a yellow cube centered in each panel. After uniformly sampling these two distributions with a variable cube size, i.e., a variable sampling volume, the estimated mean values of the number of points encountered inside the sampling volume in 100 samples randomly drawn from each population differ significantly. For the uniform distribution, the mean value fluctuates around the true mean, while it is biased for the normal distribution. Similarly, the estimated SD for each population differs markedly with an increase in the sampled volume. For a uniform distribution the SD monotonically decreases, while for the normal distribution it remains approximately constant. The results of these simplified simulations are shown in Figures 20.7 and 20.8.

CONCLUSION

I discuss a new approach for quantitative carbon analysis in soil based on INS. Although this INS method is not simple, it offers critical advantages not available with other newly emerging modalities. The key advantages of the INS system include the following:

1. It is a nondestructive method, i.e., no samples of any kind are taken. A neutron generator placed above the ground irradiates the soil, stimulating carbon characteristic gamma-ray emission that is counted by a detection system also placed above the ground.
2. The INS system can undertake multielemental analysis, expanding its usefulness.
3. It can be used either in static or scanning modes.
4. The volume sampled by the INS method is large with a large footprint; when operating in a scanning mode, the sampled volume is continuous.
5. Except for a moderate initial cost of about $100,000 for the system, no additional expenses are required for its operation over two to three years, after which a NG has to be replenished with a new tube at an approximate cost of $10,000, regardless of the number of sites analyzed.

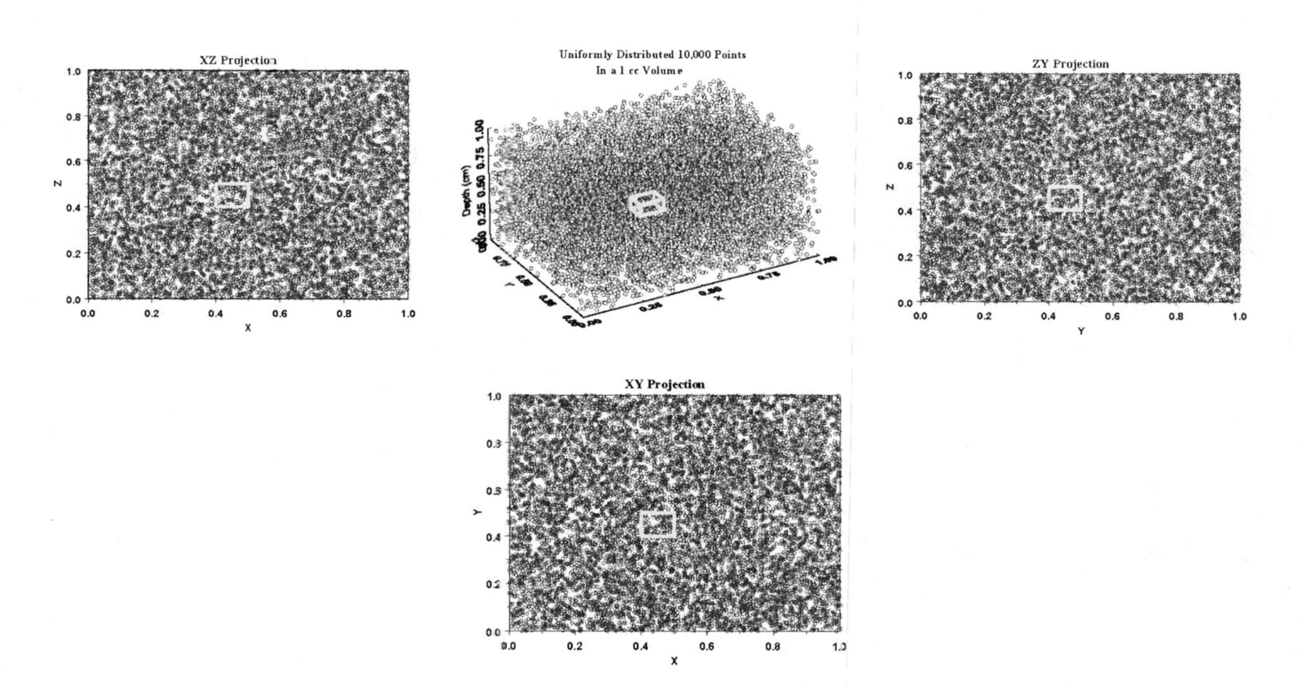

FIGURE 20.5. Ten-thousand points uniformly distributed in a unit volume (center) and projection on each plane. The sampling volume is marked in the center of each planel. (See also color gallery.)

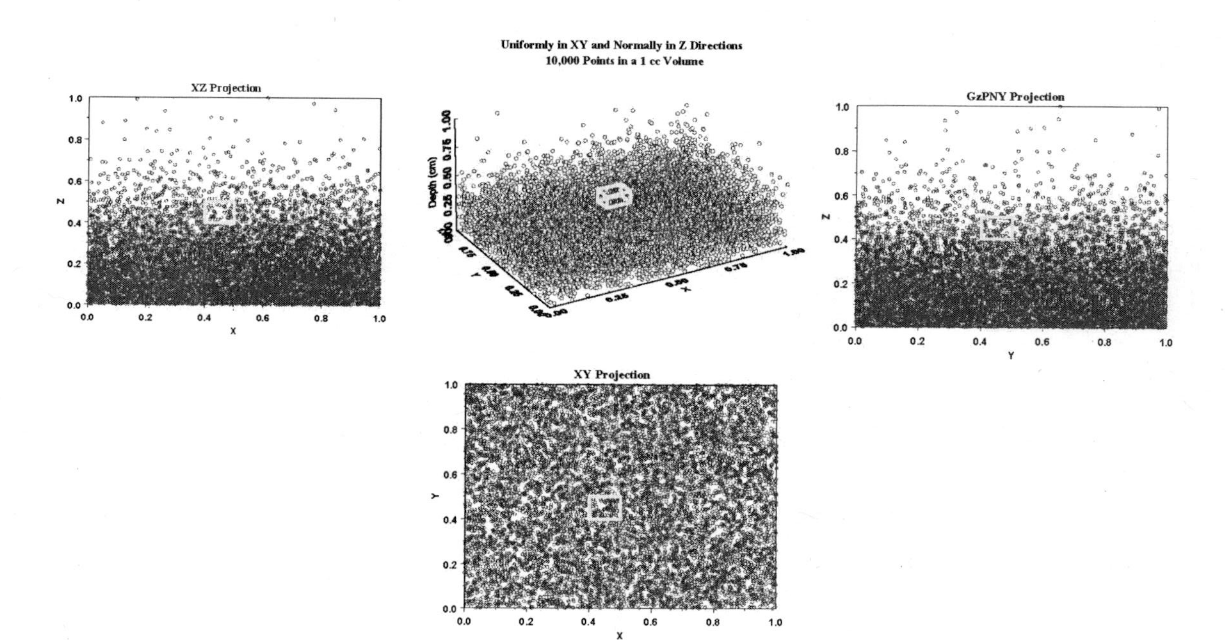

FIGURE 20.6. Ten-thousand points uniformly distributed in the lateral directions and normally distributed with depth in a unit volume (center) and projection on each plane. The sampling volume is marked in the center of each panel. (See also color gallery.)

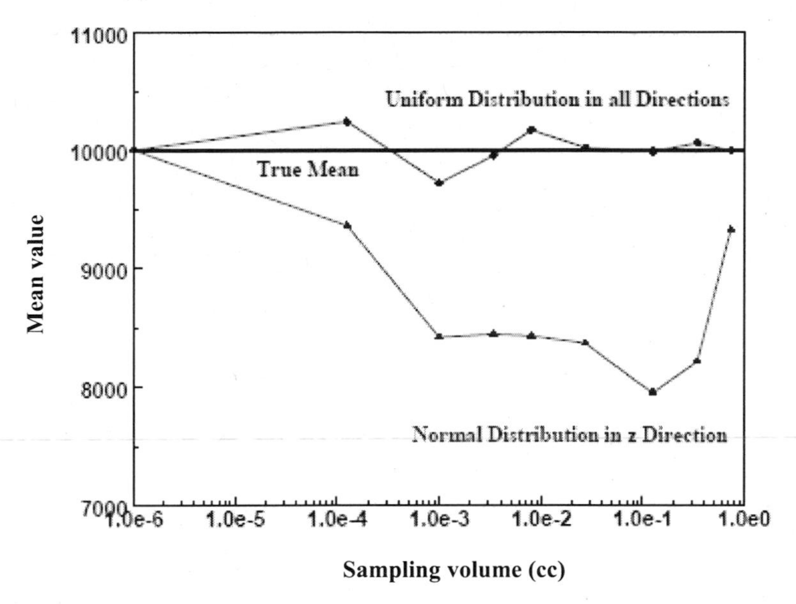

FIGURE 20.7. Variations in the estimates of the sampled mean values for uniform and normally distributed points.

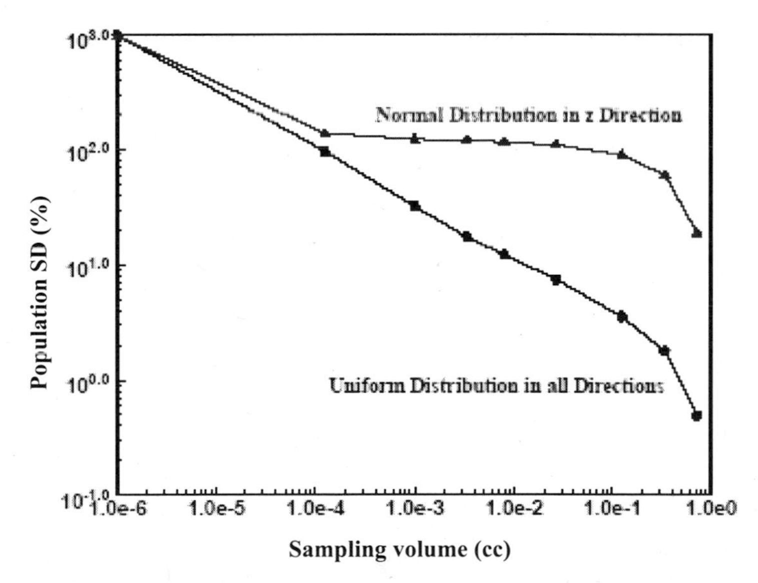

FIGURE 20.8. Variations in the estimates of the samples' standard variation (SD) values for uniform and normally distributed points.

In light of these characteristics, the INS system appears invaluable for monitoring changes in the carbon content in the field. For this purpose no calibration is required; by establishing a carbon index, changes in carbon yield can be followed with time in exactly the same location, thus giving a percent change. On the other hand, with calibration, it can be used to determine the carbon stock in the ground, thus estimating the soil's carbon inventory. However, this requires revising the standard practices for deciding upon the number of sites required to attain a given confidence level, in particular for the purposes of upward scaling. Then, geostatistical considerations should be incorporated in considering properly the averaging effects of the large volumes sampled by the INS system that would require revising standard practices in the field for determining the number of spots to be sampled.

It is highly desirable to assess properly the sampled volume for reporting the absolute value of the measured carbon. At the same time, increasing the number of detectors surrounding the NG can reduce error propagation. In the present work, only the volume irradiated by the neutrons was estimated. It should be pointed out that the carbon yield is also affected by the neutron energy spectrum that changes with depth. Thus, all these considerations must be considered carefully when evaluating the detector's configuration and the resulting counting efficiency.

In summary, INS system is a novel approach for nondestructive carbon analysis in soil with very unique features. It should contribute in assessing soil carbon inventories and assist in understanding belowground carbon processes. The complexity of carbon distribution in soil requires special attention when calibrating the INS system, and a consensus developed on the most favorable way to report carbon abundance. Clearly, this will affect the calibration procedures.

REFERENCES

Breismeister J.F., ed. (1993). MCNP-A General Purpose Monte Carlo N-Particle Transport Code Version 4A. Los Alamos National Laboratory, New Mexico, LA-12625-M.

Cramers D.A., Ebinger M.H., Breshears D.D., Unkefer P.J., Kammerdiener S.A, Ferris M.J., Catlett K.M., and Brown J.R. (2001). Measuring total soil carbon with laser-induced breakdown spectroscopy (LIBS). *J. Environ. Qual.* 30: 2202-2206.

Csikai J. (1987). *CRC handbook of fast neutron generators,* Vol. 1. CRC Press, Inc., Boca Raton, Florida.

Dane J.H., and Topp G.C., eds. (2002). *Methods of soil analysis: Part 4—physical methods.* Agron. Monogr. 5. Soil Science Society America, Inc., Madison, Wisconsin.

McCarty G.W., Reeves J.B. III, Reeves V.B., Follett R.F., and Kimble J.M. (2002). Mid-infrared and near-infrared diffuse reflectance spectroscopy for soil carbon measurement. *Soil Sci. Soc. Am. J.* 66: 640-646.

Nargolwalla S.S., and Przybylowicz E.P. (1973). *Activation analysis with neutron generators.* John Wiley & Sons, New York/London/Sydney/Toronto.

Rice C.W. (2004). Background. Available online at http://www.casmgs.colostate.edu. A CASMGS Forum, Texas A&M University, Bryan-College Station, Texas, January 20-22.

Sarmiento J.L., and Wofsy S.C. (1996). A U.S. carbon cycle science plan. A Report of the Carbon and Climate Working Group. U.S. Global Change Research Program, Washington, DC.

Schimel D., Alves D., Enting I., Heimann M., Joos E., Raynaud D., and Wigley T. (1996). CO2 and the carbon cycle. In *Climate change 1995,* eds. J.T. Houghton, L.G.M. Filbo, B.A. Callander, N. Harris, A. Kattenberg, and K. Maskell. Cambridge University Press, Cambridge, United Kingdom, pp. 76-78.

Taylor H.M., Upchurch D.R., Brown J.M., and Rogers H.H. (1991). Some methods of root investigations. In *Plants roots and their environment,* eds. B.L. McMichael and H. Persson. Elsevier Science Publisher B.V., pp. 553-564.

Wielopolski L., Mitra S., Hendrey G., Orion I., Prior S., Rogers H., Runion B., and Torbert A. (2004). Non-destructive Soil Carbon Analyzer (ND—SCA), BNL Report No.72200-2004.

Wielopolski L., Mitra S., Hendrey G., Rogers H., Torbert A., and Prior S. (2003). Non-destructive in situ soil carbon analysis: Principle and results. Proceedings of the Second Annual Conference on Carbon Sequestration, May 3-5, Article No. 225.

Wielopolski L., Orion I., Hendrey G., and Roger H. (2000). Soil carbon measurements using inelastic neutron scattering. *IEEE Trans. Nucl. Sciences* 47: 914-917.

Wielopolski L., Song Z., Orion I., Hanson A.L., and Hendrey G. (2005). Basic considerations for Monte Carlo calculations in soil. *Applied Radiation and Isotopes* 62: 97-107.

Chapter 21

Methods and Tools for Designing a Pilot Soil Carbon Sequestration Project

R. C. Izaurralde
C. W. Rice

INTRODUCTION

Carbon (C) sequestration in vegetation and soils has reached international consensus as one technology that could be used to mitigate climate change (IPCC, 2000). Brown et al. (1996) estimated that about 38 Pg C could be sequestered over a 50-year period by implementing afforestation, reforestation, and agroforestry practices over 345 Mha. Likewise, Cole et al. (1997) estimated that 40 Pg C (two-thirds of historical losses of soil organic C [SOC] from agricultural land) could be sequestered in soil over a 50- to 100-year span by applying a suite of improved agricultural practices (e.g., no-till, nutrient management, diversified crop rotations). These potentials appear small relative to the potential of other forms of sequestration (e.g., 2,900 Pg C for geologic sequestration, Dooley et al., 2004). However, terrestrial C sequestration could have an immediate application in climate change mitigation due to its availability, relatively low cost, and associated environmental benefits.

Over the past decade or so, terrestrial C sequestration has evolved from just a concept to a very active field of research and technology development. Numerous workshops and symposia have been held to discuss the issues of science and economics associated with soil C sequestration (Lal et al., 1995a,b, 1998a,b; Rosenberg and Izaurralde, 2001). Research centers

We are grateful to J. J. Dooley for his critical review of the manuscript. The manuscript was prepared with support from USDA Consortium for Agricultural Soils Mitigation of Greenhouse Gases, the USDOE–Office of Science Carbon Sequestration in Terrestrial Ecosystems program, and the NASA Office of Earth Science Applications Division.

Carbon Sequestration in Soils of Latin America

doi:10.1300/5755_22

have been organized to foster coordinated research on soil C sequestration. Such is the case of CSiTE (Carbon Sequestration in Terrestrial Ecosystems, csite.esd.ornl.gov/), sponsored since 1999 by the U.S. Department of Energy, and CASMGS (Consortium for Agricultural Soils Mitigation of Greenhouse Gases, www.casmgs.colostate.edu/), supported by the U.S. Department of Agriculture and the U.S. Environmental Protection Agency. Recent literature reviews and analyses have synthesized information on soil C sequestration rates and potential at national and global scales (Conant et al., 2001; West and Post, 2002; Lal et al., 1999; Schuman et al., 2002; VandenBygaart et al., 2003). Methods have been documented in the literature for calculating changes in soil C as a result of changes in land use and management (Ellert et al., 2002) and for conducting full C accounting of soil C sequestration (Robertson et al., 2000). A few pilot projects have been implemented in order to learn how to document soil C change at regional scales (Izaurralde, 2005). All of these examples demonstrate considerable progress in making soil C sequestration a realistic tool to mitigate climate change. However, in order to make a significant contribution to addressing climate change, soil C sequestration will have to be deployed on a global scale while meeting certain economic and environmental criteria.

In theory, learning about the spatial distribution of soil C is not that difficult. In fact, soil surveyors have been preparing maps of the distribution of soil C for many decades, although they have treated soil C content (i.e., soil organic matter content) as a static variable. Monitoring temporal changes of soil C under a given management regime has also been done on many research plots around the world. Methodologies and tools are needed to estimate, with some degree of precision, the temporal and spatial changes in soil C that would occur as a result of changes in management. These methods and tools would be essential components of any CO_2 emissions permit trading system that operates within national or international boundaries.

The objective of this chapter is to synthesize methodological advances made to detect gross changes in soil C at the field and regional scales as a result of changes in management or land use. Some of these ideas emerged from discussions about the potential of soil C sequestration in Latin America during a workshop held June 2-6, 2004, in Piracicaba, Brazil. The workshop was organized by the University of Sao Paulo and Ohio State University and sponsored by the U.S. Department of State, U.S. Department of Agriculture, and the Inter-American Institute for Global Change Research.

REGIONAL CONTEXT

With a population larger than 500 million and a land base representing about 15.2 percent (20.5×10^6 km^2) of the world's land area, Latin America is a region with dynamic demographics and land use. The focus on Latin America is timely because of the large impact, both positive and negative, that tropical deforestation and agricultural activities have on the balance of greenhouse gases. Although the rates of tropical deforestation in Brazil, Bolivia, and Venezuela appear to be decreasing (Tucker and Townshend, 2000), they still reflect a high degree of agricultural expansion and exploitation of forest resources. Concurrently, agricultural production in temperate zones of Argentina, Uruguay, Brazil, Paraguay, and Bolivia is changing significantly with the adoption of no-till practices (or direct seeding, as it is known in the region) (Figure 21.1). These significant changes in land use and management have prompted large-scale field studies to monitor the evolution of soil C and associated environmental impacts (Casas, 2002; Casas et al., 2004; Vergara Sánchez et al., 2004).

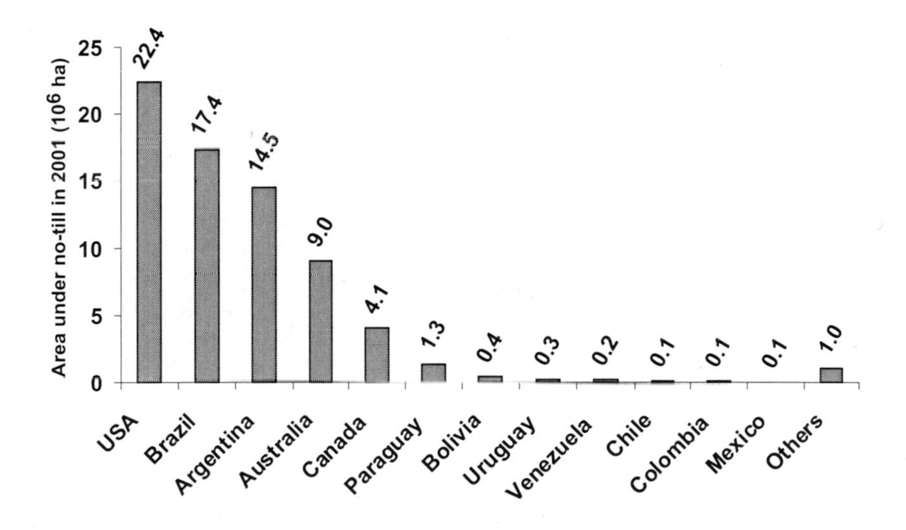

FIGURE 21.1. Estimated area under no-till worldwide in 2001. Total estimated area 71×10^6 ha. (*Sources:* Data from CTIC [Conservation Tillage Information Center, USA], AAPRESID [Asociacion Argentina de Productores de Siembra Directa, Argentina], and R. Derpsch [www.rolf-derpsch.com/startseite.htm]).

DETECTING AND SCALING CHANGES IN SOIL CARBON BY DIRECT MEASUREMENTS, SIMULATION MODELING, AND REMOTE SENSING INTERPRETATION

Base Data

A successful pilot project to test methods and tools to monitor soil C sequestration will require consideration of physical and economic aspects. The physical aspects deal mainly with (1) the selection of the land units to be measured and monitored, and (2) the acquisition and refinement of climate, soil, and management data. A select set of these factors will be treated in more detail in the following sections. The economic aspects include those associated with transactional, monitoring, and auditing costs. The reader is referred to Post et al. (2004) for an overview of the economic components for determining the cost of C sequestration as well as the value of cobenefits and discounts (e.g., additionality, uncertainty, leakage, and permanence).

Sampling Design and Data

Special attention must be given to the soil sampling design of soil C projects because the determination of short-term changes in SOC content induced by management is usually small (e.g., 0.1 to 0.5 kg $C \cdot m^{-2}$) relative to the organic C stocks present in soils (e.g., 2 to 8 kg $C \cdot m^{-2}$). There are two key issues regarding the detection of these changes. The first issue refers to the selection of controls or baseline to measure management-induced SOC changes. The second refers to the number of samples needed to establish statistically significant differences in SOC stocks not only in time but in space as well.

The first issue emerges because SOC is a dynamic property. Because of this, a time zero sampling may or may not be an appropriate control for measuring SOC changes (Izaurralde et al., 2001; McGill et al., 1996). This can be appreciated graphically in Figure 21.2 for a given set of SOC trajectories after time 0 (point O). Assuming a C sequestering practice effectively produces a C increase in soil (trajectory OD), then the only time in which SOC sequestration is measured correctly is when SOC content is at steady state (segment DB). This segment would have resulted in an underestimation of SOC sequestration had the conventional practice caused further losses in SOC content (trajectory OA, segment BA). Had the soil under the conventional practice been gaining SOC already (trajectory OC, segment

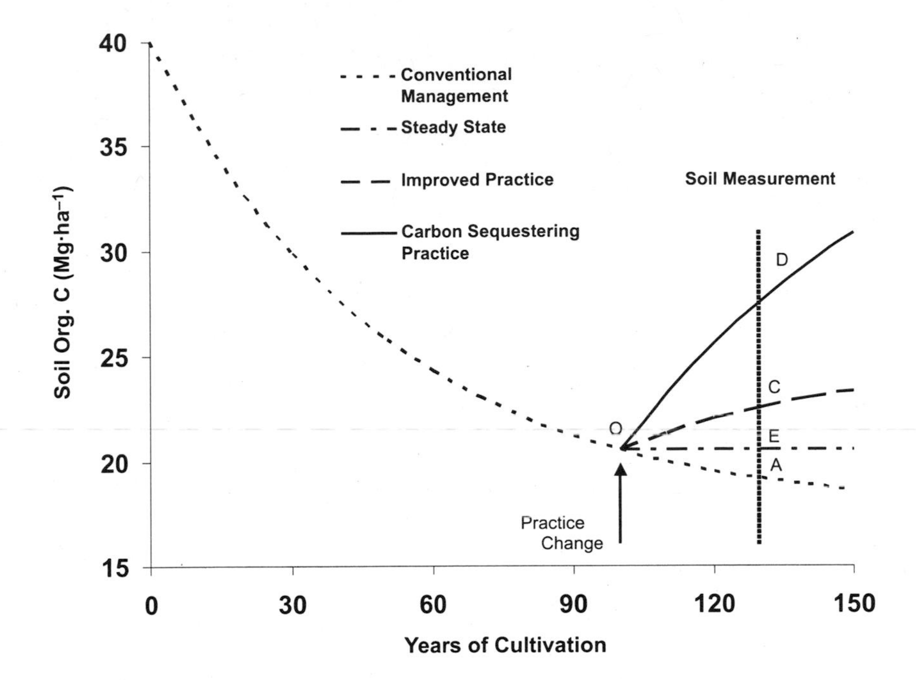

FIGURE 21.2. Possible influence of SOC dynamics on SOC sequestration rates.

CB), then the C sequestration estimate (segment DB) would have represented an overestimate of true sequestration (segment DC).

Keeping a plot under conventional practice during the sequestration period would allow for a side-by-side comparison between conventional and improved practices at time *t*. If the conventional practice kept losing SOC, then segment BA would represent a measure of the "C loss avoidance." McGill et al. (1996) observed that from the point of view of atmospheric C, both forms of scqucstration arc important. Availability of data for both types of controls appears to be the best solution to account for the impact of nonsteady-state conditions of SOC content on soil C change determinations.

An understanding of the second issue can be gained by examining the statistical formula Equation 21.1, which establishes that a given difference in SOC content (δ, kg C·m^{-2}) is directly proportional to the standard deviation of SOC content (σ, kg C·m^{-2}) and inversely proportional to the square root of the sample size. Graphically, it can be observed that a large increase in sample size is needed to detect significant statistical differences between two samples when the variance of SOC content [σ^2, (kg C·m^{-2})^2] increases (Figure 21.3) (Izaurralde, McGill, et al., 1998; Garten and Wullschleger, 1999).

$$\delta = \frac{\left(Z_{\alpha/2} + Z_{\beta}\right)\sigma}{\sqrt{\eta}} \tag{21.1}$$

Clearly, there is a need to reduce the number of samples to reduce transactional costs while maintaining the statistical power to detect changes in SOC at the field scale caused by management. Two approaches have been developed so far to address this problem. The first is based on classical statistics and uses stratified sampling techniques to locate microsite sampling units within each stratus and thus minimize soil spatial variability (Ellert et al., 2002). Soil survey knowledge is used to delineate homogeneous soil-landscape areas within fields to microsite sampling areas to be sampled at several time intervals. McConkey et al. (2000) applied this microsite ap-

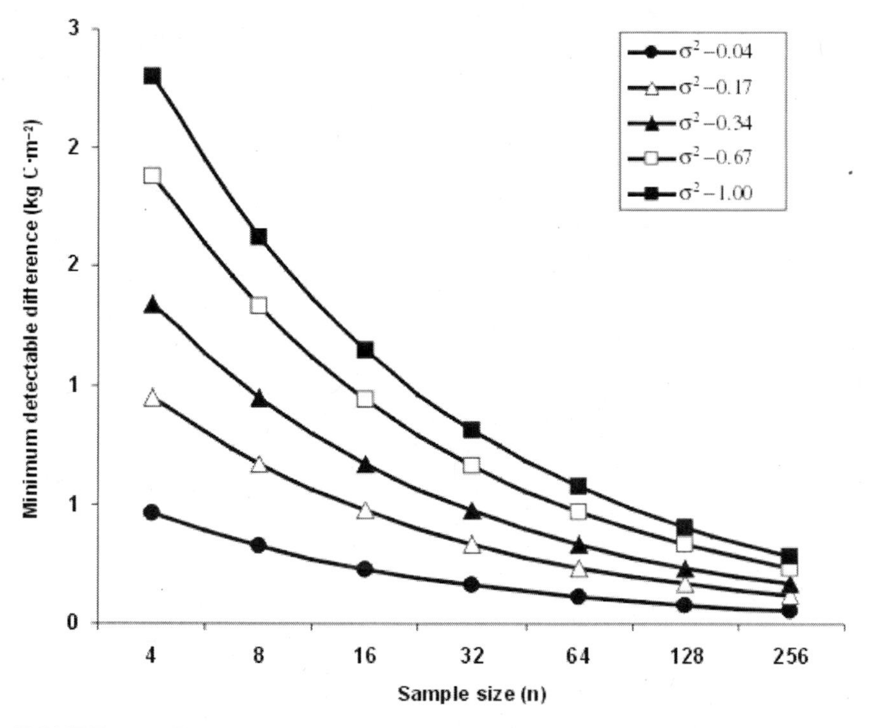

FIGURE 21.3. Relationships between sample size and minimum detectable difference in SOC density (kg C·m^{-2}) at a given level of probability (α = 0.05) and power (1 – β = 90%). (*Source:* Adapted from Garten and Wullschleger, 1999.)

proach on many conventionally tilled fields across Saskatchewan, Canada, converted to no-till (direct seeding).

The central tenet of this approach is that it is possible to reduce the spatial variability of SOC content by concentrating the sampling on a reduced area (microsite) in order to detect with accuracy the management-induced changes in SOC content and then scale-up these to the entire field. McConkey et al. (2000) adopted segmentation procedures from Pennock et al. (1987) and Pennock and Corré (2001) to describe landscapes in functional units (i.e., landform segments such as shoulder, backslope, footslope, and depression) for sampling. The remaining challenge with this method is to test the hypothesis that use of classical statistics, microsite sampling, and landscape fragmentation produces the same answer of SOC changes at the field scale as would be obtained with more intensive sampling techniques.

Confidence in the understanding of the spatial distribution of soils in the landscape has been growing over the years, especially with the development of geostatistical techniques in soil science, as well as advances in computation and information technologies (c.f., McBratney et al., 2003). Geostatistics has been increasingly used in soil science to study spatio-temporal variations in soil attributes. Odeh et al. (1994) compared several interpolation methods (e.g., multilinear regression, kriging, co-kriging, and regression kriging) in their ability to predict soil properties from landform attributes derived from a digital elevation model (DEM) and found the regression-kriging procedures the best for predicting sparsely located soil properties from dense observations of landform attributes derived from DEM data. Regression-kriging techniques have been successfully used to predict soil salinity using electromagnetic induction data (Triantafilis et al., 2001) as well as other soil properties (soil organic matter, soil pH, and topsoil depth) (Hengl et al., 2004). The remaining challenge with all these approaches is to develop methods that are accurate for estimating SOC stocks in space and time (including uncertainties) as well as inexpensive (Izaurralde, 2005).

Sampling and Processing

Once the sampling design has been selected, then the next step is to conduct the soil sampling. Several decisions will have to be made in terms of sampling depth, sampling instrumentation, and collection of ancillary information. Concerning sampling depth, researchers have been using many different depths and depth increments in SOC studies. West and Post (2002), for example, conducted a worldwide survey of long-term studies examining the impacts of crop rotations and tillage on SOC stocks. Of the

276 paired experiments reported in their Table 1, 43 percent reported SOC changes to a depth of 30 cm, 25 percent to a depth of 20 cm, and 18 percent to a depth of 15 cm. Sampling to 30 cm soil depth almost guarantees enough power to detect management-induced changes in SOC stocks. However, in order to express SOC change results to a depth of 30 cm, it is likely that an extra sampling depth (e.g., 30 to 40 cm) would have to be included so, as will be discussed later in this section, calculations for soil equivalent mass can be made. With regard to depth increments, a good scheme would be to divide the sampling depth into three or more intervals, increasing the number of samples near the soil surface to increase detection ability in the most biologically active zone.

Sampling for determination of bulk density (D_b, $Mg \cdot m^{-3}$) is another critical step for calculating SOC mass per unit area ($Mg\ C \cdot m^{-2}$) to a given soil depth (m) from SOC concentration data ($Mg\ C \cdot Mg^{-1}$). This sampling can be done with hydraulically driven probes mounted on pickup trucks, tractors, or other vehicles (Blake and Hartge, 1986). Hand-driven samplers are also appropriate, keeping in mind that the obvious goal when sampling for D_b determination is to avoid compressing the soil in the confined space of the sampler. Blake and Hartge (1986) and Culley (1993) provide detailed descriptions of the various methods that can be used to determine D_b such as the core and clod methods. Bulk density can also be determined in situ with the use of gamma radiation methods (Blake and Hartge, 1986), although instrument cost and radiation hazards may limit their use in soil C sequestration projects (Izaurralde, 2005). Lal and Kimble (2001) discuss challenges and solutions when trying to determine D_b in soils containing coarse fragments as well as soils with large swell-shrink capacity or high organic matter content.

Theory and applications of time domain reflectometry (TDR) technology have expanded vastly for in situ measurements of mass and energy in soil (Topp and Reynolds, 1998). Ren et al. (2003) used a thermo-TDR probe to make simultaneous field determinations of soil water content, temperature, electrical conductivity, thermal conductivity, thermal diffusivity, and volumetric heat capacity. With this information they further derived other soil physical parameters such as D_b, air-filled porosity, and degree of saturation. Values of D_b predicted with the thermo-TDR explained 58 percent of the variation in measured D_b. Improvements in this method could bring a way to expedite in situ measurements of D_b (Izaurralde, 2005).

Soil bulk density changes in response to soil organic matter content, soil water content, and soil compaction. Under relatively noncompacted conditions and uniform soil water content, D_b should vary inversely to SOC concentration. Figure 21.4 shows this relationship for four experimental datasets, one in Argentina and three in Canada. All four soils are loam in texture

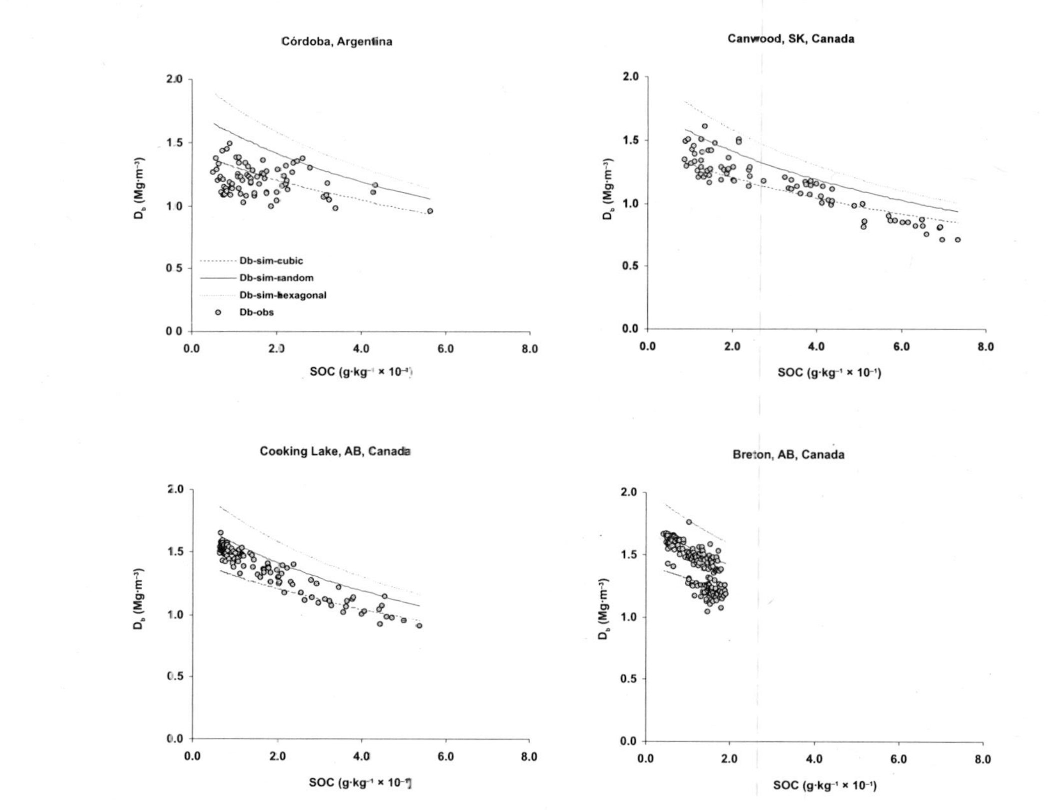

FIGURE 21.4. Predicted and observed relationship between SOC concentration and soil bulk density (D_b) for an experiment in Argentina and three in Canada. (*Sources:* Data for Córdoba are from Apezteguía, 2005; data for Canwood are from Nyborg et al., 1998; data for Cooking Lake are from Izaurralde, Nyborg, et al., 1998; and data for Breton are from Solberg et al., 1998; model predictions are from Izaurralde, 2005.)

but exhibit large variations in SOC content due to treatment effects (e.g., vegetation cover, amendment treatments, and tillage regime) and soil sampling depth. Sampling and handling errors also contribute to this variation. For comparison, each graph contains three lines describing the theoretical variation of D_b with SOC concentration for a loam soil with 20 percent clay, 40 percent silt, and 40 percent sand. These lines were predicted with the model of Adams (1973):

$$D_b = \frac{100}{\frac{\%OM}{0.244} + \frac{100-\%OM}{D_m}} \tag{21.2}$$

where %OM is percent soil organic matter, D_m is mineral bulk density ($Mg{\cdot}m^{-3}$), and the value 0.244 is the bulk density of organic matter ($Mg{\cdot}m^{-3}$). Mineral bulk density was estimated as described by Izaurralde (2005) for three kinds of three-dimensional packing of mineral particles (assumed to be spherical): cubic, random, and hexagonal. As per geometrical solutions, the packing density (*n*, fraction of a volume filled by a given collection of solids) for a cubic lattice is 0.524; it is 0.64 for a random arrangement, and 0.74 for a hexagonal close packing arrangement.

Soil Carbon Analysis

Soils contain two forms of C: organic and inorganic. Organic C is the main constituent of soil organic matter and generally is the component that responds to changes in management. Inorganic C appears largely in carbonate minerals and does not respond readily to management.

Soil organic C can be determined by either wet or dry combustion (Nelson and Sommers, 1996). In the wet combustion procedure, the soil sample is treated with acid dichromate solution in a heated vessel. The generated CO_2 due to the oxidation of organic matter is measured either titrimetrically (indirect) or gravimetrically (direct). The wet oxidation method is relatively easy to implement and has been used worldwide for many years. However, this method can result in incomplete digestion and thus a correction factor has to be used, which is generally 1.24 (Nelson and Sommers, 1996). The current preferred measurement of soil C is dry combustion. This method measures all forms of C in the soil, including inorganic C; thus, for most soil C sequestration projects where the increase in C is due to changes in soil organic C, the inorganic C must be removed by acid hydrolysis before dry combustion. Automated systems have been developed with the dry combustion methods and have become the standard in many laboratories worldwide. The automated systems are very accurate with minimal variability and low operational errors. Dry combustion instruments have a detection limit of about 10 mg $C \cdot kg^{-1}$.

Laser-induced breakdown spectroscopy (LIBS) is a method based on atomic emission spectroscopy (Radziemski and Cremers, 1989; Moenke-Blankeenburg, 1989; Rusak et al., 1997) where an intense laser pulse is focused on a sample, forming a microplasma that emits light which is then resolved for different elements. LIBS for total soil C measurement is still in its infancy, but research results have been promising. Cremers et al. (2001) evaluated the measurement of total soil C using the LIBS method on agricultural soils from Colorado and a woodland soil from New Mexico. They reported high correlation between LIBS and the dry combustion C technique. The LIBS method could be used to rapidly and efficiently measure soil C with detection limits of about 300 $mg \cdot kg^{-1}$, precision of 4 to 5 percent, and accuracy of 3 to 14 percent. Analysis time was less than one minute per sample, providing daily sample throughput much greater than that of traditional C analytical methods. A portable version is now in development. A portable LIBS instrument could allow for efficient measurement of the multitude of soil samples necessary to characterize soil heterogeneity and spatial variation of C distribution, and ultimately provide a reliable estimate of soil C inventories.

Wielopolski et al. (2000) first described the feasibility of using inelastic neutron scattering (INS) and gamma ray spectroscopy to determine soil C. This INS method is based on inelastic scattering of 14 MeV neutrons from C nuclei present in the soil and measurement of the resulting 4.44 MeV gamma ray emission. The INS system was tested, in static mode, at three different field sites (pine stand, oak forest, and sandy patch). Thirty-minute measurements were taken with the INS at each site followed by soil core collection for comparative dry combustion analysis. Results of their study indicated measurements errors of 5 to 12 percent and a minimum detection limit (MDL) of 0.018 Mg $C \cdot m^{-3}$..

Reflectance spectroscopy provides another rapid and nondestructive method for soil C measurement based on diffusely reflected radiation of illuminated soil. Both the near infrared (NIR) and mid infrared (MIR) regions have been investigated for the determination of soil C (Dalal and Henry, 1986; Ben-Dor and Banin, 1995; Chang and Laird, 2001; Reeves et al., 2001; McCarty et al., 2002). Analysis in the NIR range is the predominant means for quantitative determination of soil C, while MIR spectroscopy has been used primarily for qualitative analysis. Laboratory analyses of soils using NIR spectroscopy have been shown to successfully predict soil C, with r^2 values often reported greater than 0.80 (Chang and Laird, 2001; Chang et al., 2001; Reeves et al., 2001; McCarty et al., 2002).

Reporting: Equivalent Mass or Volumetric

Changes in soil C storage are usually reported on a volumetric or equivalent mass basis. Ellert et al. (2002) made the case for consideration of equivalent masses of soil. Reporting as equivalent mass is especially important if there has been erosion or deposition at the site. However, deeper soil sampling produces less difference between fixed volume sampling and equivalent mass. Also, samplings in small increments tend to minimize errors (Ellert et al., 2002). Usually sampling to 45 cm or greater depth is sufficient to minimize most differences, especially if this is below the management zone.

Ancillary Measurements

Estimates of crop and biomass yields are valuable inputs into soil C models to estimate and predict changes in soil C. The USDA compiles a database on crop and yield at the county level, which is reported by the National Agricultural Statistics Service (USDA-NASS, www.usda.gov/nass/pubs/pubs.htm). Technology has now advanced to use remote sensing techniques to predict crop and biomass yields (e.g., Shanahan et al., 2001). Also, yield monitors on harvesting equipment can be used to develop spatial maps of crop production (Taylor et al., 2001). Residue cover can be measured by point sampling or by remote sensing technology (McNairn et al., 2004).

Many countries have or are developing soil databases. In the United States, the National Resources Conservation Service provides several soil databases, including the National Resources Inventory and the soil survey which are now being digitized and placed into geographic information systems. These soil databases are extremely useful for providing regional and national estimates of soil C stocks and changes. The International Soil Reference and Information Centre (ISRIC) in the Netherlands has compiled a soil and terrain database for Latin America and the Caribbean (SOTERLAC) at a scale of 1:5,000,000 (lime.isric.nl/index.cfm?contentid=162). The spatial and attribute database includes about 1,800 georeferenced soil profiles. Regional, FAO (Food and Agricultural Organization), and CIAT (Centro Internacional de Agricultura Tropical) experts collaborated in the development of this major product.

Modeling and Remote Sensing

Models and remote sensing information will serve key roles in monitoring and verification activities of C sequestration projects (Post et al., 2001;

Izaurralde, 2005). Soil organic matter models or agroecosystem models with specific treatment of SOC dynamics embody our theoretical and quantitative understanding of the mechanisms regulating C transformations in soil. When initialized for specific site conditions, simulation models have been able to explain historical trends in SOM dynamics (e.g., Paustian, 2005; Parton and Rasmussen, 1994; Grant et al., 2001; Izaurralde et al., 2006). When driven with databases of climate, soil, and management, these ecosystem models have also provided insight on SOM trends at regional scales (e.g., Izaurralde et al., 2001; Falloon et al., 1998).

Two types of model complexity will be needed for project-level applications. The first type is represented by simple models, or simplified versions of complex models, because they could quickly evaluate the potential of management practices on soil C sequestration. CSTORE is an example of a simple model being developed with such an objective (Paustian, 2005). In this case, the model is intended for use in field-level prediction of soil C sequestration and also as part of a decision-support system with minimum data requirements. The CSTORE model is based on the widely used Century model (Parton et al., 1994) with a simpler representation of the SOM pools.

The second type comprises models of increased complexity, which can yield detailed understanding of the processes regulating SOC dynamics (e.g., evapotranspiration, nutrient availability, etc.). Further, these models can provide additional information about the environmental impacts of soil C sequestration on trace gas emissions (Del Grosso et al., 2000) and soil erosion (Izaurralde et al., 2006).

Models are increasingly being tested and applied with the objective of evaluating soil C sequestration practices under diverse environmental conditions. For example, in Brazil, Cerri et al. (2003) used the RothC to model soil C dynamics when converting Amazonian forests into pasture ecosystems. Apezteguía (2005) used EPIC to model soil C dynamics as affected by tillage and crop sequence in loess soils of Córdoba, Argentina. In the Pampean region of Argentina, Alvarez (2001) applied the Century model to produce regional estimates of soil C levels under grassland.

Progress is also being made to apply simulation models at landscape and regional scales. For example, the APEX model (Agricultural Policy Environmental Extender) (Williams and Izaurralde, 2005) is an implementation of the EPIC model at the watershed scale to simulate plant growth, watershed hydrology, and soil C dynamics, soil erosion, sediment transport, and sediment deposition across connected fields under diverse land-use and management practices.

Remote sensing data have also been used with simulation models to derive spatially explicit estimates of cropland net primary productivity (e.g.,

Prince et al., 2001; Lobell et al., 2002). Satellite estimates of residue cover (plant litter) would also be useful for assessing adoption of conservation practices (e.g., no-till) and calibrating models. Daughtry et al. (2004) determined spectral reflectance characteristics of dry and wet crop residues and soils over the 400 to 2400 nm wavelength region. Since crop residue cover was linearly related to a cellulose absorption index and this to NDVI, Daughtry et al. (2004) proposed a method to estimate soil tillage intensity based on these two indices. They concluded that it was possible to use advanced multispectral or hyperspectral imaging systems to conduct regional surveys of conservation practices. Recently, ETM+ data coupled with logistic regression techniques have been very successful in mapping no-till practices with a high degree of accuracy (>95 percent) for a site in Montana dominated by dryland wheat (Bricklemyer et al., 2002). South (2003) mapped no-till practices using TM data for a region in Michigan, Indiana, and Illinois with a cosine of spectral angle mapping technique (Sohn and Rebello, 2002). Validated with an intensive transect dataset, South (2003) showed that no-till mapping accuracy can be as high as 95 percent, but concluded that the time of image acquisition is critical, as no-till practices are difficult to differentiate when fields are covered with more than 30 percent crops.

Remote sensing is an excellent tool for developing soil C monitoring programs because it provides spatial data across a range of scales. Remote sensing can provide these data across scales, allowing government agencies at field and policy levels, scientists, farmers and ranchers, industry, and C market aggregators to estimate soil C levels in these ecosystems at both field and watershed levels. Of particular interest is the capacity of remote sensing to spatially characterize plant productivity and soil characteristics, two factors that impact soil C dynamics. The autonomous nature of satellite-based remote sensing makes periodic sampling easier and may help reduce expenses associated with verification and monitoring soil C sequestration programs.

CONCLUSION

Soil C sequestration has been identified as a potential mitigation practice to help attenuate the rate of increase of atmospheric CO_2, which is increasingly being demonstrated as having a major role in climate change. In anticipation of a global CO_2 emissions permit trading system in which soil C sequestration is included, methodologies and tools for the design of soil C sequestration projects have been reviewed. Aspects of direct sampling were discussed, particularly those dealing with sampling design, sampling depth, soil bulk density, methods of C determination, and calculations using equivalent mass concepts. In addition, the role of simulation models and remote sensing techniques were discussed in regard to monitoring programs, extrapolation, and auditing. The methods and tools discussed here are not meant to be prescriptive. Rather, they aim at serving the soil science community in developing methods and tools that are comparable and have the scientific rigor to make them acceptable in national and international trading mechanisms.

REFERENCES

Adams, W.A. 1973. The effect of organic matter on the bulk and true densities of some uncultivated podzolic soils. *J. Soil Sci.* 24:10-17.

Alvarez, R. 2001. Estimation of carbon losses by cultivation from soils of the Argentine Pampas using the Century model. *Soil Use Management* 17:62-66.

Apezteguia, H.P. 2005. Dinámica de la materia orgánica de los suelos de la región semiárida central de Córdoba (Argentina). PhD dissertation, Universidad Nacional de Córdoba, Córdoba, Argentina.

Ben-Dor, E., and A. Banin. 1995. Near-infrared analysis as a rapid method to simultaneously evaluate several soil properties. *Soil Sci. Soc. Am. J.* 59:364-372.

Blake, G.R., and K.H. Hartge. 1986. Bulk density. In *Methods of soil analysis,* Part I. *Physical and mineralogical methods* (pp. 363-382). Agron. Monograph no. 9 (2nd ed.). Am. Soc. Agron., Soil Sci. Soc. Am., Madison, WI.

Bricklemyer, R., R. Lawrence, and P. Miller. 2002. Documenting no-till and conventional till practices using Landsat ETM + imagery and logistic regression. *J. Soil Water Conserv.* 57:267-271.

Brown, S., J. Sathaye, M. Cannel, and P. Kauppi. 1996. Management of forests for mitigation of greenhouse gas emissions. In R.T. Watson, M.C. Zinyowera, and R.H. Moss (eds.), *Climate change 1995: Impacts, adaptations, and mitigation of climate change: Scientific-technical analyses* (pp. 773-797). Contribution of Working Group II to the Second Assessment Report of the Intergovernmental Panel on Climate Change. Cambridge University Press, Cambridge and New York.

Casas, R.R. 2002. Posibilidad de incrementar la fijación de carbono en los suelos agrícolas de la República Argentina mediante la adopción del sistema de siembra directa. Plata Basin Initiative 2002 Planning Workshop, April 8-10, Buenos Aires. Available online at www.aaas.org/international/lac/plata/casas.shtml.

Casas, R.R., M.M. Ostinelli, and G.A. Cruzate. 2004. Carbon dynamics in soil under no-till in the Argentine Pampean region. *Agron. Abs.* CD-ROM.

Cerri, C.E.P., K. Coleman, D.S. Jenkinson, M. Bernoux, R. Victoria, and C.C. Cerri. 2003. Modeling soil carbon from forest and pasture ecosystems of Amazon, Brazil. *Soil Sci. Soc. Am. J.* 67:1879-1887.

Chang, C.W., and D.A. Laird. 2001. Near-infrared reflectance spectroscopic analysis of soil C and N. *Soil Sci.* 167:110-116.

Chang, C.W., D.A. Laird, M.J. Mausbach, and C.R. Hurburgh, Jr. 2001. Near-infrared reflectance spectroscopy—Principal components regression analyses of soil properties. *Soil Sci. Soc. Am. J.* 65:480-490.

Cole, C.V., J. Duxbury, J. Freney, O. Heinemeyer, K. Minami, A. Mosier, K. Paustian, N. Rosenberg, N. Sampson, D. Sauerbeck, and Q. Zhao. 1997. Global estimates of potential mitigation of greenhouse gas emissions by agriculture. *Nutrient Cycling in Agroecosystems* 49:221-228.

Conant, R.T., K. Paustian, and E.T. Elliott. 2001. Grassland management and conversion into grassland: Effects on soil carbon. *Ecological Applic.* 11:343-355.

Cremers, D.A., M.H. Ebinger, D.D. Breshears, P.J. Unkefer, S.A. Kammerdiener, M.J. Ferris, K.M. Catlett, and J.R. Brown. 2001. Measuring total soil carbon with laser-induced breakdown spectroscopy (LIBS). *J. Environ. Qual.* 30:2202-2206.

Culley, J.L.B. 1993. Density and compressibility. In M.R. Carter (ed.), *Soil sampling and methods of analysis* (pp. 529-539). Can. Soc. Soil Sci., Lewis Publishers, Boca Raton, FL.

Dalal, R.C., and R.J. Henry. 1986. Simultaneous determination of moisture, organic carbon and total nitrogen by near infrared reflectance spectrophotometry. *Soil Sci. Soc. Am. J.* 50:120-123.

Daughtry, C.S.T., E.R. Hunt Jr., and J.E. McMurtrey III. 2004. Assessing crop residue cover using shortwave infrared reflectance. *Rem. Sens. Environ.* 90:126-134.

Del Grosso, S.J., W.J. Parton, A.R. Mosier, D.S. Ojima, A.E. Kulmala, and S. Phongpan. 2000. General model for N_2O and N_2 gas emissions from soils due to denitrification. *Global Biogeochem. Cycles* 14:1045-1060.

Dooley, J.J., S.H. Kim, J.A. Edmonds, S.J. Friedman, and M.A. Wise. 2004. A first order global geologic CO_2 storage potential supply curve and its application in a global integrated assessment model. In E.S. Rubin, D.W. Keith, and C.F. Gilboy (eds.), *Proceedings of 7th International Conference on Greenhouse Gas Control Technologies,* Volume 1: *Peer-reviewed papers and plenary presentations.* IEA Greenhouse Gas Programme, Cheltenham, UK.

Ellert, B.H., H.H. Janzen, and T. Entz. 2002. Assessment of a method to measure temporal change in soil carbon storage. *Soil Sci. Soc. Am. J.* 66:1687-1695.

Falloon, P.D., P. Smith, J.U. Smith, J. Szabó, K. Coleman, and S. Marshall. 1998. Regional estimates of carbon sequestration potential: linking the Rothamsted Carbon Model to GIS databases. *Biol. Fert. Soils* 27:236-241.

Garten, C.T., and S.D. Wullschleger. 1999. Soil carbon inventories under a bioenergy crop (switchgrass): measurement limitations. *J. Environ. Qual.* 28:1359-1365.

Grant, R.F., N.G. Juma, J.A. Robertson, R.C. Izaurralde, and W.B. McGill. 2001. Long term changes in soil C under different fertilizer, manure and rotation: Testing the mathematical model ecosys with data from the Breton Plots. *Soil Sci. Soc. Am. J.* 65:205-214.

Hengl, T., G.B.M. Heuvelink, and A. Stein. 2004. A generic framework for spatial prediction of soil variables based on regression-kriging. *Geoderma* 120:75-93.

Intergovernmental Panel on Climate Change (IPCC). 2000. *Land use, land-use change, and forestry.* Watson, R.T., I.R. Noble, B. Bolin, N.H. Ravindranath,

D.J. Verardo, and D.J. Dokken (eds.). Cambridge University Press, Cambridge, UK.

Izaurralde, R.C. 2005. Measuring and monitoring soil carbon sequestration at the project level. In R. Lal et al. (eds.), *Climate change and global food security* (pp. 467-500). CRC Press, Boca Raton, FL.

Izaurralde, R.C., K.H. Haugen-Kozyra, D.C. Jans, W.B. McGill, R.F. Grant, and J.C. Hiley. 2001. Soil organic carbon dynamics: Measurement, simulation and site to region scale-up. In R. Lal, J.M. Kimble, R.F. Follett, and B.A. Stewart (eds.), *Assessment methods for soil carbon* (pp. 553-575). Lewis Publishers, Boca Raton, FL.

Izaurralde, R.C., W.B. McGill, A. Bryden, S. Graham, M. Ward, and P. Dickey. 1998. Scientific challenges in developing a plan to predict and verify carbon storage in Canadian Prairie soils. In R. Lal, J. Kimble, R. Follett, and B.A. Stewart (eds.), *Management of carbon sequestration in soil* (pp. 433-446). Adv. Soil Sci. CRC Press, Inc., Boca Raton, FL.

Izaurralde, R.C., M. Nyborg, E.D. Solberg, H.H. Janzen, M.A. Arshad, S.S. Malhi, and M. Molina-Ayala. 1998. Carbon storage in eroded soils after five years of reclamation techniques. In R. Lal, J. Kimble, R. Follett, and B.A. Stewart (eds.), *Soil processes and the carbon cycle* (pp. 369-386). Adv. Soil Sci. CRC Press, Inc., Boca Raton, FL.

Izaurralde, R.C., J.R. Williams, W.B. McGill, N.J. Rosenberg, and M.C. Quiroga Jakas. 2006. Simulating soil C dynamics with EPIC: Model description and testing against long-term data. *Ecol. Modell.* 192:362-384.

Janik, L.J., R.H. Merry, and J.O. Skjemstad. 1998. Can mid infrared diffuse reflectance analysis replace soil extractions? *Aust. J. Exp. Agric.* 38:681-696.

Lal, R., R.F. Follett, J.M. Kimble, and C.V. Cole. 1999. Managing U.S. cropland to sequester carbon in soil. *J. Soil Water Conserv.* 54:374-381.

Lal, R., and J.M. Kimble. 2001. Importance of soil bulk density and methods of its importance. In R. Lal, J.M. Kimble, R.F. Follett, and B.A. Stewart (eds.), *Assessment methods for soil carbon* (pp. 31-44). Lewis Publishers, CRC Press, Boca Raton, FL.

Lal, R., J. Kimble, R. Follett, and B.A. Stewart (eds.). 1998a. *Management of carbon sequestration in soil.* CRC Press, Boca Raton, FL.

Lal, R., J. Kimble, R. Follett, and B.A. Stewart (eds.). 1998b. *Soil processes and the carbon cycle.* Adv. Soil Sci. CRC Press, Boca Raton, FL.

Lal, R., J. Kimble, E. Levine, and B.A. Stewart (eds.). 1995a. *Soil management and the greenhouse effect.* Lewis Publishers, CRC Press, Boca Raton, FL.

Lal, R., J. Kimble, E. Levine, and B.A. Stewart (eds.). 1995b. *Soils and global change.* Lewis Publishers, CRC Press, Boca Raton, FL.

Lobell, D.B., J.A. Hicke, G.P. Asner, C.B. Field, and C.J. Tucker. 2002. Satellite estimates of productivity and light use efficiency in United States agriculture 1982-98. *Global Change Biol.* 8:722-735.

McBratney, A.B., M.I. Mendonça Santos, and B. Minasny. 2003. On digital soil mapping. *Geoderma* 117:3-52.

McCarty, G.W., J.B. Reeves, V.B. Reeves, R.F. Follett, and J.M. Kimble. 2002. Mid-infrared and near-infrared diffuse reflectance spectroscopy for soil carbon measurement. *Soil Sci. Soc. Am. J.* 66:640-646.

McConkey, B.G., B.C. Liang, G. Padbury, and R. Heck. 2000. Prairie Soil Carbon Balance Project: Carbon sequestration from adoption of conservation cropping practices. Final Report to GEMCo. Agriculture and Agri-Food Canada, Swift Current, SK.

McGill, W.B., Y.S. Feng, and R.C. Izaurralde. 1996. Soil organic matter dynamics: from past frustrations to future expectations. Soil Biology Symposium, Solo/Suelo 1996 XIII Congresso Latino Americano de Ciéncia do Solo, August 4-8, Águas de Lindóia, São Paulo, Brazil.

McNairn, A. Smith, E. Huffman, I. Harvis, A. Pacheco, and E. Gauthier. 2004. A national inventory of land management practices: Estimating soil conservation practices using optical and radar imagery. IEEE (2004)1629-1632.

Moenke-Blankenburg, L. 1989. *Laser microanalysis.* John Wiley & Sons, New York.

Nelson, D.W., and L.E. Sommers. 1996. Total carbon, organic carbon, and organic matter. In *Methods of soil analysis.* Part 3. *Chemical methods* (pp. 961-1010). SSSA Book Series no. 5. Soil Sci. Soc. Am., Am. Soc. Agron., Madison, WI.

Nyborg, M., M. Molina-Ayala, E.D. Solberg, R.C. Izaurralde, S.S. Malhi, and H.H. Janzen. 1998. Carbon storage in grassland soil and relation to application of fertilizer. In R. Lal, J. Kimble, R. Follett, and B.A. Stewart (eds.), *Management of carbon sequestration in soil* (pp. 421-432). Adv. Soil Sci. CRC Press, Inc., Boca Raton, FL.

Odeh, I.O.A., A.B. McBratney, and D.J. Chittleborough. 1994. Spatial prediction of soil properties from landform attributes derived from a digital elevation model. *Geoderma* 63:197-214.

Parton, W.J., D.S. Ojima, C.V. Cole, and D.S. Schimel. 1994. A general model for soil organic matter dynamics: Sensitivity to litter chemistry, texture and management. In *Quantitative modeling of soil forming processes* (pp. 147-167). Spec. Public. No. 39, Soil Sci. Soc. Am., Madison, WI.

Parton, W.J., and P.E. Rasmussen. 1994. Long term effects of crop management in wheat-fallow: 2. Century model simulations. *Soil Sci. Soc. Am. J.* 58:530-536.

Paustian, K. 2005. CSTORE: A tool for monitoring soil carbon changes in agricultural systems. Available online at www.casmgs.colostate.edu/insider/vigview.asp?action=2&titleid=265, accessed April 26, 2005.

Pennock, D. J., and M.D. Corré. 2001. Development and application of landform segmentation procedures. *Soil Tillage Res.* 58:151-162.

Pennock, D.J., B.J. Zebarth, and E. De Jong. 1987. Landform classification and soil distribution in hummocky terrain, Saskatoon, Canada. *Geoderma* 40:297-315.

Post, W.M., R.C. Izaurralde, J.D. Jastrow, B.A. McCarl, J.E. Amonette, V.L. Bailey, P.M. Jardine, and J. Zhou. 2004. Carbon sequestration enhancement in U.S. soils. *BioScience* 54:895-908.

Post, W.M., R.C. Izaurralde, L.K. Mann, and N. Bliss. 2001. Monitoring and verifying changes of organic carbon in soil. *Climatic Change* 51:73-99.

Prince, S.D., J. Haskett, M. Steininger, H. Strand, and R. Wright. 2001. Net primary production of US Midwest croplands from agricultural harvest yield data. *Ecol. Applic.* 11:1194-1205.

Radziemski, L.J., and D.A. Cremers. 1989. Spectrochemical analysis using laser plasma excitation. In L.J. Radziemski and D.A. Cremers (eds.), *Application of laser-induced plasmas* (pp. 295-325). Marcel Dekker, New York.

Reeves, J.B., III, G.W. McCarty, and V.B. Reeves. 2001. Mid-infrared and diffuse reflectance spectroscopy for the quantitative analysis of agricultural soils. *J. Agric. Food Chem.* 49:766-772.

Ren, T., T.E. Ochsner, and R. Horton. 2003. Development of thermo-time domain reflectometry for vadose zone measurements. *Vadose Zone J.* 2:544-551.

Robertson, G.P., E.A. Paul, and R.R. Harwood. 2000. Greenhouse gases in intensive agriculture: Contributions of individual gases to the radiative forcing of the atmosphere. *Science* 289:1922-1925.

Rosenberg, N.J., and R.C. Izaurralde. 2001. Introductory paper to the special issue on mitigation of climate change by soil carbon sequestration. *Climatic Change* 51:1-10.

Rusak, D.A., B.C. Castle, B.W. Smith, and J.D. Winefordner. 1997. Fundamentals and application of laser-induced breakdown spectroscopy. *Crit. Rev. Anal. Chem.* 27:257-290.

Schuman, G.E., H.H. Janzen, and J.E. Herrick 2002. Soil carbon dynamics and potential carbon sequestration by rangelands. *Environmental Pollution* 116:391-396.

Shanahan, J.F., J.S. Schepers, D.D. Francis, G.E. Varvel, W.W. Wilhem, J.M. Tringe, M.R. Schlemmer, and D.J. Major. 2001. Use of remote sensing imagery to estimate corn grain yield. *Agron. J.* 93:583-589.

Sohn, Y., and N.S. Rebello. 2002. Supervised and unsupervised spectral angle classifiers. *Photogramm. Eng. Remote Sensing* 68:1271-1280.

Solberg, E.D., M. Nyborg, R.C. Izaurralde, S.S. Malhi, H.H. Janzen, and M. Molina-Ayala. 1998. Carbon storage in soils under continuous cereal grain cropping: N fertilizer and straw. In R. Lal, J. Kimble, R. Follett, and B.A. Stewart (eds.), *Management of carbon sequestration in soil* (pp. 235-254). Adv. Soil Sci. CRC Press, Inc., Boca Raton, FLA.

South, S. 2003. Identification and monitoring agricultural practices utilizing remotely sensed imagery. PhD dissertation, Michigan State University, East Lansing, Michigan.

Taylor, R.K., G.J. Klutenberg, M.D. Schrock, H. Zhang, J.P. Schimdt, and J.L. Havlin. 2001. Using yield monitor data to determine spatial crop production potential. *Trans. ASAE* 44:1409-1414.

Topp, G.C., J.L. Davis, and A.P. Annan. 1980. Electromagnetic determination of soil water content: Measurements in coaxial transmission lines. *Water Resour. Res.* 16:574-582.

Topp, G.C., and W.D. Reynolds. 1998. Time domain reflectometry: A seminal technique for measuring mass and energy in soil. *Soil & Tillage Res.* 47:125-132.

Triantafilis, J., I.O.A. Odeh, and A.B. McBratney. 2001. Five geostatistical models to predict soil salinity from electromagnetic induction data across irrigated cotton. *Soil Sci. Soc. Am. J.* 65:869-878.

Tucker, C.J., and J.R.G. Townshend. 2000. Strategies for monitoring tropical deforestation using satellite data. *Int. J. Remote Sensing* 21:1461-1471.

VandenBygaart, A.J., E.G. Gregorich, and D.A. Angers. 2003. Influence of agricultural management on soil organic carbon: A compendium and assessment of Canadian studies. *Can. J. Soil Sci.* 83:363-380.

Vergara Sánchez, M.A., J.D. Etchevers B., and M. Vargas Hernández. 2004. Variabilidad del carbono orgánico en suelos de ladera del sureste de México. *Terra* 22:359-367. Available online at www.chapingo.mx/terra/contenido/22/3/359.pdf.

West, T.O., and W.M. Post. 2002. Soil organic carbon sequestration rates by tillage and crop rotation: A global data analysis. *Soil Sci. Soc. Am. J.* 66:1930-1946.

Wielopolski, L., I. Orion, G. Hendrey, and H. Rogers. 2000. Soil carbon measurements using inelastic neutron scattering. *IEEE Trans. Nuclear Sci.* 47:914-917.

Williams, J.R., and R.C. Izaurralde. 2005. The APEX model. In V.P. Singh (ed.), *Watershed models.* CRC Press, Boca Raton, FL.

Chapter 22

Advances in Models to Measure Soil Carbon: Can Soil Carbon Really Be Measured?

John Kimble

INTRODUCTION

Soil organic carbon (SOC) is a valuable resource. It is a source of plant nutrients, it helps in forming aggregates in soils, it increases soil water and nutrient capacity, and it may even become a valuable commodity in the future. Many models are now being used to estimate changes in SOC with changes in farming practices and with potential changes from carbon sequestration.

We can start with a simple quote (Albrecht, 1938) as to why carbon is important:

> Up to the present, the policy—if it can be called a policy—has been to exhaust the supply, rather than to maintain it by regular additions according to the demands of the crops produced or the soil fertility removed. To continue very long this practice will mean a further sharp decline in crop yield.

This was written in 1938, and to be honest we have not really developed policies to deal with SOC in a totally rational way to date. The importance of soil organic matter has long been recognized, but what we still need is holistic management the does not degrade out soil resources and will allow future generations to be productive and sustainable. We need to be able to measure and report changes in SOC, and this can be done by direct methods, i.e., soil sampling, or by indirect methods, i.e., modeling. Figure 22.1 shows the general inputs to the CENTURY model; other SOC models would have similar inputs.

Carbon Sequestration in Soils of Latin America
Published by The Haworth Press, Inc., 2006.
doi:10.1300/5755_23

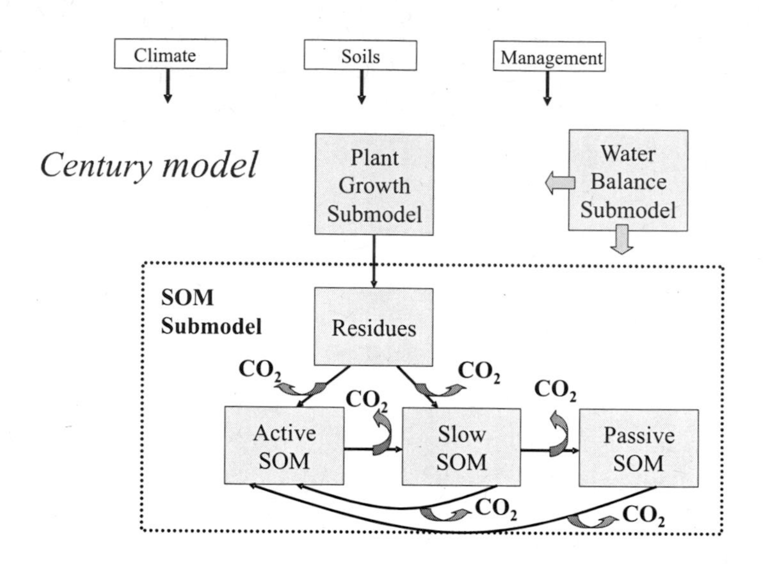

FIGURE 22.1. Input parameters that go into a model to estimate soil carbon.

Soil organic matter is linked to food security, poverty reduction, agriculture sustainability, improved nitrogen use efficiency, improvement of degraded lands, etc. We need to address farming/forestry practices that will sequester carbon, and tillage is not one of these. Tillage mixes the soil, which increases the oxidation of the SOC and leads to a degradation of soil structure and loss of the water and nutrient supply capacity of the soil; therefore, we need to put the tillage equipment out to pasture and practice conservation tillage. With a loss of SOC we lose the ability for the soil to function properly.

We need to keep in mind that there are many different carbon sinks and that all of these are important in a holistic management systems. Sinks are both above- and belowground, and the different pools are linked. We need to look at cropland, forest land, grazing land, windbreaks, wetlands, and urban landscapes to see how these systems are linked into the overall environment which is driven by carbon sequestration. There are many different strategies for the sequestration of carbon. Some of the are conservation tillage and residue management, crop rotations, use of cover crops, soil application of biosolids, water conservation (drip irrigation and subirrigation), programs that restore degraded lands, aforestation, and reforestation. The use of cover crops, for example, has many advantages: It adds carbon to the soil. It can help prevent wind and water erosion. It also allows crop residue that had in the past been returned to the soil to be used as a bioenergy feed stock without environmental degradation, which reduces the need for the use of nonrenewable energy sources.

About 40 percent of the world's soils have been degraded, and we continue to mine soils for nutrients. We have in agriculture replaced mineral elements with commercial fertilizers but in the past have not always made an effort to replace or build up SOC; however, we can do this with proper management.

Many groups are now focusing on SOC sequestration as a new income source for farmers and land managers, but the focus it just on the money that can be made directly from the carbon in the soil or vegetation. This is very short sighted, as many environmental values come from SOC sequestration that are both direct and indirect; these may be much more valuable that any money paid to land owners under a carbon credit trading system. This is not to say that carbon trading systems are not worthwhile, but they should be considered as one of many options and benefits. One major direct effect is the removal of atmospheric carbon, i.e., a reduction in one of the major greenhouse gases that is contributing to global climate change. Other direct benefits are a reduction in wind and water erosion and improving air quality. The indirect effects have a major impact in sediment detachment and transport, a reduction of dissolved loads in water, decreased leaching of chemicals into the groundwater, and reduced eutrophication of water bodies. These effects can have a very large off-site impact that can be reduced if we manage soil resources in a matter that sequesters carbon. If we do that we also may even obtain increased yields.

To cite a simple example, when infiltration rates are compared for three soils (native, no-till, and conventional tillage) the rates for 1 inch of water to enter the soil are about 10 seconds for the native forest, about 1 minute for the no-till field, and <20 minutes for the conventionally tilled field. One major difference in these soils is change in the soil structure, and this is largely controlled by SOC. If the water goes into the soil it does not run off and cause erosion, and it is also available for plant growth. Increasing SOC will also lead to a reduction in nitrogen needs in many soils, as the soil biological system functions as it should and nitrogen from plant residue is released and used by plants without as much need for chemical fertilizers. A farmer I have worked with in Illinois has increased the organic matter on his fields from about 2 percent to almost 3.5 percent and has had a 50 to 60 percent reduction in applied nitrogen with equal or improved yields.

Data from the Innovative Cropping Systems Group in Virginia have shown a 75 percent reduction in runoff when no-till is compared to conventional tillage. This translates into a 98 percent reduction in the sediment load, a 95 percent reduction in nitrogen lost per acre, and a 92 percent reduction in phosphorus loss. The major cause of this is the increased SOC and the beneficial effect this has on the soil.

Agriculture and forestry can play a role in addressing environmental issues and at the same time help to mitigate greenhouse gases, enhance soil, water, and air quality, and create agriculture sustainability. This comes from the sequestration of carbon in soils. It is a win-win strategy.

The soil and the producer benefit from increased SOC through improved soil tilth, improved water-holding capacity, improved drainage, reduced soil erosion, improved water and air quality, and long-term sustainability of production. Society benefits through reduced levels of atmospheric carbon dioxide, improved water and air quality, and increased food security.

We are interested in the pedosphere as that is our soils; this is the sphere that links the atmosphere, biosphere, lithosphere, and hydrosphere. The pedosphere is the thin mantle on the soil surface that supports most terrestrial life; water is stored and filtered through it before it moves into the groundwater. We have for a long time measured all kinds of different properties, and one of them is SOC. Land managers, farmers, foresters, and others have used measured soil properties for making many decisions related to soil fertility and many other soil properties for a very long time, so the simple answer to the question regarding SOC measurement and monitoring of changes in SOC is to just go out and do it. This will work for site-specific projects in small areas, but in larger areas or multiple areas there may be too many to measure every point. We also need to address the time of changes, and we need to compare the short and long terms (one year to several years). When looking at sampling concerns we also need to deal with natural variability in soils; they vary, as do all natural systems (soils, water bodies, vegetative pools). We also need to be concerned with laboratory measurements, as many different procedures are used to measure SOC and each gives a different result. In soil a measure of density is also needed, as a great deal of variation exists in the density of soils. We need to present our data on a volume basis, not just as a weight percent. Finally, we also need to address the type of soil with respect to carbonates, as they can affect the total carbon measurements when we use combustion for the measure technique.

This chapter will not go into details of the many different models that have been developed to estimate and scale SOC. They have been covered in other chapters and have been reviewed by others.

CAN WE MEASURE SOIL CARBON?

The simple answer is yes and we have been doing so for a very long time. SOC has always been something that was looked at in soils. In the United

States we now conduct over 2 million soil samples for fertility measurements each year, and in most, one of the factors measured is SOC (often reported as soil organic matter). There is not any real question about being able to measure SOC; we can. However, there are concerns about how to address its variability and how to scale point measurements of SOC to the field or farm level or even to large areas. If SOC is to be traded as a commodity it will need to be done on a large scale over very large areas. This is where modeling can help. It can take point measurements and the information from such points and scale these data to larger areas. Models can also estimate possible changes to carbon based on different farm management practices.

What Needs to Be Addressed?

The following are significant issues to address:

- How can scale point data to larger areas?
- Can models be used for scaling and if so how?
- What geospatial data are needed to help in scaling?
- Which human-induced activities can or should be credited?
- What is the time frame for measured change in SOC?
- What modalities, rules, and guidelines are needed?
- How verifiable are measurements?
- Which land-use changes are important?
- What are some of the uncertainties?
- What is used as a base line?
- How do we meet the need to present data on a volume basis (need for bulk density)?

The real question that needs to be addressed is, What are we trying to measure or predict and for what reason?

The major concern with SOC measurement is how to scale point data. This can be done with the use of other geospatial data (soil maps, climatic information, digital elevation models, and yield data). We have this information, but what are we really trying to measure or predict? Are we interested in absolute values or more interested in the rate of change?

When measuring carbon in agriculture and forest ecosystems the simple answer is just to take samples for the carbon determination. But that is really not possible, as many different fields and land uses must be considered. There are also long-term versus short-term changes, and many different sampling concerns (how many samples to take per field, how deep to sample, need for bulk density, etc). We need to determine which human-induced activities are of interest and the time frame for change that we are interested in. Modalities and rules are needed with clear-cut guidelines for direct measurement or for modeling. We need to consider what land uses are of interest and how they change with time. There are many uncertainties in any measurement or model, and they need to be well defined. One last concern is what base line we are going to use. The problem with setting a base line is that it is normally something form the past, when we really did not have the needed information. We can measure change from time 0 to 0 plus any number of years, but often in the Kyoto Protocol a base line of 1990 is used; this is hard to determine, as we were not sampling for that. We can, however, model this based on rates we determine experimentally and then cast back to 1990.

Measurements

What is the role that remote sensing will play? As we cannot visit every field to determine land use and land-use change or even tillage practice, we will need to use remote sensing to determine the changes taking place. As models that use remotely sensed information to estimate crop yields improve we can use them to estimate changes to the SOC based on the yield and the observed tillage system used. We also need to address the cost versus accuracy issue. We could measure accurately if we wanted to spend unlimited resources, but we know this will never be the case.

What Are Models and How Are They Used?

Why are models needed?

- To scale from point to fields to large regions
- To help predict changes and help recommend practices that will sequester carbon
- To factor in climatic variation over large areas and other variables
- To integrate data over large areas
- Not every field can be measured

Many models are in use, and both new and improved ones are being developed all the time. This chapter is not going to go into the mechanism of how the models work, or for that matter even discuss every model that has been developed. Soil scientists actually model when they are mapping. They consider the landscape and other soil forming factors and draw lines on maps; this is a model in the simplest form and it works very well. We recognizes that soil variability exists and that there may well be different soils in a given unit (these are described and a percentage given for each of them).

Models can be used to help in scaling field measurement to broader areas, integrating geospatial data using geographic information systems, integrating associated databases, and linking remote sensed.

There are empirical models or simple bookkeeping (IPCC default method with fixed input factors) and dynamic simulation models (Century, DNDC, EPIC, and RothC for example); these are described by Post et al. (1999). Bookkeeping models are useful when there is not sufficient information for the use of the more detailed dynamic process models (Metting et al., 1999).

Models Presently in Use by the USDA

The USDA reports changes to SOC, and to do this it uses models. In NRCS there is a need to look at changes different farming practices will have on SOC. For reporting, the IPCC-based default method has been used, and CENTRY has been tested in a couple states. Results of this work are reported in the *U.S. Agriculture and Forestry Greenhouse Gas Inventory 1990-2001* (USDA, 2004). When the dynamic simulation model (CENTURY) was compared (Brenner et al., 2002) to the IPCC default method for Nebraska, Iowa, and Indiana, the CENTURY model gave substantially higher values. In Nebraska the IPCC gave a value of 1.98 Tg CO_2 while the CENTURY gave 4.62 Tg CO_2 per year. Similar results were also found in Iowa and Indiana, showing that we can expect changes in different agroecological regions. Static models are useful in areas with limited data, but they do not keep up with changes to cropping systems and new technologies. Where data are available to run the dynamic models, they would seem to give much better data.

Other models used by NRCS are CQESTER and the Soil Conditioning Index. Both of these are under development and are designed to work at the farm level to support how different managements will or will not bring about carbon sequestration. Neither will give tradable credits, just an estimate of what will happen

Future Model Needs

The following are some concerns about the future:

- We need to build models from the top down and bottom up and integrate them. Most are now top down and do not work other than at broad scales.
- Models such as CENTURY are good for national and regional assessments but may not work as well at the field level without more refinements.
- The Soil Conditioning Index and CQESTR are more site-specific models (farm level).
- COMET is based on CENTURY but for local use.
- We need to link N, P, and C models together.

Many times models are run for C assuming that N and P are sufficient, yet this may not be the case. Programs to address carbon sequestration need to look at N and P concerns at the same time. The world is not made up of little boxes but is one large integrated ecosystem. All too often we try to parse it out into small sections instead of looking at it as a complex system. The use of specific models for carbon can exacerbate this problem.

GENERAL CONCERNS

In the measurement of SOC, or for that matter soil or biomass carbon, many concerns need to be addressed. We can just measure to see how much carbon there is, and this is done when harvestable timber is determined; you are really looking at a measure of carbon. Also, when soils are sampled for fertility testing SOC is normally measured, as it is an important component of fertility and has a bearing on how much herbicide and pesticide can be applied to crops. But as the possibility of carbon trading develops, what will be and needs to be measured and verified changes. We then become concerned about what human-induced activities will be credited, as this is tied to the Kyoto Protocols and is more in the political arena than the science arena. The time frame for change becomes critical, and for these modalities, rules, and guidelines are needed. We cannot expect significant changes in either biomass or SOC in a short time period; we need to be looking at four- to five-year periods. There are a couple reasons for this: (1) we are looking at small changes in large pools and (2) climate (rainfall mainly) has a major influence on biomass growth and this varies from year to year. We also need to address how verifiable measurements are and what uncertainties go into the variability. In addition, the simple question of what is to be used for a base line needs to be addressed.

Most modern soil analysis laboratories use combustion furnaces and measure the CO_2 evolved, which measures both organic and inorganic forms. Consequently, the SOC needs to be corrected for the amount of SIC. If this is not done major errors are introduced into our measurements. We also need to consider whether the data represent a point (single sample) or a regional or national estimate. The basic analytical data could be based on a point sample or a composite sample representing a field, and this needs to be known. Do the data represent a field or a regional assessment? This is a concern that must be addressed, and we need to be clear how we scale for one area to another. Models are needed for scaling, but many are top down and a combination of top down and bottom up are needed.

It is often pointed out that soils have a large amount of variability (Plates 22.1 and 22.2), but with knowledge of soil science and landscapes, variability can be described and sampling protocols developed to deal with this. One reason I feel people say that soils vary and SOC cannot be measured is that we soil scientists focus on showing variability, not on showing what we know about the variability. In soils we can go to a 100 m^2 field and sample every square meter and look at the differences we find. But if you sample every tree in a large area you would see similar variability (Cheng and Kimble, 2001). For a more information on the characterization of soil organic pools and methods of analysis, see Cheng and Kimble (2000, 2001).

On top of all these concerns we also need to address economics. We can measure very accurately and precisely, but the cost may be excessive. The overall economics of what we are doing is a concern and must be considered.

ASSESSMENT AND MEASUREMENT CONCERNS

The major concern is the need to have a measure of the density of the soil material, as there is normally an increase in density with depth. Soils may have a bulk density in the surface of close to 1 while at 50 to 100 cm the density could be 1.4 or 1.5 (a 50 percent change); this can have a major effect on the amount of carbon within a given layer. When the density of a soil is 1.1 and there is 2 percent carbon, the pool is 220 $Mg \cdot ha^{-1}$. When the density is 1.5 for the same density and thickness, the pool of SOC is 300 $Mg \cdot ha^{-1}$, or a 36 percent difference. Saying that both have the same amount based on the weight percent is not correct. When the density is used the data are reported on a volume basis. We need to consider mass versus concentration, and by measuring bulk density we can do this.

Because of variations in bulk density we really need to look at data on an equal volume projection. However, this is often not done; data are reported for a fixed depth (20 to 30 cm, for example). Figure 22.2 shows how the val-

PLATE 22.1. A soil from Florida that has very little variability. (See also color gallery.)

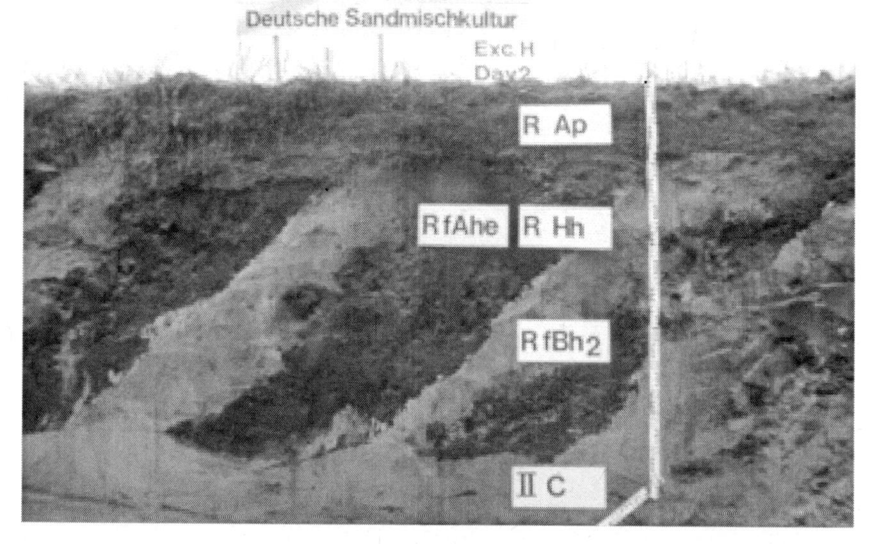

PLATE 22.2. Soil in Germany showing extreme variability but in a pattern that can be explained. (See also color gallery.)

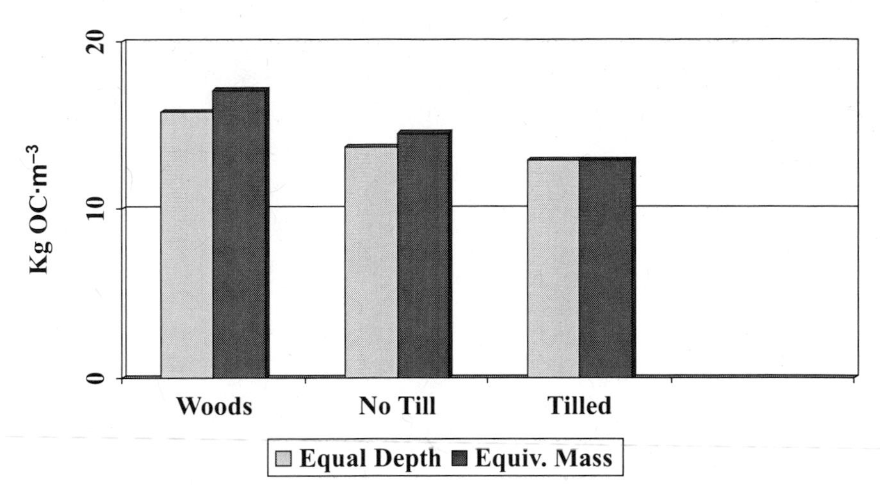

FIGURE 22.2. Comparison of KG OC·m3^{-3} for equal depth versus equivalent mass of soil.

ues reported change when we look at an equal depth versus an equivalent mass of soil. This concept is explained in detail by Ellert et al. (2001).

What are the specific concerns and/or problems with the measurement of bulk density? It has been shown that soil bulk density has a profound effect on SOC pools, and for that matter for all elemental pools in soils. One major concern is that soil bulk density is considered to be a property that varies with time. In a tilled field the bulk density will be quite different right after tillage, when the soil may be soft and loose in the tilled part, but below this the soil may be very compacted and have a high bulk density. Bulk density is a use-dependent property and will vary over time and space. This needs to be addressed. However, in many models static values are used for bulk density if they are used at all.

I will not go into detail about bulk density concerns or detail all the methods, as these are addressed by Lal and Kimble (2001) and Kimble et al. (2001). We do need to consider the methods, which include direct and indirect techniques. The direct methods are core sampling, clod sampling, and sand/water/glass beads replacement. Each of these has limitations that need to be addressed, but they do produce similar results and can be compared. Indirect methods use some type of radiation to conduct the measurement.

One major concern in bulk density no matter the method used is the amount of coarse fragments. They need to be measured and corrections made for them in any calculation of SOC. All soils are not the same and there are problems in measuring bulk density in single-grain soils, organic soils, frozen soils, wet soils, gravelly or skeletal soils, soils with high contents of salts, and soils that shrink and swell (2:1 clays). The bottom line is that the SOC pool cannot be measured without a sound knowledge of soil bulk density.

We also need to consider anthropogenic factors that can affect bulk density. This is considered to be a use-dependent property. Some of the anthropogenic factors are land use, cropping system, tillage method, residue management, and vehicular traffic. Natural factors can also affect the bulk density, such as the type of clay, the climate, and the biota.

In general, soil bulk density is a dynamic property with a wide temporal and spatial variation that is more pronounced for the surface than the subsurface horizons.

CONCLUSION

To simply answer the question, Why are we interested in modeling and measuring SOC sequestration? Why not. It is the right thing to do as it drives our agriculture/forestry systems. One caveat that needs to be addressed is the statement many make that SOC cannot be measured. Sure it can. We see it in sediments in streams, lakes, and rivers, in airborne dust, in gullies and rills in agriculture fields, and a loss of native soil fertility. Loss of SOC can easily be measured as a negative impact on the environment. And we really can and have been measuring SOC is soils for years. We have the techniques to do this, and with our knowledge of soils we can deal with variability in measurements and explain them.

We too often focus on this, worry about laboratory precision and field variation, and do not look at the real world where most things are based on averages and estimated data. We tend to focus on finding variation and not on using our knowledge of soil science to describe what we know. All systems vary, yet in soils we focus on a level of precision and accuracy that may not have any relevance to the real world because we can take so many samples and look at the variation.

We can measure SOC pools; in fact, this has been done for many years. We can also model SOC changes over time, but we should not let the models drive us. They are a tool to help us but not the answer to every question. We need to keep doing studies in the field, and this information can be used to refine and improve models.

These are some final points to consider:

- We as soil scientists need to focus on what we know about soils, explain the variability, make it part of what we do, and not use it as a crutch for not wanting to answer policy questions.

- When we address the issues of soil variability, modelers and others can use our data better.
- We may worry too much about variation at the regional and national scale.
- We try to apply field-level precision at regional and national scales, and this serves little use.
- Models need to be refined and updated, but we need to keep in mind that they are a tool to help us; for them to work we need good base line data.
- We can measure and model SOC sequestration.

We are in what many call the information age. Many say we can determine anything by developing databases and pulling the information out of them. We have many models for SOC, many reports are being written using models to prepare the numbers, and these numbers soon are ingrained in the literature. However, these numbers are only as good at the inputs used in the models, and many of them are at very broad scales, which can mask a great deal of variability. We need to keep our feet on the ground and keep the process studies going so that the needed information can be provided for large-scale modeling.

REFERENCES

Albrecht, W. A. 1938. Loss of soil organic matter and its restoration. In *Soils and men yearbook of agriculture 1938*. USDA, US Government Printing Office, Washington, DC.

Brenner, J., K Paustian, G. Bluhm, J. Cipra, M. Easter, R. Foulk, K. Killian, R. Moore, J. Schuler, P. Smith, and S. Williams. 2003. Quantifying the changes in greenhouse gas emissions due to natural resource conservation practice applications in Nebraska. Final report to the Nebraska Conservation Partnership. Colorado State University Natural Resources Ecology Laboratory and USDA Natural Conservation Service, Fort Collins, CO.

Cheng, H.H. and J.M. Kimble. 2000. Methods of analysis for soil carbon: An overview. In R. Lal, J.M. Kimble, and B.A. Stewart (Eds.), *Global climate change and tropical ecosystems*. CRC Press, Boca Raton, FL.

Cheng, H.H. and J.M. Kimble. 2001. Characterization of soil organic carbon pools. In R. Lal, J. M. Kimble, R.F. Follett, and B.A. Stewart (Eds.), *Assessment methods for soil carbon*. Lewis Publishers, Boca Raton, FL.

Ellert, B.H., H.H. Janzen, and B.G. McConkey. 2001. Measuring and comparing soil carbon storage. In R. Lal, J.M. Kimble, R.F. Follett, and B.A. Stewart (Eds.), *Assessment methods for soil carbon*. Lewis Publishers, Boca Raton, FL.

Kimble, J.M., R.B. Grossman, and S.E. Samson-Liebig. 2001. Methodology for sampling and preparation for soil carbon determination. In R. Lal, J.M. Kimble,

R.F. Follett, and B.A. Stewart (Eds.), *Assessment methods for soil carbon.* Lewis Publishers, Boca Raton, FL.

Lal, R. and J.M. Kimble. 2001. Importance of soil bulk density and methods of its measurement. In R. Lal, J.M. Kimble, R.F. Follett, and B.A. Stewart (Eds.), *Assessment methods for soil carbon.* Lewis Publishers, Boca Raton, FL.

Metting, F.B., J.L. Smith, and J.S. Amthor. 1999. Science needs new technology for carbon sequestration. In N.J. Rosenberg, R.C. Izurralde, and E.L. Malone (Eds.), *Carbon sequestration in soils: Science and monitoring, and beyond.* Proceedings of the St. Michaels Workshop, December 1998. Battelle Press, Columbus, OH.

Post, W.M., R.C. Izaurralde, L.K. Mann, and N. Bliss. 1999. Monitoring and verifying soil organic carbon sequestratiaon. In N.J. Rosenberg, R.C. Izurralde, and E.L. Malone (Eds.), *Carbon sequestration in soils: Science and monitoring, and beyond.* Proceedings of the St. Michaels Workshop, December 1998. Battelle Press, Columbus, OH.

USDA. 2004. U.S. agriculture and forestry gas inventory: 1990-2001. Global Change Program Office, Office of the Chief Economist, U.S. Department of Agriculture. Technical Bulletin No. 1907.

Chapter 23

Bulk Density Measurement for Assessment of Soil Carbon Pools

R. Lal

INTRODUCTION

Soil bulk density (ρ_b), also called apparent density, is the ratio of the mass of soil solids (M_s) to the total volume (V_t). Soil ρ_b is an important soil physical property and a strong determinant of soil quality because it affects other properties such as soil strength, water retention, available water capacity, porosity, and gaseous exchange. It has pedological, edaphological, ecological, and environmental significance (Figure 23.1). Pedologically, it affects numerous soil processes, including transport of water and dissolved/suspended substances leading to eluviation and illuviation. Edaphologically, it affects plant growth both directly and indirectly. Ecologically, ρ_b affects and is affected by activity of diverse soil fauna and flora. Environmentally, it affects quality of surface and ground waters, and fluxes of greenhouse gases (GHGs) between soils and the atmosphere.

The environmental impacts of ρ_b, water quality, and emission of GHGs have global significance. Through its impact on total porosity and pore size distribution, ρ_b influences water and transport of sediments. It is an important determinant of the magnitude and nature of non-point-source pollution (Figure 23.1). Soil ρ_b also affects air quality of both soil and atmospheric air. Gaseous composition of soil air depends on the production of CO_2, CH_4, N_2O, and other GHGs in the soil, and their diffusion from soil to the atmosphere. Both production and diffusion of gases depend on ρ_b through its effect on total porosity, pore size distribution, pore continuity, and tortuosity. Root respiration, methanogenesis, nitrification, denitrification, redox potential, and other processes affecting production and emission of GHGs in soil depend on soil ρ_b.

Carbon Sequestration in Soils of Latin America

doi:10.1300/5755_24

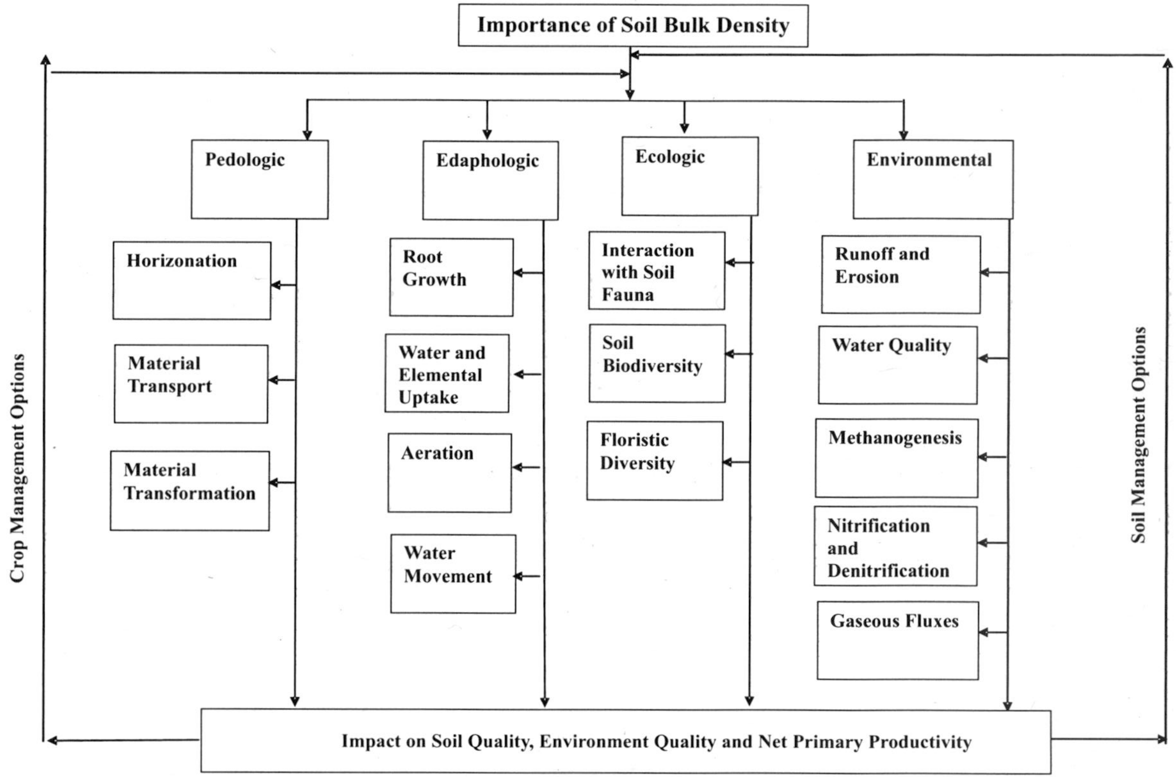

FIGURE 23.1. Soil bulk density is a key soil property with impact on numerous processes.

Knowledge of soil ρ_b is also important to understanding principal soil processes and to determination of the pool of water, nutrients (N, P, K), and soil organic carbon (SOC). Knowledge of the temporal changes in pools of these substances is important to understanding their flux as influenced by natural and managerial factors. Pool of water or SOC is computed by using Equation 23.1.

$$\text{SOC Pool} = \text{area} \times \text{depth} \times \rho_b \times \text{SOC concentration} \qquad (23.1)$$

Temporal changes in SOC pool as a measure of C flux are assessed by computing differences in SOC pool at the known time intervals. A similar procedure applies to assessment of the flux of H_2O and plant nutrients.

The objective of this chapter is to describe (1) the importance of ρ_b, (2) temporal and spatial variability due to both natural and anthropogenic factors, and (3) its impact on pools of C, N, water, and other organic and inorganic constituents in soil. It also addresses the problems involved in using data on concentrations of SOC without the supporting data on ρ_b, and practical challenges in obtaining reliable measurements of ρ_b in problem soils such as those prone to cracking and containing a large proportion of skeletal fraction.

FACTORS AFFECTING SOIL BULK DENSITY

Soil ρ_b is a highly dynamic soil property. It varies strongly over time and space. While the M_s is relatively constant, the bulk or total volume (V_t) is a highly variable entity, and temporal changes in ρ_b are affected by changes in V_t. It is the temporal changes in V_t which make it a highly variable soil physical property. Consequently, management of ρ_b is an important objective of soil surface management. Numerous factors affect V_t over time. These factors can be grouped under two categories: natural and anthropogenic (Figure 23.2). Natural factors that affect temporal changes in V_t are climate, soil properties, and biotic activities. Precipitation and temperature are important environmental factors that affect V_t through their influence on the frequency and intensity of wetting-drying and freezing-thawing cycles (Larson et al., 1980). Freezing-thawing cycles have more effect on V_t than the wetting-drying cycles. Important among soil's inherent properties are texture and clay minerals. The amount and the nature of clay content play significant roles in swell-shrink characteristics. Soils containing a large amount of high-activity clays (HAC) undergo large changes in V_t because of the swell-shrink properties (Plate 23.1). The V_t of soils containing a large proportion of skeletal material (>2 mm) (Plate 23.2) is not affected as much

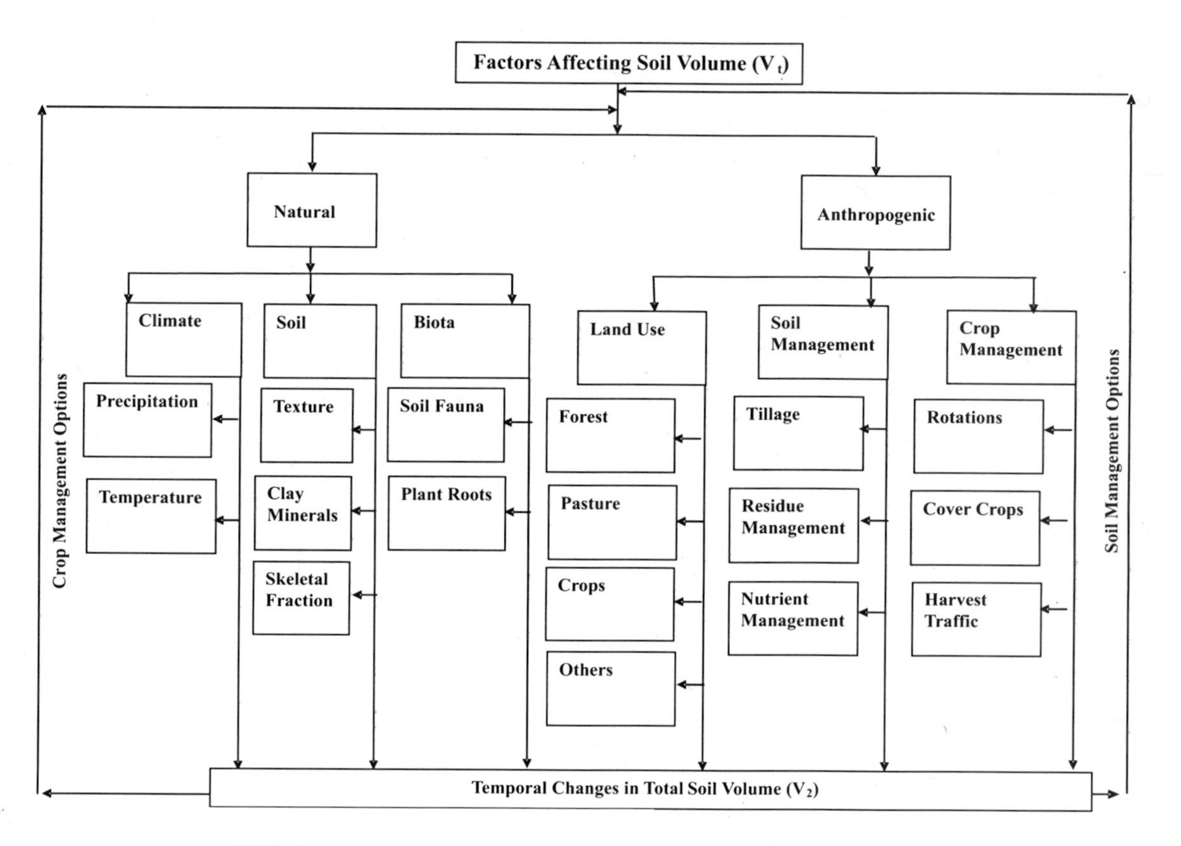

FIGURE 23.2. Factors affecting soil volume and thus the bulk density.

PLATE 23.1. Cracking clays undergo large changes in V_t. The intercrack ρ_b of the clod is high compared with that of uncracked soil.

by swell-shrink, freeze-thaw, or wet-dry cycles as that of those which contain low amounts of the skeletal material. Soil biota plays a significant role in temporal changes in ρ_b through its influence on V_t. In addition to the trampling effects of large animals, V_t is drastically influenced by the burrowing activity of soil macro fauna. Important among these are earthworms (Plate 23.3), termites (Plates 23.4 and 23.5), ants (Plate 23.6), and other soil-burrowing animals. Indeed, bioturbation (also called the drillosphere) plays an important role in both spatial and temporal changes in ρ_b through its influence on V_t. Plant roots also affect V_t by influencing macropores or biopores (Plate 23.7) because of the large pressure they exert while growing through the soil matrix. Figure 23.2 also outlines a range of anthropogenic factors that affect V_t and thus ρ_b. The impact and management of human-induced changes in V_t and ρ_b are discussed in the next section.

ANTHROPOGENIC FACTORS AFFECTING SOIL BULK DENSITY

Numerous management options affect ρ_b (Figure 23.2). Land use and land-use change influence ρ_b through canopy cover, soil perturbation, and

(a)

(b)

PLATE 23.2. It is difficult to accurately measure ρ_b of soils with a large proportion of skeletal fraction: (a) in the surface layer, (b) in the surface horizon, (c) in localized zones such as in soils affected by permafrost, and (d) in subsurface horizons.

(c)

(d)

PLATE 23.2 *(continued)*. It is difficult to accurately measure ρ_b of soils with a large proportion of skeletal fraction: (a) in the surface layer, (b) in the surface horizon, (c) in localized zones such as in soils affected by permafrost, and (d) in subsurface horizons.

PLATE 23.3. Earthworm activity profoundly affects ρ_b by influencing macropores.

the quantity and quality of biomass returned to the soils. All other factors remaining the same, especially the texture and mineralogy, ρ_b is generally in the order forest < pasture < crops. In some cases, depending on the root system of the species and management of pasture, including the stocking rate, baling, etc., soil ρ_b of pasture may be equal to or less than that of the forest plantation. Soil management, especially the tillage methods and crop residue management, has a profound impact on soil ρ_b. In general, soil ρ_b is low in systems based on crop residue mulch compared with the plow-based tillage method which incorporates crop residue. Through its influence on soil moisture and temperature regimes, residue mulching and use of conservation tillage affect activity and species diversity of soil fauna (e.g., earthworms, termites). The ρ_b is lower in mulched soils managed with conservation tillage than unmulched and plowed soils. Integrated nutrient management (INM) also affects ρ_b. Growing leguminous cover crops and green manure crops in the rotation cycle affect ρ_b through influence on the SOC pool and the root system development. Crops with a deep taproot system can penetrate dense pans, thereby increasing porosity and decreasing ρ_b. Regular use of compost, animal manure, and other biosolids also can enhance SOC pools and reduce ρ_b. Crop management, especially harvest traffic, is an im-

(a)

(b)

PLATE 23.4. Termites make numerous underground galleries between the nest (mound) (a) and the surrounding soil (b), and influence ρ_b.

PLATE 23.5. The termite-infested soil, a clod in the center, has numerous biopores caused by the termite activity and has low ρ_b.

portant factor that increases ρ_b. Heavy axle load of grain carts, with repeated traffic on a moist soil, is a principal cause of soil compaction. Mechanized harvesting can exacerbate the soil compaction hazard on tropical soils containing predominantly low-activity clays (LAC).

Therefore, knowledge about the impact of management practices on ρ_b is important to making judicious choices of appropriate land use and soil and crop management options which are crucial to the sustainable use of soil and water resources because of the important impact of ρ_b on soil quality, water quality, and flux of GHGs between soil and the atmosphere.

IMPORTANCE OF BULK DENSITY IN MEASURING POOLS OF CARBON AND OTHER CONSTITUENTS

The knowledge of ρ_b is important to understanding the impact of land-use change and soil/crop management on the assessment of elemental pools in soil. Even small changes in ρ_b, due to natural or human-induced factors, can cause drastic changes in elemental pools. The data in Table 23.1 show drastic changes in the amount of volumetric moisture content and the SOC pool due to small incremental changes in ρ_b. Thus, lack of knowledge about

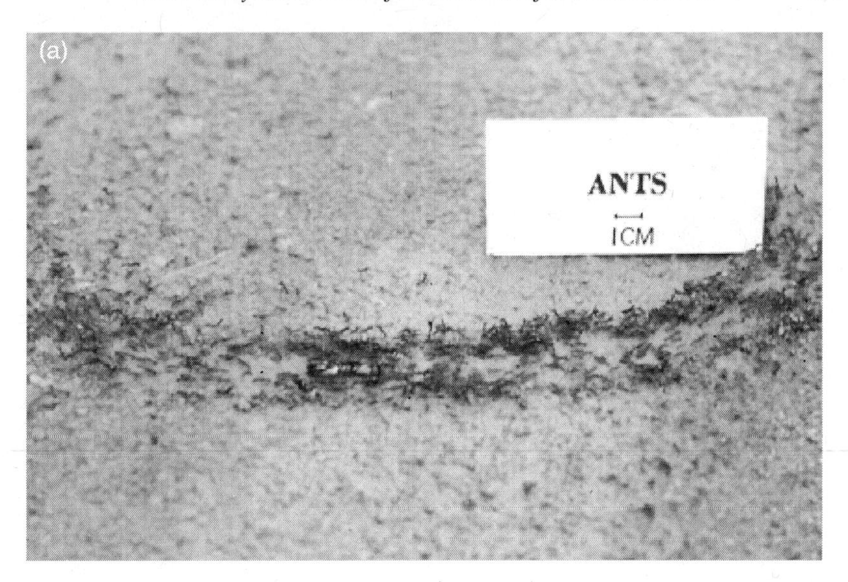

PLATE 23.6. Ants influence soil ρ_b through their burrowing activity: (a) ant colony migrating; (b) the biomass stored at the entrance of the ant nest; (c) ants can defoliate large areas in the forest; (d) the bare exposed soil can be compacted by the raindrop impact.

PLATE 23.6 *(continued)*. Ants influence soil ρ_b through their burrowing activity: (a) ant colony migrating; (b) the biomass stored at the entrance of the ant nest; (c) ants can defoliate large areas in the forest; (d) the bare exposed soil can be compacted by the raindrop impact.

PLATE 23.7. Plant roots, especially the top roots, can also decrease ρ_b by creating biopores.

TABLE 23.1. Effect of soil bulk density on mass of water and organic carbon in 0.2 m depth of a soil with gravimetric moisture content of 20 percent and soil organic carbon content of 20 g·Kg^{-1}.

Soil bulk density ($Mg \cdot m^{-3}$)	**Volumetric moisture content ($m \cdot 0.2\ m^{-1}$)**	**Soil organic carbon pool ($Mg \cdot ha^{-1}$ per 0.2 m)**
1.0	0.040	40
1.1	0.044	44
1.2	0.048	48
1.3	0.052	52
1.4	0.056	56
1.5	0.060	60
1.6	0.064	64

the ρ_b can often lead to erroneous interpretation of the management-induced changes in the SOC pool. The data in Table 23.2 provide a relevant example of a possible erroneous interpretation of the effect of tree species on SOC dynamics in soils of Costa Rica. While considering the SOC concentration, two species with a positive impact on SOC concentration were *V. guatemalensis* and *V. ferruginea* (Table 23.2). In contrast, while considering the SOC pool, two species with the maximum rate of SOC sequestration

TABLE 23.2. Effect of tree species on concentration of organic carbon in 0 to 15 cm depth of a soil at Le Selva, Costa Rica.

Tree species	SOC concentration ($g \cdot kg^{-1}$) Initial	3 years after	Change (% per year)
Control (pasture)	44.6	40.6	–8.97
Virola koschnyi	42.9	44.1	+2.80
Stryphnodendron microstachyum	43.5	44.1	+1.38
Vochysia guatemalensis	42.9	47.6	+10.96
Pithocellobium macradenium	44.1	44.7	+1.36
Pinus tecunumanii	44.7	43.5	–2.68
Hieronyma alchorneoides	45.7	46.4	+1.53
Gmelina arborea	45.2	44.4	–1.77
Vochysia ferruginea	42.5	47.0	+10.59
Inga edulis	42.9	42.9	0
Acacia mangium	42.3	44.7	+5.67
Pentaclethra macroloba	42.3	41.8	–1.18

Source: Recalculated from Fisher, 1995.

were *Pinus tecunumanii* and *Gmelina arborea*. Both *Vochysia* species had a negative impact on the SOC pool (Table 23.3). Thus, lack of data on ρ_b and inability to compute the SOC pool can lead to erroneous interpretation on the impact of management on the SOC pool and its dynamics. Similar to the SOC, lack of information on ρ_b can also cause erroneous interpretation of soil moisture conservation, water use efficiency, and nutrient use efficiency and dynamics.

REASONS FOR ACCURATE MEASUREMENT OF SOIL CARBON POOLS

The SOC concentration has been measured by soil scientists since the mid-nineteenth century (e.g., 1860). However, SOC concentration was primarily measured as an indicator of soil fertility and agronomic productivity. For most of the twentieth century, SOC concentration was measured on plot scale and mostly for the plow depth on the 0 to 20 cm layer (Table 23.4). With the realization that SOC sequestration is also essential to reducing net CO_2 emission and mitigating the climate change, the requirements for assessment of the SOC pool and flux over time have drastically changed (Ta-

TABLE 23.3. Effect of tree species on pool of organic carbon in 0 to 15 cm depth of a soil at Le Selva, Costa Rica.

	SOC concentration (mg·ha^{-1})		Change
Tree species	**Initial**	**3 years after**	**(% per year)**
Control (pasture)	43	38	–1.49
Virola koschnyi	43	37	–2.02
Stryphnodendron microstachyum	42	39	–1.13
Vochysia guatemalensis	40	39	–0.45
Pithocellobium macradenium	41	39	–0.71
Pinus tecunumanii	35	38	+0.99
Hieronyma alchorneoides	36	36	+0.18
Gmelina arborea	38	41	+1.11
Vochysia ferruginea	45	37	–2.69
Inga edulis	42	37	–1.50
Acacia mangium	41	40	–0.35
Pentaclethra macroloba	43	38	–1.63

Source: Recalculated from Fisher, 1995.

TABLE 23.4. Measurement of soil organic carbon concentration as an indicator of soil fertility.

Parameter of SOC	Measurement factors	Units of expression
Quantity	Concentration	%, g·Kg^{-1}
Depth	Plow depth	0-20 cm
Frequency	The rotation cycle	1-3 years
Precision	One decimal place	%
Scale	Plot	4 row or 8 row

ble 23.5). In this regard, the SOC pool must be measured in Mg C·ha^{-1}, over a reporting period ranging from two to ten years, and expressed on a landscape, watershed, or regional scale (Table 23.5). Assessment of SOC pool on mass basis (Mg C·ha^{-1}) cannot be made without the knowledge of soil ρ_b for different soil layers.

MEASUREMENT OF SOIL BULK DENSITY

Mathematically, soil bulk density is expressed in terms of the ratio of mass of soil solids (M_s) to the total volume (V_t), as shown in Equation 23.2,

TABLE 23.5. Measurement of soil organic carbon pool as an indicator of flux and $CO_2 - C$ from soil to the atmosphere.

Parameter of SOC	Measurement factors	Units of expression
Quantity	Amount of mass of SOC	Mg·ha, $Kg{\cdot}m^{-2}$
Depth	Soil solum	At least 1 meter
Frequency	Project duration for trading C credits	2-10 years
Precision	1 $Mg{\cdot}ha^{-1}$	Based on farm scale
Scale	Geographic unit	Farm, county, watershed, region

and appropriately corrected for the gravimetric moisture content (Equation 23.3).

$$\rho_b = M_s/V_t \qquad (23.2)$$

$$\rho_b = \rho_b'/(1 + w) \qquad (23.3)$$

Where ρ_b is the dry bulk density ($g{\cdot}cm^{-3}$, $Mg{\cdot}m^{-3}$), ρ_b' is the moist bulk density, and w is the gravimetric moisture content. There are three variables in Equations 23.2 and 23.3 which need to be measured accurately and credibly: M_s, V_t, and w. Determinations of soil mass and w can be done gravimetrically with reasonable precision and accuracy. However, there are many problems in accurate measurement of V_t. Different methods described in the literature involve a range of techniques of measuring the V_t (Figure 23.3).

Direct Methods of Measuring Soil Volume

Two direct methods of measuring V_t are the core method and the clod method. The core method is the simplest and the most direct method (Lutz, 1947; Jamison et al., 1950), and its use has been described by Veihmeyer (1929), Buchele (1961), and Vomocil (1957). Yet there are several limitations of the method, including (1) difficulty of obtaining a core in sandy or noncohesive soils, (2) soil disturbance by shattering (Baver et al., 1972), (3) compression and distortion and, especially in small cores, dependence of ρ_b value on the core size (Baver et al., 1972; Constantini, 1995), and (4) a high degree of error due to incomplete filling of the core in gravelly soils or those with coarse roots. Mechanized/motorized methods of obtaining cores have been described by Hvorslev (1949), Freitag (1971), and Buchele (1961).

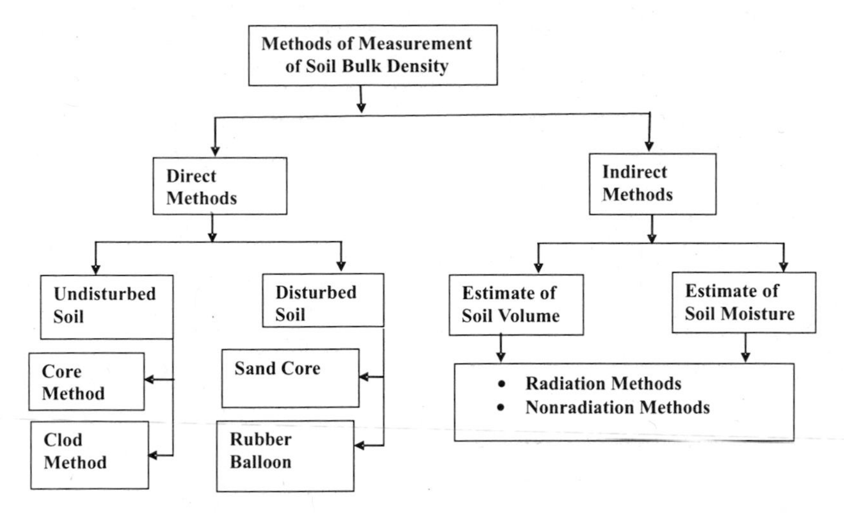

FIGURE 23.3. Methods of measuring soil bulk density.

The clod method is an alternative to the core method (Becket, 1928; Russell and Balcerek, 1944; Chepil, 1950; Gill, 1959; Brasher et al., 1966; Voorhees et al., 1966; Abrol and Palta, 1968; Zwarich and Shaykjewich, 1969; Campbell, 1973). The volume of the clod is determined by coating a clod of known weight with a water-repellent substance such as saran, paraffin, or wax. The clod method is usable under field conditions (Plate 23.8), and saran-coated clods can also be used to determine soil moisture retention characteristics.

In gravelly or other cohesionless soils where it is difficult to obtain soil cores and clods, soil is excavated and weighed. The volume of the excavated cavity is assessed by either the sand replacement method (ASTM, 1950; Blake, 1965) or the rubber balloon method (Freitag, 1971). The accuracy of this method depends on the ability to precisely measure the volume of sand or air to completely fill the soil cavity.

Indirect Methods

Indirect methods involve radiation methods, most of which are described in detail by Campbell and Henshall (1991, 2001). Most gauges involve γ-radiation emitted by a $^{137}C_s$ source, and are either the backscatter type or transmission type. For in situ use, radiation guages must be calibrated under site-specific conditions. However, calibration is influenced by several factors, including soil chemical composition, gravel concentration, and hori-

PLATE 23.8. The clod method can be used under field conditions.

zonation. Calibration is also influenced by the presence of hydrogen, which is present both in water and soil organic mater content. Lal (1974) observed significant effect of texture and gravel concentration on the calibration of the density probe. Revut and Rode (1969) proposed several indirect methods of measuring ρ_b based on assessment of penetration resistance by using a range of probes or penetrometers and also on the basis of determining the permeability of soil to water and air. Several other direct and indirect methods of measuring ρ_b are described by Blake and Hartge (1986), Culley (1993), Campbell and Henshall (2001), and Grossman and Reinsch (2002).

BULK DENSITY MEASUREMENT FOR PROBLEM SOILS

Gravelly soils pose a major challenge in measurement of ρ_b (Cunningham and Matelski, 1968). Neither radiation nor direct methods (e.g., core, clod) are applicable under these conditions. (Shipp and Matelski, 1965). If the gravels or concretions are between 2 and 10 mm size, corrections for skeletal material must be made. Corrections involve separate measurement of mass (M_{gravel}) and volume (V_{gravel}) of the gravel fraction (Equation 23.4).

$$\rho_{bc} = \frac{M_s - M_{gravel}}{V_t - V_{gravel}} \tag{23.4}$$

where ρ_{bc} is the corrected bulk density.

In addition to gravels, other problem soils include organic soils (peat, muck), wetlands, swelling soils, and salt-affected soils (Figure 23.4). Organic soils subside over time (Plate 23.9), and their ρ_b changes with *w*. The ρ_b of organic soils is low and changes with change in soil moisture content. The correction for moisture content can be simple if the change in moisture content is within the range of linear shrinkage, because decrease in soil volume equals the amount of water drained. The correction is difficult in the range of nonlinear shrinkage curve. A similar argument holds true for clayey soils of high shrink-swell capacity. However, the problem is more challenging in clayey soils prone to development of large and wide cracks. The intercrack area (Plate 23.1) is denser after a crack is developed than it would have been prior to crack development. In this regard, time of obtaining soil samples is extremely critical.

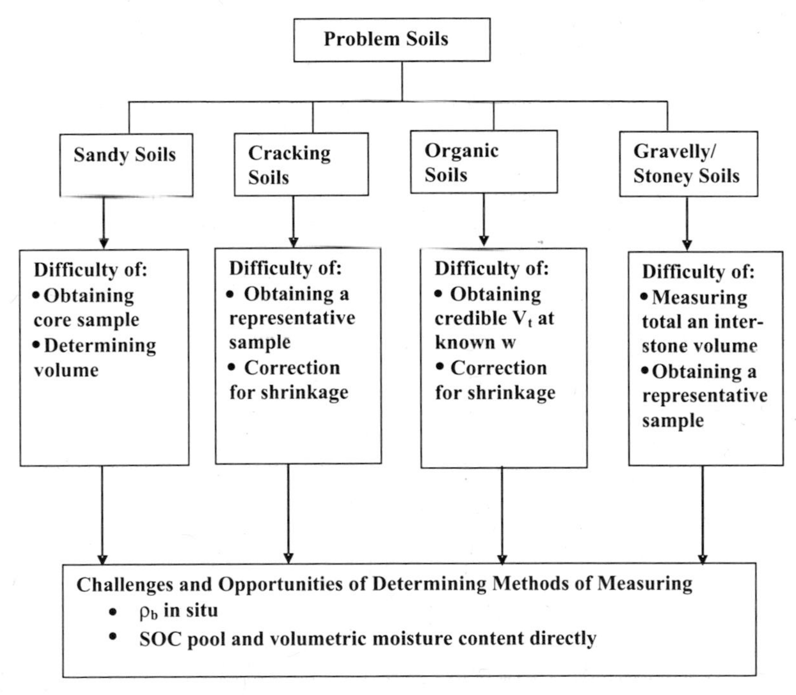

FIGURE 23.4. Problem soils that make bulk density measurements erroneous and challenging.

PLATE 23.9. Soil ρ_b or organic soils changes with time due to subsidence. This organic soil in Willard, Ohio, subsided by about 50 cm since the hand pump was installed.

When comparing SOC pool in soils with contrasting ρ_b values, it is advisable to compute the pool on equivalent soil mass basis rather than equal soil depth basis (Ellert and Bethany, 1995; Ellert et al., 2001). In such a situation, soil mass (M_s) can be computed by using Equation 23.5.

$$M_s/\text{ha} = \rho_b \times D\ 10^4\ \text{m}^2/\text{ha} \qquad (23.5)$$

where *D* is soil depth (m). For example, SOC pool is appropriately compared on equivalent mass basis in diverse treatments such as no-till, plow till, pasture, and natural forest ecosystems, because these treatments have a wide range of soil ρ_b. In such cases, SOC pool computed on equal depth basis can lead to erroneous interpretation and wrong conclusions.

PREDICTION OF SOIL BULK DENSITY

Important determinants of ρ_b are texture and SOC content (Adams, 1973; Kay, 1998, Huntington et al., 1989). Therefore, several pedotransfer functions (PTFs) have been developed relating ρ_b to these key properties. For example, Rawls (1983) related ρ_b to texture and OC content on the basis of data from 2,721 soils from the United States (Equation 23.6).

$$\rho_b = 100/\{(\%\ \text{organic matter}/\text{organic matter}\ \rho_b) + [(100 - \%\ \text{organic matter})/\text{mineral}\ \rho_b)]\} \quad (23.6)$$

They assumed ρ_b of organic matter at 0.224 $Mg{\cdot}m^{-3}$. Manrique and Jones (1991) extended the regression equation presented in Equation 3.6 to incorporate the analytical data from 12,000 soil samples. Da Silva et al. (1997) developed PTFs for comparing ρ_b among a range of tillage treatments for soils with clay content ranging from 6 to 37 percent and SOC content from 0.9 to 3.9 percent. In addition to clay and SOC contents, da Silva and colleagues developed the PTFs involving tillage method (T) and the row/interrow position (P) (Equation 23.7).

$$\rho_b = 1.573 - 0.0640T + 0.367\,P + 0.0686\,T \times P - 0.125\,C - 0.031\,Cl + 0.0021\,C \times Cl \quad (23.7)$$

where T is a tillage class range with a value of 0 and 1 for no-till and conventional tillage, respectively; P is a row/interrow position, with 0 for the row and 1 for the interrow position, respectively; and C refers to SOC content (percent), and Cl to clay content (percent). In Ohio, Calhoun et al. (2001) developed three PTFs on the basis of 937 horizon datasets. The PTF developed had an R^2 of 0.56 to 0.80. Tan et al. (2004) observed that for some soils in Ohio the model by Calhoun et al. (2001) underestimated ρ_b in soils with high SOC concentration.

Soil ρ_b is also affected by structural attributes of a soil. Important among these are aggregation, aggregate stability and strength, and aggregate size distribution. While aggregate shape, platy or prismatic, is also important, aggregation indices (e.g., WSA, MWD, GMD) provide a quantitative index which, along with clay and SOC concentrations, can be used to develop PTF for predicting ρ_b.

CONCLUSIONS

Soil bulk density is an important property, with strong impact on quality of soil and water resources. Impact of land use and soil/crop management practices on pool and flux of soil organic carbon cannot be assessed without reliable information on temporal and spatial variability (horizontal and vertical) in soil bulk density. Direct measurement of soil bulk density is done by determining the ratio of dry soil mass to the total soil volume. The latter is determined by the core and clod methods directly and by radiation transmission/scattering techniques or strength probes/penetrometers indirectly. Soil bulk density can also be estimated by developing pedotransfer functions based on empirical relationships between bulk density as a dependent variable and clay and organic carbon content as independent variables. Measurement of total soil volume is difficult for a range of problem soils, including soils with high gravel content, high organic matter content, high swell-shrink capacity, and those characterized by the formation of deep and wide cracks. Because of its significance in evaluating soil quality, pool and flux of soil carbon and water or nutrients, it is important to develop cost-effective and routine methods of measuring/predicting bulk density of a wide range of soil types. Priority must be given to in situ methods of measuring soil bulk density.

REFERENCES

Abrol, I. P. and J. P. Palta. 1968. Bulk density determination of soil clod using rubber solution as a coating material. *Soil Sci.* 106: 465-468

Adams, W. A. 1973. The effect of organic matter on the bulk and true densities of some uncultivated podzolic soils. *J. Soil Sci.* 24: 10-17.

American Society for Testing and Materials. 1950. Sand cone and rubber balloon methods. In *Procedures for testing soils.* ASTM D-18, pp. 94-110.

Baver, L. D., W. H. Gardner, and W. R. Gardner. 1972. *Soil physics,* 4th ed. Wiley, New York.

Becket, S. H. 1928. The use of high density fluids in the determination of volume weight of soils. *Soil Sci.* 25: 481-483.

Blake, G. R. 1965. Bulk density. In C. A. Bleck (ed.), *Methods of soil analysis.* ASA, Madison, WI, pp. 374-390.

Blake, G. R. and K. H. Hartge. 1986. Bulk density. In A. Klute (ed.), *Methods of soil analysis,* Part I. ASA Monograph No. 9. ASA, Madison, WI.

Brasher, B. R., D. P. Franzmeier, V. Valarsis, and S. E. Davidson. 1966. Use of Saran resin to coat natural soil and water penetration measurements. *Soil Sci.* 101: 108.

Buchele, W. F. 1961. A power sampler of undisturbed soils. *Trans. Am. Soc. Agric. Eng.* 4: 185-187, 191.

Calhoun, F. G., N. E. Smeck, B. L. Slater, J. M. Bigham, and G. F. Hall. 2001. Predicting bulk density of Ohio soils from morphology, genetic principles and laboratory characterization data. *Soil Sci. Soc. Am. J.* 65: 811-819.

Campbell, D. J. 1973. A floatation method for the rapid measurement of the wet bulk density of soil clods. *J. Soil Sci.* 24: 239-243.

Campbell, D. J. and J. K. Henshall. 1991. Bulk density. In K. A. Smith and C. E. Mullins (eds.), *Soil analysis: Physical methods.* Marcel Dekker, Inc., New York, pp. 367-397.

Campbell, D. J. and J. K. Henshall. 2001. Bulk density. In K. A. Smith and C. E. Mullins (eds.), *Soil and environmental analysis: Physical methods,* 2nd ed. Marcel Dekker, New York, pp. 315-348.

Chepil, N. S. 1950. Methods of estimating apparent density of discrete soil grains and aggregates. *Soil Sci.* 70: 351-362.

Constantini, A. 1995. Soil sampling bulk density in the coastal lowlands of southeast Queensland. *Aust. J. Soil Res.* 33: 11-18.

Culley, J. L. B. 1993. Density and compressibility. In. M.R. Carter (ed.), *Soil sampling and methods of analysis.* Lewis Publishers, Boca Raton, FL, pp. 529-640.

Cunningham, R. L. and R. P. Matelski. 1968. Bulk density measurements on certain soils high in coarse fragments. *Soil Sci. Soc. Am. Proc.* 32: 108-111.

da Silva, A., B. D. Kay, and E. Perfect. 1997. Management versus inherent soil properties. *Soil Tillage Res.* 44: 81-93.

Ellert, B. H. and J. R. Bethany. 1995. Calculation of organic matter and nutrients stored in soils under contrasting management regimes. *Can. J. Soil Sci.* 75: 529-538.

Ellert, B. H., H. H. Janzen, and B. G. McConkey. 2001. Measuring and comparing soil C storage. In R. Lal, J. M. Kimble, R. F. Follett, and B. A. Stewart (eds.), *Assessment methods for soil carbon.* Lewis Publishers, Boca Raton, FL, pp. 131-146.

Fisher, R. F. 1995. Amelioration of degraded rain forest soils by plantations of native trees. *Soil Sci. Am. J.* 59: 544-549.

Freitag, D. R. 1971. Methods of measuring soil compaction. In. K. K. Barnes et al. (eds.), *Compaction of agric. soils.* Am. Soc. Agric. Eng., St. Joseph, MI, pp. 47-103.

Gill, W. R. 1959. Soil bulk density changes due to moisture changes in soil. *Trans Am. Soc. Agric. Eng.* 2: 104-105.

Grossman, R. B. and T. G. Reinsch 2002. Bulk density and linear extensibility. In J. H. Dane and G. C. Topp (eds.), *Methods of soil analysis,* Part 4. Soil Sci. Soc. Am., Madison WI, pp. 201-228.

Huntington, T. G., C. E. Johnson, A. H. Johnson, T. G. Siccana, and D. F. Ryan. 1989. Carbon, organic matter and bulk density relationships in a forested Spodosol. *Soil Sci.* 137: 177-181.

Hvorslev, M. J. 1949. Sub-surface exploration and sampling of soils for civil engineering purposes. Rep. on Res. Proj. on Am. Soc. of Civil Eng., U.S. Army Engrs. Waterways Expt. Sta., Vicksburg, MS.

Jamison, V. C., H. H. Weaver, and I. F. Reed. 1950. A hammer-driven soil core sampler. *Soil Sci.* 69: 487-496.

Kay, B. D. 1998. Soil structure and organic carbon: A review. In R. Lal, J. M. Kimble, R. F. Follett, and B. A. Stewart (eds.), *Soil processes and the carbon cycle.* CRC Press, Boca Raton, FL, pp. 169-198.

Lal, R. 1974. Effect of soil texture and density on the neutron and density probe calibration for some tropical soils. *Soil Sci.* 117: 183-190.

Larson, W. E., S. C. Gupta, and R. A. Useche. 1980. Compression of agricultural soils from eight soil orders. *Soil Sci. Soc. Am. J.* 44: 450-457.

Lutz, J. F. 1947. Apparatus for collecting undisturbed soil samples. *Soil Sci.* 64: 399-401.

Manrique, L. A. and C. A. Jones. 1991. Bulk density of soils in relation to soil physical and chemical properties. *Soil Sci. Soc. Am. J.* 55: 476-481.

Rawls, W. J. 1983. Estimating soil bulk density from particle size analysis and organic matter content. *Soil Sci.* 135: 123-124.

Revut, I. B. and A. A. Rode (eds.). 1969. *Experimental methods of studying soil structure.* USDA-NSF, Amerind Publishing Co., New Delhi, India (translated in 1981).

Russell, E. W. and W. Balcerek. 1944. The determination of volume and airspace of soil clods. *J. Agric. Sci., Camb.* 34: 123-132.

Shipp, R. F. and R. P. Matelski. 1965. Bulk density and coarse-fragment determination on some Pennsylvania soils. *Soil Sci.* 99: 392-397.

Tan, Z. X., R. Lal, N. E. Smeck, F. G. Calhoun, Brian K. Slater, B. Parkinson, and R. M. Gehring. 2004. Taxonomic and geographic distribution of soil organic carbon pools in Ohio. *Soil Sci. Soc. Am. J.* 68: 1896-1904.

Veihmeyer, F. J. 1929. An improved soil sampling tube. *Soil Sci.* 27: 147-152.

Vomocil, J. A. 1957. Measurement of soil bulk density and penetrability and review of methods. *Adv. Agron.* 9: 159-175.

Voorhees, W. B., R. R. Allmaras, and W. E. Larson. 1966. Porosity of surface soil aggregates at various moisture contents. *Soil Sci. Soc. Am. Proc.* 30: 163-167.

Zwarich, M. A. and C. F. Shaykjewich. 1969. An evaluation of several methods of measuring bulk density of soils. *Can. U. Soil Sci.* 49(2): 241-245.

PART IV: CONCLUSION

Chapter 24

Recommendations for Research and Development

R. Lal
C. Cerri
M. Bernoux
J. Etchevers

A principal reason for considering Latin America as an important region in soil/terrestrial C sequestration in any future scenarios for mitigating climate change is the fact that it encompasses 60 percent of the world's tropical rain forest (TRF). The conversion of TRF to agricultural or urban ecosystems will exacerbate the rates of CO_2 emissions and pose a large environmental challenge to future generations because of the projected climate change. A close examination of the global C pools shows that the key manageable pool is the terrestrial, comprising both soil and vegetation components. Therefore, it is important to identify ways and means by which atmospheric CO_2 can be incorporated into soil and biotic pools.

Several recommendations emerged from the workshop. Important among these are the following.

ESTABLISHMENT OF THE LATIN AMERICAN SOIL CARBON NETWORK

The workshop led to the development of the Latin American Soil Carbon Network (LASCANet). The objective of the LASCANet is to develop a research, extension, outreach, and training program in Latin America on soil carbon sequestration to improve soil quality, increase agronomic productivity, decrease non-point-source pollution, and reduce the rate of enrichment of atmospheric concentration of CO_2. The LASCANet facilities exchange of scientists and students collates and disseminates scientific information,

The help received from all rapporteurs and chairs of the discussion panels in summarizing the salient points is gratefully acknowledged.

Carbon Sequestration in Soils of Latin America

doi:10.1300/5755_25

creates awareness about the importance of soil carbon sequestration to sustainable management of natural resources, liaises with land managers and policymakers, and establishes cross-linkages among stakeholders in the Latin America region. The LASCANet is identifying ways to minimize carbon loss and maximize retaining carbon in land, reducing the effects of CO_2 and other trace gases (CH_4, N_2O, NO_x) on global climate change, and improving environment quality.

RATES OF SOIL CARBON SEQUESTRATION

Inventory of the SOC pool for different ecoregions is given in Table 24.1. Work Group 1 developed a matrix of the first approximation of rates of soil C sequestration. Surprisingly, the available data show that measured rates of SOC sequestration are higher for Latin America than North America. Most rates are in the range of 0.3 to 1.5 Mg C·ha^{-1} per year. The potential of SOC sequestration for Latin America is 0.1 to 0.2 Pg C per year, with the large potential of afforestation in the Amazonia, and adoption of recommended management practices in the Cerrados, Llanos, and Pampas ecoregions. Practices for SOC sequestration include land-use conversion to natural fallowing, afforestation, fast-growing timber plantation, and agro-

TABLE 24.1. Inventory of soil carbon pools for the soil ecoregions of Latin America.

			Agricultural impacted area			Mosaic of native and agricultural impacted area		
Soil ecoregion	Area (1000 km^2)	C pool, 0-30 cm (Tg C)	1000 km^2	%	C pool[a] (Tg C)	1000 km^2	%	C pool[a] (Tg C)
Amazonia	6734.4	33711	196.3	2.9	881	277.4	4.1	1367
Andean Equatorial	630.8	4136	98.5	15.6	603	198.9	31.5	1290
Brazilian Atlantic	1133.0	6267	363.0	32.0	1740	345.2	30.5	1879
Caribbean	655.6	3370	141.9	21.6	759	129.6	19.8	658
Central America	1137.0	7140	415.5	36.5	2378	4.2	0.4	32
Central Andes	1806.5	7822	14.2	0.8	78	120.7	6.7	685
Central Brazilian	1882.1	7986	697.8	37.1	2963	320.0	17.0	1291
East Brazilian	1012.3	3806	155.7	15.4	631	348.5	34.4	1262
Eastern Pampas	825.8	5419	235.1	28.5	1624	205.4	24.9	1338
Mexican	1367.6	5793	55.8	4.1	316	0.0	0.0	0
Preandean	1763.2	8125	148.8	8.4	768	154.9	8.8	799
South Argentina	961.5	3185	1.5	0.2	9	6.5	0.7	7
South Chilean	346.3	3743	18.7	5.4	213	12.8	3.7	119
Total	20256.1	100503	2542.8		12963	2124.1		10727

Source: Based on the FAO 1:5,000,000 vector map and the estimates from Batjes et al., 1996.
[a]Corresponding C pool (0-30 cm) under native condition that was impacted by land-use activity.

forestry. It is important to avoid deforestation by intensification of agriculture on existing land through adoption of recommended management practices and restoration of degraded lands. The latter includes conversion of plow tillage to no-till with cover crops and mulching, elimination of burning, integrated nutrient management, use of agroforestry systems in place of slash-and-burn agriculture, and implementation of a silvopastoral system with improved species. The SOC sequestration rates are available for land uses in each ecoregion but need to be validated under on-farm conditions.

MEASUREMENT, MONITORING, AND VERIFICATION

Principal issues discussed with regard to soil sampling included depth, frequency, intensity of sampling, replications, and cost-effectiveness of different monitoring techniques. The choice of depth of sampling depends on the objectives and should be to 1 m, although 0.30 cm depth can be adequate in some cases. The frequency of sampling can be once in three years, and one sample per ha may be needed on a farm or landscape scale. Bulk density must be measured for each layer and corrected for large roots and gravels. Monitoring technologies include destructive sampling and non-invasive methods, which are encouraged where point measurements or volumetric requirements are needed. The aboveground biomass can be measured by remote sensing and geoprocessing techniques, and it is useful to establish the relation between C in the biomass and that in the soil. In addition to C, it is also important to measure N and P concentrations, aggregation, and soil erosion in managed and reference ecosystems.

POLICIES AND PROGRAMS

Policies and programs to promote soil C sequestration include the following:

- Promoting SOC sequestration by implementing policies that are linked to
  —international treaties and
  —ancillary benefits of SOC sequestration such as increased productivity, socioeconomic benefits, and environmental services, including erosion control, reduction in siltation and non-point-source pollution, increase in biodiversity, and enhancement of the material and aesthetic value of the land.
- Developing country-specific policy instruments such as
  —improving policies for soil management and incentives to comply with them,

—creating economic incentives for farmers to adopt recommended practices, and
—supporting demand-driven research in soil C sequestration with farmer participation.

- Enhancing interaction between scientists and policymakers, and integrating farmers and the private sector by
  —identifying influential groups within each country from the public, private, and research sectors and facilitating meetings among them,
  —establishing a network for Latin America that includes policymakers, biophysical scientists, social scientists, farmers/land managers, and producer associations, and
  —promoting interministerial cooperation within countries (e.g., agriculture, environment ministries) and among countries.
- Facilitating information/technology transfer to the private sector (farmers, land owners, agribusiness, associations) by
  —identifying appropriate practices/activities for specific land-use systems within each ecoregion,
  —encouraging research on the economic benefits of these activities and practices (on-farm demonstration plots, pilot programs), and
  —identifying information delivery strategies (e.g., extension services, producer associations, NGOs).

HOW GOVERNMENTS CAN ENCOURAGE SOIL CARBON SEQUESTRATION

Government policies can be extremely important in providing incentives to farmers and land managers for soil C sequestration. Specifically, national governments can do the following:

- Develop policies that are linked to international treaties (e.g., climate change, biodiversity, desertification) and ancillary benefits of soil C sequestration. Important among these are increased productivity, socioeconomic benefits, and environmental services or societal benefits (e.g., reduced erosion and sedimentation, improved soil quality, improved water availability and quality, enhanced air quality, improved biodiversity, and aesthetic value of the land).
- Develop country-specific policy instruments with regard to the following:
  —Improve policies for soil/land management and incentives to comply with them.

- —Provide economic incentives (e.g., credit lines/loans, tax exemptions, yield guarantees/insurance) for farmers engaged in approved activities and management practices.
- —Pay for environmental services without interfering with international carbon trading.
- —Create and support government-funded research and extension on carbon sequestration and related topics.

- Learn from existing programs/projects that have been effective in promoting carbon sequestration (e.g., Carbo Europe) and develop policies accordingly.
- Create information mechanisms at all levels, but with specific reference to the following:
  - —For the general public and different society groups to be aware of the global warming problem, potential benefits of carbon sequestration, and other environmental services.
  - —For the donor community to be aware of the range of social benefits implied when supporting research program/projects on carbon sequestration and other environmental services.

COMMUNICATION AMONG SCIENTISTS, POLICYMAKERS, AND LAND MANAGERS

There is a strong need to establish and strengthen channels of communication among scientists, government agencies, and land managers. Scientists need to provide the following information to the government:

- Relevant scientific information, in a short and clear format they can use, provided in timely manner
- Scope of the socioeconomic and political implications of decisions they may make based on this information

Similarly, government agencies need to provide the following information to the scientific community so that the latter can develop a demand-driven and mission-oriented research program.

- Relevant and timely information on
  - —policy decisions that support scientific research on carbon sequestration and related topics,
  - —financial support strategies,
  - —additional information on how to apply to C funds for research projects, and

—activities that may enhance communication between policymakers and scientists.

- Information about the government's needs for collaboration from scientists in the decision-making process at all levels

Furthermore, scientists and policymakers must closely interact with farmers and the private sector to create favorable environments for restoration of degraded soils and adoption of recommended land use and soil/crop management practices. Integration of farmers with the private sector is also important to trading C credits and creating synergistic effects. Specific needs toward this goal are the following:

- Identifying influential groups within each country from the public, private, and research sectors and facilitating meetings between them.
- Establishing a network for the Americas (Web page or Listserv) that includes policymakers, biophysical scientists, social scientists, farmers/land managers, and producer associations.
- Promoting interministerial cooperation within countries (e.g., agriculture, environment ministries, etc.) and between countries (e.g., PROCISUR).

Soil C sequestration is a relatively new concept with regard to trading C credits and providing environmental services for mitigating climate exchange and improving water quality. Therefore, it is important to facilitate information and technology transfer to the private sector, including farmers, land owners, agribusiness, commodity growers' associations, and other stakeholders. Technology transfers can be achieved through the following:

- Identifying appropriate activities/management practices for specific land-use systems within each ecoregion
- Encouraging research of the economic benefits of these activities and practices such as on-farm demonstration plots or pilot programs
- Identifying information delivery strategies such as extension services, producer associations, and nongovernmental organizations

SPECIFIC RESEARCH PRIORITIES

Several specific research priorities were identified. Notable among these were the following:

1. Recovery of degraded land and degraded pastures for all of Latin America
2. Identification of hot spots of production/consumption of other GHGs
3. Improvement of model parameters (mineralogy, hydrology, interface aquatic/terrestrial ecosystems)
4. Participatory approaches to enhance adoption of best management practices
5. Impacts of climate change on the different ecosystems especially for the dry ecosystem
6. The magnitude of SOC lost due to agricultural practices versus how much lost is really C dislocation due to erosion
7. Making current cropping systems more "C friendly," especially those that are already under no-till
8. Identifying the mechanisms of SOM stabilization in tropical soils
9. Determining the net C sequestration when all the greenhouse gases (CO_2, CH_4, and N_2O) are considered
10. Establishing and updating databases
11. Effectively establishing and monitoring the "quality of systems" (e.g., what is the minimum residue amount needed for each ag system and agroecoregion to attain sequestration)
12. Developing effective scientific networks for addressing critical issues

REFERENCE

Batjes, N. H., G. Fischer, F. O. Nachtergaele, V. S. Stolboroy, and H. T. Van Velthuizen. 1997. *Soil data derived from WISE for use in global and regional AEZ studies,* Vol. 10. FAO/IISA/ISRIC, Luxembourg.

Index

Page numbers followed by the letter "f" indicate figures; those followed by the letter "t" indicate tables.

doi:10.1300/5755_26

Order a copy of this book with this form or online at:
http://www.haworthpress.com/store/product.asp?sku=5755

CARBON SEQUESTRATION IN SOILS OF LATIN AMERICA

________in hardbound at $69.95 (ISBN-13: 978-1-56022-136-4; ISBN-10: 1-56022-136-4)

________in softbound at $49.95 (ISBN-13: 978-1-56022-137-1; ISBN-10: 1-56022-137-2)

524 pages plus index • Includes photos and illustrations

Or order online and use special offer code HEC25 in the shopping cart.

COST OF BOOKS________

POSTAGE & HANDLING________
(US: $4.00 for first book & $1.50 for each additional book)
(Outside US: $5.00 for first book & $2.00 for each additional book)

SUBTOTAL________

IN CANADA: ADD 7% GST________

STATE TAX________
(NJ, NY, OH, MN, CA, IL, IN, PA, & SD residents, add appropriate local sales tax)

FINAL TOTAL________
(If paying in Canadian funds, convert using the current exchange rate, UNESCO coupons welcome)

☐ **BILL ME LATER:** (Bill-me option is good on US/Canada/Mexico orders only; not good to jobbers, wholesalers, or subscription agencies.)

☐ Check here if billing address is different from shipping address and attach purchase order and billing address information.

Signature________________________

☐ **PAYMENT ENCLOSED:** $________________

☐ **PLEASE CHARGE TO MY CREDIT CARD.**

☐ Visa ☐ MasterCard ☐ AmEx ☐ Discover ☐ Diner's Club ☐ Eurocard ☐ JCB

Account # ________________________

Exp. Date________________________

Signature________________________

Prices in US dollars and subject to change without notice.

NAME________________________

INSTITUTION________________________

ADDRESS________________________

CITY________________________

STATE/ZIP________________________

COUNTRY________________ COUNTY (NY residents only)________________

TEL________________ FAX________________

E-MAIL________________________

May we use your e-mail address for confirmations and other types of information? ☐ Yes ☐ No
We appreciate receiving your e-mail address and fax number. Haworth would like to e-mail or fax special discount offers to you, as a preferred customer. **We will never share, rent, or exchange your e-mail address or fax number.** We regard such actions as an invasion of your privacy.

Order From Your Local Bookstore or Directly From

The Haworth Press, Inc.

10 Alice Street, Binghamton, New York 13904-1580 • USA
TELEPHONE: 1-800-HAWORTH (1-800-429-6784) / Outside US/Canada: (607) 722-5857
FAX: 1-800-895-0582 / Outside US/Canada: (607) 771-0012
E-mail to: orders@haworthpress.com

For orders outside US and Canada, you may wish to order through your local sales representative, distributor, or bookseller.
For information, see http://haworthpress.com/distributors

(Discounts are available for individual orders in US and Canada only, not booksellers/distributors.)

PLEASE PHOTOCOPY THIS FORM FOR YOUR PERSONAL USE.

http://www.HaworthPress.com BOF06